College Preparatory Mathematics

Custom Edition for Stark State College

Modules 13–18 Only

Taken from:

Prealgebra & Introductory Algebra, Third Edition
by Margaret L. Lial, Diana L. Hestwood, John Hornsby, and Terry McGinnis

Basic College Mathematics, Eighth Edition
by Margaret L. Lial, Stanley A. Salzman, Diana L. Hestwood

Intermediate Algebra, Ninth Edition
by Margaret L. Lial, John Hornsby, and Terry McGinnis

Learning Solutions

New York Boston San Francisco
London Toronto Sydney Tokyo Singapore Madrid
Mexico City Munich Paris Cape Town Hong Kong Montreal

Cover Art: *Regatta Single*, by Angela Sciaraffa

Taken from:

Prealgebra & Introductory Algebra, Third Edition
by Margaret L. Lial, Diana L. Hestwood, John Hornsby, and Terry McGinnis

Published by Addison Wesley
Boston, MA 02116

Basic College Mathematics, Eighth Edition
by Margaret L. Lial, Stanley A. Salzman, Diana L. Hestwood

Published by Addison Wesley

Intermediate Algebra, Ninth Edition
by Margaret L. Lial, John Hornsby, and Terry McGinnis

Published by Addison Wesley

This special edition published in cooperation with Pearson Learning Solutions.

Pearson Learning Solutions, 501 Boylston Street, Suite 900, Boston, MA 02116
A Pearson Education Company
www.pearsoned.com

Printed in the United States of America

1 2 3 4 5 6 7 8 9 10 V0NA 15 14 13 12 11 10

000200010270651731

MT

PEARSON

ISBN 10: 0-558-92660-6
ISBN 13: 978-0-558-92660-1

CONTENTS

13

Solving Inequalities/Absolute Values/Intro to Functions

13.1 Basic Concepts

OBJECTIVES

1. Write sets using set notation.
2. Use number lines.
3. Know the common sets of numbers.
4. Find additive inverses.
5. Use absolute value.
6. Use inequality symbols.

OBJECTIVE 1 Write sets using set notation. A **set** is a collection of objects called the **elements,** or **members,** of the set. In algebra, the elements of a set are usually numbers. Set braces, $\{\ \}$, are used to enclose the elements. For example, 2 is an element of the set $\{1, 2, 3\}$. Since we can count the number of elements in the set $\{1, 2, 3\}$ and the counting process comes to an end, it is a **finite set.**

In algebra, we refer to certain sets of numbers by name. The set

$$N = \{1, 2, 3, 4, 5, 6, \ldots\} \quad \text{Natural (counting) numbers}$$

is called the **natural numbers,** or the **counting numbers.** The three dots (called *ellipsis points*) show that the list continues in the same pattern indefinitely. We cannot list all of the elements of the set of natural numbers, so it is an **infinite set.**

When 0 is included with the set of natural numbers, we have the set of **whole numbers,** written

$$W = \{0, 1, 2, 3, 4, 5, 6, \ldots\}. \quad \text{Whole numbers}$$

A set containing no elements, such as the set of whole numbers less than 0, is called the **empty set,** or **null set,** usually written $\emptyset$.

CAUTION
Do *not* write $\{\emptyset\}$ for the empty set; $\{\emptyset\}$ is a set with one element, $\emptyset$. Use only the notation $\emptyset$ for the empty set.

◀ Work Problem 1 at the Side.

In algebra, letters called **variables** are often used to represent numbers or to define sets of numbers. For example,

$$\{x \mid x \text{ is a natural number between 3 and 15}\}$$

(read "the set of all elements x such that x is a natural number between 3 and 15") defines the set

$$\{4, 5, 6, 7, \ldots, 14\}.$$

The notation $\{x \mid x$ is a natural number between 3 and $15\}$ is an example of **set-builder notation.**

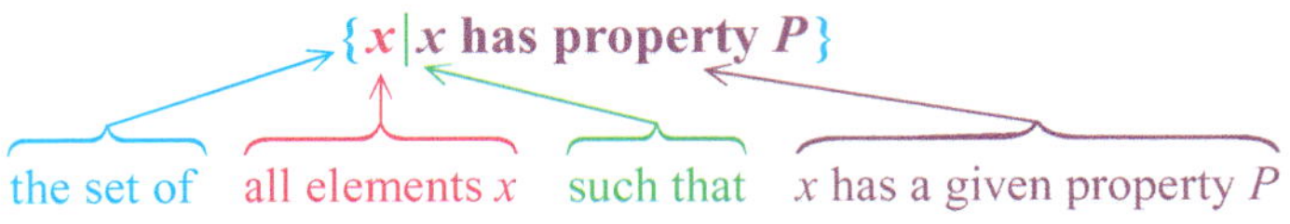

Example 1 Listing the Elements in Sets

List the elements in each set.

(a) $\{x \mid x$ is a natural number less than $4\}$
The natural numbers less than 4 are 1, 2, and 3. This set is $\{1, 2, 3\}$.

(b) $\{y \mid y$ is one of the first five even natural numbers$\}$ is $\{2, 4, 6, 8, 10\}$.

(c) $\{z \mid z$ is a natural number greater than or equal to $7\}$
The set of natural numbers greater than or equal to 7 is an infinite set, written with ellipsis points as $\{7, 8, 9, 10, \ldots\}$.

◀ Work Problem 2 at the Side.

---

1 Consider the set
$$\left\{0, 10, \frac{3}{10}, 52, 98.6\right\}.$$

(a) Which elements of the set are natural numbers?

(b) Which elements of the set are whole numbers?

2 List the elements in each set.

(a) $\{x \mid x$ is a whole number less than $5\}$

(b) $\{y \mid y$ is a whole number greater than $12\}$

ANSWERS

1. (a) 10 and 52 (b) 0, 10, and 52
2. (a) $\{0, 1, 2, 3, 4\}$ (b) $\{13, 14, 15, \ldots\}$

Example 2 Using Set-Builder Notation to Describe Sets

Use set-builder notation to describe each set.

(a) $\{1, 3, 5, 7, 9\}$

There are often several ways to describe a set with set-builder notation. One way is $\{y \mid y$ is one of the first five odd natural numbers$\}$.

(b) $\{5, 10, 15, \ldots\}$

This set can be described as $\{x \mid x$ is a multiple of 5 greater than 0$\}$.

Work Problem 3 at the Side. ▶

3 Use set-builder notation to describe each set.

(a) $\{0, 1, 2, 3, 4, 5\}$

(b) $\{7, 14, 21, 28, \ldots\}$

OBJECTIVE 2 Use number lines. A good way to get a picture of a set of numbers is by using a **number line.** To construct a number line, choose any point on a horizontal line and label it 0. Next, choose a point to the right of 0 and label it 1. The distance from 0 to 1 establishes a scale that can be used to locate more points, with positive numbers to the right of 0 and negative numbers to the left of 0. The number 0 is neither positive nor negative. A number line is shown in Figure 1.

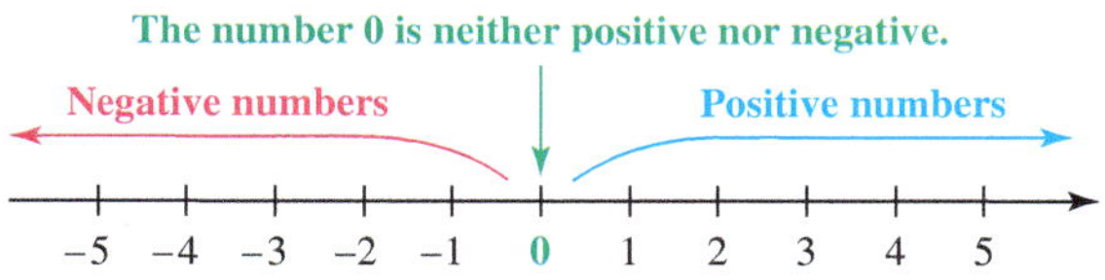

Figure 1

The set of numbers identified on the number line in Figure 1, including positive and negative numbers and 0, is part of the set of **integers,** written

$$I = \{\ldots, -3, -2, -1, 0, 1, 2, 3, \ldots\}. \quad \text{Integers}$$

Each number on a number line is called the **coordinate** of the point that it labels, while the point is the **graph** of the number. Figure 2 shows a number line with several selected points graphed on it.

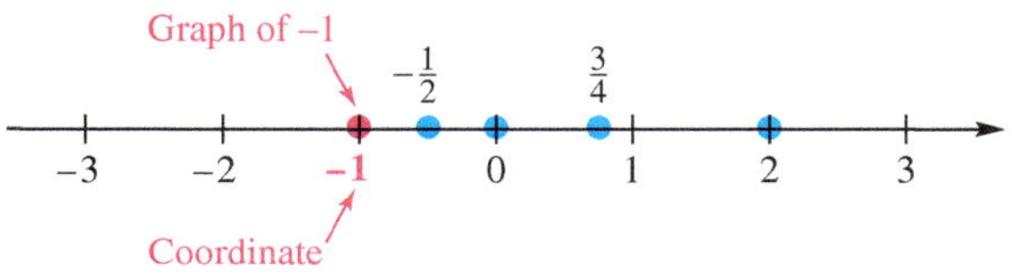

Figure 2

Work Problem 4 at the Side. ▶

4 Graph the elements of each set.

(a) $\{-4, -2, 0, 2, 4, 6\}$

(b) $\left\{-1, 0, \frac{2}{3}, 2.5\right\}$

(c) $\left\{5, \frac{16}{3}, 6, \frac{13}{2}, 7, \frac{29}{4}\right\}$

The fractions $-\frac{1}{2}$ and $\frac{3}{4}$, graphed on the number line in Figure 2, are examples of *rational numbers*. A **rational number** can be expressed as the quotient of two integers, with denominator not 0. The set of all rational numbers is written

$$\left\{\frac{p}{q} \,\middle|\, p \text{ and } q \text{ are integers, } q \neq 0\right\}. \quad \text{Rational numbers}$$

The set of rational numbers includes the natural numbers, whole numbers, and integers, since these numbers can be written as fractions. For example, $14 = \frac{14}{1}$, $-3 = \frac{-3}{1}$, and $0 = \frac{0}{1}$. A rational number written as a fraction, such as $\frac{1}{8}$ or $\frac{2}{3}$, can also be expressed as a decimal by dividing the numerator by the denominator, as shown on the next page.

ANSWERS

3. (a) One answer is $\{x \mid x$ is a whole number less than 6$\}$. (b) One answer is $\{x \mid x$ is a multiple of 7 greater than 0$\}$.

4. (a)
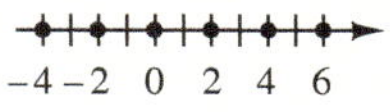

(b)

(c)
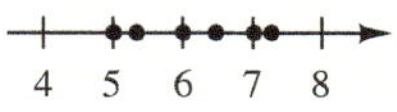

$8\overline{)1.000}$ gives 0.125 ← Terminating decimal	$3\overline{)2.000\ldots}$ gives 0.666... ← Repeating decimal
8; 20; 16; 40; 40; 0 ← Remainder is 0.	18; 20; 18; 20; 18; 2 ← Remainder is never 0.
$\frac{1}{8} = 0.125$	$\frac{2}{3} = 0.\overline{6}$ ← A bar is written over the repeating digit(s).

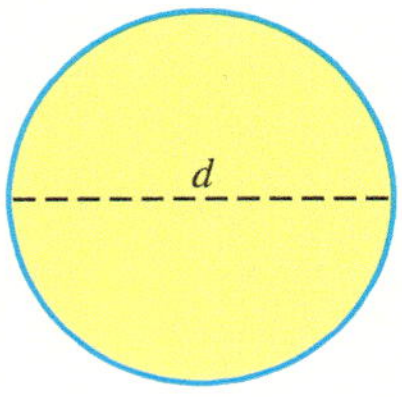

$\pi = \frac{C}{d}$

Figure 3

Thus, terminating decimals, such as $0.125 = \frac{1}{8}$, $0.8 = \frac{4}{5}$, and $2.75 = \frac{11}{4}$, and repeating decimals, such as $0.\overline{6} = \frac{2}{3}$ and $0.\overline{27} = \frac{3}{11}$, are rational numbers.

Decimal numbers that neither terminate nor repeat are *not* rational, and thus are called **irrational numbers.** Many square roots are irrational numbers; for example, $\sqrt{2} = 1.4142135\ldots$ and $-\sqrt{7} = -2.6457513\ldots$ repeat indefinitely without pattern. (Some square roots *are* rational: $\sqrt{16} = 4$, $\sqrt{100} = 10$, and so on.) Another irrational number is π (pronounced "pie"), the ratio of the distance around, or circumference, C of a circle to its diameter d. See Figure 3.

Some of the rational and irrational numbers discussed above are graphed on the number line in Figure 4. The rational numbers together with the irrational numbers make up the set of **real numbers.** ***Every point on a number line corresponds to a real number, and every real number corresponds to a point on the number line.***

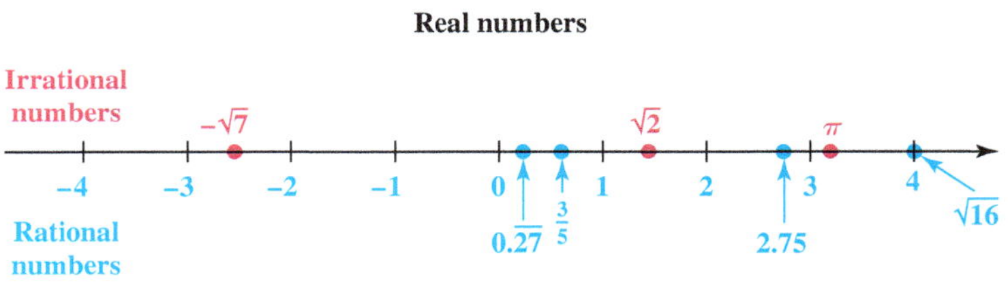

Figure 4

OBJECTIVE 3 Know the common sets of numbers. The sets of numbers listed below will be used throughout the rest of this text.

Sets of Numbers

Natural numbers	$\{1, 2, 3, 4, 5, 6, \ldots\}$
Whole numbers	$\{0, 1, 2, 3, 4, 5, 6, \ldots\}$
Integers	$\{\ldots, -3, -2, -1, 0, 1, 2, 3, \ldots\}$
Rational numbers	$\left\{\frac{p}{q} \middle\vert p \text{ and } q \text{ are integers}, q \neq 0\right\}$ *Examples:* $\frac{4}{1}$, 1.3, $-\frac{9}{2}$, $\frac{16}{8}$ or 2, $\sqrt{9}$ or 3, $0.\overline{6}$
Irrational numbers	$\{x \vert x$ **is a real number that is not rational**$\}$ *Examples:* $\sqrt{3}$, $-\sqrt{2}$, π
Real numbers	$\{x \vert x$ **is a rational number or an irrational number**$\}$*

*An example of a number that is not real is $\sqrt{-1}$. This number is part of the *complex number system*.

Figure 5 shows that the set of real numbers includes both the rational and irrational numbers. ***Every real number is either rational or irrational.*** Also, notice that the integers are elements of the set of rational numbers, and that the whole numbers and natural numbers are elements of the set of integers.

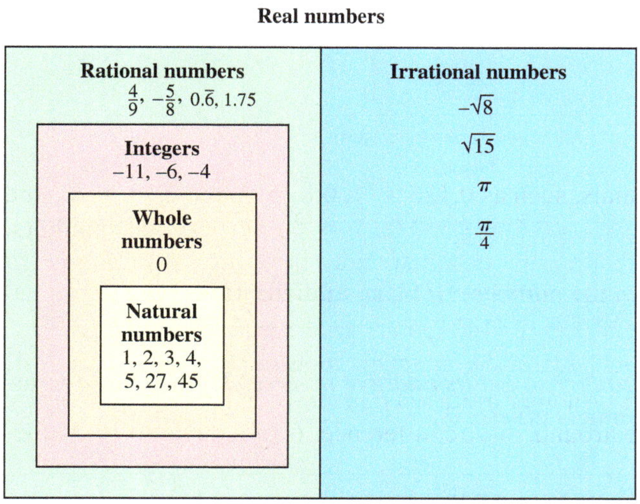

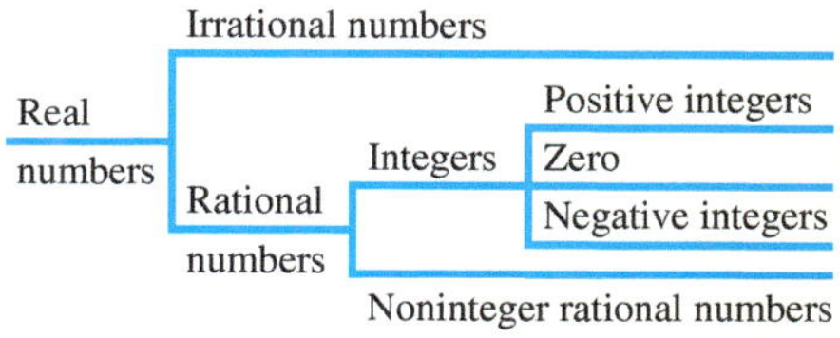

Figure 5 The Real Numbers

Example 3 Identifying Examples of Number Sets

Which numbers in

$$\left\{-8, -\sqrt{2}, -\frac{9}{64}, 0, 0.5, \frac{2}{3}, 1.\overline{12}, \sqrt{3}, 2\right\}$$

are elements of each set?

(a) Integers
$-8, 0,$ and 2 are integers.

(b) Rational numbers
$-8, -\frac{9}{64}, 0, 0.5, \frac{2}{3}, 1.\overline{12},$ and 2 are rational numbers.

(c) Irrational numbers
$-\sqrt{2}$ and $\sqrt{3}$ are irrational numbers.

(d) Real numbers
All the numbers in the given set are real numbers.

Work Problem **5** at the Side. ▶

Example 4 Determining Relationships between Sets of Numbers

Decide whether each statement is *true* or *false*.

(a) All irrational numbers are real numbers.
This is true. As shown in Figure 5, the set of real numbers includes all irrational numbers.

(b) Every rational number is an integer.
This statement is false. Although some rational numbers are integers, other rational numbers, such as $\frac{2}{3}$ and $-\frac{1}{4}$, are not.

Work Problem **6** at the Side. ▶

5 Select all the sets from the following list that apply to each number.
Whole number
Rational number
Irrational number
Real number

(a) -6 **(b)** 12

(c) $0.\overline{3}$ **(d)** $-\sqrt{15}$

(e) π **(f)** $\frac{22}{7}$

(g) 3.14

6 Decide whether the statement is *true* or *false*. If *false,* tell why.

(a) All whole numbers are integers.

(b) Some integers are whole numbers.

(c) Every real number is irrational.

ANSWERS

5. **(a)** rational, real **(b)** whole, rational, real **(c)** rational, real **(d)** irrational, real **(e)** irrational, real **(f)** rational, real **(g)** rational, real

6. **(a)** true **(b)** true **(b)** false; Some real numbers are irrational, but others are rational numbers.

OBJECTIVE 4 Find additive inverses. Look at the number line in Figure 6. For each positive number, there is a negative number on the opposite side of 0 that lies the same distance from 0. These pairs of numbers are called *additive inverses, negatives,* or *opposites* of each other. For example, 3 is the additive inverse of -3, and -3 is the additive inverse of 3.

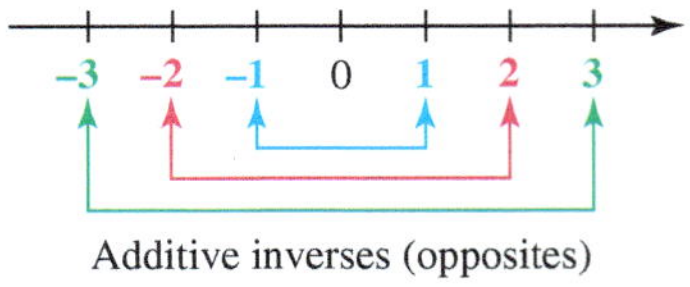

Figure 6

> **Additive Inverse**
>
> For any real number a, the number $-a$ is the **additive inverse** of a.

Change the sign of a number to get its additive inverse. The sum of a number and its additive inverse is always 0.

> **Uses of the Symbol −**
>
> The symbol "−" can be used to indicate any of the following:
>
> **1.** a negative number, such as -9 or -15;
>
> **2.** the additive inverse of a number, as in "-4 is the additive inverse of 4";
>
> **3.** subtraction, as in $12 - 3$.

In the expression $-(-5)$, the symbol "−" is being used in two ways: the first − indicates the additive inverse (or opposite) of -5, and the second indicates a negative number, -5. Since the additive inverse of -5 is 5, then

$$-(-5) = 5.$$

This example suggests the following property.

> **$-(-a)$**
>
> For any real number a, $\qquad -(-a) = a.$

Numbers written with positive or negative signs, such as $+4$, $+8$, -9, and -5, are called **signed numbers.** A positive number can be called a signed number even though the positive sign is usually left off. The following table shows the additive inverses of several signed numbers. The number 0 is its own additive inverse.

Number	Additive Inverse
6	-6
-4	4
$\frac{2}{3}$	$-\frac{2}{3}$
-8.7	8.7
0	0

◀ Work Problem 7 at the Side.

7 Give the additive inverse of each number.

(a) 9

(b) -12

(c) $-\frac{6}{5}$

(d) 0

(e) 1.5

ANSWERS

7. (a) -9 **(b)** 12 **(c)** $\frac{6}{5}$ **(d)** 0 **(e)** -1.5

OBJECTIVE 5 Use absolute value. Geometrically, the **absolute value** of a number a, written $|a|$, is the distance on the number line from 0 to a. For example, the absolute value of 5 is the same as the absolute value of -5 because each number lies five units from 0. See Figure 7. That is,

$$|5| = 5 \quad \text{and} \quad |-5| = 5.$$

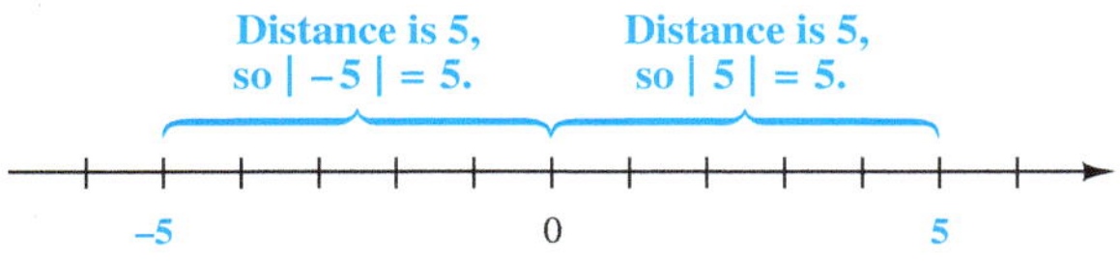

Figure 7

> **CAUTION**
> ***Because absolute value represents distance, and distance is never negative, the absolute value of a number is always positive or 0.***

The formal definition of absolute value follows.

> **Absolute Value**
>
> $$|a| = \begin{cases} a & \textbf{if } a \textbf{ is positive or 0} \\ -a & \textbf{if } a \textbf{ is negative} \end{cases}$$

The second part of this definition, $|a| = -a$ if a is negative, requires careful thought. If a is a *negative* number, then $-a$, the additive inverse or opposite of a, is a positive number. Thus, $|a|$ is positive. For example, if $a = -3$, then

$$|a| = |-3| = -(-3) = 3. \qquad |a| = -a \text{ if } a \text{ is negative.}$$

Example 5 Finding Absolute Value

Simplify by finding each absolute value.

(a) $|13| = 13$ **(b)** $|-2| = -(-2) = 2$

(c) $|0| = 0$ **(d)** $|-0.75| = 0.75$

(e) $-|8|$

Evaluate the absolute value first. Then find the additive inverse.

$$-|8| = -(8) = -8$$

(f) $-|-8|$

Work as in part (e): $|-8| = 8$, so

$$-|-8| = -(8) = -8.$$

(g) $|-2| + |5|$

Evaluate each absolute value first, and then add.

$$|-2| + |5| = 2 + 5 = 7$$

(h) $-|5 - 2| = -|3| = -3$

Work Problem 8 at the Side. ▶

8 Find the value of each expression.

(a) $|6|$

(b) $|-3|$

(c) $-\left|\frac{1}{4}\right|$

(d) $-|-2|$

(e) $-|-7.25|$

(f) $|-6| + |-3|$

(g) $|-9| - |-4|$

(h) $-|9 - 4|$

ANSWERS

8. (a) 6 **(b)** 3 **(c)** $-\frac{1}{4}$ **(d)** -2 **(e)** -7.25 **(f)** 9 **(g)** 5 **(h)** -5

Absolute value is useful when comparing size without regard to sign.

9 Refer to the table in Example 6. Of the software publishers and fabric mills industries, which will show the greater change (without regard to sign)?

Example 6 **Comparing Rates of Change in Industries**

The projected annual rates of change in employment (in percent) in some of the fastest-growing and in some of the most rapidly declining industries from 2002 through 2012 are shown in the table.

Industry (2002–2012)	Annual Rate of Change (in percent)
Software publishers	5.3
Care services for the elderly	4.5
Child day-care services	3.6
Cut-and-sew apparel manufacturing	−12.2
Fabric mills	−5.9
Metal ore mining	−4.8

Source: U.S. Bureau of Labor Statistics.

What industry in the list is expected to see the greatest change? the least change?

We want the greatest *change,* without regard to whether the change is an increase or a decrease. Look for the number in the list with the largest absolute value. That number is found in cut-and-sew apparel manufacturing, since

$$|-12.2| = 12.2.$$

Similarly, the least change is in the child day-care services industry:

$$|3.6| = 3.6.$$

◀ Work Problem **9** at the Side.

OBJECTIVE 6 **Use inequality symbols.** The statement

$$4 + 2 = 6$$

is an **equation**—a statement that two quantities are equal. The statement

$$4 \neq 6$$

(read "4 is not equal to 6") is an **inequality**—a statement that two quantities are *not* equal. When two numbers are not equal, one must be less than the other. The symbol $<$ means "is less than." For example,

$$8 < 9, \quad -6 < 15, \quad -6 < -1, \quad 0.5 < 0.9, \quad \text{and} \quad 0 < \frac{4}{3}.$$

The symbol $>$ means "is greater than." For example,

$$12 > 5, \quad 9 > -2, \quad -4 > -6, \quad 1.25 > 1.2, \quad \text{and} \quad \frac{6}{5} > 0.$$

In each case, the symbol "points" toward the lesser number.

ANSWER

9. Fabric mills

The number line in Figure 8 shows the graphs of the numbers 4 and 9. We know that $4 < 9$. On the graph, 4 is to the left of 9. ***The lesser of two numbers is always to the left of the other on a number line.***

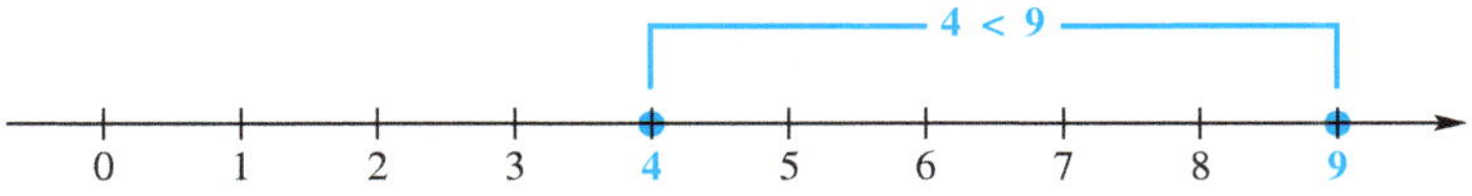

Figure 8

Inequalities on a Number Line

On a number line,

$a < b$ if a is to the left of b; $a > b$ if a is to the right of b.

We can use a number line to determine order. As shown on the number line in Figure 9, -6 is located to the left of 1. For this reason, $-6 < 1$. Also, $1 > -6$. From the same number line, $-5 < -2$, or $-2 > -5$.

Figure 9

CAUTION

Be careful when ordering negative numbers. Since -5 is to the left of -2 on the number line in Figure 9, $-5 < -2$, or $-2 > -5$. In each case, the symbol points to -5, the lesser number.

Work Problem **10** *at the Side.* ▶

The following table summarizes results about positive and negative numbers in both words and symbols.

Words	Symbols
Every negative number is less than 0.	If a is negative, then $a < 0$.
Every positive number is greater than 0.	If a is positive, then $a > 0$.
0 is neither positive nor negative.	

In addition to the symbols $\neq$, $<$, and $>$, the symbols $\leq$ and $\geq$ are often used.

INEQUALITY SYMBOLS

Symbol	Meaning	Example
$\neq$	is not equal to	$3 \neq 7$
$<$	is less than	$-4 < -1$
$>$	is greater than	$3 > -2$
$\leq$	is less than or equal to	$6 \leq 6$
$\geq$	is greater than or equal to	$-8 \geq -10$

10 Insert $<$ or $>$ in each blank to make a true statement.

(a) 3 ______ 7

(b) 9 ______ 2

(c) -4 ______ -8

(d) -2 ______ -1

(e) 0 ______ -3.5

(f) $\frac{5}{8}$ ______ $\frac{3}{4}$

(g) -0.3 ______ -0.5

Answers

10. (a) $<$ **(b)** $>$ **(c)** $>$ **(d)** $<$ **(e)** $>$ **(f)** $<$ **(g)** $>$

The following table shows several inequalities and why each is true.

Inequality	Why It Is True
$6 \leq 8$	$6 < 8$
$-2 \leq -2$	$-2 = -2$
$-9 \geq -12$	$-9 > -12$
$-3 \geq -3$	$-3 = -3$
$6 \cdot 4 \leq 5(5)$	$24 < 25$

Notice the reason why $-2 \leq -2$ is true. ***With the symbol ≤, if either the < part or the = part is true, then the inequality is true. This is also the case with the ≥ symbol.***

In the last row of the table, recall that the dot in $6 \cdot 4$ indicates the product 6×4, or 24, and $5(5)$ means 5×5, or 25. Thus, the inequality $6 \cdot 4 \leq 5(5)$ becomes $24 \leq 25$, which is true.

◀ *Work Problem* **11** *at the Side.*

11 Answer *true* or *false*.

(a) $-2 \leq -3$

(b) $0.5 \leq 0.5$

(c) $-9 \geq -1$

(d) $5 \cdot 8 \leq 7 \cdot 7$

(e) $3(4) > 2(6)$

Answers

11. (a) false **(b)** true **(c)** false **(d)** true **(e)** false

13.1 Exercises

FOR EXTRA HELP — MyMathLab — MathXL PRACTICE — WATCH — DOWNLOAD — READ — REVIEW

Write each set by listing its elements. See Example 1.

1. $\{x \mid x$ is a natural number less than $6\}$

2. $\{m \mid m$ is a natural number less than $9\}$

3. $\{z \mid z$ is an integer greater than $4\}$

4. $\{y \mid y$ is an integer greater than $8\}$

5. $\{a \mid a$ is an even integer greater than $8\}$

6. $\{k \mid k$ is an odd integer less than $1\}$

7. $\{x \mid x$ is an irrational number that is also rational$\}$

8. $\{r \mid r$ is a number that is both positive and negative$\}$

9. $\{p \mid p$ is a number whose absolute value is $4\}$

10. $\{w \mid w$ is a number whose absolute value is $7\}$

Write each set using set-builder notation. See Example 2. (More than one description is possible.)

11. $\{2, 4, 6, 8\}$

12. $\{11, 12, 13, 14\}$

13. $\{4, 8, 12, 16, \ldots\}$

14. $\{\ldots, -6, -3, 0, 3, 6, \ldots\}$

Graph the elements of each set on a number line.

15. $\{-3, -1, 0, 4, 6\}$

16. $\{-4, -2, 0, 3, 5\}$

17. $\left\{-\frac{2}{3}, 0, \frac{4}{5}, \frac{12}{5}, \frac{9}{2}, 4.8\right\}$

18. $\left\{-\frac{6}{5}, -\frac{1}{4}, 0, \frac{5}{6}, \frac{13}{4}, 5.2, \frac{11}{2}\right\}$

*Which elements of each set are **(a)** natural numbers, **(b)** whole numbers, **(c)** integers, **(d)** rational numbers, **(e)** irrational numbers, **(f)** real numbers? See Example 3.*

19. $\left\{-8, -\sqrt{5}, -0.6, 0, \frac{3}{4}, \sqrt{3}, \pi, 5, \frac{13}{2}, 17, \frac{40}{2}\right\}$

20. $\left\{-9, -\sqrt{6}, -0.7, 0, \frac{6}{7}, \sqrt{7}, 4.\overline{6}, 8, \frac{21}{2}, 13, \frac{75}{5}\right\}$

Decide whether each statement is true *or* false. *If* false, *tell why. See Example 4.*

21. Every rational number is an integer.

22. Every natural number is an integer.

23. Every irrational number is an integer.

24. Every integer is a rational number.

25. Every natural number is a whole number.

26. Some rational numbers are irrational.

27. Some rational numbers are whole numbers.

28. Some real numbers are integers.

29. The absolute value of any number is the same as the absolute value of its additive inverse.

30. The absolute value of any nonzero number is positive.

Give **(a)** *the additive inverse and* **(b)** *the absolute value of each number. See the discussion of additive inverses and Example 5.*

31. 6

32. 8

33. -12

34. -15

35. $\frac{6}{5}$

36. 0.13

Find the value of each expression. See Example 5.

37. $|-8|$

38. $|-11|$

39. $\left|\frac{3}{2}\right|$

40. $\left|\frac{7}{4}\right|$

41. $-|5|$

42. $-|17|$

43. $-|-2|$

44. $-|-8|$

45. $-|4.5|$

46. $-|12.6|$

47. $|-2| + |3|$

48. $|-16| + |12|$

49. $|-9| - |-3|$

50. $|-10| - |-5|$

51. $|-1| + |-2| - |-3|$

52. $|-6| + |-4| - |-10|$

Solve each problem. See Example 6.

53. The table shows the percent change in population from 2000 through 2006 for selected states.

State	Percent Change
Alabama	3.4
Iowa	1.9
Louisiana	−4.1
Michigan	1.6
North Dakota	−1.0
West Virginia	0.6

Source: U.S. Census Bureau.

(a) Which state had the greatest change in population? What was this change? Was it an increase or a decrease?

(b) Which state had the least change in population? What was this change? Was it an increase or a decrease?

54. The table gives the net trade balance, in millions of dollars, for selected U.S. trade partners for January 2006.

Country	Trade Balance (in millions of dollars)
India	−1257
China	−17,911
Netherlands	756
France	−85
Turkey	−78

Source: U.S. Census Bureau.

A negative balance means that imports to the United States exceeded exports from the United States, while a positive balance means that exports exceeded imports.

(a) Which country had the greatest discrepancy between exports and imports? Explain.

(b) Which country had the least discrepancy between exports and imports? Explain.

Sea level refers to the surface of the ocean. The depth of a body of water such as an ocean or sea can be expressed as a negative number, representing average depth in feet below sea level. On the other hand, the altitude of a mountain can be expressed as a positive number, indicating its height in feet above sea level. The table gives selected depths and heights.

Body of Water	Average Depth in Feet (as a negative number)	Mountain	Altitude in Feet (as a positive number)
Pacific Ocean	−12,925	McKinley	20,320
South China Sea	−4,802	Point Success	14,158
Gulf of California	−2,375	Matlalcueyetl	14,636
Caribbean Sea	−8,448	Rainier	14,410
Indian Ocean	−12,598	Steele	16,644

Source: World Almanac and Book of Facts.

55. List the bodies of water in order, starting with the deepest and ending with the shallowest.

56. List the mountains in order, starting with the shortest and ending with the tallest.

57. *True* or *false:* The absolute value of the depth of the Pacific Ocean is greater than the absolute value of the depth of the Indian Ocean.

58. *True* or *false:* The absolute value of the depth of the Gulf of California is greater than the absolute value of the depth of the Caribbean Sea.

Use order on a number line to answer true *or* false *to each statement.*

59. $-6 < -2$

60. $-4 < -3$

61. $-4 > -3$

62. $-2 > -1$

63. $3 > -2$

64. $5 > -3$

65. $-3 \geq -3$

66. $-4 \leq -4$

Use an inequality symbol to write each statement.

67. 7 is greater than y.

68. -4 is less than 12.

69. 5 is greater than or equal to 5.

70. -3 is less than or equal to -3.

71. $3t - 4$ is less than or equal to 10.

72. $5x + 4$ is greater than or equal to 19.

73. $5x + 3$ is not equal to 0.

74. $6x + 7$ is not equal to -3.

First simplify each side of the inequality. Then tell whether the resulting statement is true *or* false.

75. $-6 < 7 + 3$

76. $-7 < 4 + 2$

77. $2 \cdot 5 \geq 4 + 6$

78. $8 + 7 \leq 3 \cdot 5$

79. $-|-3| \geq -3$

80. $-|-5| \leq -5$

81. $-8 > -|-6|$

82. $-9 > -|-4|$

The graph shows egg production in millions of eggs in selected states for 2005 and 2006. Use this graph to work Exercises 83–87.

83. In 2005, was egg production in Iowa (IA) less than or greater than egg production in California (CA)?

84. In 2006, which states had production greater than 6000 million eggs?

85. In which states was 2006 egg production less than 2005 production?

86. If x represents 2006 egg production for Texas (TX) and y represents 2006 egg production for California (CA), which is true: $x < y$ or $x > y$?

87. If x represents 2005 egg production for Ohio (OH) and y represents 2006 egg production for Ohio (OH), which is true: $x < y$ or $x > y$?

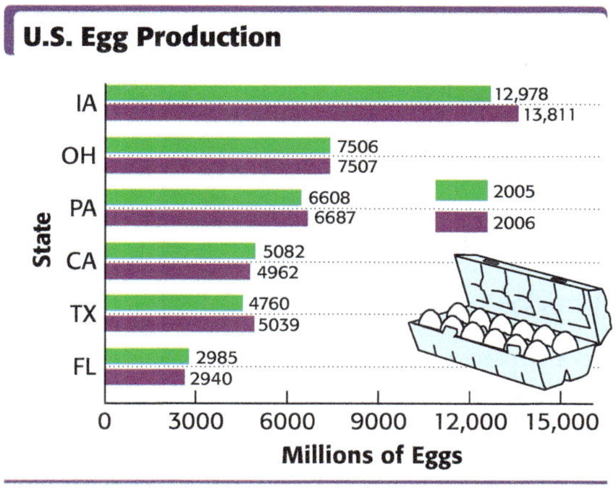

Source: U.S. Department of Agriculture.

13.2 Linear Inequalities in One Variable

OBJECTIVES

1. **Graph intervals on a number line.**
2. **Solve linear inequalities using the addition property.**
3. **Solve linear inequalities using the multiplication property.**
4. **Solve linear inequalities with three parts.**
5. **Solve applied problems using linear inequalities.**

Solving inequalities is closely related to solving equations. In this section we introduce properties for solving inequalities.

Inequalities are algebraic expressions related by

$<$ "is less than,"	$\leq$ "is less than or equal to,"
$>$ "is greater than,"	$\geq$ "is greater than or equal to."

We solve an inequality by finding all real number solutions for it. For example, the solution set of $x \leq 2$ includes *all* real numbers that are less than or equal to 2, not just the integers less than or equal to 2. For example, -2.5, -1.7, -1, $\frac{1}{2}$, $\sqrt{2}$, $\frac{7}{4}$, and 2 are real numbers less than or equal to 2 and are therefore solutions of $x \leq 2$.

OBJECTIVE 1 Graph intervals on a number line. A good way to show the solution set of an inequality is by graphing. We graph all the real numbers satisfying $x \leq 2$ by placing a square bracket at 2 on a number line and drawing an arrow extending from the bracket to the left (to represent the fact that all numbers less than 2 are also part of the graph). The graph is shown in Figure 1.

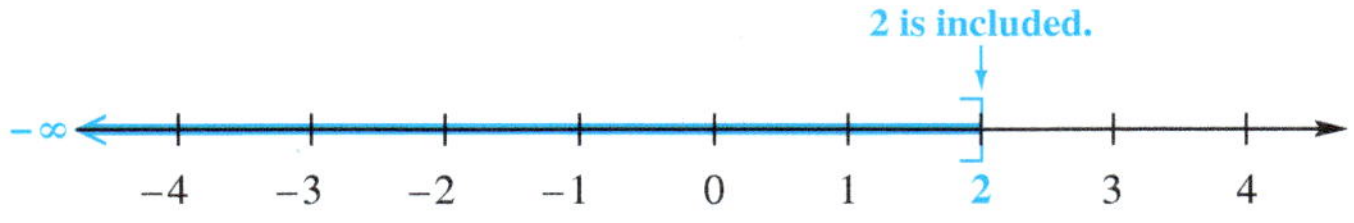

Figure 1

The set of numbers less than or equal to 2 is an example of an **interval** on the number line. To write intervals, we use **interval notation.** For example, the interval of all numbers less than or equal to 2 is written $(-\infty, 2]$. The negative infinity symbol $-\infty$ does not indicate a number. It is used to show that the interval includes all real numbers less than 2. As on the number line, the square bracket indicates that 2 is included in the solution set. ***A parenthesis is always used next to the infinity symbol. The set of real numbers is written in interval notation as*** $(-\infty, \infty)$.

Example 1 Graphing Intervals Written in Interval Notation on Number Lines

Write each inequality in interval notation and graph it.

(a) $x > -5$

The statement $x > -5$ says that x can represent any number greater than -5, but x cannot equal -5. This interval is written $(-5, \infty)$. We show this solution set on a number line by placing a parenthesis at -5 and drawing an arrow to the right, as in Figure 2. The parenthesis at -5 shows that -5 is *not* part of the graph.

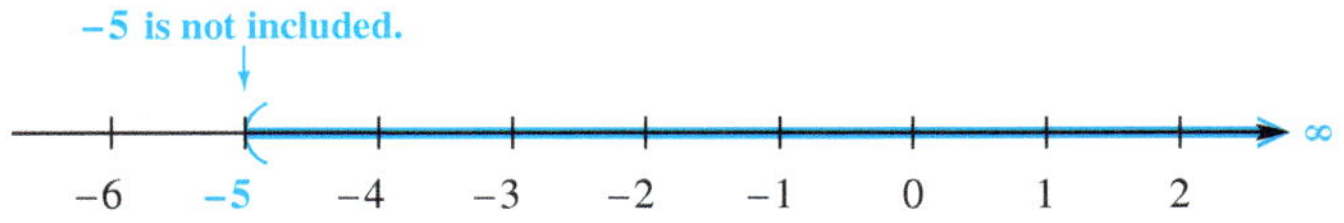

Figure 2

Continued on Next Page

(b) $-1 \leq x < 3$

This statement is read "-1 is less than or equal to x *and* x is less than 3." Thus, we want the set of numbers that are *between* -1 and 3, with -1 included and 3 excluded. In interval notation, we write the solution set as $[-1, 3)$, using a square bracket at -1 because it is part of the graph and a parenthesis at 3 because it is not part of the graph. See Figure 3.

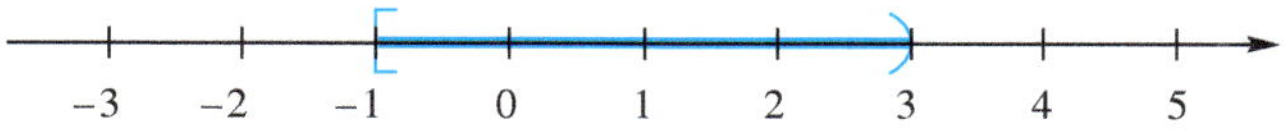

Figure 3

Work Problem **1** at the Side. ▶

We now summarize the various types of intervals.

Type of Interval	Set-Builder Notation	Interval Notation	Graph
Open interval	$\{x \mid a < x < b\}$	(a, b)	
Closed interval	$\{x \mid a \leq x \leq b\}$	$[a, b]$	
Half-open (or half-closed) interval	$\{x \mid a \leq x < b\}$	$[a, b)$	
	$\{x \mid a < x \leq b\}$	$(a, b]$	
Disjoint interval*	$\{x \mid x < a \text{ or } x > b\}$	$(-\infty, a) \cup (b, \infty)$	
Infinite interval	$\{x \mid x > a\}$	(a, ∞)	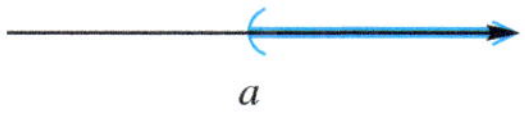
	$\{x \mid x \geq a\}$	$[a, \infty)$	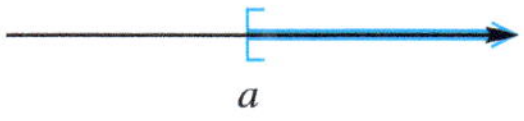
	$\{x \mid x < a\}$	$(-\infty, a)$	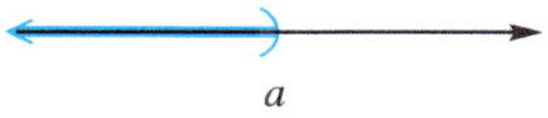
	$\{x \mid x \leq a\}$	$(-\infty, a]$	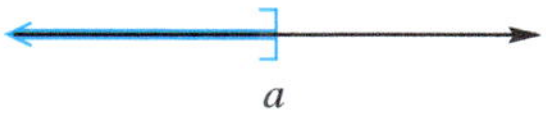
	$\{x \mid x \text{ is a real number}\}$	$(-\infty, \infty)$	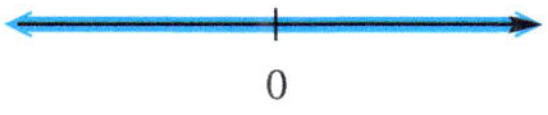

An **inequality** is a statement with algebraic expressions related by $<$, $\leq$, $>$, or $\geq$.

Linear Inequality

A **linear inequality in one variable** can be written in the form

$$Ax + B < C,$$

where A, B, and C are real numbers, with $A \neq 0$.

(While we give definitions and rules only for $<$, they are also valid for $>$, $\leq$, and $\geq$.)

*We will work with disjoint intervals in **Section 13.3** when we study *set operations* and *compound inequalities*.

1 Write each inequality in interval notation and graph it.

(a) $x < -1$

(b) $x \geq -3$

(c) $-4 \leq x < 2$

(d) $0 < x < 3.5$

Answers

1. (a) $(-\infty, -1)$

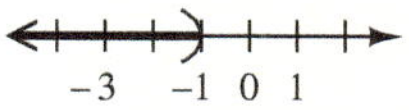

(b) $[-3, \infty)$

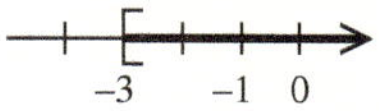

(c) $[-4, 2)$

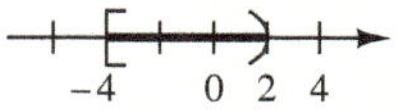

(d) $(0, 3.5)$

2 Solve each inequality, check your solutions, and graph the solution set.

(a) $x - 3 < -9$

(b) $p + 6 < 8$

Examples of linear inequalities include

$$x + 5 < 2, \quad t - 3 \geq 5, \quad \text{and} \quad 2k + 5 \leq 10. \qquad \text{Linear inequalities}$$

OBJECTIVE 2 Solve linear inequalities using the addition property. We solve an inequality by finding all numbers that make the inequality true. Usually, an inequality has an infinite number of solutions. These solutions, like solutions of equations, are found by producing a series of simpler equivalent inequalities. **Equivalent inequalities** are inequalities with the same solution set. We use the *addition property of inequality* to produce equivalent inequalities.

Addition Property of Inequality

For all real numbers A, B, and C, the inequalities

$$A < B \quad \text{and} \quad A + C < B + C$$

are equivalent.

In words, adding the same number to each side of an inequality does not change the solution set.

Example 2 Using the Addition Property of Inequality

Solve $x - 7 < -12$, and graph the solution set.

$$x - 7 < -12$$
$$x - 7 + 7 < -12 + 7 \qquad \text{Add 7.}$$
$$x < -5$$

Check Substitute -5 for x in the *equation* $x - 7 = -12$.

$$x - 7 = -12$$
$$-5 - 7 \stackrel{?}{=} -12 \qquad \text{Let } x = -5.$$
$$-12 = -12 \qquad \text{True}$$

The result, a true statement, shows that -5 is the boundary point. Now we test a number on each side of -5 to verify that numbers *less than* -5 make the *inequality* true. We choose -4 and -6.

$$x - 7 < -12$$

$-4 - 7 \stackrel{?}{<} -12$ Let $x = -4$.	$-6 - 7 \stackrel{?}{<} -12$ Let $x = -6$.
$-11 < -12$ False	$-13 < -12$ True
-4 is *not* in the solution set.	-6 is in the solution set.

Thus, $(-\infty, -5)$, the infinite interval graphed in Figure 4, is the solution set.

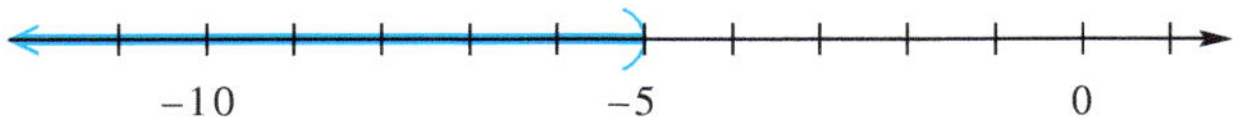

Figure 4

As with equations, the addition property of inequality can be used to *subtract* the same number from each side of an inequality. For example, to solve the inequality $x + 4 > 10$, we subtract 4 from each side to get $x > 6$.

◀ Work Problem **2** at the Side.

Answers

2. (a) $(-\infty, -6)$

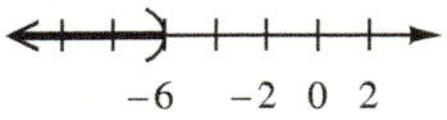

(b) $(-\infty, 2)$

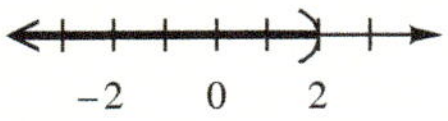

Example 3 **Using the Addition Property of Inequality**

Solve $14 + 2m \leq 3m$, and graph the solution set.

$$14 + 2m \leq 3m$$

$$14 + 2m - 2m \leq 3m - 2m \qquad \text{Subtract } 2m.$$

$$14 \leq m \qquad \text{Combine like terms.}$$

$$m \geq 14 \qquad \text{Rewrite.}$$

Pay careful attention here.

The inequality $14 \leq m$ (14 is less than or equal to m) can also be written $m \geq 14$ (m is greater than or equal to 14). ***Notice that in each case, the inequality symbol points to the lesser number,* 14.**

Check

$$14 + 2m = 3m$$

$$14 + 2(14) \stackrel{?}{=} 3(14) \qquad \text{Let } m = 14.$$

$$42 = 42 \qquad \text{True}$$

So 14 satisfies the equality part of $\leq$. Choose 10 and 15 as test points.

$$14 + 2m < 3m$$

$14 + 2(10) \stackrel{?}{<} 3(10)$ Let $m = 10$.	$14 + 2(15) \stackrel{?}{<} 3(15)$ Let $m = 15$.
$34 < 30$ False	$44 < 45$ True
10 is not in the solution set.	15 is in the solution set.

The check confirms that $[14, \infty)$ is the solution set. See Figure 5.

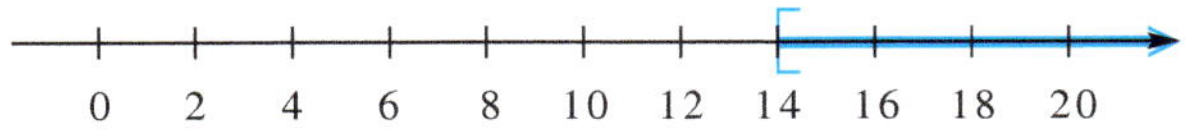

Figure 5

Work Problem **3** *at the Side.* ▶

CAUTION
To avoid errors, rewrite an inequality such as $14 \leq m$ as $m \geq 14$ so that the variable is on the left, as in Example 3.

OBJECTIVE 3 Solve linear inequalities using the multiplication property. Solving an inequality such as $3x \leq 15$ requires dividing each side by 3 using the *multiplication property of inequality*. Consider

$$-2 < 5,$$

a true statement. Multiply each side by 8.

$$-2(8) < 5(8) \qquad \text{Multiply by 8.}$$

$$-16 < 40 \qquad \text{True}$$

The result is true. Start again with $-2 < 5$, and multiply each side by -8.

$$-2(-8) < 5(-8) \qquad \text{Multiply by } -8.$$

$$16 < -40 \qquad \text{False}$$

The result, $16 < -40$, is false. To make it true, we must change the direction of the inequality symbol to get

$$16 > -40. \qquad \text{True}$$

Work Problem **4** *at the Side.* ▶

3 Solve $2k - 5 \geq 1 + k$, check, and graph the solution set.

4 Multiply both sides of each inequality by -5. Then insert the correct symbol, either $<$ or $>$, in the first blank, and fill in the other blanks in part (b).

(a) $7 < 8$

-35 ______ -40

(b) $-1 > -4$

5 ______ ______

ANSWERS

3. $[6, \infty)$

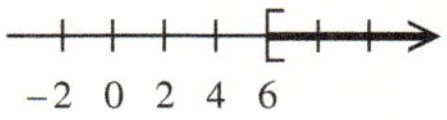

4. **(a)** $>$ **(b)** $<$; 20

5 Solve, check, and graph the solution set of each inequality.

(a) $2x < -10$

(b) $-7k \geq 8$

(c) $-9m < -81$

As these examples suggest, multiplying each side of an inequality by a *negative* number reverses the direction of the inequality symbol. The same is true for dividing by a negative number since division is defined in terms of multiplication.

Multiplication Property of Inequality

For all real numbers A, B, and C, with $C \neq 0$,

(a) the inequalities

$$A < B \quad \text{and} \quad AC < BC \quad \text{are equivalent } \mathbf{if}\ C > 0;$$

(b) the inequalities

$$A < B \quad \text{and} \quad AC > BC \quad \text{are equivalent } \mathbf{if}\ C < 0.$$

In words, each side of an inequality may be multiplied (or divided) by a *positive* number without changing the direction of the inequality symbol. ***Multiplying (or dividing) by a negative number requires that we reverse the direction of the inequality symbol.***

Example 4 Using the Multiplication Property of Inequality

Solve each inequality, and graph the solution set.

(a) $5m \leq -30$

Use the multiplication property to divide each side by 5. ***Since 5 is positive, do not reverse the direction of inequality symbol.***

$$5m \leq -30$$

$$\frac{5m}{5} \leq \frac{-30}{5} \qquad \text{Divide by 5.}$$

$$m \leq -6$$

Check that the solution set is the interval $(-\infty, -6]$, graphed in Figure 6.

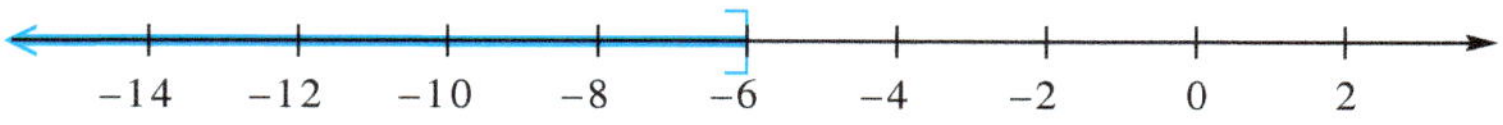

Figure 6

(b) $-4k \leq 32$

Divide each side by -4. ***Since -4 is negative, reverse the direction of inequality symbol.***

$$-4k \leq 32$$

$$\frac{-4k}{-4} \geq \frac{32}{-4} \qquad \text{Divide by } -4\text{; reverse the direction of the symbol.}$$

$$k \geq -8$$

Reverse the inequality symbol when dividing by a negative number.

Check the solution set. Figure 7 shows the graph of the solution set, $[-8, \infty)$.

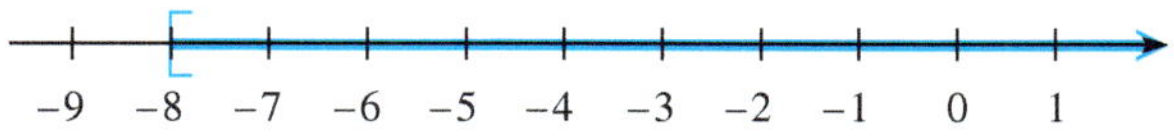

Figure 7

◀ *Work Problem* **5** *at the Side.*

Answers

5. (a) $(-\infty, -5)$

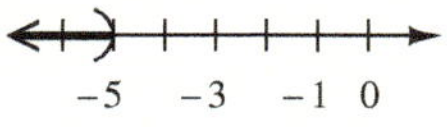

(b) $\left(-\infty, -\frac{8}{7}\right]$

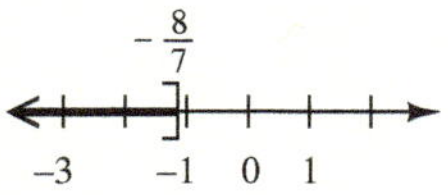

(c) $(9, \infty)$

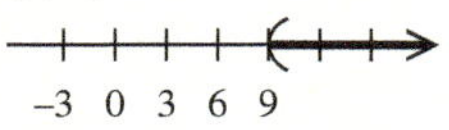

Solving a Linear Inequality

Step 1 **Simplify each side separately.** Clear parentheses, fractions, and decimals using the distributive property as needed, and combine like terms.

Step 2 **Isolate the variable terms on one side.** Use the addition property of inequality to get all terms with variables on one side of the inequality and all numbers on the other side.

Step 3 **Isolate the variable.** Use the multiplication property of inequality to change the inequality to the form $x < k$ or $x > k$.

CAUTION

Reverse the direction of the inequality symbol only when multiplying or dividing each side of an inequality by a negative number.

Example 5 Solving a Linear Inequality

Solve $-3(x + 4) + 2 \geq 7 - x$, and graph the solution set.

Step 1
$$-3(x+4) + 2 \geq 7 - x$$
$$-3x - 12 + 2 \geq 7 - x \quad \text{Distributive property}$$
$$-3x - 10 \geq 7 - x \quad \text{Combine like terms.}$$

Step 2
$$-3x - 10 + x \geq 7 - x + x \quad \text{Add } x.$$
$$-2x - 10 \geq 7 \quad \text{Combine like terms.}$$
$$-2x - 10 + 10 \geq 7 + 10 \quad \text{Add 10.}$$
$$-2x \geq 17 \quad \text{Combine like terms.}$$

Step 3
$$\frac{-2x}{-2} \leq \frac{17}{-2} \quad \text{Divide by } -2\text{; change } \geq \text{ to } \leq.$$
$$x \leq -\frac{17}{2}$$

Be sure to reverse the inequality symbol.

Figure 8 shows the graph of the solution set, $(-\infty, -\frac{17}{2}]$.

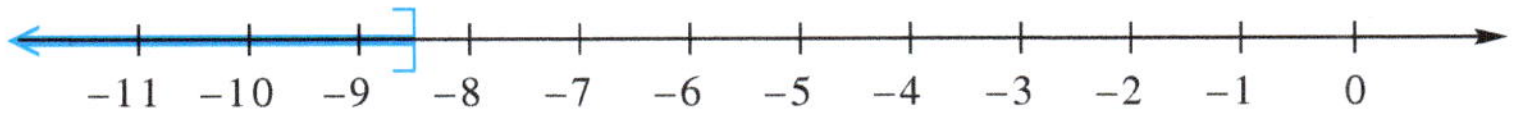

Figure 8

Work Problem 6 at the Side. ▶

6 Solve, check, and graph the solution set of each inequality.

(a) $x + 4(2x - 1) \geq x + 2$

(b) $m - 2(m - 4) \leq 3m$

Note

In Step 2 of Example 5, if we add $3x$ (instead of x) to both sides, we get

$$-3x - 10 + 3x \geq 7 - x + 3x \quad \text{Add } 3x.$$
$$-10 \geq 2x + 7$$
$$-10 - 7 \geq 2x + 7 - 7 \quad \text{Subtract 7.}$$
$$-17 \geq 2x$$
$$-\frac{17}{2} \geq x, \quad \text{or} \quad x \leq -\frac{17}{2}. \quad \text{Divide by 2; rewrite.}$$

The result "$-\frac{17}{2}$ is greater than or equal to x" means the same thing as "x is less than or equal to $-\frac{17}{2}$." Thus, the solution set is the same.

ANSWERS

6. (a) $\left[\frac{3}{4}, \infty\right)$

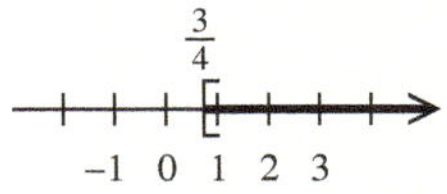

(b) $[2, \infty)$

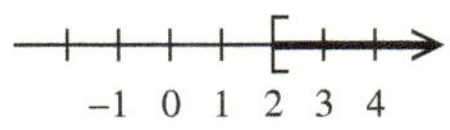

Example 6 **Solving a Linear Inequality with Fractions**

Solve $-\frac{2}{3}(r-3)-\frac{1}{2}<\frac{1}{2}(5-r)$, and graph the solution set.

Step 1 To clear fractions, multiply by the least common denominator, 6.

$$-\frac{2}{3}(r-3)-\frac{1}{2}<\frac{1}{2}(5-r)$$

$$6\left[-\frac{2}{3}(r-3)-\frac{1}{2}\right]<6\left[\frac{1}{2}(5-r)\right] \quad \text{Multiply by 6, the LCD.}$$

Be careful here.

$$6\left[-\frac{2}{3}(r-3)\right]-6\left(\frac{1}{2}\right)<6\left[\frac{1}{2}(5-r)\right] \quad \text{Distributive property}$$

$$-4(r-3)-3<3(5-r) \quad \text{Multiply.}$$

$$-4r+12-3<15-3r \quad \text{Distributive property}$$

$$-4r+9<15-3r$$

Step 2

$$-4r+9+3r<15-3r+3r \quad \text{Add } 3r.$$

$$-r+9<15$$

$$-r+9-9<15-9 \quad \text{Subtract 9.}$$

$$-r<6$$

Step 3

$$-1(-r)>-1(6) \quad \text{Multiply by } -1\text{; change } < \text{ to } >.$$

$$r>-6$$

Check that the solution set is $(-6, \infty)$. See Figure 9.

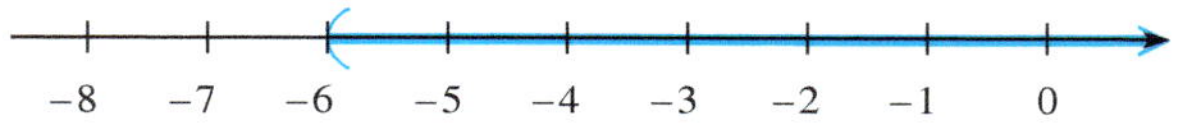

Figure 9

◀ Work Problem **7** at the Side.

OBJECTIVE 4 **Solve linear inequalities with three parts.** For some applications, it is necessary to work with a **three-part inequality** such as

$$3<x+2<8,$$

where $x+2$ is *between* 3 and 8. We solve this inequality as follows.

$$3-2<x+2-2<8-2 \quad \text{Subtract 2 from } each \text{ part.}$$

$$1<x<6$$

Thus, x must be between 1 and 6 so that $x+2$ will be between 3 and 8. The solution set, the open interval $(1, 6)$, is graphed in Figure 10.

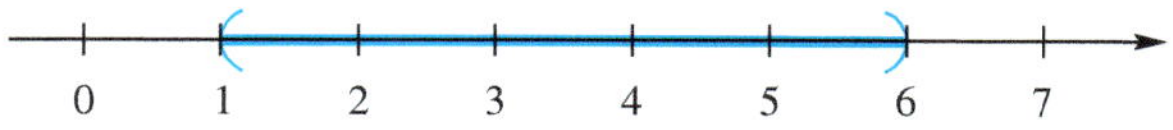

Figure 10

CAUTION

In three-part inequalities, the order of the parts is important. For example, do *not* write $8<x+2<3$, since this implies that $8<3$, a false statement. ***We write three-part inequalities so that the symbols point in the same direction, and both point toward the lesser number.***

◀ Work Problem **8** at the Side.

7 Solve, check, and graph the solution set of each inequality.

(a) $5-3(m-1) \leq 2(m+3)+1$

(b) $\frac{1}{4}(m+3)+2 \leq \frac{3}{4}(m+8)$

8 Rewrite each three-part inequality using the order in which the numbers appear on the number line.

(a) $1>x>-1$

(b) $16 \geq p \geq 11$

(c) $-2>t \geq -8$

ANSWERS

7. (a) $\left[\frac{1}{5}, \infty\right)$

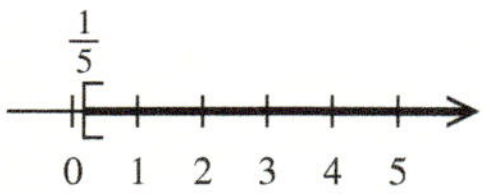

(b) $\left[-\frac{13}{2}, \infty\right)$

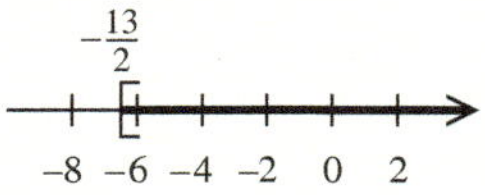

8. (a) $-1<x<1$
(b) $11 \leq p \leq 16$
(c) $-8 \leq t<-2$

Example 7 **Solving a Three-Part Inequality**

Solve $-2 \le -3k - 1 \le 5$, and graph the solution set.

Begin by adding 1 to each of the three parts to isolate the variable term in the middle.

$$-2 + 1 \le -3k - 1 + 1 \le 5 + 1 \quad \text{Add 1 to each part.}$$

$$-1 \le -3k \le 6$$

$$\frac{-1}{-3} \ge \frac{-3k}{-3} \ge \frac{6}{-3} \quad \text{Divide each part by } -3\text{; reverse the direction of the inequality symbols.}$$

$$\frac{1}{3} \ge k \ge -2$$

$$-2 \le k \le \frac{1}{3} \quad \text{Rewrite in order based on the number line.}$$

Check that the solution set is the closed interval $[-2, \frac{1}{3}]$. See Figure 11.

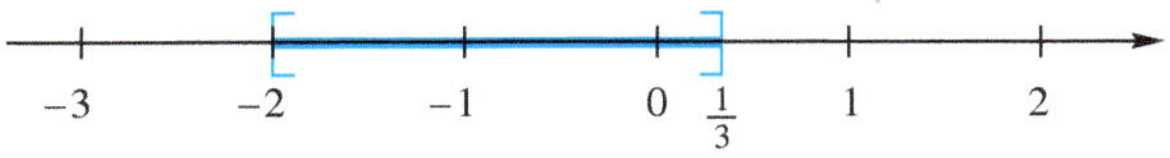

Figure 11

Work Problem 9 at the Side. ▶

9 Solve, check, and graph the solution set of each inequality.

(a) $-3 \le x - 1 \le 7$

(b) $5 < 3x - 4 < 9$

Examples of the types of solution sets to be expected from solving linear equations and linear inequalities are shown below.

SOLUTIONS OF LINEAR EQUATIONS AND INEQUALITIES

Equation or Inequality	Typical Solution Set	Graph of Solution Set
Linear equation $5x + 4 = 14$	$\{2\}$	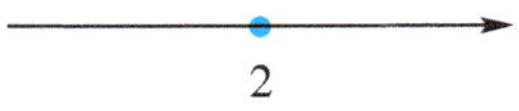
Linear inequality $5x + 4 < 14$	$(-\infty, 2)$	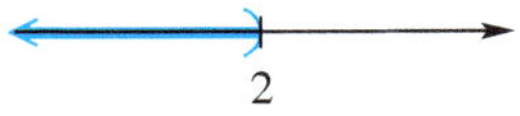
$5x + 4 > 14$	$(2, \infty)$	
Three-part inequality $-1 \le 5x + 4 \le 14$	$[-1, 2]$	

OBJECTIVE 5 **Solve applied problems using linear inequalities.** Besides the familiar "is less than" and "is greater than," other expressions also indicate inequalities, as shown in the table below.

Word Expression	Interpretation
a exceeds *b*	$a > b$
a is at least *b*	$a \ge b$
a is no less than *b*	$a \ge b$
a is at most *b*	$a \le b$
a is no more than *b*	$a \le b$

Answers

9. (a) $[-2, 8]$

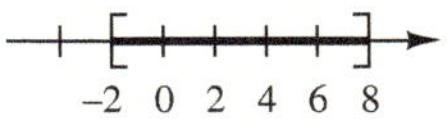

(b) $\left(3, \frac{13}{3}\right)$

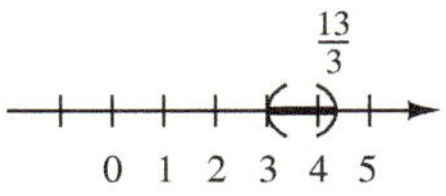

In Examples 8 and 9, we use the six problem-solving steps from **Section 2.3,** changing Step 3 from "Write an equation" to "Write an inequality."

10 Solve the problem.

A rental company charges \$10 to rent a leaf blower, plus \$7.50 per hr. Marge Ruhberg can spend no more than \$40 to blow leaves from her driveway and pool deck. What is the *maximum* amount of time she can use the rented leaf blower?

Example 8 **Using a Linear Inequality to Solve a Rental Problem**

A rental company charges \$20 to rent a chain saw, plus \$9 per hr. Tom Ruhberg can spend no more than \$65 to clear some logs from his yard. What is the *maximum* amount of time he can use the rented saw?

Step 1 **Read** the problem again.

Step 2 **Assign a variable.** Let h = the number of hours he can rent the saw.

Step 3 **Write an inequality.** He must pay \$20, plus \$9h, to rent the saw for h hours, and this amount must be *no more than* \$65.

$$\underbrace{20 + 9h}_{\text{Cost of renting}} \quad \underbrace{\leq}_{\text{is no more than}} \quad \underbrace{65}_{\text{65 dollars.}}$$

Step 4 **Solve.**

$$9h \leq 45 \qquad \text{Subtract 20.}$$

$$h \leq 5 \qquad \text{Divide by 9.}$$

Step 5 **State the answer.** He can use the saw for a maximum of 5 hr. (He may use it for less time, as indicated by the inequality $h \leq 5$.)

Step 6 **Check.** If Tom uses the saw for **5** hr, he will spend

$$20 + 9(5) = 65 \text{ dollars,} \quad \text{the maximum amount.}$$

◀ *Work Problem* **10** *at the Side.*

11 Solve the problem.

Alex Lose has grades of 92, 90, and 84 on his first three history tests. What grade must he make on his fourth test in order to keep an average of at least 90?

Example 9 **Finding an Average Test Score**

Emma Saska has scores of 88, 86, and 90 on her first three algebra tests. An average score of at least 90 will earn an A in the class. What possible scores on her fourth test will earn her an A average?

Let x represent the score on the fourth test. Her average score must be at least 90. To find the average of four numbers, add them and then divide by 4.

$$\underbrace{\frac{88 + 86 + 90 + x}{4}}_{\text{Average}} \quad \underbrace{\geq}_{\text{is at least}} \quad \underbrace{90}_{90.}$$

$$\frac{264 + x}{4} \geq 90 \qquad \text{Add the scores.}$$

$$264 + x \geq 360 \qquad \text{Multiply by 4.}$$

$$x \geq 96 \qquad \text{Subtract 264.}$$

She must score **96** or more on her fourth test.

Check $\dfrac{88 + 86 + 90 + 96}{4} = \dfrac{360}{4} = 90,$ the minimum score.

A score of 96 or more will give an average of at least 90, as required.

◀ *Work Problem* **11** *at the Side.*

Answers

10. 4 hr
11. at least 94

13.2 Exercises

FOR EXTRA HELP

MyMathLab | Math XL PRACTICE | WATCH | DOWNLOAD | READ | REVIEW

Match each inequality in Column I with the correct graph or interval in Column II.

I	II
1. $x \le 3$	**A.** 
2. $x > 3$	**B.** 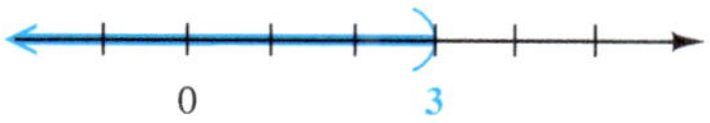
3. $x < 3$	**C.** $(3, \infty)$
4. $x \ge 3$	**D.** $(-\infty, 3]$
5. $-3 \le x \le 3$	**E.** $(-3, 3)$
6. $-3 < x < 3$	**F.** $[-3, 3]$

7. Refer to the graph, and write an inequality or a three-part inequality for each description.

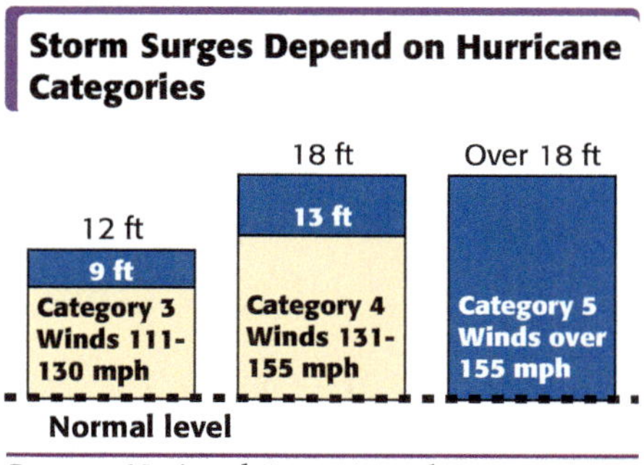

Source: National Oceanic and Atmospheric Administration.

(a) The wind speed s (in miles per hour) of a Category 4 hurricane

(b) The wind speed s (in miles per hour) of a Category 5 hurricane

(c) The storm surge x (in feet) from a Category 3 hurricane

(d) The storm surge x (in feet) from a Category 5 hurricane

8. Dr. Paul Donohue writes a syndicated column in which readers question him on a variety of health topics. Reader C. J. wrote, "Many people say they can weigh more because they have a large frame. How is frame size determined?" Here is Dr. Donohue's response:

> *"For a man, a wrist circumference between* 6.75 *and* 7.25 in. *[inclusive] indicates a medium frame. Anything above is a large frame and anything below, a small frame."*

Using x to represent wrist circumference in inches, write an inequality or a three-part inequality that represents wrist circumference for a male with the indicated frame size.

(a) Small frame

(b) Medium frame

(c) Large frame

(*Source: The Gazette,* Cedar Rapids, Iowa, October 4, 2004.)

9. A student solved the following inequality as shown.

$$4x \ge -64$$

$$\frac{4x}{4} \le \frac{-64}{4}$$

$$x \le -16$$

Solution set: $(-\infty, -16]$

WHAT WENT WRONG? Give the correct solution set.

10. Explain how you will determine whether to use parentheses or brackets when graphing the solution set of an inequality.

Solve each inequality, giving solution sets in both interval and graph forms. Check your answers. See Examples 1–6.

11. $x - 4 \leq 3$

12. $t - 3 \leq 1$

13. $4x + 1 \geq 21$

14. $5t + 2 \geq 52$

15. $5x > -25$

16. $7x < -28$

17. $-4x < 16$

18. $-2m > 10$

19. $-\frac{3}{4}r \geq 30$

20. $-\frac{2}{3}x \leq 12$

21. $-1.3m \geq -5.2$

22. $-2.5x \leq -1.25$

23. $\frac{3k - 1}{4} > 5$

24. $\frac{5z - 6}{8} < 8$

25. $\frac{2k - 5}{-4} > 5$

26. $\frac{3z - 2}{-5} < 6$

27. $3k + 1 < -20$

28. $5z + 6 > -29$

29. $x + 4(2x - 1) \geq x$

30. $m - 2(m - 4) \leq 3m$

31. $-(4 + r) + 2 - 3r < -14$

32. $-(9 + k) - 5 + 4k \geq 4$

33. $-3(z - 6) > 2z - 2$

34. $-2(x + 4) \leq 6x + 16$

35. $\frac{2}{3}(3k - 1) \geq \frac{3}{2}(2k - 3)$

36. $\frac{7}{5}(10m - 1) < \frac{2}{3}(6m + 5)$

37. $-\frac{1}{4}(p + 6) + \frac{3}{2}(2p - 5) < 10$

38. $\frac{3}{5}(k - 2) - \frac{1}{4}(2k - 7) \leq 3$

Relating Concepts (Exercises 39–43) For Individual or Group Work

Work Exercises 39–43 in order.

39. Solve the linear equation

$$5(x + 3) - 2(x - 4) = 2(x + 7),$$

and graph the solution set on a number line.

40. Solve the linear inequality

$$5(x + 3) - 2(x - 4) > 2(x + 7),$$

and graph the solution set on a number line.

41. Solve the linear inequality

$$5(x + 3) - 2(x - 4) < 2(x + 7),$$

and graph the solution set on a number line.

42. Graph all the solution sets of the equation and inequalities in Exercises 39–41 on the same number line. What set do you obtain?

43. Based on the results of Exercises 39–41, complete the following using a conjecture (educated guess): The solution set of

$$-3(x + 2) = 3x + 12$$

is $\{-3\}$, and the solution set of

$$-3(x + 2) < 3x + 12$$

is $(-3, \infty)$. Therefore the solution set of

$$-3(x + 2) > 3x + 12$$

is ____________.

44. Which is the graph of $-2 < x$?

A. 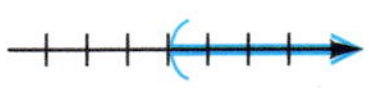

B. 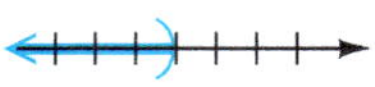

C. 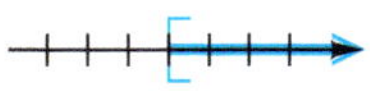

D. 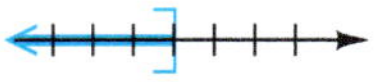

Solve each inequality, giving solution sets in both interval and graph forms. Check your answers. See Example 7.

45. $-4 < x - 5 < 6$

46. $-1 < x + 1 < 8$

47. $-9 \leq k + 5 \leq 15$

48. $-4 \leq m + 3 \leq 10$

49. $-6 \leq 2(z + 2) \leq 16$

50. $-15 < 3(p + 2) < 24$

51. $-16 < 3t + 2 < -10$

52. $-19 < 3x - 5 \leq 1$

53. $4 < -9x + 5 \leq 8$

54. $4 < -2x + 3 \leq 8$

55. $-1 \leq \frac{2x - 5}{6} \leq 5$

56. $-3 < \frac{3m + 1}{4} \leq 3$

The weather forecast by time of day for the U.S. Olympic Track and Field Trials, in Sacramento, California, is shown in the figure. Use this graph to work Exercises 57–60.

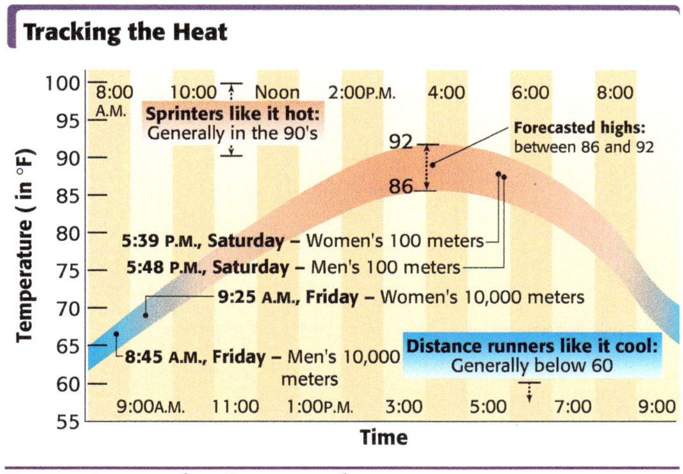

***Source*:** Accuweather, Bee research.

57. Sprinters prefer Fahrenheit temperatures in the 90's. Using the upper boundary of the forecast, in what time period is the temperature expected to be at least 90°F?

58. Distance runners prefer cool temperatures. During what time period are temperatures predicted to be no more than 70°F? Use the lower forecast boundary.

59. What range of temperatures is predicted for the Women's 100-m event?

60. What range of temperatures is forecast for the Men's 10,000-m event?

Solve each problem. See Examples 8 and 9.

61. Amber is signing up for cell phone service. She is trying to decide between Plan A, which costs \$54.99 a month with a free phone included, and Plan B, which costs \$49.99 a month, but would require her to buy a phone for \$129. Under either plan, Amber does not expect to go over the included number of monthly minutes. After how many months would Plan B be a better deal?

62. Stuart and Tracy Sudak need to rent a truck to move their belongings to their new apartment. They can rent a truck of the size they need from U-Haul for \$29.95 a day plus 28 cents per mile or from Budget Truck Rentals for \$34.95 a day plus 25 cents per mile. After how many miles would the Budget rental be a better deal than the U-Haul one?

63. Bonnie Boehme earned scores of 90 and 82 on her first two tests in English Literature. What score must she make on her third test to keep an average of 84 or greater?

64. Scott Barnett scored 92 and 96 on his first two tests in Methods in Teaching Mathematics. What score must he make on his third test to keep an average of 90 or greater?

A product will produce a profit only when the revenue (R) from selling the product exceeds the cost (C) of producing it. Find the least whole number of units x that must be sold for each business to show a profit for the item described.

65. Peripheral Visions, Inc. finds that the cost to produce x studio-quality DVDs is

$$C = 20x + 100,$$

while the revenue produced from them is $R = 24x$ (C and R in dollars).

66. Speedy Delivery finds that the cost to make x deliveries is

$$C = 3x + 2300,$$

while the revenue produced from them is $R = 5.50x$ (C and R in dollars).

67. A BMI (body mass index) between 19 and 25 is considered healthy. Use the formula

$$\text{BMI} = \frac{704 \times (\text{weight in pounds})}{(\text{height in inches})^2}$$

to find the weight range w, to the nearest pound, that gives a healthy BMI for each height. (*Source: Washington Post.*)

(a) 72 in. **(b)** Your height in inches

68. To achieve the maximum benefit from exercising, the heart rate in beats per minute should be in the target heart rate zone (*THR*). For a person aged A, the formula is

$$0.7(220 - A) \leq THR \leq 0.85(220 - A).$$

Find the *THR* to the nearest whole number for each age. (*Source:* Hockey, Robert V., *Physical Fitness: The Pathway to Healthful Living*, Times Mirror/Mosby College Publishing, 1989.)

(a) 35 **(b)** Your age

Find the unknown numbers in each description.

69. Six times a number is between -12 and 12.

70. Half a number is between -3 and 2.

71. When 1 is added to twice a number, the result is greater than or equal to 7.

72. If 8 is subtracted from a number, then the result is at least 5.

73. One third of a number is added to 6, giving a result of at least 3.

74. Three times a number, minus 5, is no more than 7.

Study Skills

USING STUDY CARDS

You may have used "flash cards" in other classes. In math, "study cards" can be helpful. The main things to remember in math besides terms and definitions are *sets of steps to follow* to solve problems (and how to know which set of steps to follow) and *concepts about how math works* (principles). So, the cards may look different but will be just as useful.

In this two-part activity, you will find four types of study cards to use in math. Look carefully at what kinds of information to put on them and where to put it. Then use them the way you would any flash card:

- to quickly review when you have a few minutes,
- for daily reviews,
- to review before a quiz or test.

Remember, the most helpful thing about study cards is making them. After each card description you will find an assignment to try, marked **Now Try This.**

OBJECTIVES

1 **Create study cards for new terms.**

2 **Create study cards for new procedures.**

New Vocabulary Cards

To make a new vocabulary card, put the word (spelled correctly) and the page number where it is found on the front of the card. On the back, write:

- the definition (in your own words if possible),
- an example, including any exceptions or other special information,
- any related words, and
- a sample problem (if appropriate).

Front of Card

Interval notation *p. 128*

Back of Card

Definition: Using symbols to describe an interval on a number line.

Symbols: ∞ $-\infty$ ( ) [ ] ( ] [ )

Use interval notation to tell what numbers are in the solution set for an inequality.

Examples: $(-5, \infty)$ *All numbers greater than* -5, *not including* -5

$[-5, 5)$ *All numbers between* -5 *and* 5, *including* -5, *excluding* 5

Now Try This

List 4 new vocabulary words/concepts you need to learn right now. Make a card for each one.

__________ __________ __________ __________

Procedure ("Steps") Cards

To make a procedure (steps) card, write the name of the procedure at the top on the front of the card. Then write each step *in words*. If you need to learn abbreviations for some words, include them along with the whole words written out. On the back of the card, put an example of the procedure, showing each step you need to take. You can review by looking at the front and practicing a new worked example, or by looking at the back and remembering the procedure and its steps.

Front of Card

Solving a Linear Inequality

1. *Simplify each side separately. (Clear parentheses and combine like terms.)*
2. *Isolate variable terms on one side. (Add or subtract the same number from both sides.)*
3. *Isolate the variable. (Divide both sides by the same number; if dividing by a negative number, reverse direction of inequality.)*

Back of Card

Solve $-3(x+4)+2 \geq 7-x$ *and graph the solution set.*

$-3(x+4)+2 \geq 7-x$	*Clear parentheses.*
$-3x-12+2 \geq 7-x$	*Combine like terms.*
$-3x-10 \geq 7-x$	*Both sides are simplified.*
$-3x-10+x \geq 7-x+x$	*Add x to both sides.*
$-2x-10 \geq 7$	*Variable term sill not isolated.*
$-2x-10+10 \geq 7+10$	*Add 10 to both sides.*
$\frac{-2x}{-2} \geq \frac{17}{-2}$	*Divide both sides by* -2*; dividing by negative, reverse direction of inequality symbol.*
$x \leq -\frac{17}{2}$	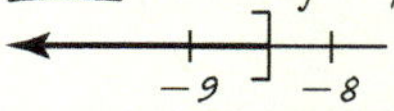 $-\frac{17}{2} = -8\frac{1}{2}$

Now Try This ▶▶▶

What procedure are you learning right now? Write below the steps that you will put on your study card.

Procedure: ______________________

Step 1. ______________________

Step 2 ______________________

Step 3 ______________________

Step 4 ______________________

Step 5 ______________________

13.3 Set Operations and Compound Inequalities

The table shows symptoms of an overactive thyroid and an underactive thyroid.

Underactive Thyroid	Overactive Thyroid
Sleepiness, s	Insomnia, i
Dry hands, d	Moist hands, m
Intolerance of cold, c	Intolerance of heat, h
Goiter, g	Goiter, g

Source: The Merck Manual of Diagnosis and Therapy, 16th Edition, Merck Research Laboratories, 1992.

Let N be the set of symptoms for an underactive thyroid, and let O be the set of symptoms for an overactive thyroid. Suppose we are interested in the set of symptoms that are found in *both* sets N *and* O. In this section, we discuss the use of the words *and* and *or* as they relate to sets and inequalities.

OBJECTIVES

1. **Find the intersection of two sets.**
2. **Solve compound inequalities with the word *and*.**
3. **Find the union of two sets.**
4. **Solve compound inequalities with the word *or*.**

OBJECTIVE 1 Find the intersection of two sets. The intersection of two sets is defined using the word *and*.

Intersection of Sets

For any two sets A and B, the **intersection** of A and B, symbolized $A \cap B$, is defined as follows:

$$A \cap B = \{x \mid x \text{ is an element of } A \textbf{ and } x \text{ is an element of } B\}.$$

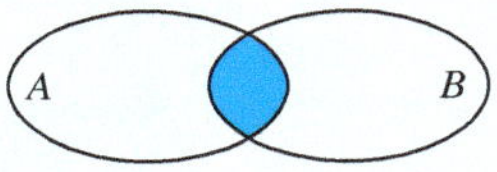

Example 1 Finding the Intersection of Two Sets

Let $A = \{1, 2, 3, 4\}$ and $B = \{2, 4, 6\}$. Find $A \cap B$.

The set $A \cap B$ contains those elements that belong to both A *and* B.

$$A \cap B = \{1, 2, 3, 4\} \cap \{2, 4, 6\}$$
$$= \{2, 4\}$$

Work Problem 1 at the Side. ▶

1 List the elements in each set.

(a) $A \cap B$, if $A = \{3, 4, 5, 6\}$ and $B = \{5, 6, 7\}$

(b) $N \cap O$ (Refer to the thyroid table.)

A **compound inequality** consists of two inequalities linked by a connective word such as *and* or *or*. Examples of compound inequalities are

$$x + 1 \leq 9 \quad \textbf{and} \quad x - 2 \geq 3$$

and

$$2x > 4 \quad \textbf{or} \quad 3x - 6 < 5.$$

Compound inequalities

OBJECTIVE 2 Solve compound inequalities with the word *and*.

Solving a Compound Inequality with *and*

Step 1 Solve each inequality individually.

Step 2 Since the inequalities are joined with *and*, the solution set of the compound inequality will include all numbers that satisfy both inequalities in Step 1 (the intersection of the solution sets).

ANSWERS

1. **(a)** $\{5, 6\}$ **(b)** $\{g\}$

2 Solve each compound inequality, and graph the solution set.

(a) $x < 10$ and $x > 2$

(b) $x + 3 \leq 1$ and $x - 4 \geq -12$

Example 2 Solving a Compound Inequality with *and*

Solve the compound inequality $x + 1 \leq 9$ and $x - 2 \geq 3$.

Step 1 Solve each inequality individually.

$$x + 1 \leq 9 \quad \text{and} \quad x - 2 \geq 3$$
$$x + 1 - 1 \leq 9 - 1 \quad \text{and} \quad x - 2 + 2 \geq 3 + 2$$
$$x \leq 8 \quad \text{and} \quad x \geq 5$$

Step 2 Because the inequalities are joined with the word *and,* the solution set will include all numbers that satisfy *both* inequalities in Step 1 at the same time. Thus, the compound inequality is true whenever $x \leq 8$ and $x \geq 5$ are both true. The top graph in Figure 12 shows $x \leq 8$, and the bottom graph shows $x \geq 5$.

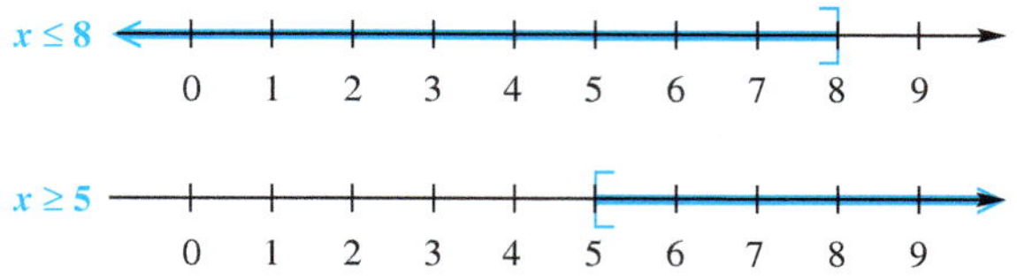

Figure 12

Find the intersection of the two graphs in Figure 12 to get the solution set of the compound inequality. The intersection of the two graphs in Figure 13 shows that the solution set is the closed interval [5, 8].

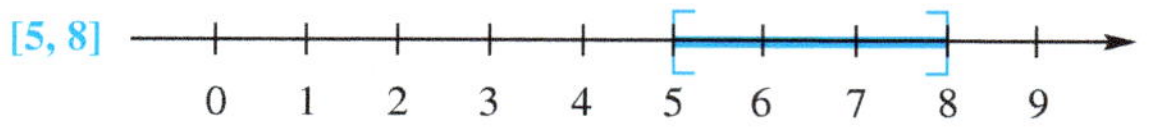

Figure 13

◀ Work Problem **2** at the Side.

3 Solve

$2x \geq x - 1$ and $3x \geq 3 + 2x$,

and graph the solution set.

Example 3 Solving a Compound Inequality with *and*

Solve the compound inequality $-3x - 2 > 5$ and $5x - 1 \leq -21$.

Step 1 Solve each inequality individually.

$$-3x - 2 > 5 \quad \text{and} \quad 5x - 1 \leq -21$$
$$-3x > 7 \quad \text{and} \quad 5x \leq -20$$
$$x < -\frac{7}{3} \quad \text{and} \quad x \leq -4$$

Remember to reverse the inequality symbol.

The graphs of $x < -\frac{7}{3}$ and $x \leq -4$ are shown in Figure 14.

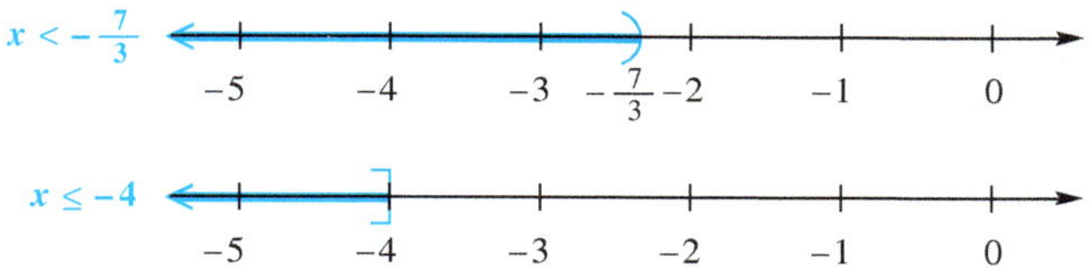

Figure 14

Step 2 Now find all values of x that satisfy both conditions; that is, the real numbers that are less than $-\frac{7}{3}$ and also less than or equal to -4. See Figure 15. The solution set is the infinite interval $(-\infty, -4]$.

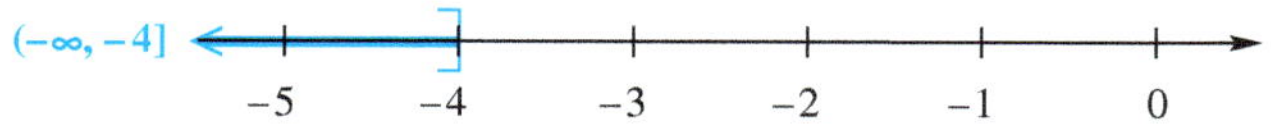

Figure 15

◀ Work Problem **3** at the Side.

Answers

2. (a) (2, 10)

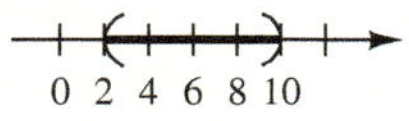

(b) [−8, −2]

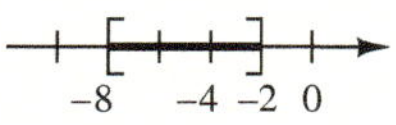

3. [3, ∞)

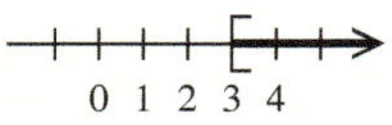

Example 4 Solving a Compound Inequality with *and*

Solve $x + 2 < 5$ and $x - 10 > 2$.

First solve each inequality individually.

$$x + 2 < 5 \quad \text{and} \quad x - 10 > 2$$
$$x < 3 \quad \text{and} \quad x > 12$$

The graphs of $x < 3$ and $x > 12$ are shown in Figure 16.

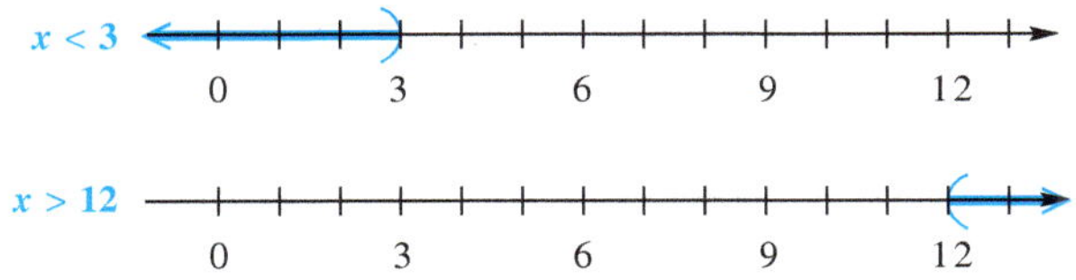

Figure 16

There is no number that is both less than 3 *and* greater than 12, so the given compound inequality has no solution. The solution set is $\emptyset$. See Figure 17.

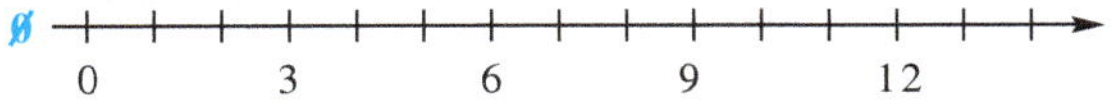

Figure 17

Work Problem 4 at the Side. ▶

OBJECTIVE 3 Find the union of two sets. The union of two sets is defined using the word *or*.

Union of Sets

For any two sets A and B, the **union** of A and B, symbolized $A \cup B$, is defined as follows:

$$A \cup B = \{x \mid x \text{ is an element of } A \textbf{ or } x \text{ is an element of } B\}.$$

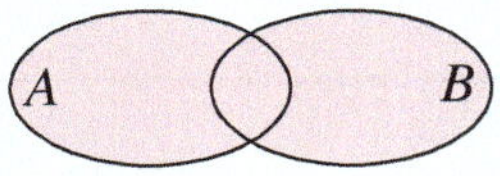

Example 5 Finding the Union of Two Sets

Let $A = \{1, 2, 3, 4\}$ and $B = \{2, 4, 6\}$. Find $A \cup B$.

Begin by listing all the elements of set A: 1, 2, 3, 4. Then list any additional elements from set B. In this case the elements 2 and 4 are already listed, so the only additional element is 6. Therefore,

$$A \cup B = \{1, 2, 3, 4\} \cup \{2, 4, 6\}$$
$$= \{1, 2, 3, 4, 6\}.$$

The union consists of all elements in either A *or* B (or both).

Note

Although the elements 2 and 4 appeared in both sets A and B, they are written only once in $A \cup B$.

Work Problem 5 at the Side. ▶

4 Solve.

(a) $x < 5$ and $x > 5$

(b) $x + 2 > 3$ and $2x + 1 < -3$

5 List the elements in each set.

(a) $A \cup B$, if $A = \{3, 4, 5, 6\}$ and $B = \{5, 6, 7\}$

(b) $N \cup O$ from the thyroid table at the beginning of this section

Answers

4. (a) $\emptyset$ **(b)** $\emptyset$

5. (a) $\{3, 4, 5, 6, 7\}$ **(b)** $\{s, d, c, g, i, m, h\}$

OBJECTIVE 4 Solve compound inequalities with the word *or*. Use the following steps.

Solving a Compound Inequality with *or*

Step 1 Solve each inequality individually.

Step 2 Since the inequalities are joined with *or*, the solution set includes all numbers that satisfy either one of the two inequalities in Step 1 (the union of the solution sets).

Example 6 Solving a Compound Inequality with *or*

Solve $6x - 4 < 2x$ or $-3x \le -9$.

Step 1 Solve each inequality individually.

$$6x - 4 < 2x \quad \text{or} \quad -3x \le -9$$
$$4x < 4$$
$$x < 1 \quad \text{or} \quad x \ge 3$$

Remember to reverse the inequality symbol.

The graphs of these two inequalities are shown in Figure 18.

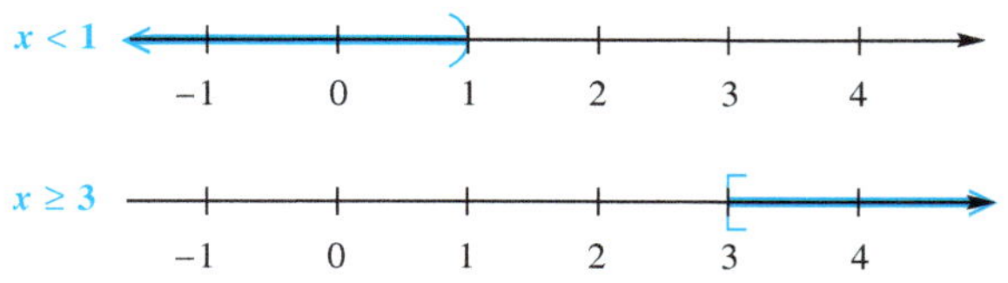

Figure 18

Step 2 Since the inequalities are joined with *or*, we find the union of the two solution sets, as shown in Figure 19. The solution set is the disjoint interval

$$(-\infty, 1) \cup [3, \infty).$$

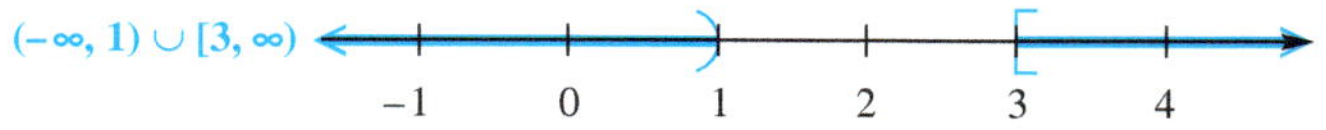

Figure 19

CAUTION

When inequalities are used to write the solution set in Example 6, it *must* be written as

$$x < 1 \quad \text{or} \quad x \ge 3,$$

which keeps the numbers 1 and 3 in their order on the number line. Writing $3 \le x < 1$ would imply that $3 \le 1$, which is ***FALSE.*** There is no other way to write the solution set of such a union.

◀ *Work Problem* 6 *at the Side.*

6 Solve. Give each solution set in both interval and graph forms.

(a) $x + 2 > 3$ or
$2x + 1 < -3$

(b) $x - 1 > 2$ or
$3x + 5 < 2x + 6$

ANSWERS

6. (a) $(-\infty, -2) \cup (1, \infty)$

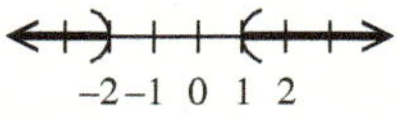

(b) $(-\infty, 1) \cup (3, \infty)$

Example 7 Solving a Compound Inequality with *or*

Solve $-4x + 1 \geq 9$ or $5x + 3 \leq -12$.

First, solve each inequality individually.

$$\begin{aligned} -4x + 1 &\geq 9 &\quad \text{or} \quad 5x + 3 &\leq -12 \\ -4x &\geq 8 &\quad \text{or} \quad 5x &\leq -15 \\ x &\leq -2 &\quad \text{or} \quad x &\leq -3 \end{aligned}$$

The graphs of these two inequalities are shown in Figure 20.

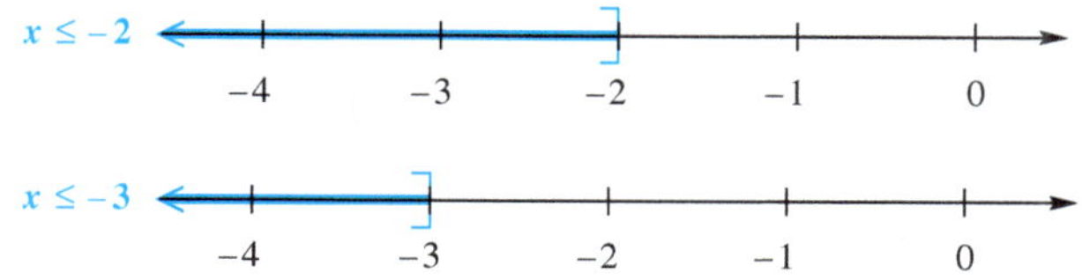

Figure 20

By taking the union, we obtain the interval $(-\infty, -2]$. See Figure 21.

Figure 21

Work Problem 7 at the Side. ▶

Example 8 Solving a Compound Inequality with *or*

Solve $-2x + 5 \geq 11$ or $4x - 7 \geq -27$.

Solve each inequality individually.

$$\begin{aligned} -2x + 5 &\geq 11 &\quad \text{or} \quad 4x - 7 &\geq -27 \\ -2x &\geq 6 &\quad \text{or} \quad 4x &\geq -20 \\ x &\leq -3 &\quad \text{or} \quad x &\geq -5 \end{aligned}$$

The graphs of these two inequalities are shown in Figure 22.

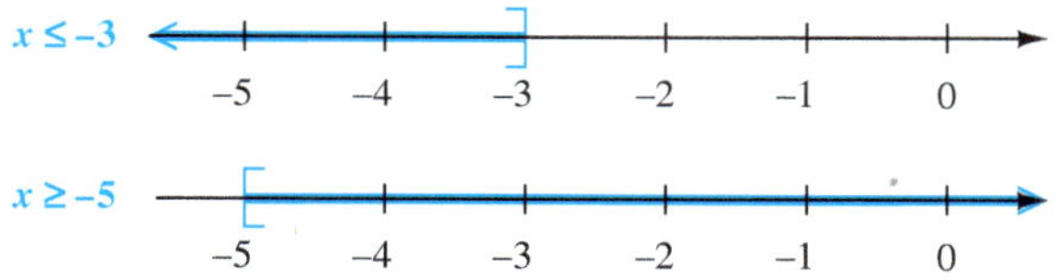

Figure 22

By taking the union, we obtain every real number as a solution, since every real number satisfies at least one of the two inequalities. The set of all real numbers is written in interval notation as $(-\infty, \infty)$ and graphed as in Figure 23.

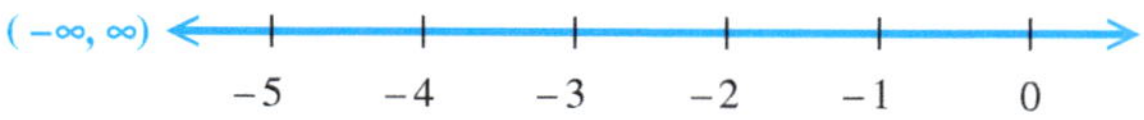

Figure 23

Work Problem 8 at the Side. ▶

7 Solve. Give each solution set in both interval and graph forms.

(a) $2x + 1 \leq 9$ or $2x + 3 \leq 5$

(b) $3x - 4 > 2$ or $-2x + 5 < 3$

8 Solve

$3x - 2 \leq 13$ or $x + 5 \geq 7$.

Give the solution set in both interval and graph forms.

ANSWERS

7. (a) $(-\infty, 4]$

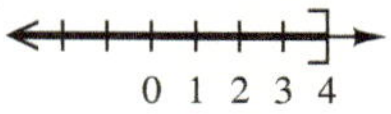

(b) $(1, \infty)$

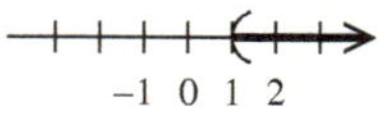

8. $(-\infty, \infty)$

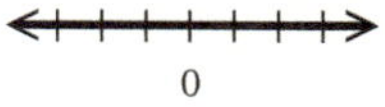

9 From Example 9, list the elements that satisfy each set.

(a) The set of films with admissions greater than 130,000,000 and gross income less than $800,000,000

(b) The set of films with admissions greater than 130,000,000 or gross income less than $800,000,000

Example 9 **Applying Intersection and Union**

The five highest-grossing domestic films (adjusted for inflation) as of July, 2005, are listed in the table.

FIVE ALL-TIME HIGHEST GROSSING DOMESTIC FILMS

Film	Admissions	Gross Income
Gone with the Wind	202,044,569	$1,293,085,000
Star Wars	178,119,595	$1,139,965,000
The Sound of Music	142,415,376	$911,458,000
E.T.	141,925,359	$908,322,298
The Ten Commandments	131,000,000	$838,400,000

Source: Exhibitor Relations Co., Inc.

List the elements of the following sets.

(a) The set of top-five films with admissions greater than 180,000,000 *and* gross income greater than $1,000,000,000

The only film that satisfies both conditions is *Gone with the Wind,* so the set is

{*Gone with the Wind*}.

(b) The set of top-five films with admissions less than 170,000,000 *or* gross income greater than $1,000,000,000

Here, a film that satisfies at least one of the conditions is in the set. This set includes all five films:

{*Gone with the Wind, Star Wars, The Sound of Music, E.T., The Ten Commandments*}.

◀ Work Problem **9** at the Side.

Answers

9. **(a)** ∅ **(b)** {*Gone with the Wind, Star Wars, The Sound of Music, E.T., The Ten Commandments*}

13.3 Exercises

Decide whether each statement is true *or* false. *If it is* false, *explain why.*

1. The union of the solution sets of $2x + 1 = 3$, $2x + 1 > 3$, and $2x + 1 < 3$ is $(-\infty, \infty)$.

2. The intersection of the sets $\{x \mid x \geq 5\}$ and $\{x \mid x \leq 5\}$ is $\emptyset$.

3. The union of the sets $(-\infty, 6)$ and $(6, \infty)$ is $\{6\}$.

4. The intersection of the sets $[6, \infty)$ and $(-\infty, 6]$ is $\{6\}$.

Let $A = \{1, 2, 3, 4, 5, 6\}$, $B = \{1, 3, 5\}$, $C = \{1, 6\}$, *and* $D = \{4\}$. *Specify each set. See Examples 1 and 5.*

5. $A \cap D$ **6.** $B \cap C$ **7.** $B \cap \emptyset$ **8.** $A \cap \emptyset$

9. $A \cup B$ **10.** $B \cup D$ **11.** $B \cup C$ **12.** $C \cup B$

Two sets are specified by graphs. Graph the intersection of the two sets.

13.
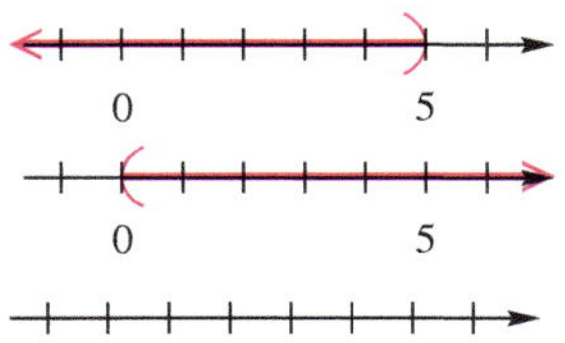

14.
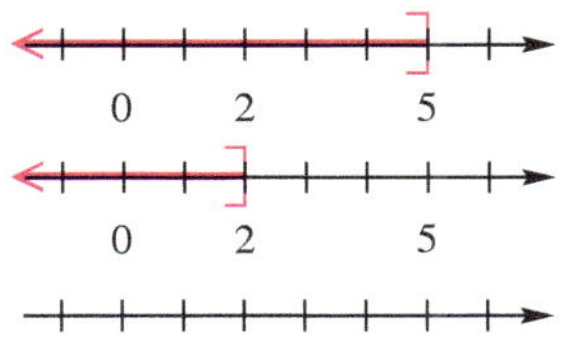

15.
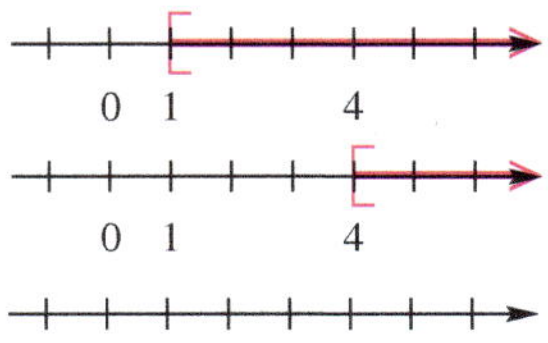

Two sets are specified by graphs. Graph the union of the two sets.

16.
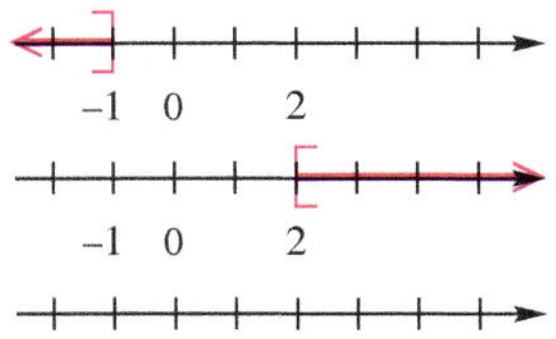

17.
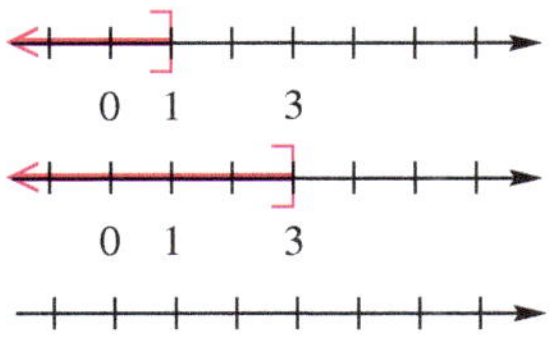

18.
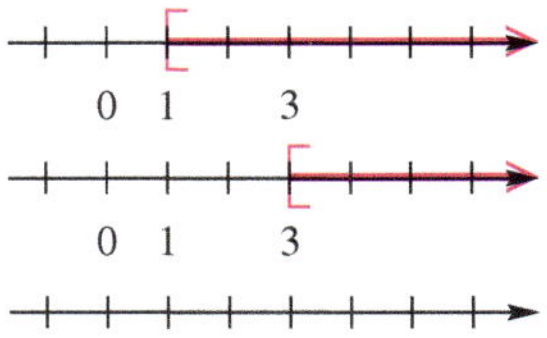

19. Give an example of intersection applied to a real-life situation.

20. A compound inequality uses one of the words *and* or *or*. Explain how you will determine whether to use *intersection* or *union* when graphing the solution set.

For each compound inequality, give the solution set in both interval and graph forms. See Examples 2–4.

21. $x < 2$ and $x > -3$

22. $x < 5$ and $x > 0$

23. $x \leq 2$ and $x \leq 5$

24. $x \geq 3$ and $x \geq 6$

25. $x \leq 3$ and $x \geq 6$

26. $x \leq -1$ and $x \geq 3$

27. $x - 3 \leq 6$ and $x + 2 \geq 7$

28. $x + 5 \leq 11$ and $x - 3 \geq -1$

29. $3x - 4 \leq 8$ and $4x - 1 \leq 15$

30. $7x + 6 \leq 48$ and $-4x \geq -24$

For each compound inequality, give the solution set in both interval and graph forms. See Examples 6–8.

31. $x \leq 1$ or $x \leq 8$

32. $x \geq 1$ or $x \geq 8$

33. $x \geq -2$ or $x \geq 5$

34. $x \leq -2$ or $x \leq 6$

35. $x + 3 \geq 1$ or $x - 8 \leq -4$

36. $x + 6 \geq 11$ or $x - 4 \leq 3$

37. $x + 2 > 7$ or $1 - x > 6$

38. $x + 1 > 3$ or $x + 4 < 2$

39. $x + 1 > 3$ or $-4x + 1 \geq 5$

40. $3x < x + 12$ or $x + 1 > 10$

41. $4x - 8 > 0$ or $4x - 1 < 7$

42. $3x < x + 12$ or $3x - 8 > 10$

Express each set in the simplest interval form.

43. $(-\infty, -1] \cap [-4, \infty)$

44. $[-1, \infty) \cap (-\infty, 9]$

45. $(-\infty, -6] \cap [-9, \infty)$

46. $(5, 11] \cap [6, \infty)$

47. $(-\infty, 3) \cup (-\infty, -2)$

48. $[-9, 1] \cup (-\infty, -3)$

49. $[3, 6] \cup (4, 9)$

50. $[-1, 2] \cup (0, 5)$

For each compound inequality, state whether intersection or union should be used. Then give the solution set in both interval and graph forms. See Examples 2–4 and 6–8.

51. $x < -1$ and $x > -5$

52. $x > -1$ and $x < 7$

53. $x < 4$ or $x < -2$

54. $x < 5$ or $x < -3$

55. $x + 1 \geq 5$ and $x - 2 \leq 10$

56. $2x - 6 \leq -18$ and $2x \geq -18$

57. $-3x \leq -6$ or $-3x \geq 0$

58. $-8x \leq -24$ or $-5x \geq 15$

Relating Concepts (Exercises 59–64) For Individual or Group Work

The figures represent the backyards of neighbors Luigi, Mario, Than, and Joe. Find the area and the perimeter of each yard. Suppose that each resident has 150 ft *of fencing and enough sod to cover* 1400 ft^2 *of lawn.*

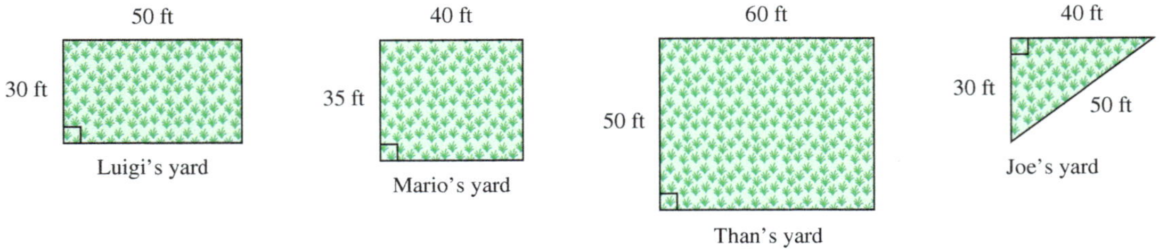

Give the name or names of the residents whose yards satisfy each description.

59. The yard can be fenced *and* the yard can be sodded.

60. The yard can be fenced *and* the yard cannot be sodded.

61. The yard cannot be fenced *and* the yard can be sodded.

62. The yard cannot be fenced *and* the yard cannot be sodded.

63. The yard can be fenced *or* the yard can be sodded.

64. The yard cannot be fenced *or* the yard can be sodded.

Average expenses for full-time college students at 2-year institutions during the 2005–2006 academic year are shown in the table.

COLLEGE EXPENSES (IN DOLLARS), 2-YEAR INSTITUTIONS

Type of Expense	Public Schools	Private Schools
Tuition and fees	1935	12,450
Board rates	2306	4726
Dormitory charges	2251	3994

Source: National Center for Education Statistics, U.S. Department of Education.

Use the table to list the elements of each set. See Example 9.

65. The set of expenses that are less than \$2500 for public schools *and* are greater than \$5000 for private schools

66. The set of expenses that are greater than \$2300 for public schools *and* are less than \$12,000 for private schools

67. The set of expenses that are less than \$2300 for public schools *or* are greater than \$10,000 for private schools

68. The set of expenses that are greater than \$12,000 *or* are less than \$2000

Study Skills

USING STUDY CARDS REVISITED

This is the second part of the Study Cards activity. As you get further into a chapter, you can choose particular problems that will serve as a good test review. Here are two more types of study cards that will help you.

OBJECTIVES

1 **Create study cards for difficult problems.**

2 **Create study cards of quiz problems.**

Tough Problems Card

When you are doing your homework and find yourself saying, "This is really hard," or "I'm having trouble with this," make a tough problem study card. On the front, write out the procedure to work the type of problem *in words*. If there are special notes (like what *not* to do), include them. On the back, work at least one example. Make sure you label what you are doing.

Solving a Linear Inequality with Fractions
First step: Clear the inequality of fractions.
— *Find a common denominator.*
— *Multiply each term by the common denominator.*

Front of Card

Solve $\frac{3}{4}(m-3)+2 \le \frac{1}{2}(m+8)$ *and graph the solution set*

$$\frac{4}{1}\left[\frac{3}{4}(m-3)\right]+4(2) \le \frac{4}{1}\left[\frac{1}{2}(m+8)\right]$$ *Common denom. is 4. Multiply every term by 4.*

$$3(m-3)+8 \le 2(m+8)$$
$$3m-9+8 \le 2m+16$$
$$3m-1 \le 2m+16$$ *Simplify each side.*

$$3m-1-2m \le 2m+16-2m$$ *Subtract 2m.*

$$m-1 \le 16$$
$$m-1+1 \le 16+1$$
$$m \le 17$$

Back of Card

> *Choose three types of difficult problems, and work them out on study cards. Be sure to put the words for solving the problem on one side and the worked problem on the other side.*

Now Try This

Practice Quiz Cards

Quiz study cards cover each type of problem you learn. They are useful when you prepare for a test. To make a quiz card, put the problem with the direction words (like *solve, simplify, estimate*) on the front of a card, and work the problem on the back. If you like, include the page number from the text. When you review, work the problem on a separate paper and check it by looking at the back of your quiz card.

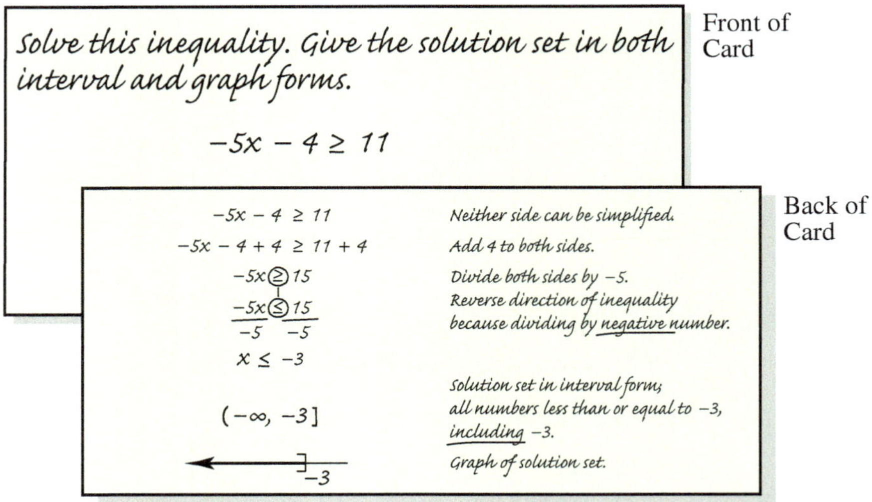

Now Try This ▶▶▶

Choose three problems from different sections of this chapter, and work them on study cards. Be sure you don't just choose the easiest problems.

Why Are Study Cards Brain Friendly?

First, **making the study cards is an active technique.** You have to make decisions about what is most important and how to put it on a card. This kind of thinking is more involved than just memorizing, and as a result, you will understand the concepts better and remember them longer.

Second, **the cards are visually appealing** (if you write neatly and try some color). You will remember a visual image longer and may even be able to "picture in your mind" how your cards look. This will help you during tests.

Third, because **study cards are small and portable,** you can review them easily whenever you have a few minutes. Even while you're waiting for a bus or have a few minutes between classes, you can take out your cards and read over them. After a while, the information will become automatic and easier to remember.

13.4 Absolute Value Equations and Inequalities

In a production line, quality is controlled by randomly choosing items from the line and checking to see how selected measurements vary from the optimum measure. These differences are sometimes positive and sometimes negative, so they are expressed with absolute value. For example, a machine that fills quart milk cartons might be set to release 1 qt (32 oz) plus or minus 2 oz per carton. Then the number of ounces in each carton should satisfy the *absolute value inequality* $|x - 32| \le 2$, where x is the number of ounces.

OBJECTIVES

1. **Use the distance definition of absolute value.**
2. **Solve equations of the form $|ax + b| = k$, for $k > 0$.**
3. **Solve inequalities of the form $|ax + b| < k$ and of the form $|ax + b| > k$, for $k > 0$.**
4. **Solve absolute value equations that involve rewriting.**
5. **Solve equations of the form $|ax + b| = |cx + d|$.**
6. **Solve special cases of absolute value equations and inequalities.**

OBJECTIVE 1 Use the distance definition of absolute value.

In **Section 1.1,** we saw that the absolute value of a number x, written $|x|$, represents the distance from x to 0 on the number line. For example, the solutions of $|x| = 4$ are 4 and -4, as shown in Figure 24.

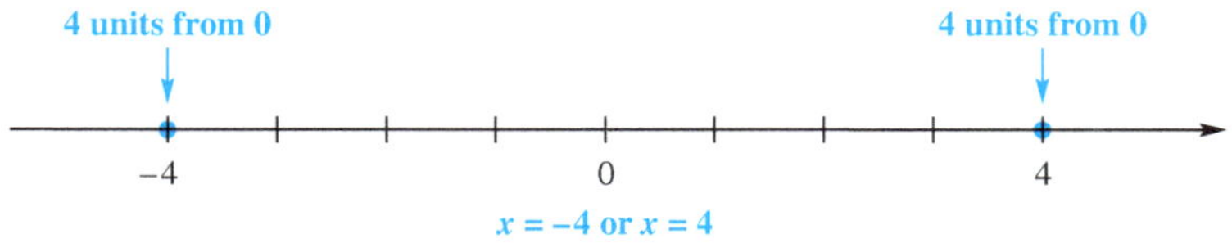

Figure 24

Because absolute value represents distance from 0, it is reasonable to interpret the solutions of $|x| > 4$ to be all numbers that are *more* than 4 units from 0. The set $(-\infty, -4) \cup (4, \infty)$ fits this description. Figure 25 shows the graph of the solution set of $|x| > 4$. Because the graph consists of two separate intervals, the solution set is described using *or* as $x < -4$ or $x > 4$.

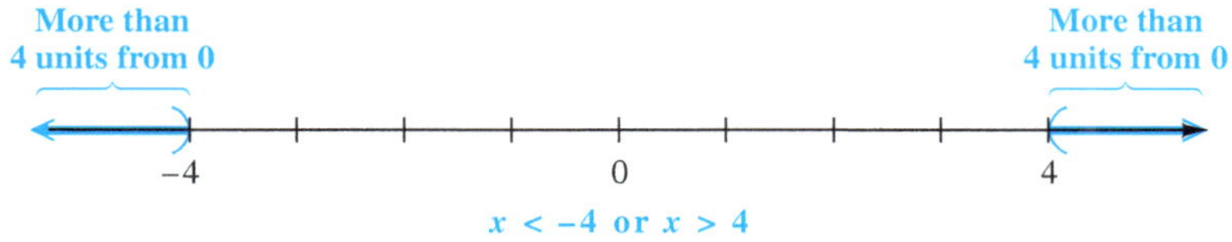

Figure 25

The solution set of $|x| < 4$ consists of all numbers that are *less* than 4 units from 0 on the number line. Another way of thinking of this is to think of all numbers *between* -4 and 4. This set of numbers is given by $(-4, 4)$, as shown in Figure 26. Here, the graph shows that $-4 < x < 4$, which means $x > -4$ *and* $x < 4$.

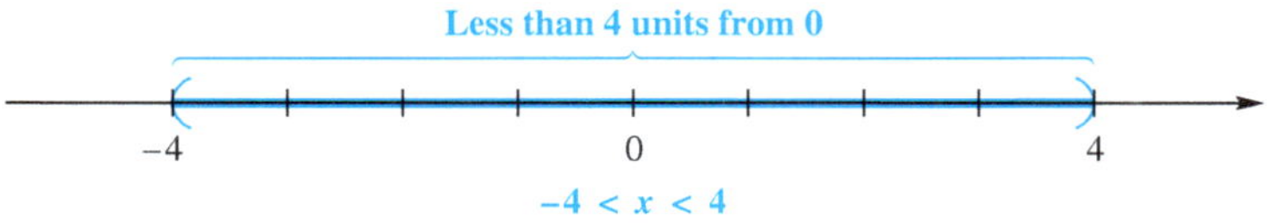

Figure 26

Work Problem **1** *at the Side.* ▶

The equation and inequalities just described are examples of **absolute value equations and inequalities.** They involve the absolute value of a variable expression and generally take the form

$$|ax + b| = k, \quad |ax + b| > k, \quad \text{or} \quad |ax + b| < k,$$

where k is a positive number. From Figures 24–26, we see that

$|x| = 4$ has the same solution set as $x = -4$ **or** $x = 4$,

$|x| > 4$ has the same solution set as $x < -4$ **or** $x > 4$,

$|x| < 4$ has the same solution set as $x > -4$ **and** $x < 4$.

1 Graph the solution set of each equation or inequality.

(a) $|x| = 3$

(b) $|x| > 3$

(c) $|x| < 3$

ANSWERS

1. (a)
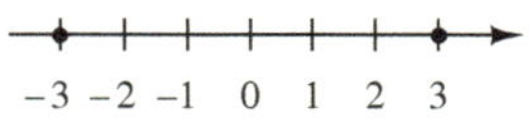

(b)
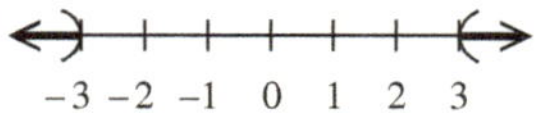

(c)
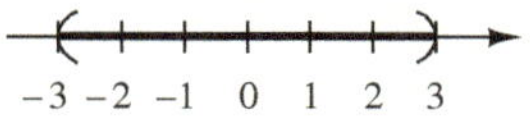

Thus, we can solve an absolute value equation or inequality by solving the appropriate compound equation or inequality.

2 Solve, check, and graph the solution set of each equation.

(a) $|x + 2| = 3$

(b) $|3x - 4| = 11$

Solving Absolute Value Equations and Inequalities

Let k be a positive real number, and p and q be real numbers.

Case 1 To solve $|ax + b| = k$, solve the compound equation

$$ax + b = k \quad \text{or} \quad ax + b = -k.$$

The solution set is usually of the form $\{p, q\}$, which includes two numbers.

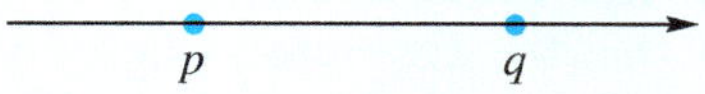

Case 2 To solve $|ax + b| > k$, solve the compound inequality

$$ax + b > k \quad \text{or} \quad ax + b < -k.$$

The solution set is of the form $(-\infty, p) \cup (q, \infty)$, which is a disjoint interval.

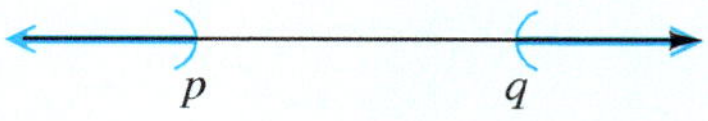

Case 3 To solve $|ax + b| < k$, solve the three-part inequality

$$-k < ax + b < k.$$

The solution set is of the form (p, q), an open interval.

OBJECTIVE 2 Solve equations of the form $|ax + b| = k$, for $k > 0$. ***Remember that because absolute value refers to distance from the origin, an absolute value equation will have two parts.***

Example 1 Solving an Absolute Value Equation

Solve $|2x + 1| = 7$.

For $|2x + 1|$ to equal 7, $2x + 1$ must be 7 units from 0 on the number line. This can happen only when $2x + 1 = 7$ or $2x + 1 = -7$. This is Case 1 in the preceding box. Solve this compound equation as follows.

$$2x + 1 = 7 \quad \text{or} \quad 2x + 1 = -7$$
$$2x = 6 \quad \text{or} \quad 2x = -8$$
$$x = 3 \quad \text{or} \quad x = -4$$

Check by substituting 3 and then -4 in the original absolute value equation to verify that the solution set is $\{-4, 3\}$. The graph is shown in Figure 27.

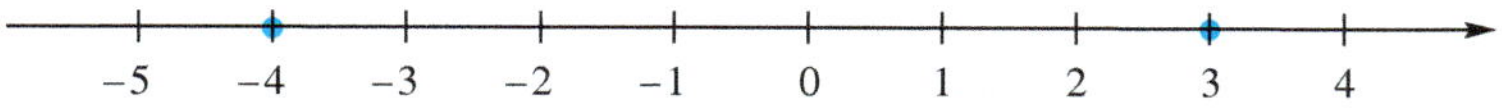

Figure 27

◀ Work Problem **2** at the Side.

ANSWERS

2. (a) $\{-5, 1\}$

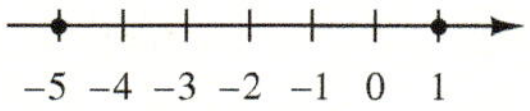

(b) $\left\{-\frac{7}{3}, 5\right\}$

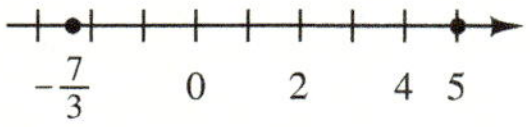

Note

Some people prefer to write the compound statements in Cases 1 and 2 of the box on the previous page as the equivalent forms

$$ax + b = k \quad \text{or} \quad -(ax + b) = k$$

and

$$ax + b > k \quad \text{or} \quad -(ax + b) > k.$$

These forms produce the same results.

OBJECTIVE 3 Solve inequalities of the form $|ax + b| < k$ and of the form $|ax + b| > k$, for $k > 0$.

Example 2 Solving an Absolute Value Inequality with >

Solve $|2x + 1| > 7$.

By Case 2 in the box on the preceding page, this absolute value inequality is rewritten as

$$2x + 1 > 7 \quad \text{or} \quad 2x + 1 < -7,$$

because $2x + 1$ must represent a number that is *more* than 7 units from 0 on either side of the number line. Now, solve the compound inequality.

$$2x + 1 > 7 \quad \text{or} \quad 2x + 1 < -7$$
$$2x > 6 \quad \text{or} \quad 2x < -8$$
$$x > 3 \quad \text{or} \quad x < -4$$

Check these solutions. The solution set is $(-\infty, -4) \cup (3, \infty)$, a disjoint interval. See the graph in Figure 28.

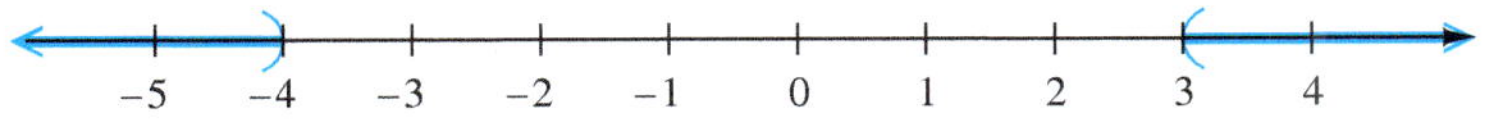

Figure 28

Work Problem **3** at the Side. ▶

Example 3 Solving an Absolute Value Inequality with <

Solve $|2x + 1| < 7$.

The expression $2x + 1$ must represent a number that is less than 7 units from 0 on either side of the number line. Another way of thinking of this is to realize that $2x + 1$ must be between -7 and 7. As Case 3 in the box on the preceding page shows, this is written as the three-part inequality

$$-7 < 2x + 1 < 7.$$

$-8 < 2x < 6$ Subtract 1 from each part.

$-4 < x < 3$ Divide each part by 2.

Check that the solution set is $(-4, 3)$, so the graph consists of the open interval shown in Figure 29.

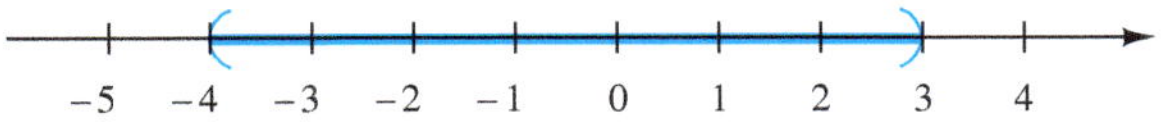

Figure 29

Work Problem **4** at the Side. ▶

3 Solve, check, and graph the solution set of each inequality.

(a) $|x + 2| > 3$

(b) $|3x - 4| \geq 11$

4 Solve, check, and graph the solution set of each inequality.

(a) $|x + 2| < 3$

(b) $|3x - 4| \leq 11$

ANSWERS

3. (a) $(-\infty, -5) \cup (1, \infty)$

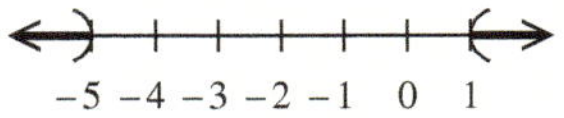

(b) $\left(-\infty, -\frac{7}{3}\right] \cup [5, \infty)$

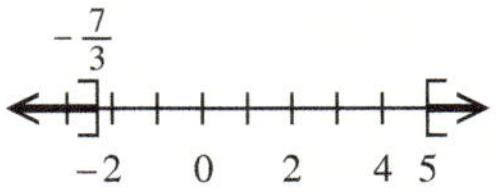

4. (a) $(-5, 1)$

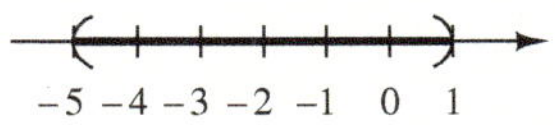

(b) $\left[-\frac{7}{3}, 5\right]$

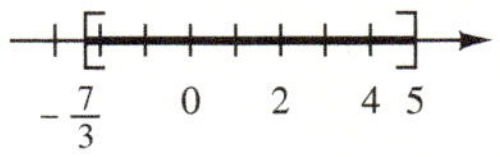

5 Solve $|5x + 2| - 9 = -7$. Check your solutions.

Look back at Figures 27, 28, and 29, with the graphs of $|2x + 1| = 7$, $|2x + 1| > 7$, and $|2x + 1| < 7$. If we find the union of the three sets, we get the set of all real numbers. This is because, for any value of x, $|2x + 1|$ will satisfy one and only one of the following: It is either equal to 7, greater than 7, or less than 7.

CAUTION

When solving absolute value equations and inequalities of the types in Examples 1, 2, and 3, remember the following.

1. The methods described apply when the constant is alone on one side of the equation or inequality and is *positive*.
2. Absolute value equations and absolute value inequalities of the form $|ax + b| > k$ translate into "or" compound statements.
3. Absolute value inequalities of the form $|ax + b| < k$ translate into "and" compound statements, which may be written as three-part inequalities.
4. An "or" statement *cannot* be written in three parts. It would be incorrect to use $-7 > 2x + 1 > 7$ in Example 2, because this would imply that $-7 > 7$, which is *false*.

OBJECTIVE 4 Solve absolute value equations that involve rewriting. Sometimes an absolute value equation or inequality requires some rewriting before it can be set up as a compound statement.

Example 4 Solving an Absolute Value Equation That Requires Rewriting

Solve $|x + 3| + 5 = 12$.

First, get the absolute value alone on one side of the equals sign.

$$|x + 3| + 5 = 12$$

$$|x + 3| + 5 - 5 = 12 - 5 \quad \text{Subtract 5.}$$

$$|x + 3| = 7$$

Now use the method shown in Example 1.

$$x + 3 = 7 \quad \text{or} \quad x + 3 = -7$$

$$x = 4 \quad \text{or} \quad x = -10$$

Check these solutions by substituting each one in the original equation.

Check $|x + 3| + 5 = 12$

$\lvert 4 + 3\rvert + 5 \stackrel{?}{=} 12$	Let $x = 4$.	$\lvert -10 + 3\rvert + 5 \stackrel{?}{=} 12$	Let $x = -10$.
$\lvert 7\rvert + 5 \stackrel{?}{=} 12$		$\lvert -7\rvert + 5 \stackrel{?}{=} 12$	
$12 = 12$	True	$12 = 12$	True

The check confirms that the solution set is $\{-10, 4\}$.

◀ Work Problem **5** at the Side.

CAUTION

When solving an equation like the one in Example 4, do *not* simply drop the absolute value bars.

Answer

5. $\left\{-\frac{4}{5}, 0\right\}$

We use a method similar to that used in Example 4 to solve an absolute value *inequality* that requires rewriting:

$\lvert x+3\rvert+5\geq 12$	$\lvert x+3\rvert+5\leq 12$
$\lvert x+3\rvert\geq 7$	$\lvert x+3\rvert\leq 7$
$x+3\geq 7$ or $x+3\leq -7$	$-7\leq x+3\leq 7$
$x\geq 4$ or $x\leq -10.$	$-10\leq x\leq 4.$
Solution set: $(-\infty,-10]\cup[4,\infty)$	Solution set: $[-10,4]$

Work Problem **6** at the Side. ▶

6 Solve each inequality, and graph the solution set.

(a) $\lvert x+2\rvert-3>2$

(b) $\lvert 3x+2\rvert+4\leq 15$

OBJECTIVE 5 Solve equations of the form $\lvert ax+b\rvert=\lvert cx+d\rvert$. By definition, for two expressions to have the same absolute value, they must either be equal or be negatives of each other.

Solving $\lvert ax+b\rvert=\lvert cx+d\rvert$

To solve an absolute value equation of the form

$$\lvert ax+b\rvert=\lvert cx+d\rvert,$$

solve the compound equation

$$ax+b=cx+d \quad\text{or}\quad ax+b=-(cx+d).$$

Example 5 Solving an Equation with Two Absolute Values

Solve $\lvert z+6\rvert=\lvert 2z-3\rvert$.

This equation is satisfied either if $z+6$ and $2z-3$ are equal to each other, or if $z+6$ and $2z-3$ are negatives of each other. Thus,

$$\begin{aligned} z+6&=2z-3 &&\text{or}& z+6&=-(2z-3)\\ z+9&=2z &&\text{or}& z+6&=-2z+3\\ 9&=z &&\text{or}& 3z&=-3\\ z&=9 &&\text{or}& z&=-1. \end{aligned}$$

Check that the solution set is $\{9,-1\}$.

Work Problem **7** at the Side. ▶

7 Solve each equation.

(a) $\lvert k-1\rvert=\lvert 5k+7\rvert$

(b) $\lvert 4r-1\rvert=\lvert 3r+5\rvert$

OBJECTIVE 6 Solve special cases of absolute value equations and inequalities. When a typical absolute value equation or inequality involves a *negative* constant or 0 alone on one side, we use the properties of absolute value to solve. Keep the following in mind.

Special Cases of Absolute Value

Case 1 The absolute value of an expression can never be negative—that is, $\lvert a\rvert\geq 0$ for all real numbers a.

Case 2 The absolute value of an expression equals 0 only when the expression is equal to 0.

ANSWERS

6. (a) $(-\infty,-7)\cup(3,\infty)$

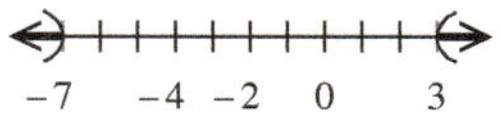

(b) $\left[-\frac{13}{3},3\right]$

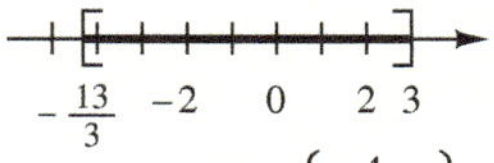

7. (a) $\{-1,-2\}$ **(b)** $\left\{-\frac{4}{7},6\right\}$

8 Solve each equation.

(a) $|6x + 7| = -5$

(b) $\left|\frac{1}{4}x - 3\right| = 0$

9 Solve.

(a) $|x| > -1$

(b) $|x| < -5$

(c) $|x + 2| \leq 0$

(d) $|t - 10| - 2 \leq -3$

Example 6 Solving Special Cases of Absolute Value Equations

Solve each equation.

(a) $|5r - 3| = -4$

See Case 1 in the box on the preceding page. ***The absolute value of an expression can never be negative,*** so there are no solutions for this equation. The solution set is $\emptyset$.

(b) $|7x - 3| = 0$

See Case 2 in the box on the preceding page. The expression $7x - 3$ will equal 0 *only* if

$$7x - 3 = 0$$

$$7x = 3 \quad \text{Add 3.}$$

$$x = \frac{3}{7}. \quad \text{Divide by 7.}$$

Thus, the solution set of the original equation is $\{\frac{3}{7}\}$, with just one element. Check this solution by substituting it in the original equation.

◀ Work Problem **8** at the Side.

Example 7 Solving Special Cases of Absolute Value Inequalities

Solve each inequality.

(a) $|x| \geq -4$

The absolute value of a number is always greater than or equal to 0. Thus, $|x| \geq -4$ is true for *all* real numbers. The solution set is $(-\infty, \infty)$.

(b) $|x + 6| - 3 < -5$

$$|x + 6| < -2 \quad \text{Add 3 to each side.}$$

There is no number whose absolute value is less than -2, so this inequality has no solution. The solution set is $\emptyset$.

(c) $|x - 7| + 4 \leq 4$

$$|x - 7| \leq 0 \quad \text{Subtract 4 from each side.}$$

The value of $|x - 7|$ will never be less than 0. However, $|x - 7|$ will *equal* 0 when $x = 7$. Therefore, the solution set is $\{7\}$.

◀ Work Problem **9** at the Side.

Answers

8. **(a)** $\emptyset$ **(b)** $\{12\}$
9. **(a)** $(-\infty, \infty)$ **(b)** $\emptyset$ **(c)** $\{-2\}$ **(d)** $\emptyset$

13.4 Exercises

FOR EXTRA HELP MyMathLab

Math XL PRACTICE · WATCH · DOWNLOAD · READ · REVIEW

Match each absolute value equation or inequality in Column I with the graph of its solution set in Column II.

I	II
1. $\lvert x \rvert = 5$	**A.** 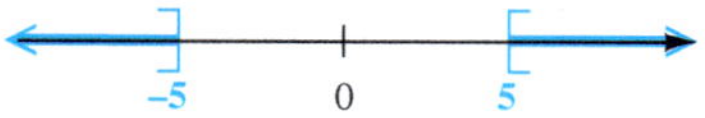
$\lvert x \rvert < 5$	**B.** 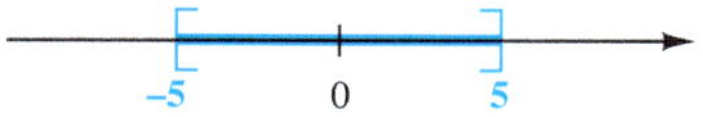
$\lvert x \rvert > 5$	**C.** 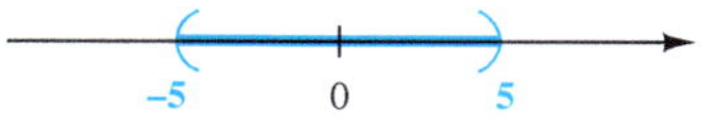
$\lvert x \rvert \le 5$	**D.** 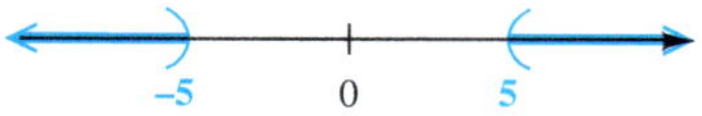
$\lvert x \rvert \ge 5$	**E.** 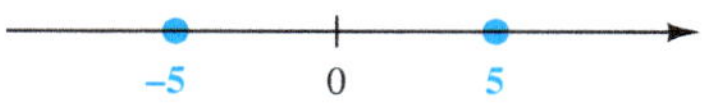

I	II
2. $\lvert x \rvert = 9$	**A.** 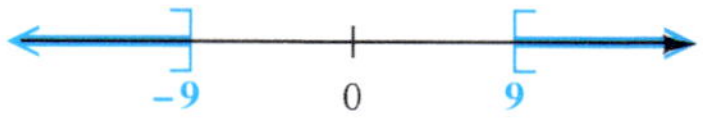
$\lvert x \rvert > 9$	**B.** 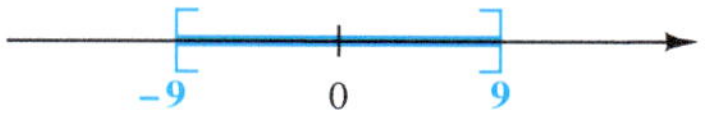
$\lvert x \rvert \ge 9$	**C.** 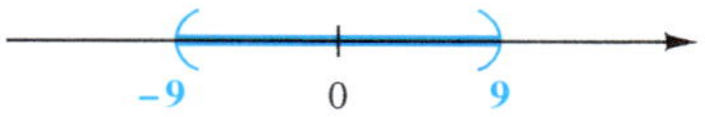
$\lvert x \rvert < 9$	**D.** 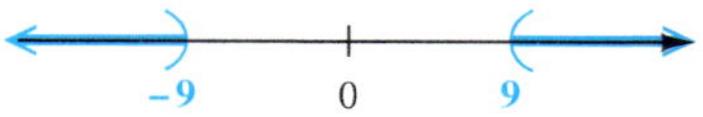
$\lvert x \rvert \le 9$	**E.** 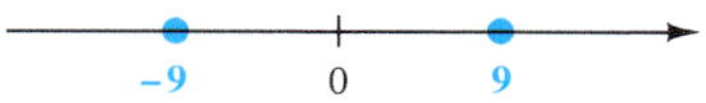

3. How many solutions will $\lvert ax + b \rvert = k$ have if
(a) $k = 0$; **(b)** $k > 0$; **(c)** $k < 0$?

4. Explain when to use *and* and when to use *or* if you are solving an absolute value equation or inequality of the form $\lvert ax + b \rvert = k$, $\lvert ax + b \rvert < k$, or $\lvert ax + b \rvert > k$, where k is a positive number.

Solve each equation. See Example 1.

5. $\lvert x \rvert = 12$

6. $\lvert x \rvert = 14$

7. $\lvert 4x \rvert = 20$

8. $\lvert 5x \rvert = 30$

9. $\lvert x - 3 \rvert = 9$

10. $\lvert p - 5 \rvert = 13$

11. $\lvert 2x + 1 \rvert = 9$

12. $\lvert 2x + 3 \rvert = 19$

13. $\lvert 4r - 5 \rvert = 17$

14. $\lvert 5t - 1 \rvert = 21$

15. $\lvert 2x + 5 \rvert = 14$

16. $\lvert 2x - 9 \rvert = 18$

17. $\left\lvert \frac{1}{2}x + 3 \right\rvert = 2$

18. $\left\lvert \frac{2}{3}q - 1 \right\rvert = 5$

19. $\left\lvert 1 - \frac{3}{4}k \right\rvert = 7$

20. $\left\lvert 2 - \frac{5}{2}m \right\rvert = 14$

Solve each inequality, and graph the solution set. See Example 2.

21. $|x| > 3$

22. $|x| > 2$

23. $|k| \geq 4$

24. $|r| \geq 1$

25. $|t + 2| > 8$

26. $|r + 5| > 20$

27. $|3x - 1| \geq 8$

28. $|4x + 1| \geq 21$

29. $|3 - x| > 5$

30. $|5 - x| > 3$

31. The graph of the solution set of $|2x + 1| = 9$ is given here.

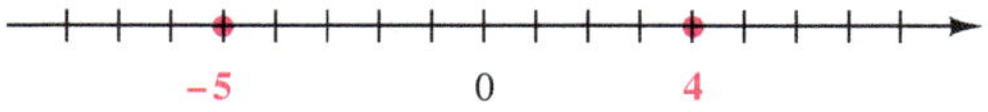

Without actually doing the algebraic work, graph the solution set of each inequality, referring to the graph above.

(a) $|2x + 1| < 9$

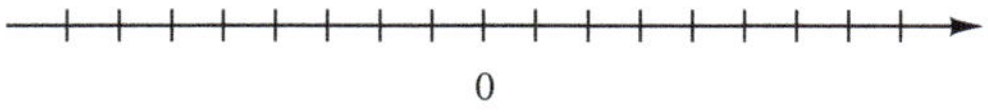

(b) $|2x + 1| > 9$

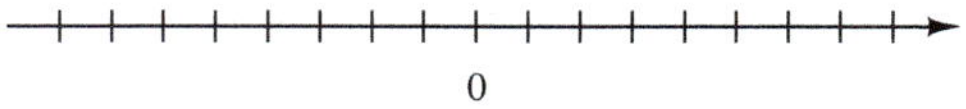

32. The graph of the solution set of $|3x - 4| < 5$ is given here.

Without actually doing the algebraic work, graph the solution set of the equation and the inequality, referring to the graph above.

(a) $|3x - 4| = 5$

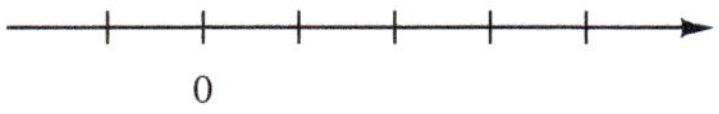

(b) $|3x - 4| > 5$

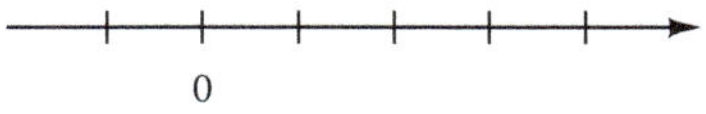

Solve each inequality, and graph the solution set. See Example 3. (Hint: Compare your answers to those in Exercises 21–30.)

33. $|x| \leq 3$

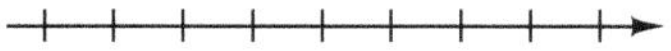

34. $|x| \leq 2$

35. $|k| < 4$

36. $|r| < 1$

37. $|t + 2| \leq 8$

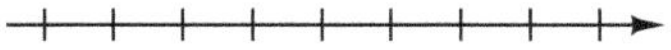

38. $|r + 5| \leq 20$

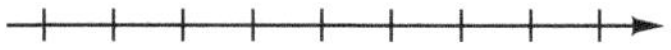

39. $|3x - 1| < 8$

40. $|4x + 1| < 21$

41. $|3 - x| \leq 5$

42. $|5 - x| \leq 3$

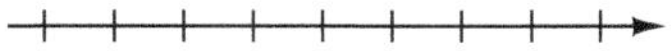

Exercises 43–50 represent a sampling of the various types of absolute value equations and inequalities. Decide which method of solution applies, find the solution set, and graph. See Examples 1–3.

43. $|-4 + k| > 6$

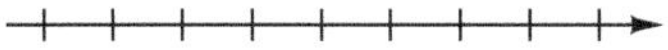

44. $|-3 + t| > 5$

45. $|7 + 2z| = 5$

46. $|9 - 3p| = 3$

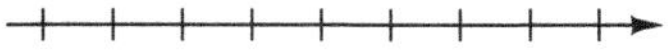

47. $|3r - 1| \leq 11$

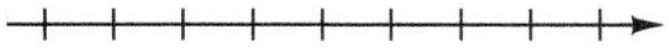

48. $|2s - 6| \leq 6$

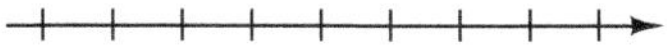

49. $|-3x - 8| \leq 4$

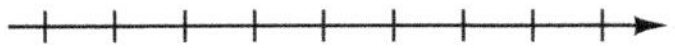

50. $|-2x - 6| \leq 5$

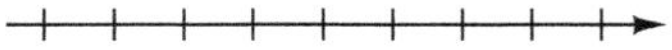

Solve each equation or inequality. Give the solution set using set notation for equations and interval notation for inequalities. See Example 4.

51. $|x| - 1 = 4$

52. $|x| + 3 = 10$

53. $|x + 4| + 1 = 2$

54. $|x + 5| - 2 = 12$

55. $|2x + 1| + 3 > 8$

56. $|6x - 1| - 2 > 6$

57. $|x + 5| - 6 \leq -1$

58. $|r - 2| - 3 \leq 4$

Solve each equation. See Example 5.

59. $|3x + 1| = |2x + 4|$

60. $|7x + 12| = |x - 8|$

61. $\left|m - \frac{1}{2}\right| = \left|\frac{1}{2}m - 2\right|$

62. $\left|\frac{2}{3}r - 2\right| = \left|\frac{1}{3}r + 3\right|$

63. $|6x| = |9x + 1|$

64. $|13x| = |2x + 1|$

65. $|2p - 6| = |2p + 11|$

66. $|3x - 1| = |3x + 9|$

Solve each equation or inequality. See Examples 6 and 7.

67. $|x| \geq -10$

68. $|x| \geq -15$

69. $|12t - 3| = -8$

70. $|13w + 1| = -3$

71. $|4x + 1| = 0$

72. $|6r - 2| = 0$

73. $|2q - 1| < -6$

74. $|8n + 4| < -4$

75. $|x + 5| > -9$

76. $|x + 9| > -3$

77. $|7x + 3| \leq 0$

78. $|4x - 1| \leq 0$

79. $|5x - 2| \geq 0$

80. $|4 + 7x| \geq 0$

81. $|10z + 7| > 0$

82. $|4x + 1| > 0$

83. $|x - 2| + 3 \geq 2$

84. $|k - 4| + 5 \geq 4$

85. The 2007 recommended daily intake (RDI) of calcium for females aged 19–50 is 1000 mg/day. Actual vitamin needs vary from person to person. Write an absolute value inequality to express the RDI plus or minus 100 mg and solve it. (*Source:* Food and Nutrition Board, National Academy of Sciences Institute of Medicine.)

86. The average clotting time of blood is 7.45 sec with a variation of plus or minus 3.6 sec. Write this statement as an absolute value inequality and solve it.

Relating Concepts (Exercises 87–90) For Individual or Group Work

The 10 tallest buildings in Kansas City, Missouri, are listed along with their heights.

Building	Height (in feet)
One Kansas City Place	632
Town Pavilion	591
Hyatt Regency Crown Center	504
Kansas City Power and Light	481
Fidelity Bank and Trust Building	454
City Hall	443
1201 Walnut	427
Federal Office Building	413
Commerce Tower	407
City Center Square	404

Source: World Almanac and Book of Facts.

Use this information to ***work Exercises 87–90 in order.***

87. To find the average of a group of numbers, we add the numbers and then divide by the number of items added. Use a calculator to find the average of the heights.

88. Let k represent the average height of these buildings. If a height x satisfies the inequality

$$|x - k| < t,$$

then the height is said to be within t feet of the average. Using your result from Exercise 87, list the buildings that are within 50 ft of the average.

89. Repeat Exercise 88, but find the buildings that are within 75 ft of the average.

90. **(a)** Write an absolute value inequality that describes the height of a building that is *not* within 75 ft of the average.

(b) Solve the inequality you wrote in part (a).

(c) Use the result of part (b) to find the buildings that are not within 75 ft of the average.

(d) Confirm that your answer to part (c) makes sense by comparing it with your answer to Exercise 89.

Summary Exercises on Solving Linear and Absolute Value Equations and Inequalities

This section of miscellaneous equations and inequalities provides practice in solving all the types introduced in **Chapters 2 and 3.** You might wish to refer to the boxes in these chapters that summarize the various methods of solution.

Solve each equation or inequality. Give the solution set using set notation for equations and interval notation for inequalities.

1. $4z + 1 = 49$

2. $|m - 1| = 6$

3. $6q - 9 = 12 + 3q$

4. $3p + 7 = 9 + 8p$

5. $|a + 3| = -4$

6. $2m + 1 \leq m$

7. $8r + 2 \geq 5r$

8. $4(a - 11) + 3a = 20a - 31$

9. $2q - 1 = -7$

10. $|3q - 7| - 4 = 0$

11. $6z - 5 \leq 3z + 10$

12. $|5z - 8| + 9 \geq 7$

13. $9x - 3(x + 1) = 8x - 7$

14. $|x| \geq 8$

15. $9x - 5 \geq 9x + 3$

16. $13p - 5 > 13p - 8$

17. $|q| < 5.5$

18. $4z - 1 = 12 + z$

19. $\frac{2}{3}x + 8 = \frac{1}{4}x$

20. $-\frac{5}{8}x \geq -20$

21. $\frac{1}{4}p < -6$

22. $7z - 3 + 2z = 9z - 8z$

23. $\frac{3}{5}q - \frac{1}{10} = 2$

24. $|r - 1| < 7$

25. $r + 9 + 7r = 4(3 + 2r) - 3$

26. $6 - 3(2 - p) < 2(1 + p) + 3$

27. $|2p - 3| > 11$

28. $\frac{x}{4} - \frac{2x}{3} = -10$

29. $|5a + 1| \leq 0$

30. $5z - (3 + z) \geq 2(3z + 1)$

31. $-2 \leq 3x - 1 \leq 8$

32. $-1 \leq 6 - x \leq 5$

33. $|7z - 1| = |5z + 3|$

34. $|p + 2| = |p + 4|$

35. $|1 - 3x| \geq 4$

36. $\frac{1}{2} \leq \frac{2}{3}r \leq \frac{5}{4}$

37. $-(m + 4) + 2 = 3m + 8$

38. $\frac{p}{6} - \frac{3p}{5} = p - 86$

39. $-6 \leq \frac{3}{2} - x \leq 6$

40. $|5 - x| < 4$

41. $|x - 1| \geq -6$

42. $|2r - 5| = |r + 4|$

43. $8q - (1 - q) = 3(1 + 3q) - 4$

44. $8x - (x + 3) = -(2x + 1) - 12$

45. $|r - 5| = |r + 9|$

46. $|r + 2| < -3$

47. $2x + 1 > 5$ or $3x + 4 < 1$

48. $1 - 2x \geq 5$ and $7 + 3x \geq -2$

13.5 Linear Inequalities in Two Variables

OBJECTIVES

1. Graph linear inequalities in two variables.
2. Graph the intersection of two linear inequalities.
3. Graph the union of two linear inequalities.

OBJECTIVE 1 Graph linear inequalities in two variables. In **Section 3.1** we graphed linear inequalities in one variable on the number line. We now graph linear inequalities in two variables on a rectangular coordinate system.

Linear Inequality in Two Variables

An inequality that can be written as

$$Ax + By < C \quad \text{or} \quad Ax + By > C,$$

where A, B, and C are real numbers and A and B are not both 0, is a **linear inequality in two variables.**

The symbols $\le$ and $\ge$ may replace $<$ and $>$ in the definition.

Consider the graph in Figure 30. The graph of the line $x + y = 5$ divides the points in the rectangular coordinate system into three sets:

1. Those points that lie on the line itself and satisfy the equation $\boldsymbol{x + y = 5}$ [like (0, 5), (2, 3), and (5, 0)];
2. Those that lie in the half-plane above the line and satisfy the inequality $x + y > 5$ [like (5, 3) and (2, 4)];
3. Those that lie in the half-plane below the line and satisfy the inequality $x + y < 5$ [like (0, 0) and (−3, −1)].

The graph of the line $x + y = 5$ is called the **boundary line** for the inequalities $x + y > 5$ and $x + y < 5$. Graphs of linear inequalities in two variables are *regions* in the real number plane that may or may not include boundary lines.

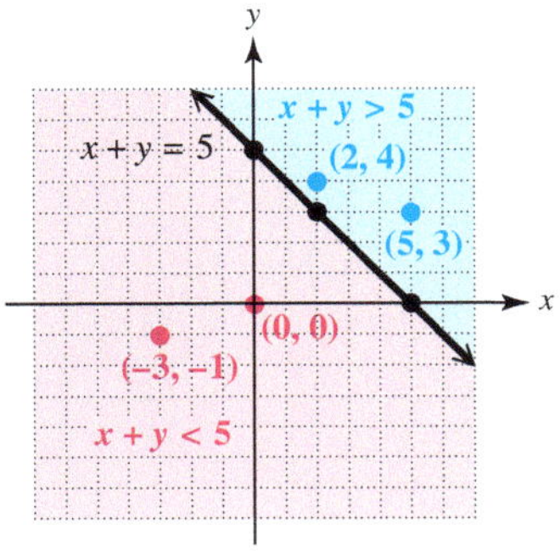

Figure 30

To graph a linear inequality in two variables, follow these steps.

Graphing a Linear Inequality

Step 1 **Draw the graph of the straight line that is the boundary.** Make the line solid if the inequality involves $\le$ or $\ge$; make the line dashed if the inequality involves $<$ or $>$.

Step 2 **Choose a test point.** Choose any point not on the line, and substitute the coordinates of this point in the inequality.

Step 3 **Shade the appropriate region.** Shade the region that includes the test point if it satisfies the original inequality; otherwise, shade the region on the other side of the boundary line.

1 Graph each inequality.

(a) $x + y \leq 4$

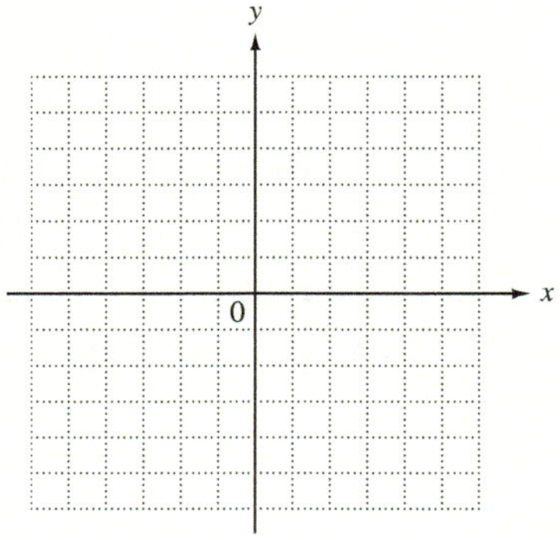

(b) $3x + y \geq 6$

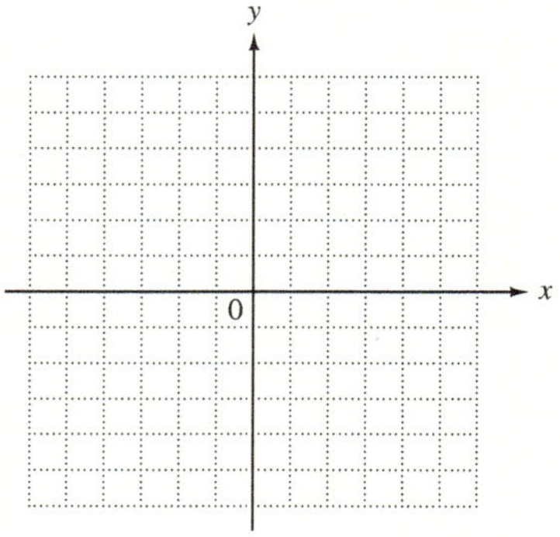

Example 1 **Graphing a Linear Inequality**

Graph $3x + 2y \geq 6$.

Step 1 First graph the line $3x + 2y = 6$. The graph of this line, the boundary of the graph of the inequality, is shown in Figure 31.

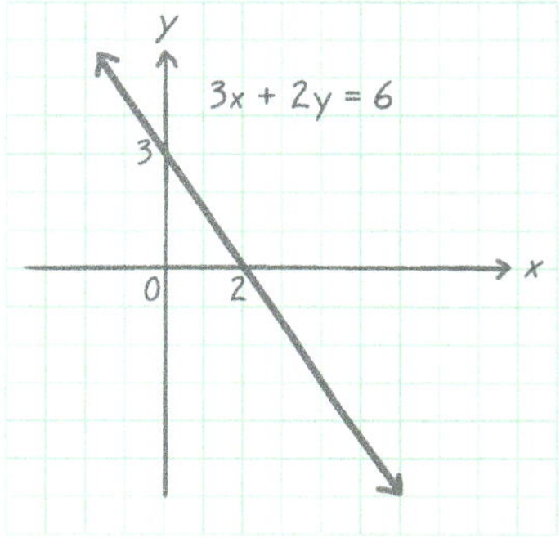

Figure 31

Step 2 The graph of the inequality $3x + 2y \geq 6$ includes the points of the boundary line $3x + 2y = 6$ and either the points *above* the line $3x + 2y = 6$ or the points *below* that line. To decide which, select any point not on the line $3x + 2y = 6$ as a test point. The origin, (0, 0), is often a good choice. Substitute the values from the test point (0, 0) for x and y in the inequality.

$$3x + 2y > 6$$

$$3(0) + 2(0) \overset{?}{>} 6 \quad \text{Let } x = 0 \text{ and } y = 0.$$

$$0 > 6 \quad \text{False}$$

Step 3 Because the result is false, (0, 0) does *not* satisfy the inequality, and so the solution set includes all points on the other side of the line. This region is shaded in Figure 32.

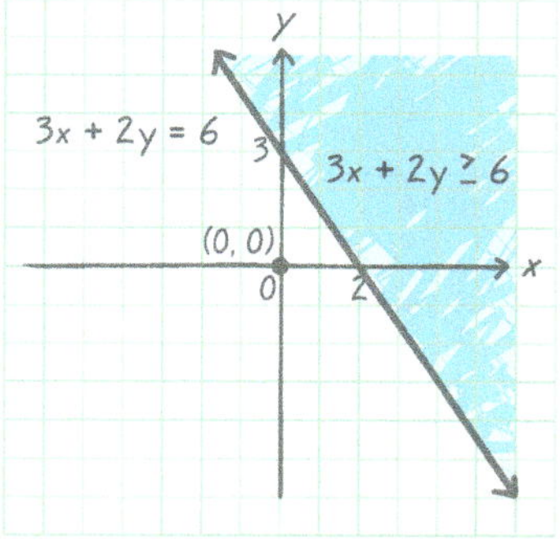

Figure 32

◀ *Work Problem* **1** *at the Side.*

If the inequality is written in the form $y > mx + b$ or $y < mx + b$, the inequality symbol indicates which half-plane to shade.

If $y > mx + b$, shade **above** the boundary line.

If $y < mx + b$, shade **below** the boundary line.

This method works only if the inequality is solved for y.

ANSWERS

1. (a)

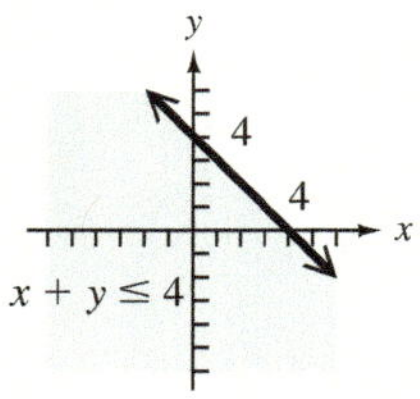

(b)

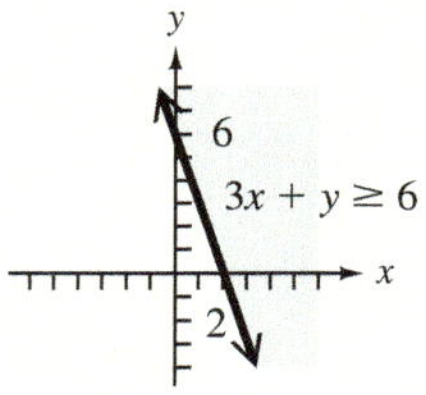

Example 2 **Graphing a Linear Inequality with Boundary Passing through the Origin**

Graph $3x - 4y > 0$.

First graph the boundary line. If $x = 0$, then $y = 0$. Thus, this line passes through the origin. Two other points on the line are $(4, 3)$ and $(-4, -3)$. The points of the boundary line do not belong to the inequality $3x - 4y > 0$ (because inequality symbol does not include equality). For this reason, the line is dashed. Now solve the inequality for y.

$$3x - 4y > 0$$

$$-4y > -3x \quad \text{Subtract } 3x.$$

$$y < \frac{3}{4}x \quad \text{Divide by } -4\text{; change } > \text{ to } <.$$

Because of the *is less than* symbol, shade *below* the line. As a check, choose a test point not on the line. Because the origin is on the line, we must choose a different point, such as $(2, -1)$. Substitute for x and y in the original inequality.

$$3x - 4y > 0$$

$$3(2) - 4(-1) \overset{?}{>} 0 \quad \text{Let } x = 2 \text{ and } y = -1.$$

$$6 + 4 \overset{?}{>} 0$$

$$10 > 0 \quad \text{True}$$

This result agrees with the decision to shade below the line. The solution set, graphed in Figure 33, includes only those points in the shaded half-plane (not those on the line).

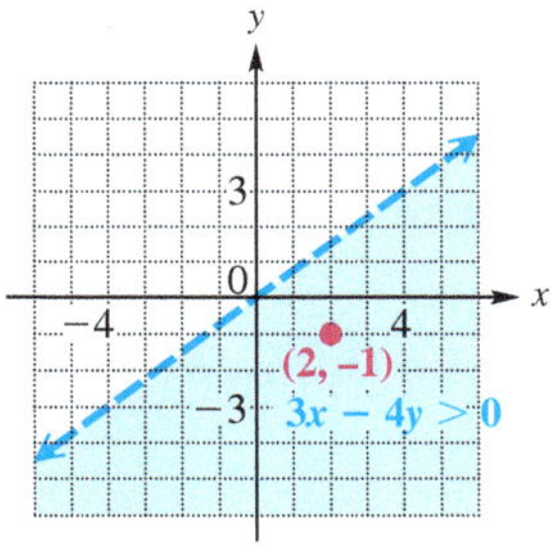

Figure 33

Work Problem **2** *at the Side.* ▶

2 Graph each inequality.

(a) $x + y > 0$

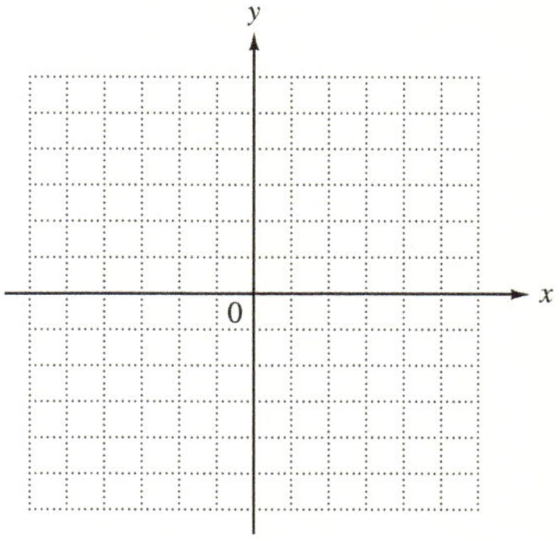

(b) $3x - 2y > 0$

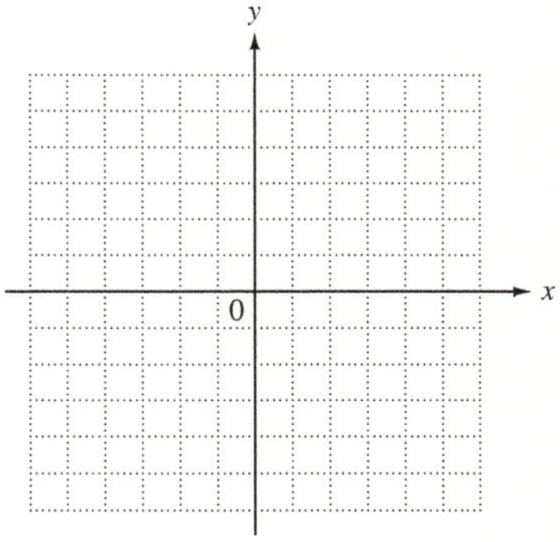

OBJECTIVE 2 **Graph the intersection of two linear inequalities.** In **Section 3.2** we discussed how the words *and* and *or* are used with compound inequalities. In that section, the inequalities had one variable. Those ideas can be extended to include inequalities in two variables.

A pair of inequalities joined with the word *and* is interpreted as the intersection of the solution sets of the inequalities. ***The graph of the intersection of two or more inequalities is the region of the plane where all points satisfy all of the inequalities at the same time.***

ANSWERS

2. (a)

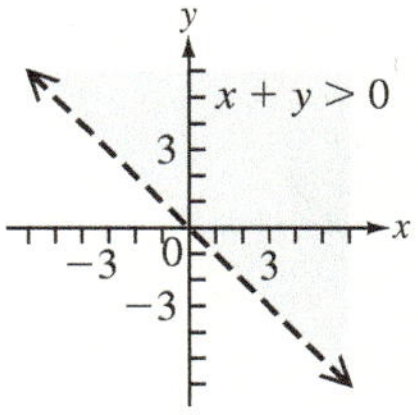

(b)

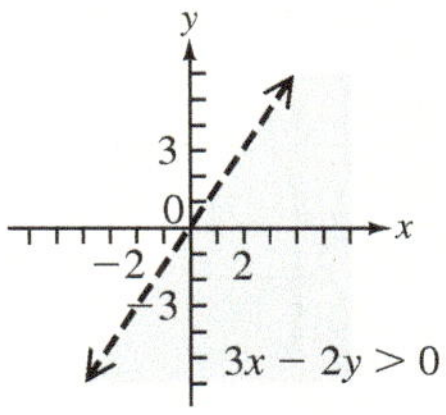

3 Graph $x - y \leq 4$ and $x \geq -2$.

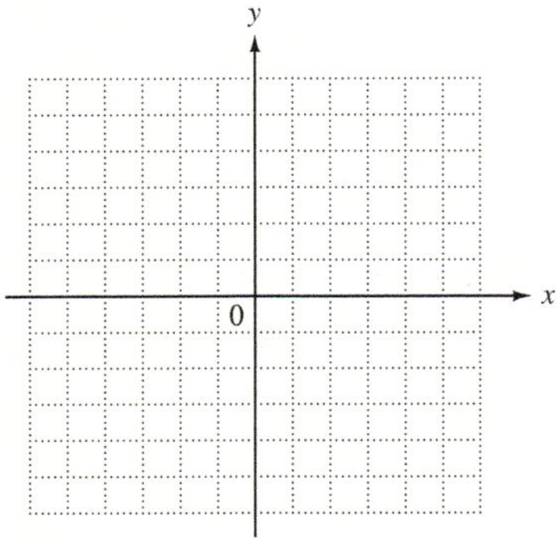

Example 3 Graphing the Intersection of Two Inequalities

Graph $2x + 4y \geq 5$ and $x \geq 1$.

To begin, we graph each of the two inequalities $2x + 4y \geq 5$ and $x \geq 1$ separately. The graph of $2x + 4y \geq 5$ is shown in Figure 34(a), and the graph of $x \geq 1$ is shown in Figure 34(b).

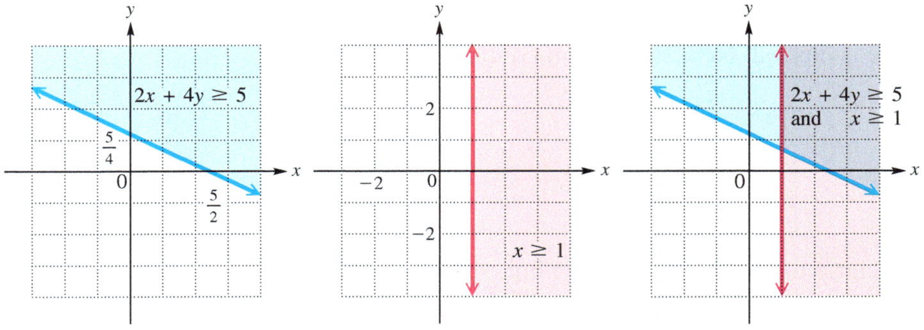

(a) (b) (c)

Figure 34

In practice, the two graphs in Figures 34(a) and 34(b) are graphed on the same axes. Then we use heavy shading to identify the intersection of the graphs, as shown in Figure 34(c). To check, we can use a test point from each of the four regions formed by the intersection of the boundary lines. Verify that only ordered pairs in the heavily shaded region satisfy both inequalities.

◀ Work Problem **3** at the Side.

4 Graph $7x - 3y < 21$ or $x > 2$.

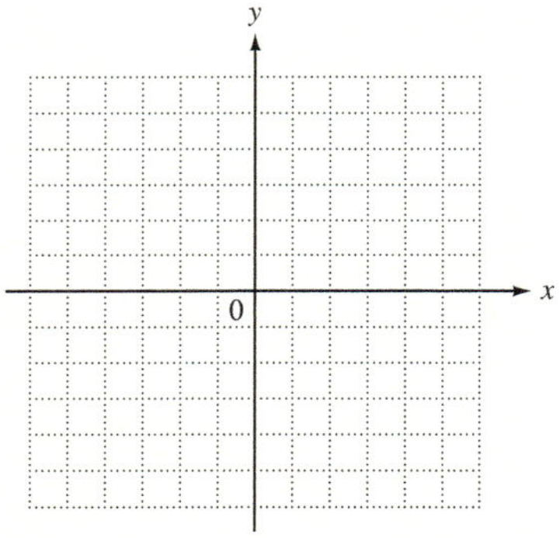

OBJECTIVE 3 Graph the union of two linear inequalities. When two inequalities are joined by the word *or*, we must find the union of the graphs of the inequalities. ***The graph of the union of two inequalities includes all of the points that satisfy either inequality.***

Example 4 Graphing the Union of Two Inequalities

Graph $2x + 4y \geq 5$ or $x \geq 1$.

The graphs of the two inequalities are shown in Figures 34(a) and 34(b) in Example 3. The graph of the union is shown in Figure 35.

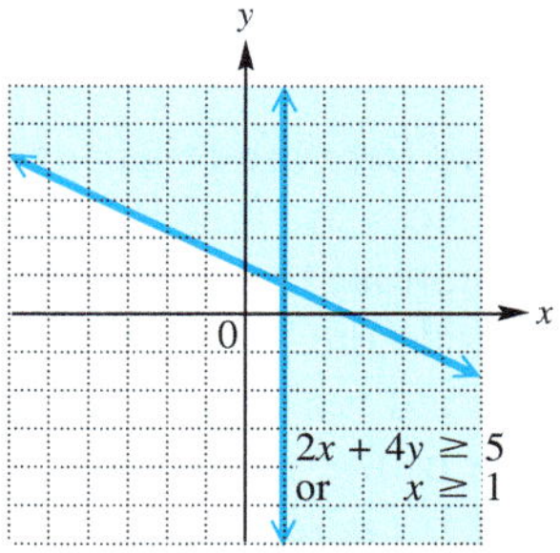

Figure 35

◀ Work Problem **4** at the Side.

ANSWERS

3.

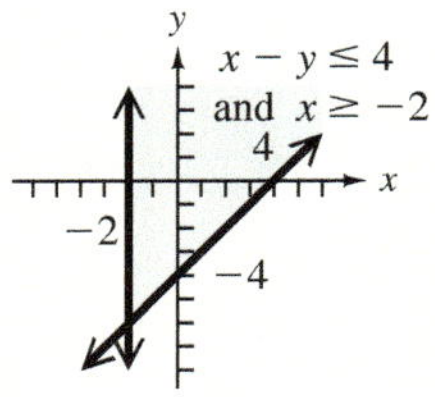

4.

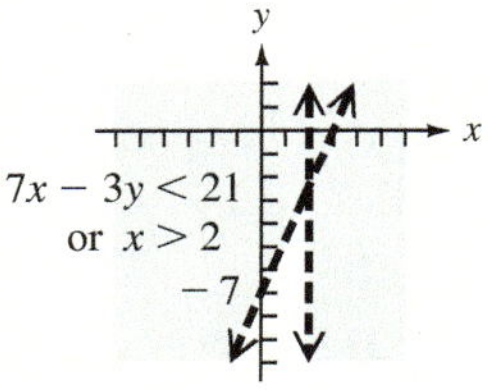

13.5 Exercises

FOR EXTRA HELP

MyMathLab

MathXL PRACTICE WATCH DOWNLOAD READ REVIEW

In each statement, fill in the first blank with one of the words solid *or* dashed. *Fill in the second blank with one of the words* above *or* below.

1. The boundary of the graph of $y \leq -x + 2$ will be a ________ line, and the shading will be ________ the line.

2. The boundary of the graph of $y < -x + 2$ will be a ________ line, and the shading will be ________ the line.

3. The boundary of the graph of $y > -x + 2$ will be a ________ line, and the shading will be ________ the line.

4. The boundary of the graph of $y \geq -x + 2$ will be a ________ line, and the shading will be ________ the line.

5. How is the boundary line $Ax + By = C$ used in graphing either $Ax + By < C$ or $Ax + By > C$?

6. Describe the two methods discussed in the text for deciding which region is the solution set of a linear inequality in two variables.

Graph each linear inequality. See Examples 1 and 2.

7. $x + y \leq 2$

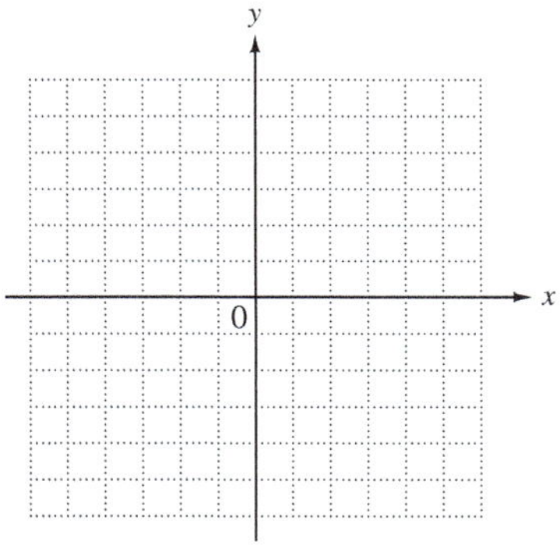

8. $x + y \leq -3$

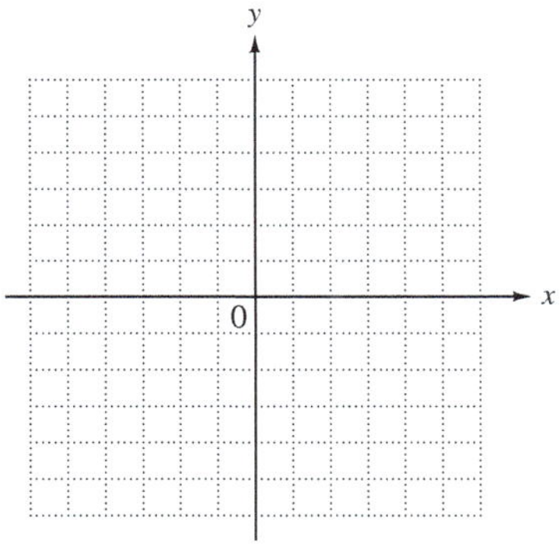

9. $4x - y < 4$

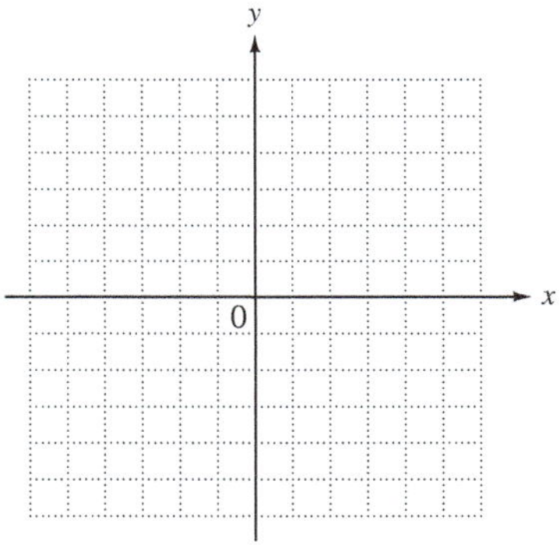

10. $3x - y < 3$

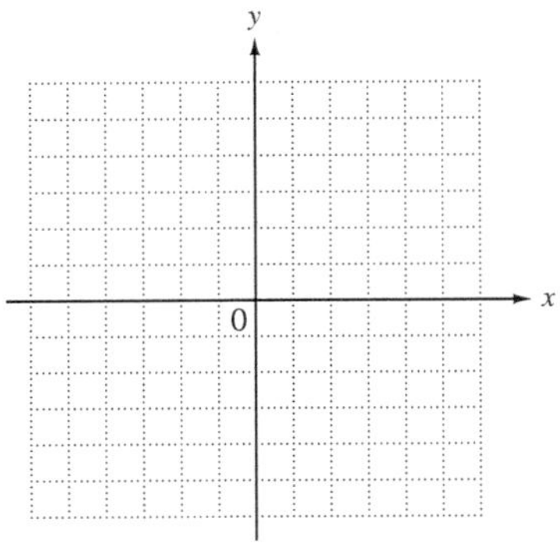

11. $x + 3y \geq -2$

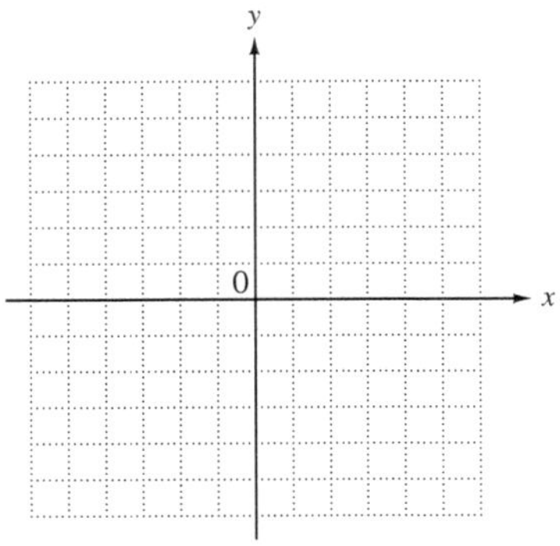

12. $x + 4y \geq -3$

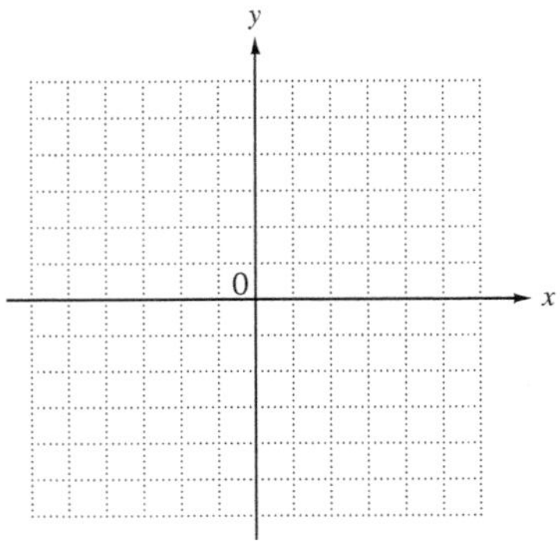

13. $x + y > 0$

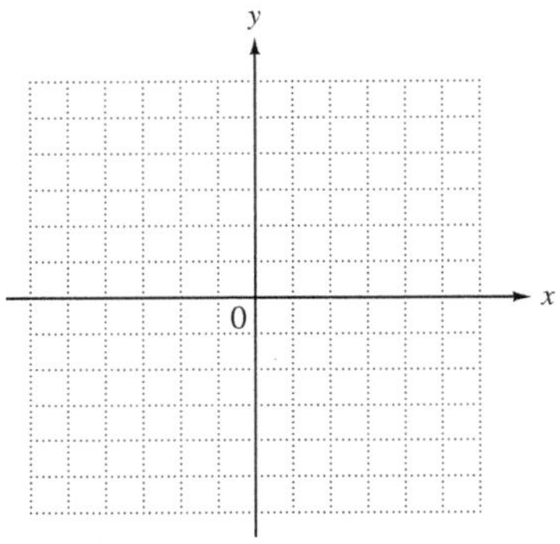

14. $x + 2y > 0$

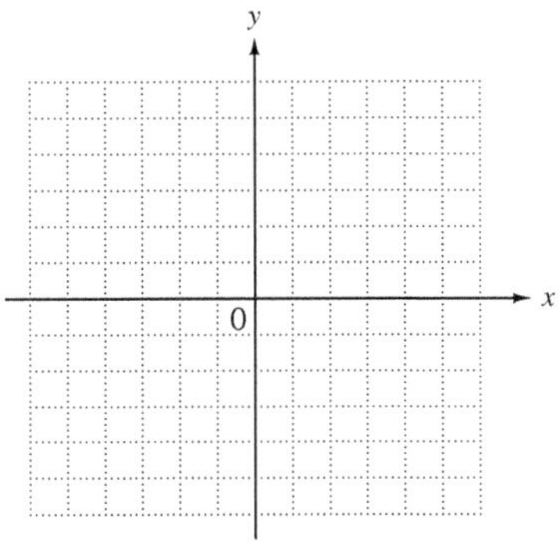

15. $x - 3y \leq 0$

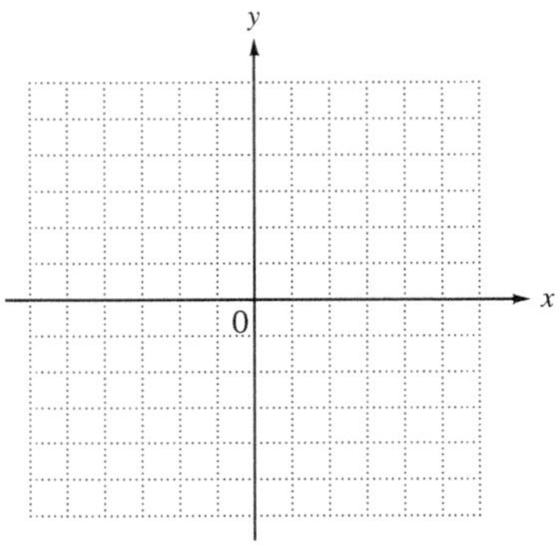

16. $x - 5y \leq 0$

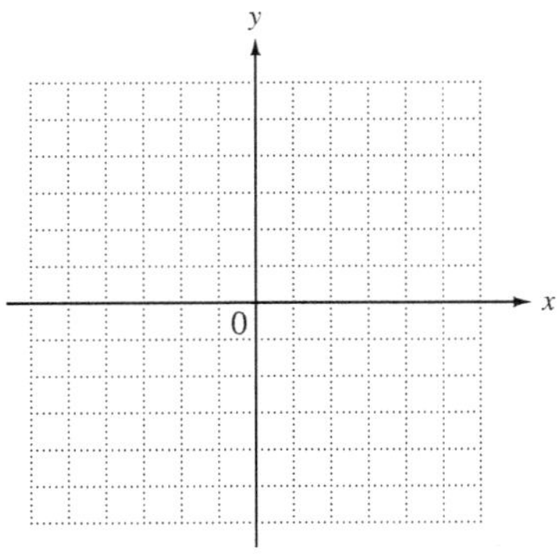

17. $y < x$

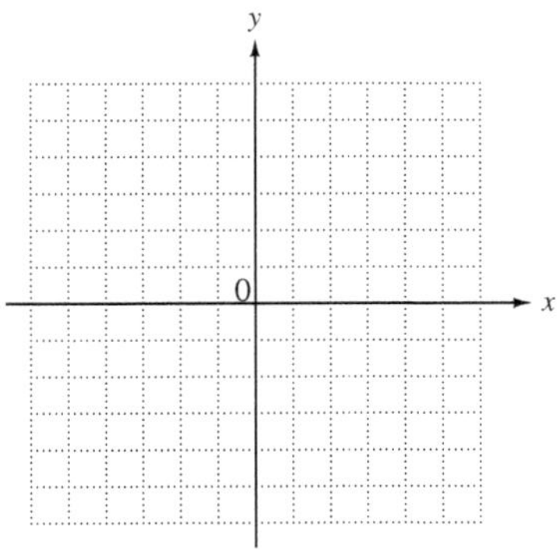

18. $y \leq 4x$

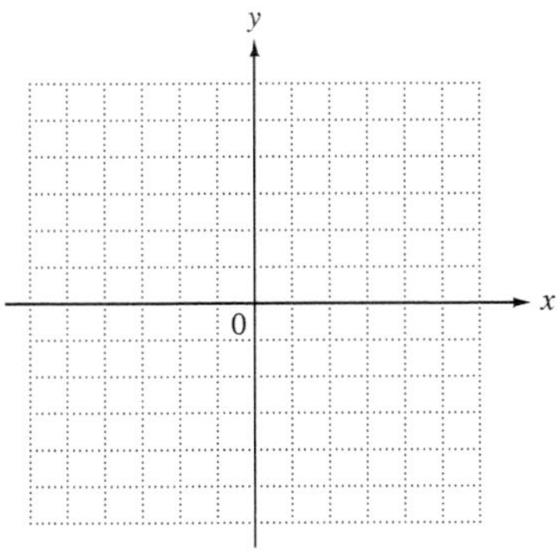

Graph the intersection of each pair of inequalities. See Example 3.

19. $x + y \leq 1$ and $x \geq 1$

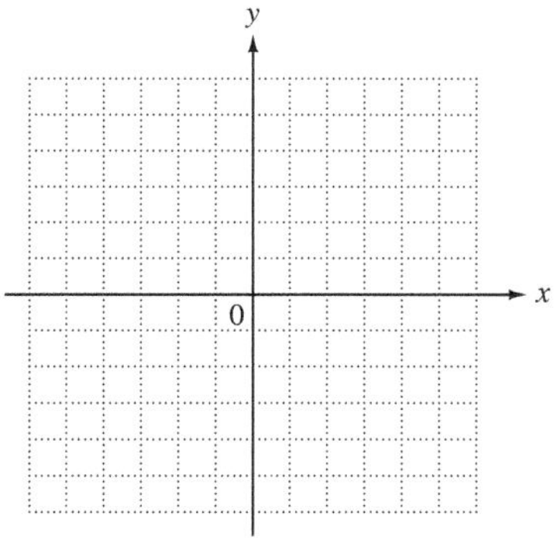

20. $x - y \geq 2$ and $x \geq 3$

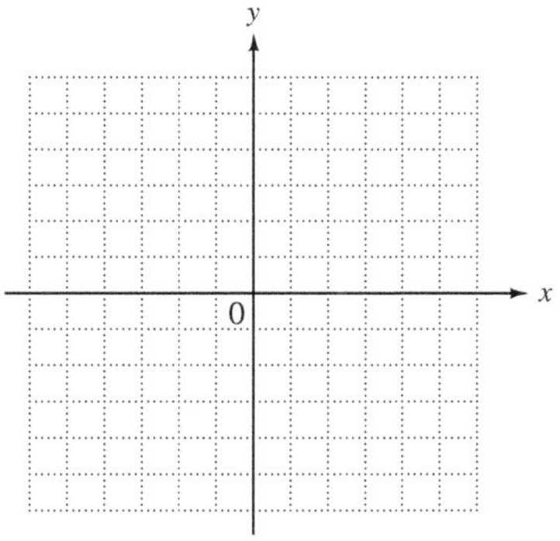

21. $2x - y \geq 2$ and $y < 4$

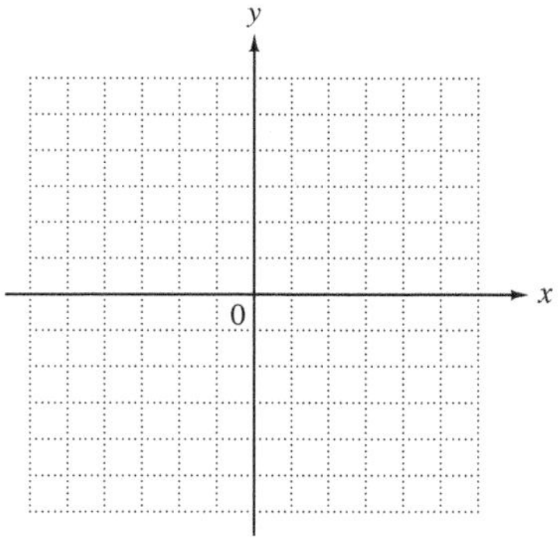

22. $3x - y \geq 3$ and $y < 3$

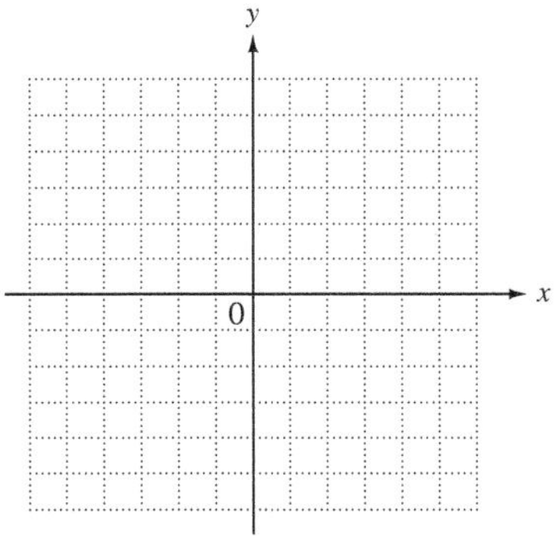

23. $x + y > -5$ and $y < -2$

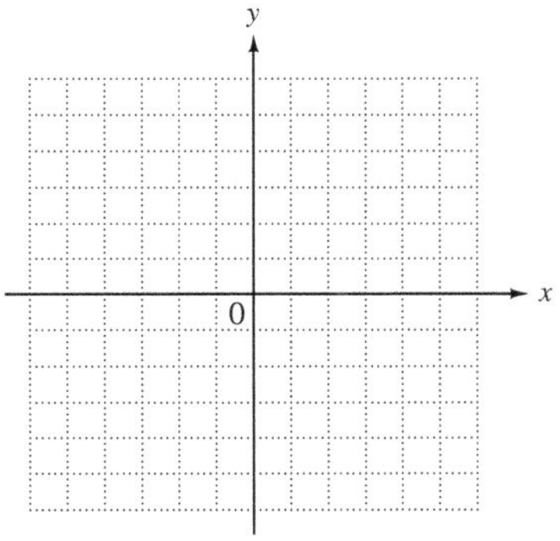

24. $6x - 4y < 10$ and $y > 2$

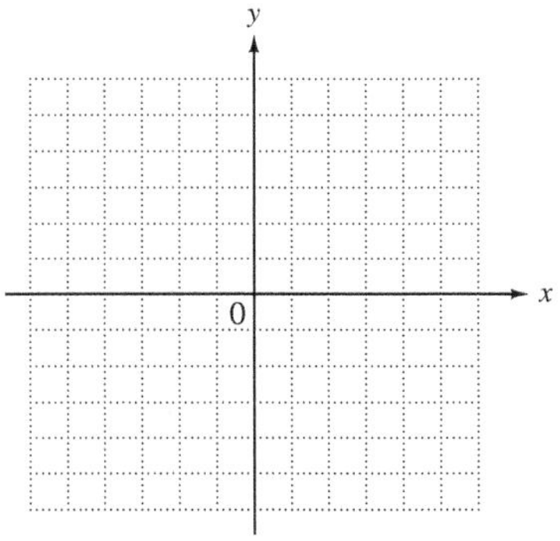

*Use the method described in **Section 3.3** to write each inequality as a compound inequality, and graph its solution set in the rectangular coordinate plane.*

25. $|x| \geq 3$

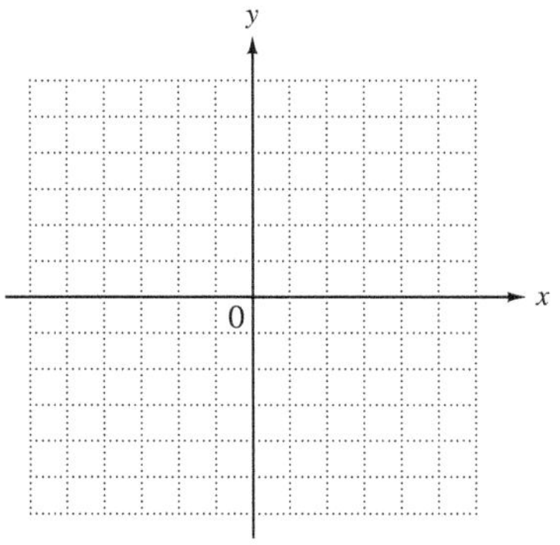

26. $|y| < 5$

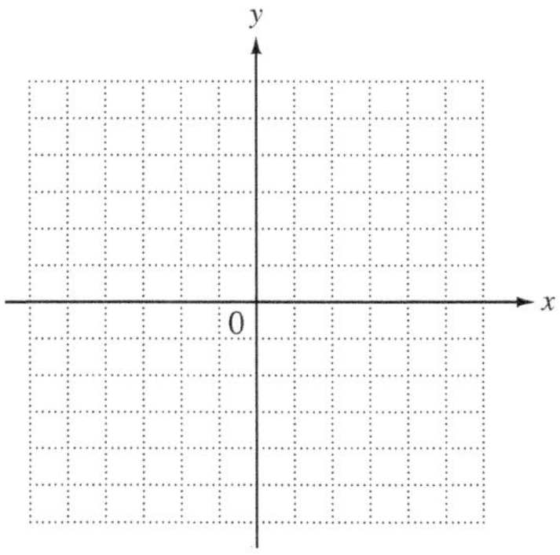

27. $|y + 1| < 2$

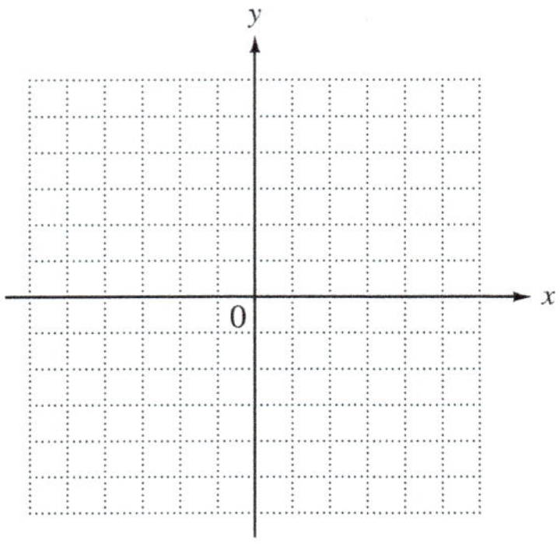

28. $|x - 2| \geq 1$

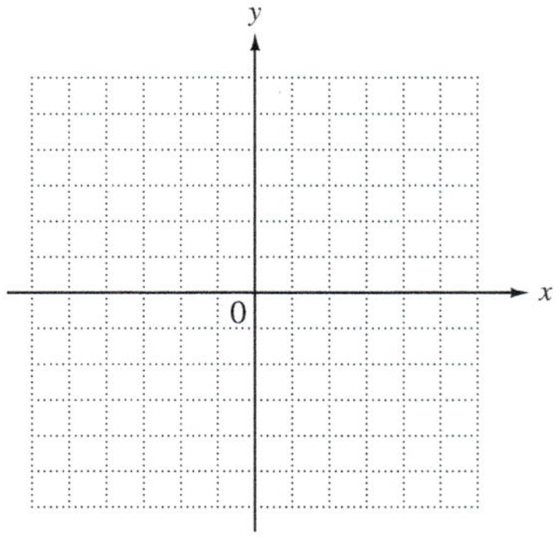

Graph the union of each pair of inequalities. See Example 4.

29. $x - y \geq 1$ or $y \geq 2$

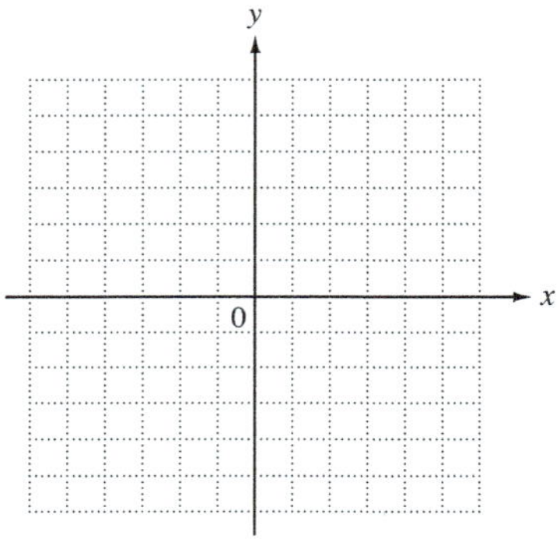

30. $x + y \leq 2$ or $y \geq 3$

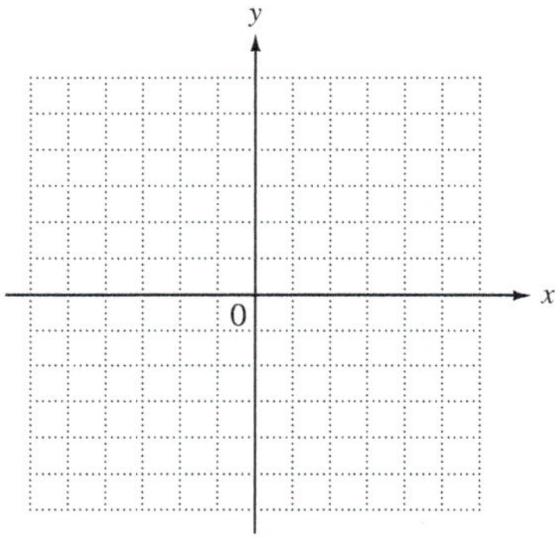

31. $x - 2 > y$ or $x < 1$

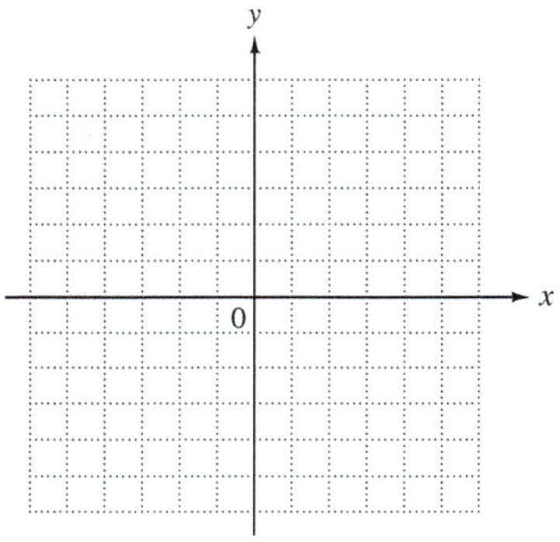

32. $x + 3 < y$ or $x > 3$

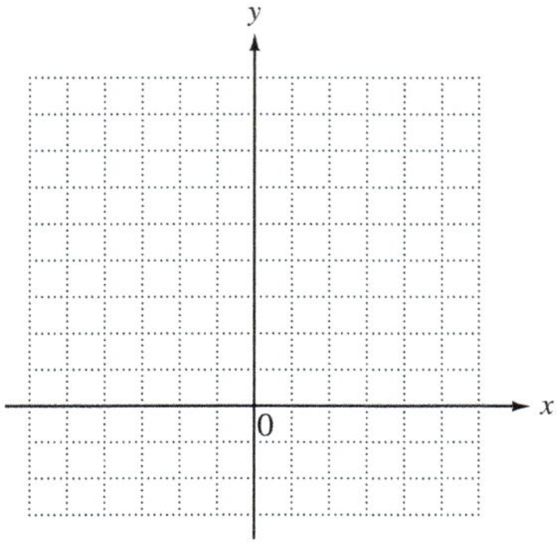

33. $3x + 2y < 6$ or $x - 2y > 2$

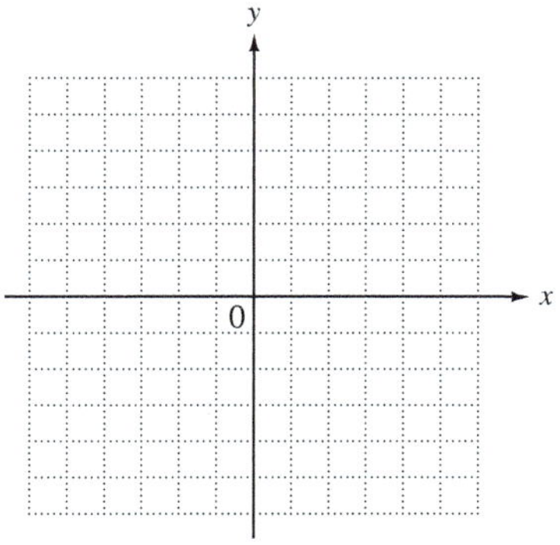

34. $x - y \geq 1$ or $x + y \leq 4$

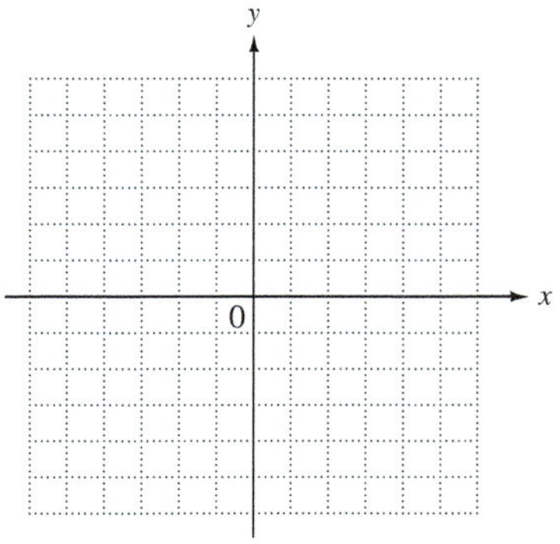

13.6 Introduction to Functions

OBJECTIVES

1. Define and identify relations and functions.
2. Find domain and range.
3. Identify functions defined by graphs and equations.
4. Use function notation.
5. Graph linear and constant functions.

We often describe one quantity in terms of another. Consider the following.

- The amount of your paycheck if you are paid hourly depends on the number of hours you worked.
- The cost at the gas station depends on the number of gallons of gas you pumped into your car.
- The distance traveled by a car moving at a constant speed depends on the time traveled.

We can use ordered pairs to represent these corresponding quantities. For example, we indicate the relationship between the amount of your paycheck and hours worked by writing ordered pairs in which the first number represents hours worked and the second number represents paycheck amount in dollars. Then the ordered pair (5, 40) indicates that when you work 5 hr, your paycheck is \$40. Similarly, the ordered pairs (10, 80) and (20, 160) show that working 10 hr results in an \$80 paycheck and working 20 hr results in a \$160 paycheck.

Work Problem 1 at the Side. ▶

1 What would the ordered pair (40, 320) in the correspondence between number of hours worked and paycheck amount (in dollars) indicate?

Since the amount of your paycheck *depends* on the number of hours worked, your paycheck amount is called the *dependent variable,* and the number of hours worked is called the *independent variable*. Generalizing, if the value of the variable y depends on the value of the variable x, then y is the **dependent variable** and x is the **independent variable.**

Independent variable → x; Dependent variable → y

$$(x, y)$$

OBJECTIVE 1 Define and identify relations and functions. Since we can write related quantities using ordered pairs, a set of ordered pairs such as

$$\{(5, 40), (10, 80), (20, 160), (40, 320)\}$$

is called a *relation*.

> **Relation**
>
> A **relation** is any set of ordered pairs.

A special kind of relation, called a *function,* is very important in mathematics and its applications.

> **Function**
>
> A **function** is a relation in which, for each value of the first component of the ordered pairs, there is *exactly one value* of the second component.

ANSWER

1. It indicates that when you work 40 hr, your paycheck is \$320.

2 Determine whether each relation defines a function.

(a) $\{(0, 3), (-1, 2), (-1, 3)\}$

(b) $\{(2, -2), (4, -4), (6, -6)\}$

(c) $\{(-1, 5), (0, 5)\}$

Example 1 Determining Whether Relations Are Functions

Tell whether each relation defines a function.

$$F = \{(1, 2), (-2, 4), (3, -1)\}$$
$$G = \{(-2, -1), (-1, 0), (0, 1), (1, 2), (2, 2)\}$$
$$H = \{(-4, 1), (-2, 1), (-2, 0)\}$$

Relations F and G are functions, because for each different x-value there is exactly one y-value. Notice that in G, the last two ordered pairs have the same y-value (1 is paired with 2, and 2 is paired with 2). This does not violate the definition of function, since the first components (x-values) are different and each is paired with only one second component (y-value).

In relation H, however, the last two ordered pairs have the *same* x-value paired with *two different* y-values (-2 is paired with both 1 and 0), so H is a relation but not a function. ***In a function, no two ordered pairs can have the same first component and different second components.***

Different y-values

$$H = \{(-4, 1), (-2, 1), (-2, 0)\} \quad \text{Not a function}$$

Same x-value

◀ Work Problem **2** at the Side.

In a function, there is* exactly one *value of the dependent variable, the second component, for each value of the independent variable, the first component. This is what makes functions so important in applications.

Relations and functions can also be expressed as a correspondence or *mapping* from one set to another, as shown in Figure 36 for function F and relation H from Example 1. The arrow from 1 to 2 indicates that the ordered pair (1, 2) belongs to F—each first component is paired with exactly one second component. In the mapping for set H, which is not a function, the first component -2 is paired with two different second components, 1 and 0.

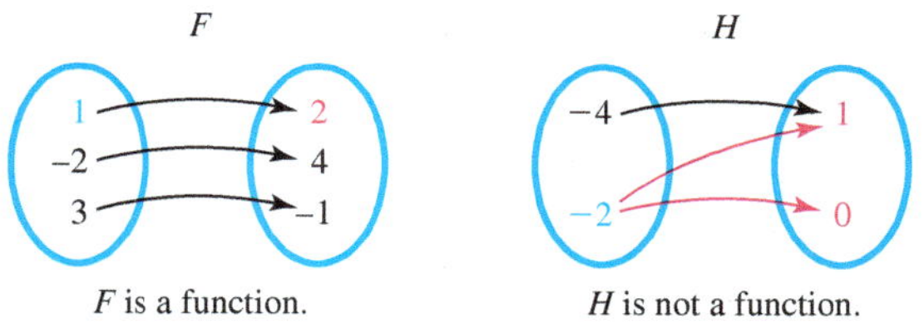

Figure 36

Since relations and functions are sets of ordered pairs, we can represent them using tables and graphs. A table and graph for function F is shown in Figure 37.

x	y
1	2
-2	4
3	-1

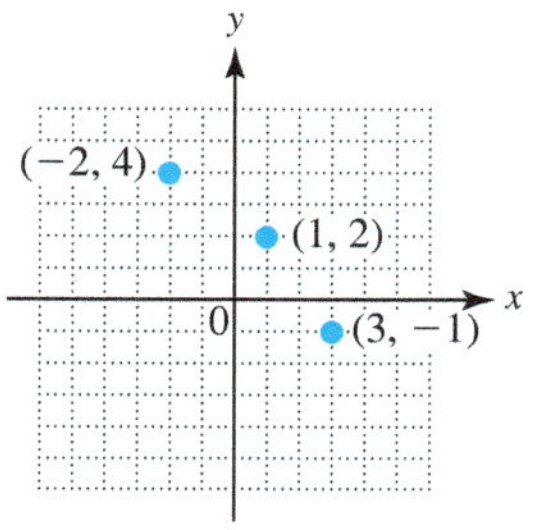

Graph of F

Figure 37

Answers

2. (a) not a function **(b)** function **(c)** function

Finally, we can describe a relation or function using a rule that tells how to determine the dependent variable for a specific value of the independent variable. The rule may be given in words, such as "the dependent variable is twice the independent variable." Usually, however, the rule is given as an equation:

$$y = 2x.$$

↑ Dependent variable ↑ Independent variable

An equation is the most efficient way to define a relation or function.

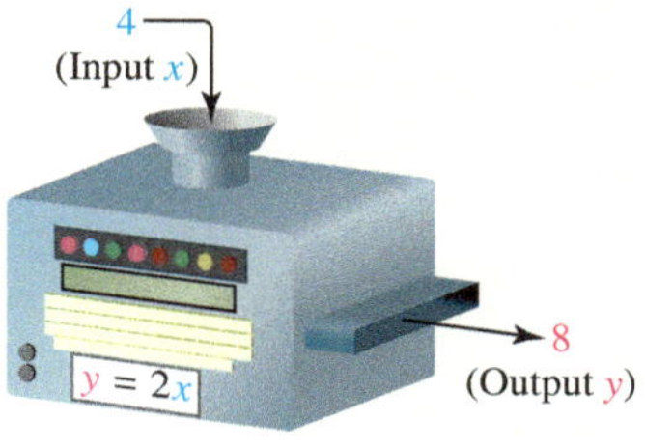

Function machine

> **Note**
>
> Another way to think of a function relationship is to think of the independent variable as an input and the dependent variable as an output. This is illustrated by the input-output (function) machine in the margin for the function defined by $y = 2x$.

OBJECTIVE 2 Find domain and range. For every relation, there are two important sets of elements called the *domain* and *range*.

> **Domain and Range**
>
> In a relation, the set of all values of the independent variable (x) is the **domain.** The set of all values of the dependent variable (y) is the **range.**

Example 2 Finding Domains and Ranges of Relations

Give the domain and range of each relation. Tell whether the relation defines a function.

(a) $\{(3, -1), (4, 2), (4, 5), (6, 8)\}$

The domain, the set of x-values, is $\{3, 4, 6\}$; the range, the set of y-values, is $\{-1, 2, 5, 8\}$. This relation is not a function because the same x-value 4 is paired with two different y-values, 2 and 5.

(b)

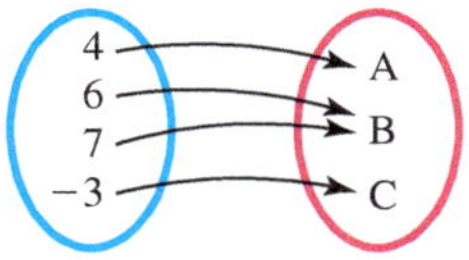

The domain of this relation is

$$\{4, 6, 7, -3\}.$$

The range is

$$\{A, B, C\}.$$

This mapping defines a function—each x-value corresponds to exactly one y-value.

(c)

x	y
−5	2
0	2
5	2

This is a table of ordered pairs, so the domain is the set of x-values, $\{-5, 0, 5\}$, and the range is the set of y-values, $\{2\}$. The table defines a function because each different x-value corresponds to exactly one y-value (even though it is the same y-value).

Work Problem 3 at the Side. ▶

3 Give the domain and range of each relation. Does the relation define a function?

(a) $\{(4, 0), (4, 1), (4, 2)\}$

(b)

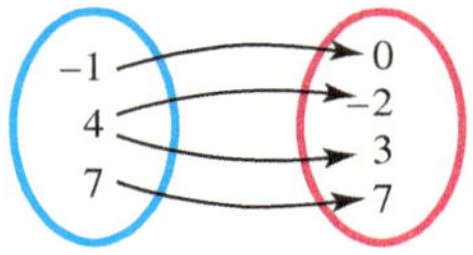

(c)

Year	Cell Phone Subscribers (in thousands)
2002	140,766
2003	158,722
2004	182,140
2005	207,896
2006	233,041

Source: CTIA-The Wireless Association.

Answers

3. (a) domain: $\{4\}$; range: $\{0, 1, 2\}$; No, the relation does not define a function.
(b) domain: $\{-1, 4, 7\}$; range: $\{0, -2, 3, 7\}$; No, the relation does not define a function.
(c) domain: $\{2002, 2003, 2004, 2005, 2006\}$; range: $\{140{,}766, 158{,}722, 182{,}140, 207{,}896, 233{,}041\}$; Yes, the relation defines a function.

4 Give the domain and range of each relation.

(a)

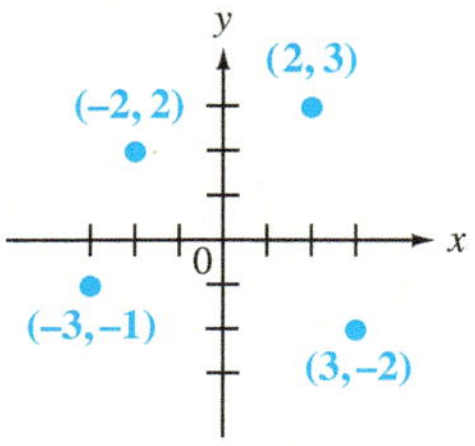

(b)

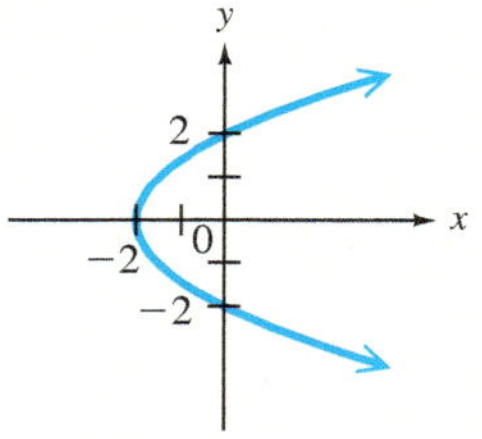

(c)

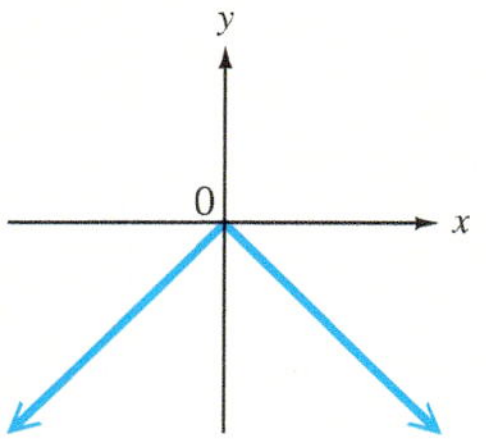

The graph of a relation gives a picture of the relation, which can be used to determine its domain and range.

Example 3 Finding Domains and Ranges from Graphs

Give the domain and range of each relation.

(a)

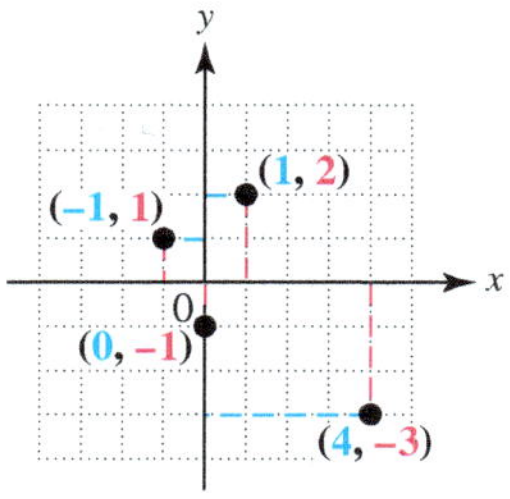

The domain is the set of x-values,

$$\{-1, 0, 1, 4\}.$$

The range is the set of y-values,

$$\{-3, -1, 1, 2\}.$$

(b)

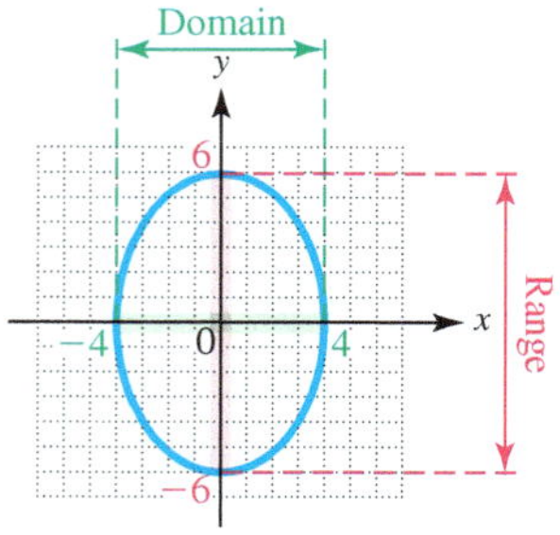

The x-values of the points on the graph include all numbers between -4 and 4, inclusive. The y-values include all numbers between -6 and 6, inclusive. Using interval notation,

the domain is $[-4, 4]$;

the range is $[-6, 6]$.

(c)

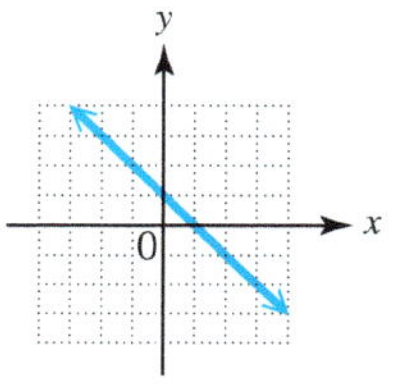

The arrowheads indicate that the line extends indefinitely left and right, as well as up and down. Therefore, both the domain and the range include all real numbers, written $(-\infty, \infty)$.

(d)

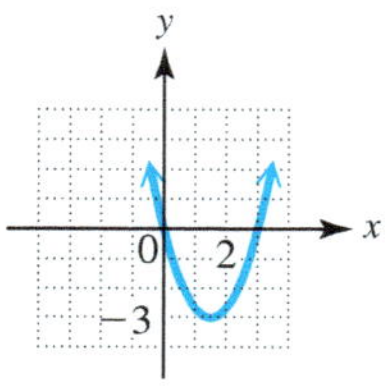

The arrowheads indicate that the graph extends indefinitely left and right, as well as upward. The domain is $(-\infty, \infty)$. Because there is a least y-value, -3, the range includes all numbers greater than or equal to -3, written $[-3, \infty)$.

◀ Work Problem 4 at the Side.

Since relations are often defined by equations, such as $y = 2x + 3$ and $y^2 = x$, we must sometimes determine the domain of a relation from its equation. We assume the following agreement on the domain of a relation.

> **Agreement on Domain**
> The domain of a relation is assumed to be all real numbers that produce real numbers when substituted for the independent variable.

To illustrate this agreement, since any real number can be used as a replacement for x in $y = 2x + 3$, the domain of this function is the set of all real numbers. The function defined by $y = \frac{1}{x}$ has all real numbers except 0 as domain, since y is undefined if $x = 0$. ***In general, the domain of a function defined by an algebraic expression is all real numbers, except those numbers that lead to division by 0 or an even root of a negative number.***

Answers

4. **(a)** domain: $\{-3, -2, 2, 3\}$; range: $\{-2, -1, 2, 3\}$
(b) domain: $[-2, \infty)$; range: $(-\infty, \infty)$
(c) domain: $(-\infty, \infty)$; range: $(-\infty, 0]$

OBJECTIVE 3 Identify functions defined by graphs and equations. Since each value of x in a function corresponds to only one value of y, any vertical line drawn through the graph of a function must intersect the graph in at most one point. This is the *vertical line test* for a function.

Vertical Line Test

If every vertical line intersects the graph of a relation in no more than one point, then the relation represents a function.

For example, the graph shown in Figure 38(a) is not the graph of a function since a vertical line intersects the graph in more than one point. The graph in Figure 38(b) does represent a function.

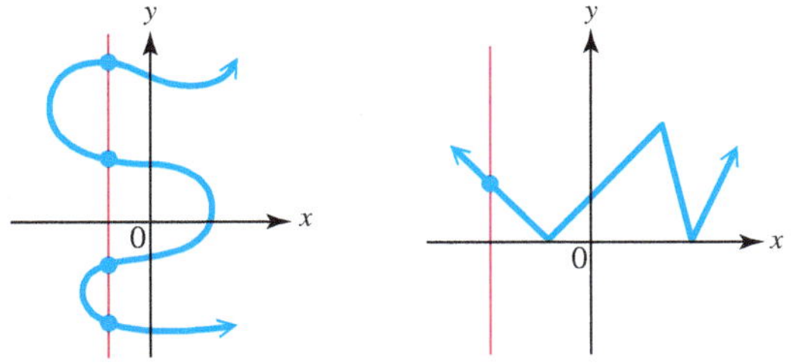

Not a function – the same x-value corresponds to four different y-values.

(a)

Function – each x-value corresponds to only one y-value.

(b)

Figure 38

Example 4 Using the Vertical Line Test

Use the vertical line test to determine whether each relation graphed in Example 3 is a function.

(a)

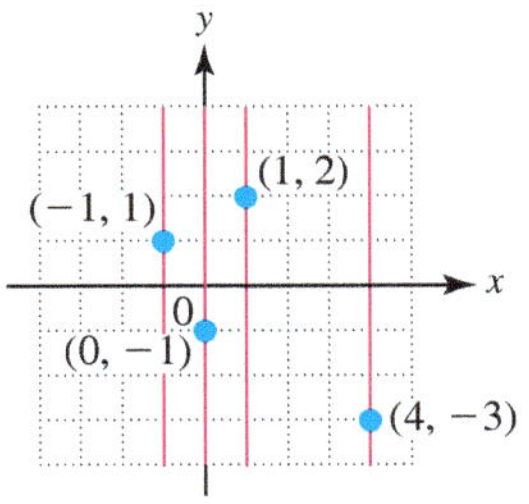

Function

(b)

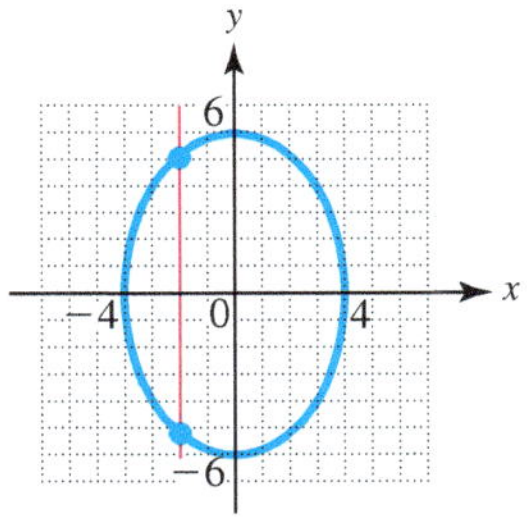

Not a function

(c)

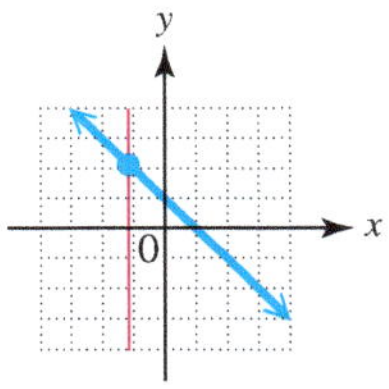

Function

(d)

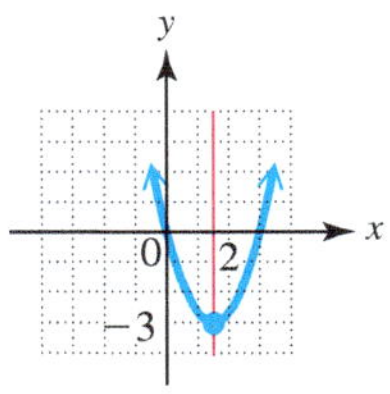

Function

The graphs in (a), (c), and (d) represent functions. The graph of the relation in (b) fails the vertical line test, since the same x-value corresponds to two different y-values; therefore, it is not the graph of a function.

Work Problem 5 at the Side. ▶

5 Use the vertical line test to decide which graphs represent functions.

A.

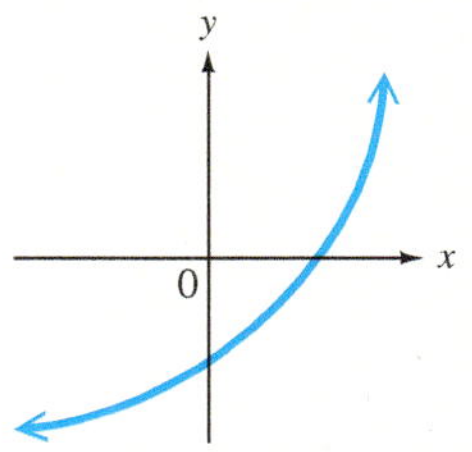

B.

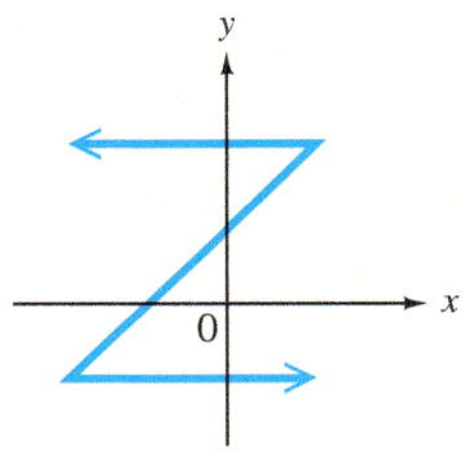

C.

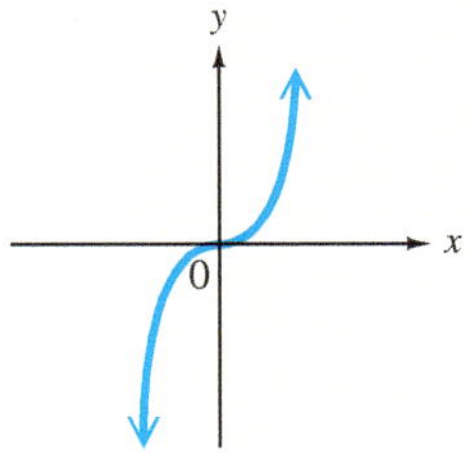

ANSWER

5. A and C are graphs of functions.

6 Decide whether each relation defines a function, and give the domain.

(a) $y = 6x + 12$

(b) $y \leq 4x$

(c) $y = -\sqrt{3x - 2}$

(d) $y^2 = 25x$

Note

Graphs that do not represent functions are still relations. ***Remember that all equations and graphs represent relations and that all relations have a domain and range.***

It can be more difficult to decide whether a relation defined by an equation or an inequality is a function. The next example gives some hints that may help.

Example 5 Identifying Functions from Their Equations

Decide whether each relation defines a function and give the domain.

(a) $y = x + 4$

In the defining equation, $y = x + 4$, y is always found by adding 4 to x. Thus, each value of x corresponds to just one value of y and the relation defines a function; x can be any real number, so the domain is $(-\infty, \infty)$.

(b) $y = \sqrt{2x - 1}$

For any choice of x in the domain, there is exactly one corresponding value for y (the radical is a nonnegative number), so this equation defines a function. Since the equation involves a square root, the quantity under the radical sign cannot be negative. Thus,

$$2x - 1 \geq 0$$
$$2x \geq 1$$
$$x \geq \frac{1}{2},$$

and the domain of the function is $[\frac{1}{2}, \infty)$.

(c) $y^2 = x$

The ordered pairs (16, 4) and (16, −4) both satisfy this equation. Since one value of x, 16, corresponds to two values of y, 4 and −4, this equation does not define a function. Because x is equal to the square of y, the values of x must always be nonnegative. The domain of the relation is $[0, \infty)$.

(d) $y \leq x - 1$

By definition, y is a function of x if every value of x leads to exactly one value of y. Here a particular value of x, say 1, corresponds to many values of y. The ordered pairs (1, 0), (1, −1), (1, −2), (1, −3), and so on, all satisfy the inequality. Thus, this relation does not define a function. Any number can be used for x, so the domain is the set of real numbers, $(-\infty, \infty)$.

(e) $y = \frac{5}{x - 1}$

Given any value of x in the domain, we find y by subtracting 1, then dividing the result into 5. This process produces exactly one value of y for each value in the domain, so this equation defines a function. The domain includes all real numbers except those that make the denominator 0. We find these numbers by setting the denominator equal to 0 and solving for x.

$$x - 1 = 0$$
$$x = 1$$

The domain includes all real numbers *except* 1, written $(-\infty, 1) \cup (1, \infty)$.

◀ Work Problem 6 at the Side.

ANSWERS

6. (a) yes; $(-\infty, \infty)$ **(b)** no; $(-\infty, \infty)$ **(c)** yes; $\left[\frac{2}{3}, \infty\right)$ **(d)** no; $[0, \infty)$

In summary, three variations of the definition of function are given here.

Variations of the Definition of Function

1. A **function** is a relation in which, for each value of the first component of the ordered pairs, there is exactly one value of the second component.
2. A **function** is a set of ordered pairs in which no first component is repeated.
3. A **function** is a rule or correspondence that assigns exactly one range value to each distinct domain value.

OBJECTIVE 4 Use function notation. When a function f is defined with a rule or an equation using x and y for the independent and dependent variables, we say "y is a function of x" to emphasize that *y depends on x*. We use the notation

$$y = f(x),$$

called **function notation,** to express this and read $f(x)$ as "f of x." (In this special notation the parentheses do not indicate multiplication.) The letter f stands for *function*. For example, if $y = 9x - 5$, we can name this function f and write

$$f(x) = 9x - 5.$$

Note that ***$f(x)$ is just another name for the dependent variable y.*** For example, if $y = f(x) = 9x - 5$ and $x = 2$, then we find y, or $f(2)$, by replacing x with 2.

$$\begin{aligned} y &= f(2) \\ &= 9 \cdot 2 - 5 \\ &= 18 - 5 \\ &= 13. \end{aligned}$$

For function f, the statement "if $x = 2$, then $y = 13$" is represented by the ordered pair $(2, 13)$ and is abbreviated with function notation as

$$f(2) = 13.$$

Read $f(2)$ as "f of 2" or "f at 2." Also,

$$f(0) = 9 \cdot 0 - 5 = -5 \qquad \text{and} \qquad f(-3) = 9(-3) - 5 = -32.$$

These ideas can be illustrated as follows.

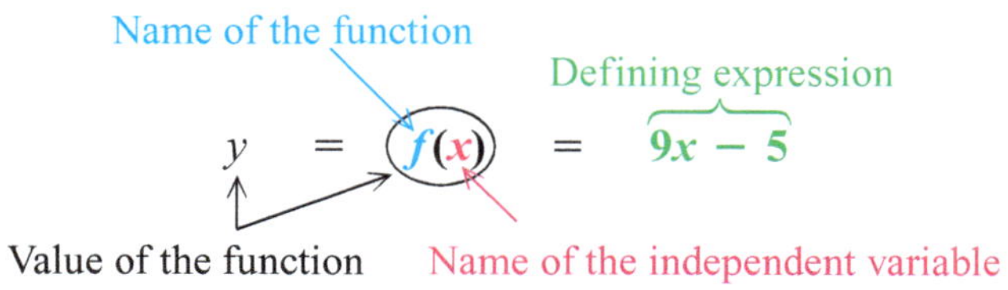

CAUTION
The symbol $f(x)$ *does not* indicate "f times x," but represents the y-value for the indicated x-value. As just shown, $f(2)$ is the y-value that corresponds to the x-value 2.

7 Find $f(-3)$, $f(p)$, and $f(m+1)$.

(a) $f(x) = 6x - 2$

(b) $f(x) = \dfrac{-3x + 5}{2}$

(c) $f(x) = \dfrac{1}{6}x - 1$

Example 6 Using Function Notation

Let $f(x) = -x^2 + 5x - 3$. Find the following.

(a) $f(2)$

Do not read this as "f times 2." Read it as "f of 2."

$f(x) = -x^2 + 5x - 3$ — The base in $-x^2$ is x, not $(-x)$.

$f(2) = -2^2 + 5 \cdot 2 - 3$ — Replace x with 2.

$f(2) = -4 + 10 - 3$ — Apply the exponent; multiply.

$f(2) = 3$ — Add and subtract.

Since $f(2) = 3$, the ordered pair (2, 3) belongs to f.

(b) $f(q)$

$f(x) = -x^2 + 5x - 3$

$f(q) = -q^2 + 5q - 3$ — Replace x with q.

The replacement of one variable with another is important in later courses.

Sometimes letters other than f, such as g, h, or capital letters F, G, and H, are used to name functions.

Example 7 Using Function Notation

Let $g(x) = 2x + 3$. Find and simplify $g(a + 1)$.

$g(x) = 2x + 3$

$g(a+1) = 2(a+1) + 3$ — Replace x with $a + 1$.

$g(a+1) = 2a + 2 + 3$

$g(a+1) = 2a + 5$

◀ Work Problem **7** at the Side.

Functions can be evaluated in a variety of ways, as shown in Example 8.

Example 8 Using Function Notation

For each function, find $f(3)$.

(a) $f(x) = 3x - 7$

$f(3) = 3(3) - 7$

$f(3) = 9 - 7$

$f(3) = 2$

(b) $f = \{(-3, 5), (0, 3), (3, 1), (6, -1)\}$

We want $f(3)$, the y-value of the ordered pair where $x = 3$. As indicated by the ordered pair (3, 1), when $x = 3$, $y = 1$, so $f(3) = 1$.

(c)

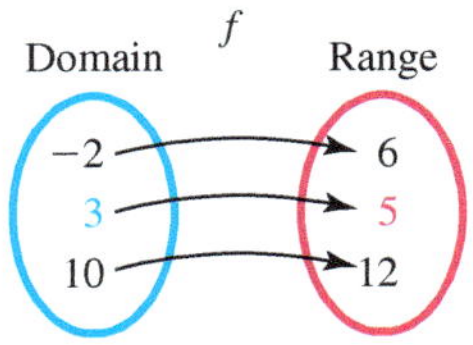

The domain element 3 is paired with 5 in the range, so $f(3) = 5$.

Continued on Next Page

Answers

7. (a) -20; $6p - 2$; $6m + 4$

(b) 7; $\dfrac{-3p + 5}{2}$; $\dfrac{-3m + 2}{2}$

(c) $-\dfrac{3}{2}$; $\dfrac{1}{6}p - 1$; $\dfrac{1}{6}(m + 1) - 1$, or $\dfrac{1}{6}m - \dfrac{5}{6}$

(d)

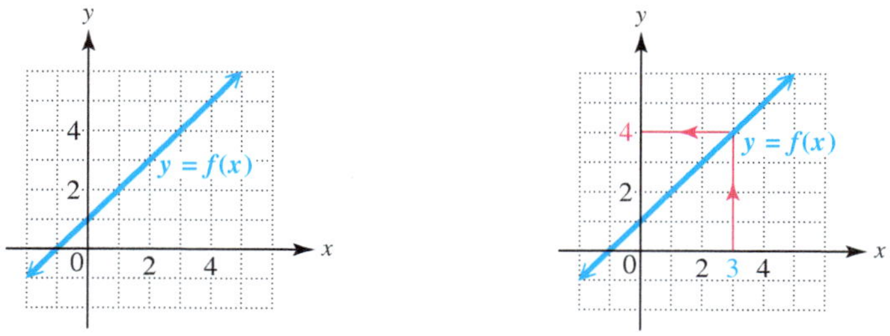

Figure 39

To evaluate $f(3)$, find 3 on the x-axis. See Figure 39. Then move up until the graph of f is reached. Moving horizontally to the y-axis gives 4 for the corresponding y-value. Thus, $f(3) = 4$.

Work Problem **8** *at the Side.* ▶

If a function f is defined by an equation with x and y instead of with function notation, use the following steps to find $f(x)$.

Writing an Equation Using Function Notation

Step 1 Solve the equation for y.

Step 2 Replace y with $f(x)$.

Example 9 **Writing Equations Using Function Notation**

Rewrite each equation using function notation $f(x)$. Then find $f(-2)$ and $f(a)$.

(a) $y = x^2 + 1$

This equation is already solved for y. Since $y = f(x)$,

$$f(x) = x^2 + 1.$$

To find $f(-2)$, let $x = -2$.

$$f(-2) = (-2)^2 + 1$$
$$f(-2) = 4 + 1$$
$$f(-2) = 5$$

Find $f(a)$ by letting $x = a$: $f(a) = a^2 + 1.$

(b) $x - 4y = 5$

First solve $x - 4y = 5$ for y. Then replace y with $f(x)$.

$$x - 4y = 5$$
$$x - 5 = 4y \quad \text{Add } 4y\text{; subtract 5.}$$
$$y = \frac{x-5}{4}, \quad \text{so} \quad f(x) = \frac{1}{4}x - \frac{5}{4}$$

Now find $f(-2)$ and $f(a)$.

$$f(-2) = \frac{1}{4}(-2) - \frac{5}{4} = -\frac{7}{4} \quad \text{Let } x = -2.$$
$$f(a) = \frac{1}{4}a - \frac{5}{4} \quad \text{Let } x = a.$$

Work Problem **9** *at the Side.* ▶

8 For each function, find $f(-2)$.

(a) $f(x) = -4x - 8$

(b) $f = \{(0, 5), (-1, 3), (-2, 1)\}$

(c)

x	$f(x)$
-4	16
-2	4
0	0
2	4
4	16

9 Rewrite each equation using function notation $f(x)$. Then find $f(-1)$.

(a) $y = \sqrt{x + 2}$

(b) $x^2 - 4y = 3$

Answers

8. **(a)** 0 **(b)** 1 **(c)** 4

9. **(a)** $f(x) = \sqrt{x + 2}$; 1

(b) $f(x) = \frac{x^2 - 3}{4}$, or $f(x) = \frac{1}{4}x^2 - \frac{3}{4}$; $-\frac{1}{2}$

OBJECTIVE 5 Graph linear and constant functions. Our first two-dimensional graphing was of straight lines. Linear equations (except for vertical lines with equations $x = a$) define *linear functions.*

Linear Function

A function that can be defined by

$$f(x) = ax + b,$$

for real numbers a and b, is a **linear function.** The value of a is the slope m of the graph of the function.

A linear function defined by $f(x) = b$ (whose graph is a horizontal line) is sometimes called a **constant function.** The domain of any linear function is $(-\infty, \infty)$. The range of a nonconstant linear function is $(-\infty, \infty)$, while the range of the constant function defined by $f(x) = b$ is $\{b\}$.

Example 10 Graphing Linear and Constant Functions

Graph each function. Give the domain and range.

(a) $f(x) = \frac{1}{4}x - \frac{5}{4}$

Recall from **Section 4.3** that m is the slope of the line and $(0, b)$ is the y-intercept. In Example 9(b), we wrote the equation $x - 4y = 5$ as the linear function defined by

$$f(x) = \frac{1}{4}x - \frac{5}{4}.$$

Slope (pointing to $\frac{1}{4}$); y-intercept is $(0, -\frac{5}{4})$.

To graph this function, plot the y-intercept $(0, -\frac{5}{4})$ and use the definition of slope as $\frac{\text{rise}}{\text{run}}$ to find a second point on the line. Since the slope is $\frac{1}{4}$, move 1 unit up from $(0, -\frac{5}{4})$ and 4 units to the right to the point $(4, -\frac{1}{4})$. Draw the straight line through these points to obtain the graph shown in Figure 40. The domain and range are both $(-\infty, \infty)$.

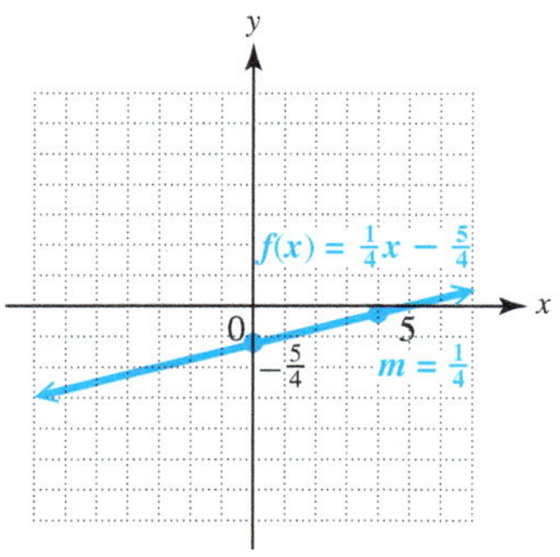

Figure 40

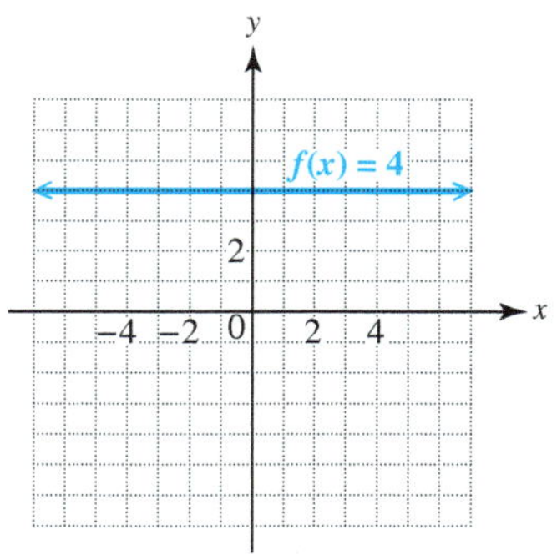

Figure 41

(b) $f(x) = 4$

This is a constant function. Its graph is the horizontal line containing all points with y-coordinate equal to 4. See Figure 41. The domain is $(-\infty, \infty)$ and the range is $\{4\}$.

◀ Work Problem **10** at the Side.

10 Graph each linear function. Give the domain and range.

(a) $f(x) = \frac{3}{4}x - 2$

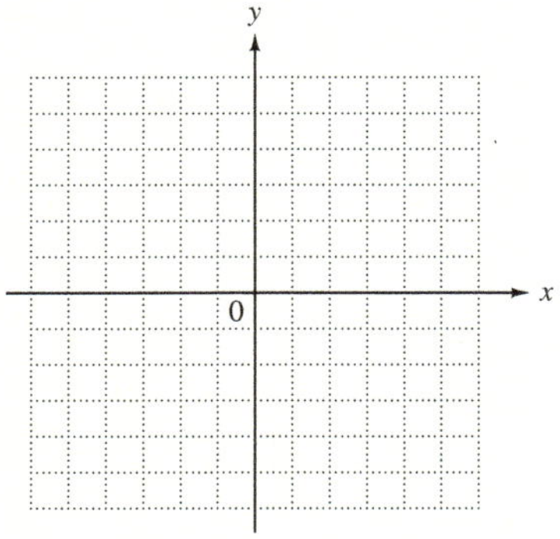

(b) $g(x) = 3$

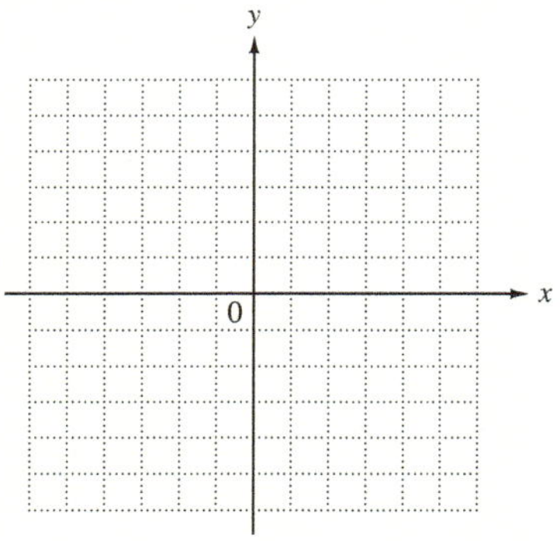

ANSWERS

10. (a)

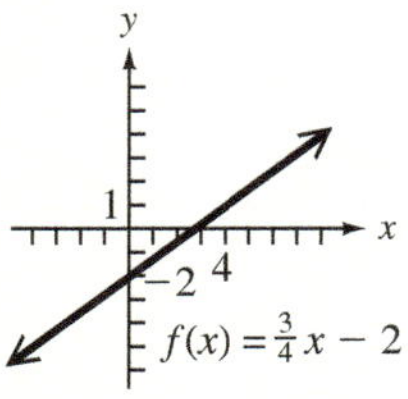

domain: $(-\infty, \infty)$; range: $(-\infty, \infty)$

(b)

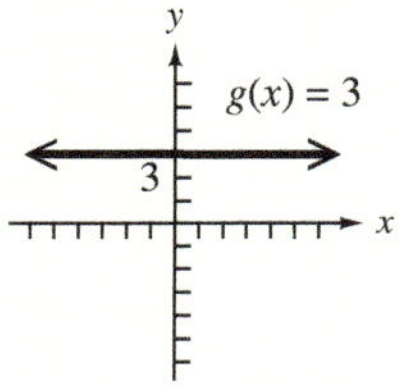

domain: $(-\infty, \infty)$; range: $\{3\}$

13.6 Exercises

1. In an ordered pair of a relation, is the first element the independent or the dependent variable?

2. Give an example of a relation that is not a function, having domain $\{-3, 2, 6\}$ and range $\{4, 6\}$. (There are many possible correct answers.)

3. Explain what is meant by each term.

(a) Relation **(b)** Domain of a relation

(c) Range of a relation **(d)** Function

4. Describe the use of the vertical line test.

Decide whether each relation is a function, and give the domain and the range. Use the vertical line test in Exercises 17–22. See Examples 1–4.

5. $\{(5, 1), (3, 2), (4, 9), (7, 3)\}$

6. $\{(8, 0), (5, 4), (9, 3), (3, 9)\}$

7. $\{(2, 4), (0, 2), (2, 6)\}$

8. $\{(9, -2), (-3, 5), (9, 1)\}$

9. $\{(-3, 1), (4, 1), (-2, 7)\}$

10. $\{(-12, 5), (-10, 3), (8, 3)\}$

11. $\{(1, 1), (1, -1), (0, 0), (2, 4), (2, -4)\}$

12. $\{(2, 5), (3, 7), (4, 9), (5, 11)\}$

13.

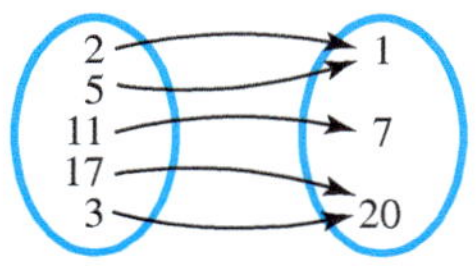

14.

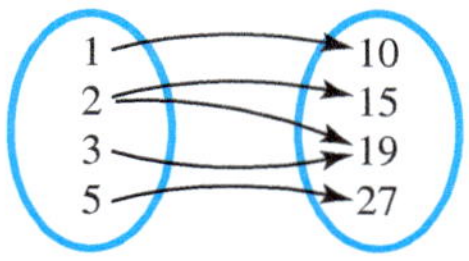

15.

x	y
1	5
1	2
1	-1
1	-4

16.

x	y
4	-3
2	-3
0	-3
-2	-3

17.

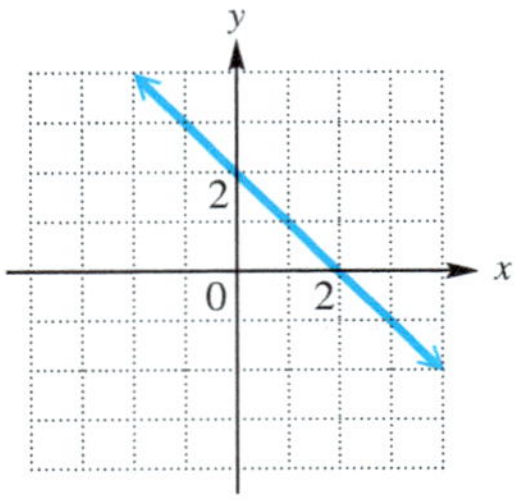

18.

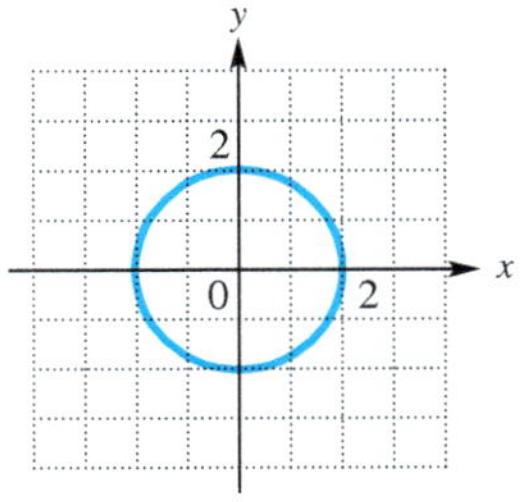

19.

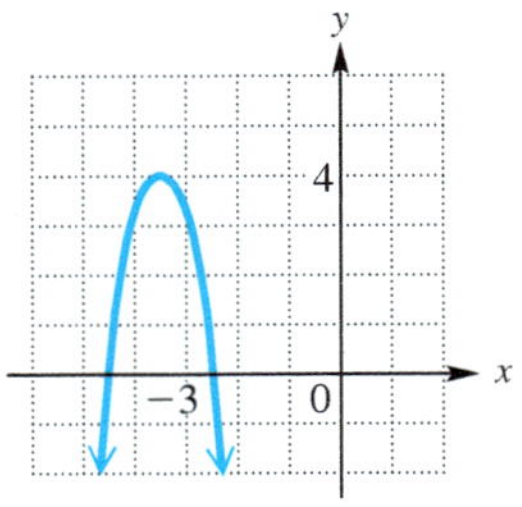

20.

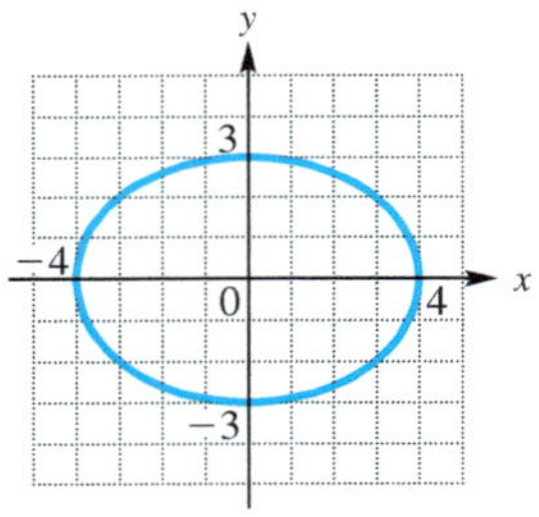

21.

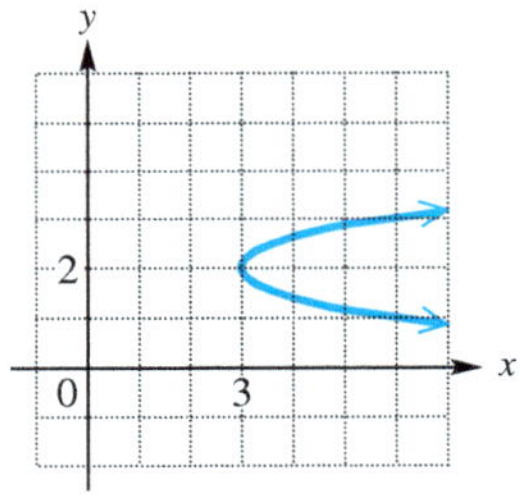

22.

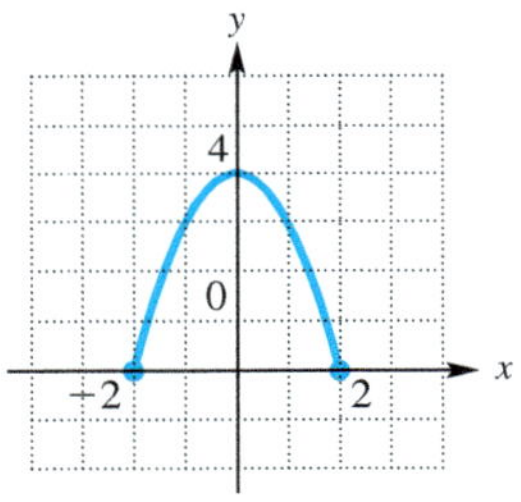

Decide whether each relation defines y as a function of x. Give the domain. See Example 5.

23. $y = x^2$

24. $y = x^3$

25. $x = y^6$

26. $x = y^4$

27. $y = 2x - 6$

28. $y = -6x + 8$

29. $x + y < 4$

30. $x - y < 3$

31. $y = \sqrt{x}$

32. $y = -\sqrt{x}$

33. $xy = 1$

34. $xy = -3$

35. $y = \sqrt{4x + 2}$

36. $y = \sqrt{9 - 2x}$

37. $y = \dfrac{2}{x - 9}$

38. $y = \dfrac{-7}{x - 16}$

39. Refer to the graph to answer the questions.

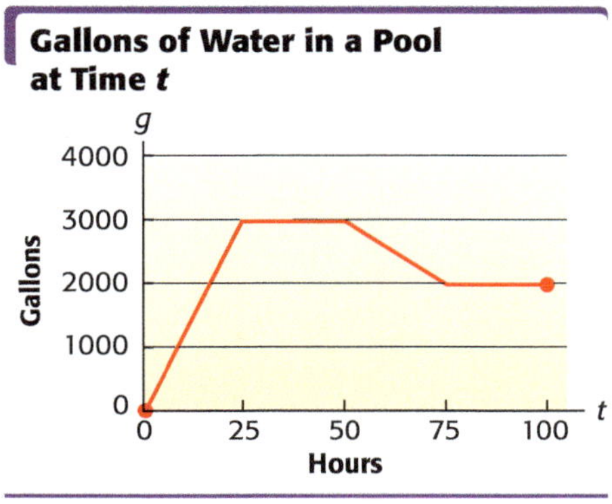

(a) What numbers are possible values of the dependent variable?

(b) For how long is the water level increasing? decreasing?

(c) How many gallons are in the pool after 90 hr?

(d) Call this function *g*. What is $g(0)$? What does it mean in this example?

40. The graph shows the daily megawatts of electricity used on a record-breaking summer day in Sacramento, California.

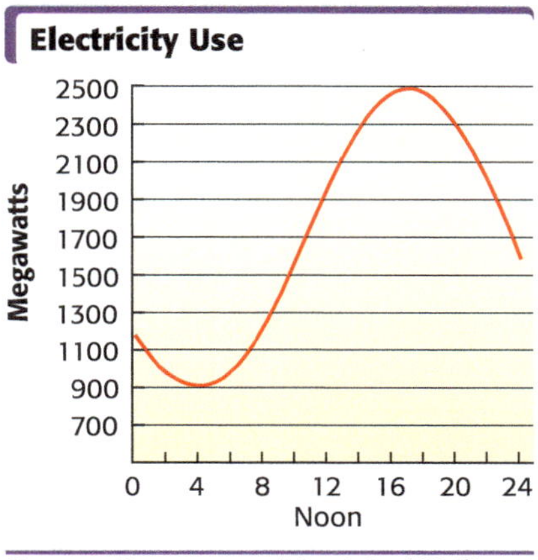

Source: Sacramento Municipal Utility District.

(a) Is this the graph of a function?

(b) What is the domain?

(c) Estimate the number of megawatts used at 8 A.M.

(d) At what time was the most electricity used? the least electricity?

41. Give an example of a function from everyday life. (*Hint:* Fill in the blanks: _______ depends on _______, so _______ is a function of _______.)

42. Choose the correct response: The notation $f(3)$ means

A. the variable f times 3 or $3f$

B. the value of the dependent variable when the independent variable is 3

C. the value of the independent variable when the dependent variable is 3

D. f equals 3.

Let $f(x) = -3x + 4$ *and* $g(x) = -x^2 + 4x + 1$. *Find the following. See Examples 6 and 7.*

43. $f(0)$

44. $f(-3)$

45. $g(-2)$

46. $g(10)$

47. $f(p)$

48. $g(k)$

49. $f(-x)$

50. $g(-x)$

51. $f(x + 2)$

52. $g\left(-\frac{1}{x}\right)$

53. $g\left(\frac{p}{3}\right)$

54. $f(3t - 2)$

For each function, find ***(a)*** $f(2)$ *and* ***(b)*** $f(-1)$. *See Example 8.*

55. $f = \{(-1, 3), (4, 7), (0, 6), (2, 2)\}$

56. $f = \{(2, 5), (3, 9), (-1, 11), (5, 3)\}$

57.

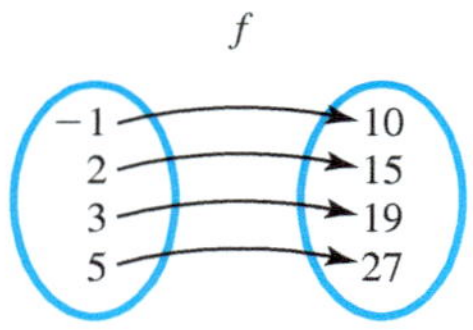

58.

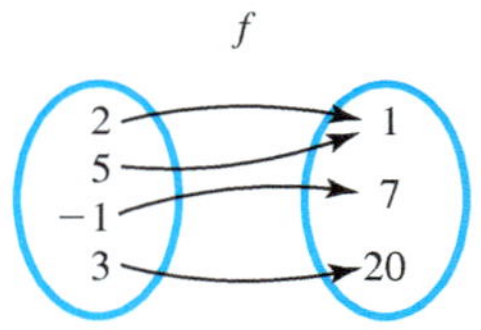

59.

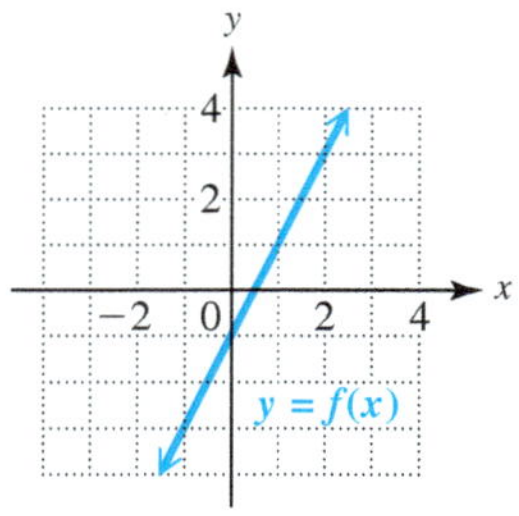

60.

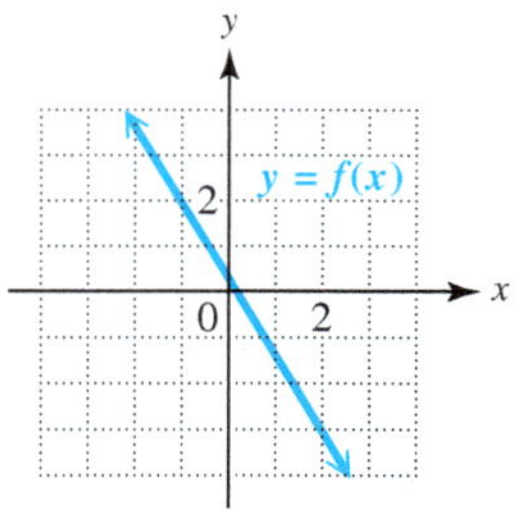

61. Fill in each blank with the correct response.

The equation $2x + y = 4$ has a straight ______ as its graph. One point that lies on the graph is (3, ______). If we solve the equation for y and use function notation, we have a ______ function defined by $f(x) =$ ______. For this function, $f(3) =$ ______, meaning that the point (______, ______) lies on the graph of the function.

62. Which of the following defines a linear function?

A. $y = \frac{2}{5}x - 3$ **B.** $y = \frac{1}{x}$

C. $y = x^2$ **D.** $y = \sqrt{x}$

An equation that defines y as a function f of x is given. ***(a)*** *Solve for y in terms of x, and replace y with the function notation f(x).* ***(b)*** *Find f(3). See Example 9.*

63. $x + 3y = 12$ **64.** $x - 4y = 8$ **65.** $y + 2x^2 = 3$

66. $y - 3x^2 = 2$ **67.** $4x - 3y = 8$ **68.** $-2x + 5y = 9$

Graph each linear or constant function. Give the domain and range. See Example 10.

69. $f(x) = -2x + 5$

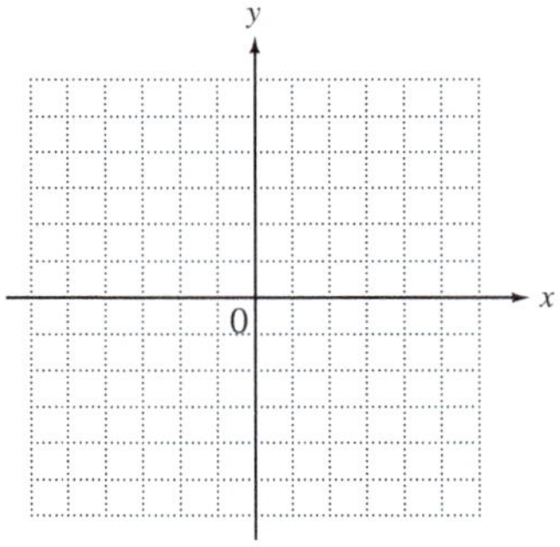

70. $g(x) = 4x - 1$

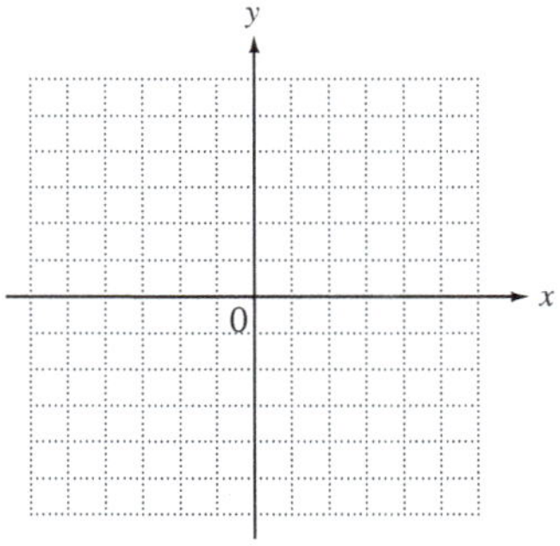

71. $h(x) = \frac{1}{2}x + 2$

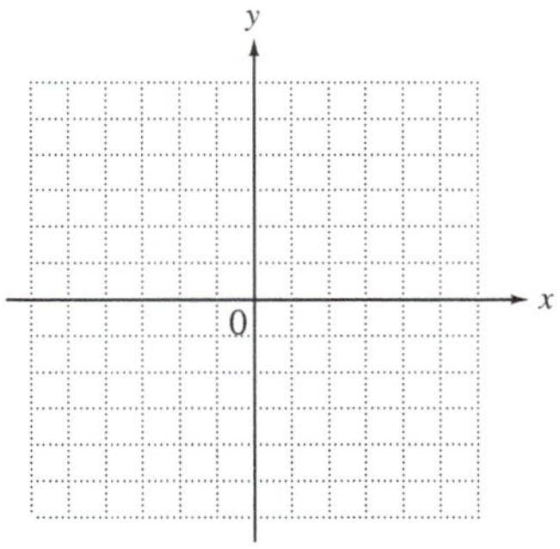

72. $F(x) = -\frac{1}{4}x + 1$

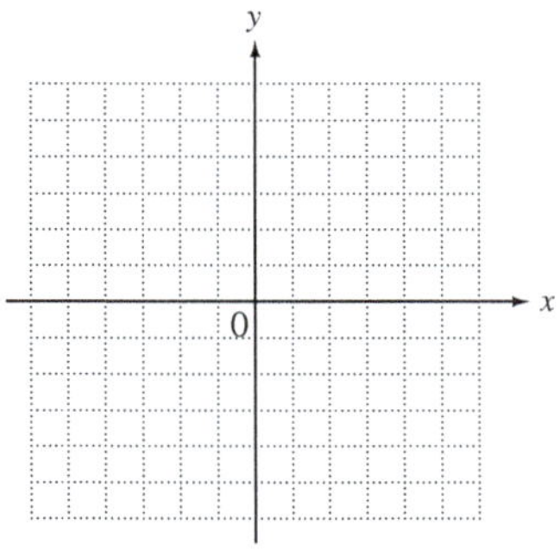

73. $g(x) = -4$

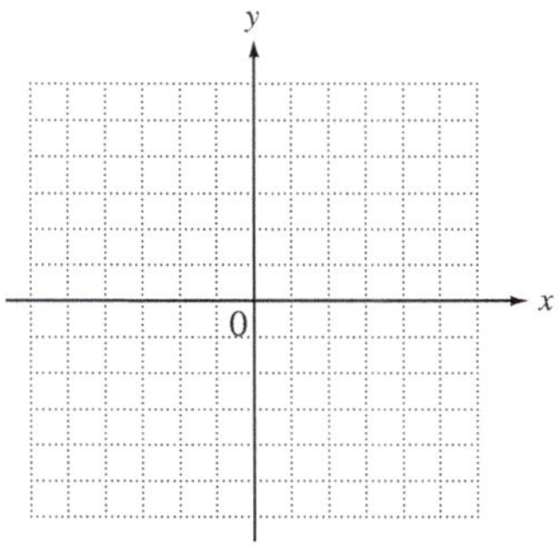

74. $f(x) = 5$

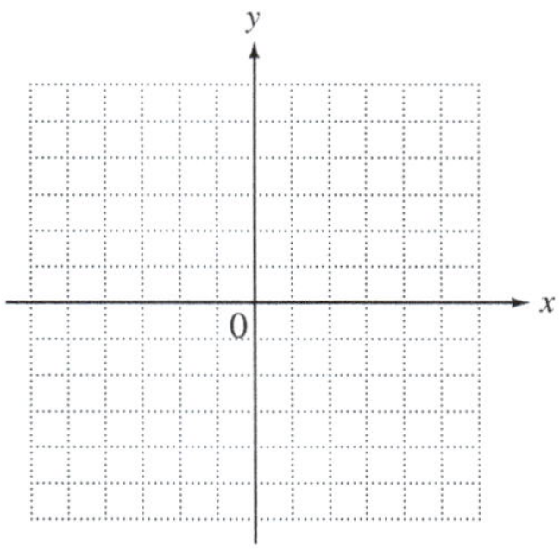

Solve each problem.

75. Suppose that a taxicab driver charges \$2.50 per mi.

(a) Fill in the table with the correct response for the price $f(x)$ he charges for a trip of x miles.

x	$f(x)$
0	
1	
2	
3	

(b) The linear function that gives a rule for the amount charged is $f(x) =$ ______.

(c) Graph this function for the domain $\{0, 1, 2, 3\}$.

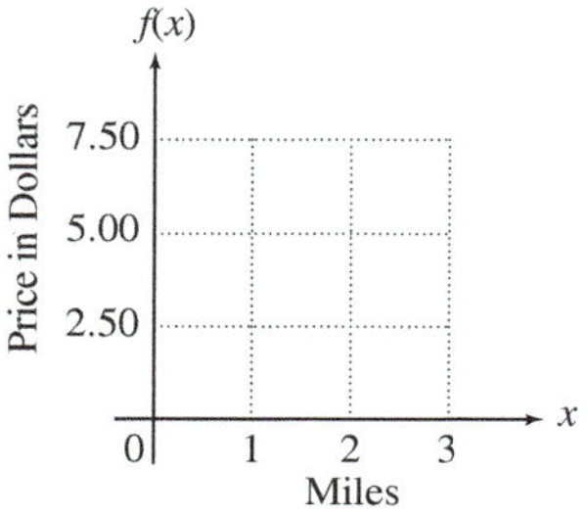

76. Suppose that a package weighing x pounds costs $f(x)$ dollars to ship to a given location, where $f(x) = 3.75x$.

(a) What is the value of $f(3)$?

(b) In your own words, describe what 3 and the value $f(3)$ mean in part (a), using the terms *independent variable* and *dependent variable.*

(c) How much would it cost to mail a 5-lb package? Write the answer using function notation.

Forensic scientists use the lengths of certain bones to calculate the height of a person. Two bones often used are the tibia (t), the bone from the ankle to the knee, and the femur (r), the bone from the knee to the hip socket. A person's height (h) is determined from the lengths of these bones using functions defined by the following formulas. All measurements are in centimeters.

Functions for men: $h(r) = 69.09 + 2.24r$ or $h(t) = 81.69 + 2.39t$

Functions for women: $h(r) = 61.41 + 2.32r$ or $h(t) = 72.57 + 2.53t$

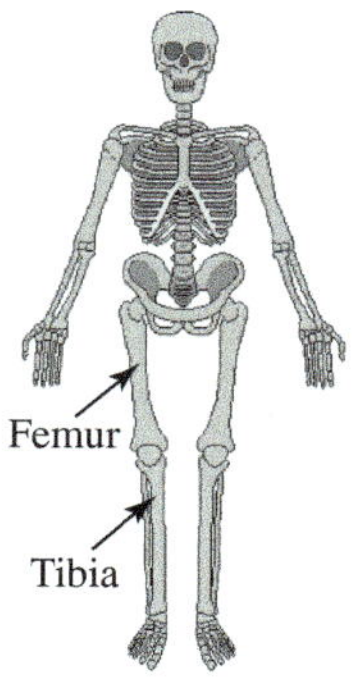

77. Find the height of a man with a femur measuring 56 cm.

78. Find the height of a man with a tibia measuring 40 cm.

79. Find the height of a woman with a femur measuring 50 cm.

80. Find the height of a woman with a tibia measuring 36 cm.

14

Systems of Equations

14.1 Solving Systems of Linear Equations by Graphing

OBJECTIVES

1. Decide whether a given ordered pair is a solution of a system.
2. Solve linear systems by graphing.
3. Solve special systems by graphing.
4. Identify special systems without graphing.

A **system of linear equations,** often called a **linear system,** consists of two or more linear equations with the same variables.

$$\begin{aligned}2x + 3y &= 4\\3x - y &= -5\end{aligned}\qquad\begin{aligned}x + 3y &= 1\\-y &= 4 - 2x\end{aligned}\qquad\begin{aligned}x - y &= 1\\y &= 3\end{aligned}\qquad \text{Linear systems}$$

In the system on the right, think of $y = 3$ as an equation in two variables by writing it as $0x + y = 3$.

OBJECTIVE 1 Decide whether a given ordered pair is a solution of a system. A **solution of a system** of linear equations is an ordered pair that makes both equations true at the same time. A solution of an equation is said to *satisfy* the equation.

EXAMPLE 1 Determining Whether an Ordered Pair Is a Solution

Is $(4, -3)$ a solution of each system?

(a) $x + 4y = -8$
$3x + 2y = 6$

To decide whether $(4, -3)$ is a solution of the system, substitute 4 for x and -3 for y in each equation.

$x + 4y = -8$		$3x + 2y = 6$	
$4 + 4(-3) \stackrel{?}{=} -8$		$3(4) + 2(-3) \stackrel{?}{=} 6$	
$4 + (-12) \stackrel{?}{=} -8$	Multiply.	$12 + (-6) \stackrel{?}{=} 6$	Multiply.
$-8 = -8$	True	$6 = 6$	True

Because $(4, -3)$ satisfies both equations, it is a solution of the system.

(b) $2x + 5y = -7$
$3x + 4y = 2$

Again, substitute 4 for x and -3 for y in both equations.

$2x + 5y = -7$		$3x + 4y = 2$	
$2(4) + 5(-3) \stackrel{?}{=} -7$		$3(4) + 4(-3) \stackrel{?}{=} 2$	
$8 + (-15) \stackrel{?}{=} -7$	Multiply.	$12 + (-12) \stackrel{?}{=} 2$	Multiply.
$-7 = -7$	True	$0 = 2$	False

The ordered pair $(4, -3)$ is not a solution of this system because it does not satisfy the second equation.

◀ Work Problem 1 at the Side.

OBJECTIVE 2 Solve linear systems by graphing. The set of all ordered pairs that are solutions of a system is its **solution set.** One way to find the solution set of a system of two linear equations is to graph both equations on the same axes. The graph of each line shows points whose coordinates satisfy the equation of that line. Any intersection point would be on both lines and would therefore be a solution of *both* equations. ***Thus, the coordinates of any point where the lines intersect give a solution of the system.***

1 Fill in the blanks, and decide whether the given ordered pair is a solution of the system.

(a) $(2, 5)$

$3x - 2y = -4$
$5x + y = 15$

$3x - 2y = -4$
$3(\_\_\_) - 2(\_\_\_) \stackrel{?}{=} -4$

$5x + y = 15$
$5(2) + \_\_\_ \stackrel{?}{=} \_\_\_$

$(2, 5)$ ______ a solution. (is/is not)

(b) $(1, -2)$

$x - 3y = 7$
$4x + y = 5$

$(1, -2)$ ______ a solution. (is/is not)

Answers

1. (a) 2; 5; 5; 15; is (b) is not

The graph in Figure 1 shows that the solution of the system in Example 1(a) is the intersection point $(4, -3)$. Because *two different* straight lines can intersect at no more than one point, there can never be more than one solution for such a system.

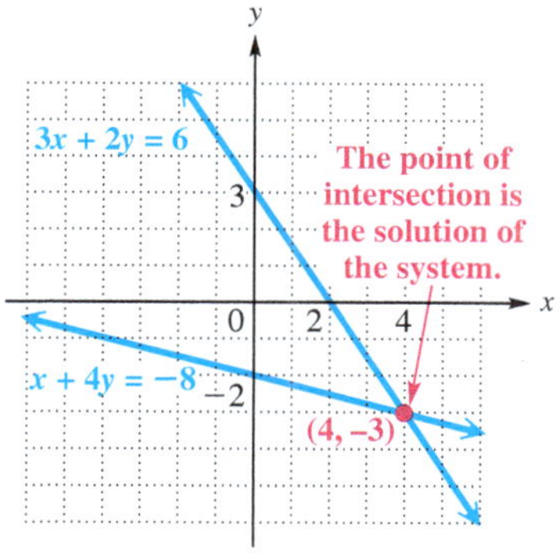

Figure 1

2 Solve each system of equations by graphing both equations on the same axes. Check your solutions.

(a) $5x - 3y = 9$

$x + 2y = 7$

(One of the lines is already graphed.)

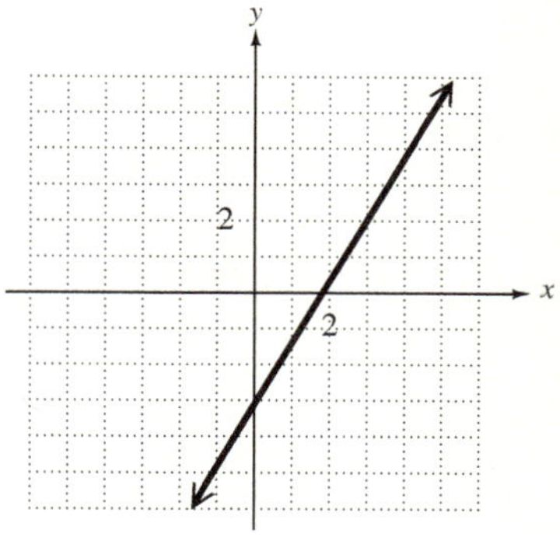

(b) $x + y = 4$

$2x - y = -1$

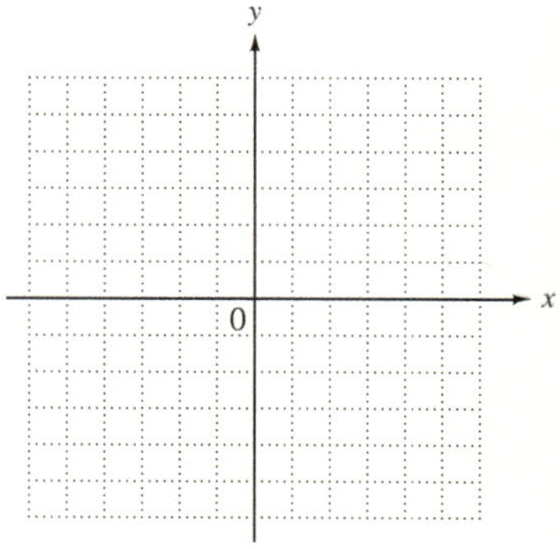

EXAMPLE 2 Solving a System by Graphing

Solve the system of equations by graphing both equations on the same axes.

$$2x + 3y = 4$$
$$3x - y = -5$$

We graph these two equations by plotting several points for each line. Recall from **Section 11.2** that the intercepts are often convenient choices.

$2x + 3y = 4$

x	y
0	$\frac{4}{3}$
2	0
−2	$\frac{8}{3}$

Find a third ordered pair as a check.

$3x - y = -5$

x	y
0	5
$-\frac{5}{3}$	0
−2	−1

The lines in Figure 2 suggest that the graphs intersect at the point $(-1, 2)$. We check this by substituting -1 for x and 2 for y in both equations.

Check

$$2x + 3y = 4$$
$$2(-1) + 3(2) \stackrel{?}{=} 4$$
$$4 = 4 \quad \text{True}$$
$$3x - y = -5$$
$$3(-1) - 2 \stackrel{?}{=} -5$$
$$-5 = -5 \quad \text{True}$$

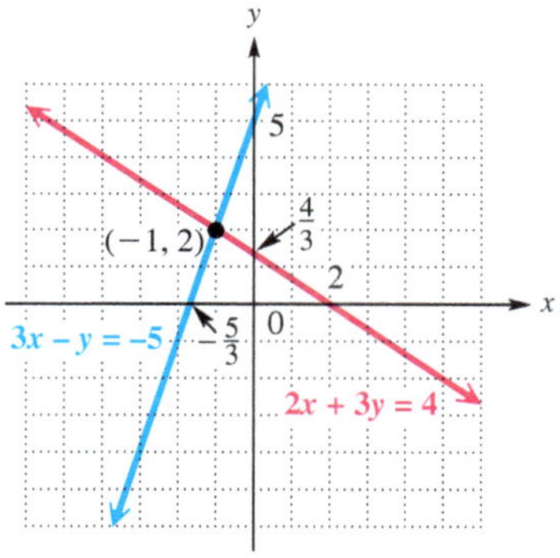

Figure 2

Because $(-1, 2)$ satisfies both equations, the *solution set* of this system is $\{(-1, 2)\}$. We write the ordered pair in the solution set between set braces.

Work Problem **2** at the Side. ▶

Note

We can also graph a linear system by writing each equation in the system in slope-intercept form and using the slope and y-intercept to graph each line. For Example 2,

$2x + 3y = 4$ becomes $y = -\frac{2}{3}x + \frac{4}{3}$ $\quad$ y-intercept $(0, \frac{4}{3})$; slope $-\frac{2}{3}$

$3x - y = -5$ becomes $y = 3x + 5.$ $\quad$ y-intercept $(0, 5)$; slope 3, or $\frac{3}{1}$

Confirm that graphing these equations results in the same lines and the same solution shown in Figure 2.

Answers

2. **(a)** $\{(3, 2)\}$ **(b)** $\{(1, 3)\}$

To solve a linear system by graphing, follow these steps.

Solving a Linear System by Graphing

Step 1 **Graph each equation** of the system on the same coordinate axes.

Step 2 **Find the coordinates of the point of intersection** of the graphs if possible. This is the solution of the system.

Step 3 **Check** the solution in *both* of the original equations. Then write the solution set.

CAUTION

A difficulty with the graphing method of solution is that it may not be possible to determine from the graph the exact coordinates of the point that represents the solution, particularly if these coordinates are not integers. For this reason, algebraic methods of solution are explained later in this chapter. The graphing method does, however, show geometrically how solutions are found and is useful when approximate answers will do.

OBJECTIVE 3 Solve special systems by graphing. Sometimes the graphs of the two equations in a system either do not intersect at all or are the same line, as in the systems in Example 3.

EXAMPLE 3 Solving Special Systems

Solve each system by graphing.

(a) $2x + y = 2$
$2x + y = 8$

The graphs of these lines are shown in Figure 3. The two lines are parallel and have no points in common. For such a system, **there is no solution.** Its solution set is the **empty set**, or **null set,** symbolized **$\emptyset$.**

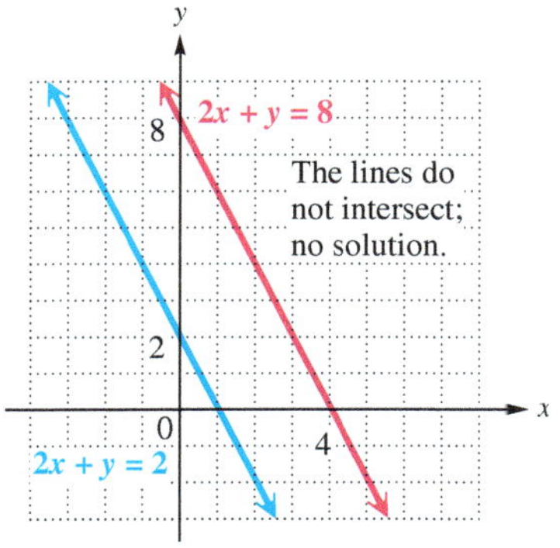

Figure 3

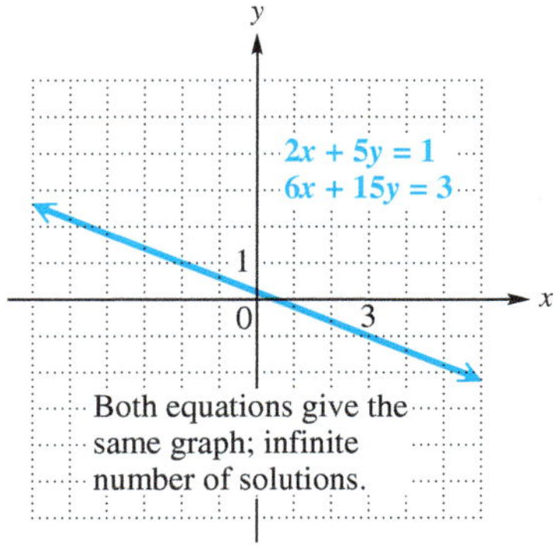

Figure 4

(b) $2x + 5y = 1$
$6x + 15y = 3$

The graphs of these two equations are the same line. See Figure 4. The second equation can be obtained by multiplying each side of the first equation by 3. In this case, every point on the line is a solution of the system, and *the solution set contains an infinite number of ordered pairs* that satisfy the equations.

Continued on Next Page

We write the solution set as

$$\{(x, y) \mid 2x + 5y = 1\},$$

This is the first equation in the system. See the Note below.

read "the set of ordered pairs (x, y) such that $2x + 5y = 1$." This notation is called **set-builder notation.** It is convenient to use this notation when it is not possible to list all the elements of a set

Note

When a system has an infinite number of solutions, as in Example 3(b), either equation of the system could be used to write the solution set. ***We prefer to use the equation in standard form with coefficients that are integers having greatest common factor 1.*** If neither of the given equations of the system is in this form, use an *equivalent* equation that is in standard form with coefficients that are integers having greatest common factor 1 to write the solution set with set-builder notation.

*Work Problem **3** at the Side.* ▶

The system in Example 2 has exactly one solution. A system with at least one solution is called a **consistent system.** A system of equations with no solution, such as the one in Example 3(a), is called an **inconsistent system.** The equations in Example 2 are **independent equations** with different graphs. The equations of the system in Example 3(b) have the same graph and are equivalent. Because they are different forms of the same equation, these equations are called **dependent equations.**

Examples 2 and 3 show the three cases that may occur when solving a system of two equations with two variables.

Three Cases for Solutions of Systems

1. The graphs intersect at exactly one point, which gives the (single) ordered-pair solution of the system. The **system is consistent** and the **equations are independent.** See Figure 5(a).
2. The graphs are parallel lines, so there is no solution and the solution set is $\emptyset$. The **system is inconsistent** and the **equations are independent.** See Figure 5(b).
3. The graphs are the same line. There is an infinite number of solutions, and the solution set is written in set-builder notation as $\{(x, y) \mid$ ___________$\}$, where one of the equations is written after the $\mid$ symbol. The **system is consistent** and the **equations are dependent.** See Figure 5(c).

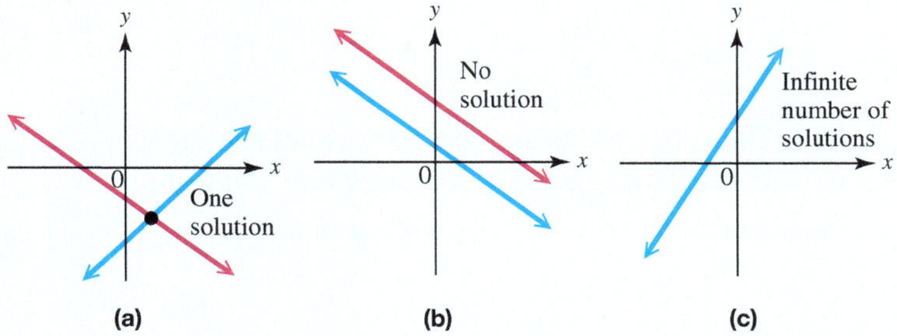

Figure 5

3 Solve each system of equations by graphing both equations on the same axes.

(a) $3x - y = 4$
$6x - 2y = 12$

(One of the lines is already graphed.)

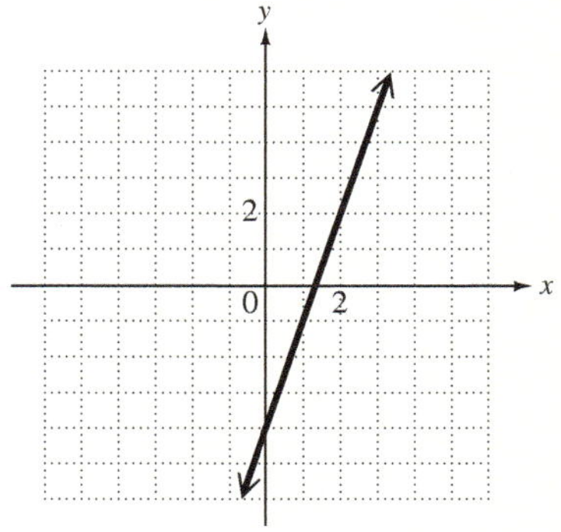

(b) $x - 3y = -2$
$2x - 6y = -4$

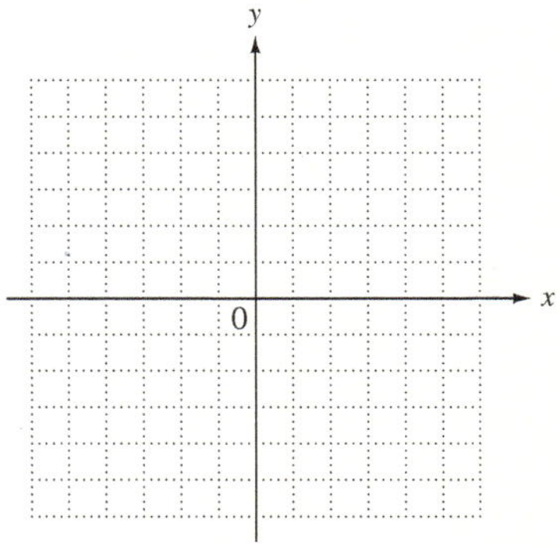

Answers

3. **(a)** There is *no solution,* so write the solution set as $\emptyset$.
(b) There is an *infinite number of solutions,* so use set-builder notation. $\{(x, y) \mid x - 3y = -2\}$

OBJECTIVE 4 Identify special systems without graphing. Example 3 showed that the graphs of an inconsistent system are parallel lines and the graphs of a system of dependent equations are the same line. We can recognize these special kinds of systems without graphing by using slopes.

EXAMPLE 4 Identifying the Three Cases by Using Slopes

Describe each system without graphing. State the number of solutions.

(a) $3x + 2y = 6$

$-2y = 3x - 5$

Write each equation in slope-intercept form, $y = mx + b$, by solving for y.

$$3x + 2y = 6$$
$$2y = -3x + 6 \quad \text{Subtract } 3x.$$
$$y = -\frac{3}{2}x + 3 \quad \text{Divide by 2.}$$

$$-2y = 3x - 5$$
$$y = -\frac{3}{2}x + \frac{5}{2} \quad \text{Divide by } -2.$$

Both equations have slope $-\frac{3}{2}$ but they have different y-intercepts, 3 and $\frac{5}{2}$. In **Section 11.3,** we found that lines with the same slope are parallel, so these equations have graphs that are parallel lines. Thus, the system has no solution.

(b) $2x - y = 4$

$x = \frac{y}{2} + 2$

Again, write the equations in slope-intercept form.

$$2x - y = 4$$
$$-y = -2x + 4$$
$$y = 2x - 4$$

$$x = \frac{y}{2} + 2$$
$$\frac{y}{2} + 2 = x$$
$$\frac{y}{2} = x - 2$$
$$y = 2x - 4$$

The equations are exactly the same—their graphs are the same line. Thus, the system has an infinite number of solutions.

(c) $x - 3y = 5$

$2x + y = 8$

In slope-intercept form, the equations are as follows.

$$x - 3y = 5$$
$$-3y = -x + 5$$
$$y = \frac{1}{3}x - \frac{5}{3}$$

$$2x + y = 8$$
$$y = -2x + 8$$

The graphs of these equations are neither parallel nor the same line, since the slopes are different. This system has exactly one solution.

◀ Work Problem 4 at the Side.

4 Describe each system without graphing. State the number of solutions.

(a) $2x - 3y = 5$

$3y = 2x - 7$

(b) $-x + 3y = 2$

$2x - 6y = -4$

(c) $6x + y = 3$

$2x - y = -11$

ANSWERS

4. **(a)** The equations represent parallel lines. The system has no solution.
(b) The equations represent the same line. The system has an infinite number of solutions.
(c) The equations represent lines that are neither parallel nor the same line. The system has exactly one solution.

14.1 Exercises

1. Which ordered pair could be a solution of the system graphed? Why is it the only valid choice?

A. (2, 2)
B. (−2, 2)
C. (−2, −2)
D. (2, −2)

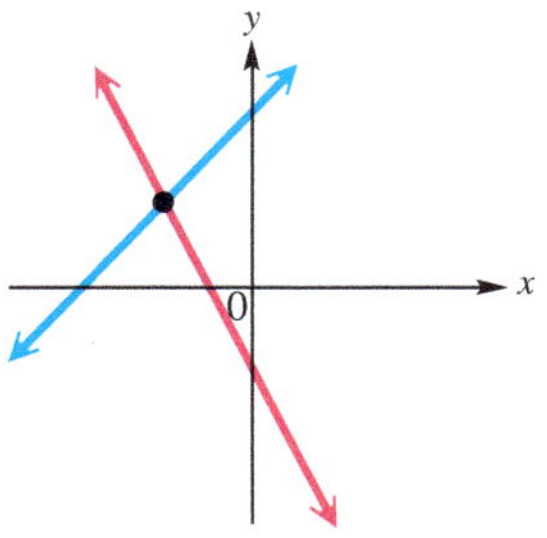

2. Which ordered pair could be a solution of the system graphed? Why is it the only valid choice?

A. (2, 0)
B. (0, 2)
C. (−2, 0)
D. (0, −2)

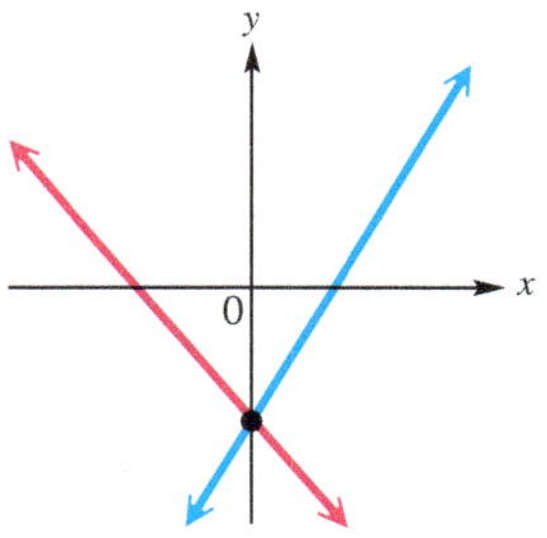

3. How can you tell without graphing that this system has no solution?

$$x + y = 2$$
$$x + y = 4$$

4. Explain why a system of two linear equations cannot have exactly two solutions.

Decide whether the given ordered pair is a solution of the given system. See Example 1.

5. (2, −3)
$x + y = -1$
$2x + 5y = 19$

6. (4, 3)
$x + 2y = 10$
$3x + 5y = 3$

7. (−1, −3)
$3x + 5y = -18$
$4x + 2y = -10$

8. (−9, −2)
$2x - 5y = -8$
$3x + 6y = -39$

9. (7, −2)
$4x = 26 - y$
$3x = 29 + 4y$

10. (9, 1)
$2x = 23 - 5y$
$3x = 24 + 3y$

11. (6, −8)
$-2y = x + 10$
$3y = 2x + 30$

12. (−5, 2)
$5y = 3x + 20$
$3y = -2x - 4$

Solve each system of equations by graphing. If the system is inconsistent or the equations are dependent, say so. See Examples 2 and 3.

13. $x - y = 2$
$x + y = 6$

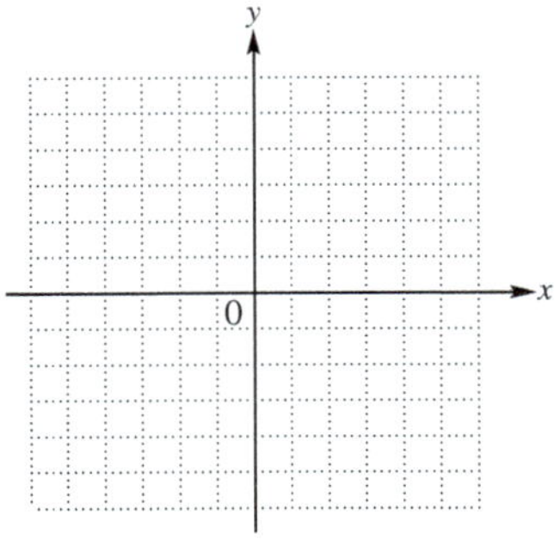

14. $x - y = 3$
$x + y = -1$

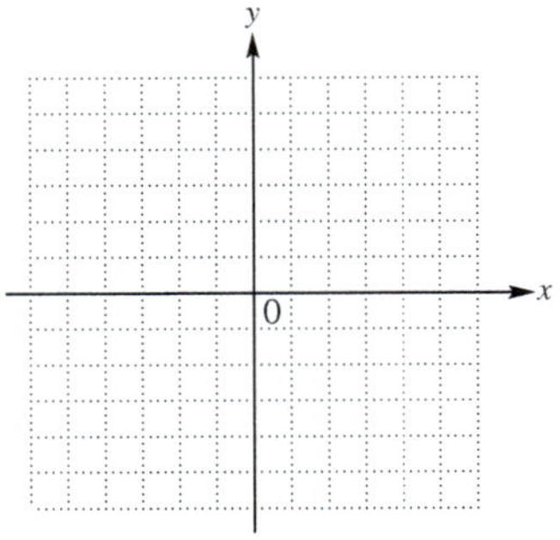

15. $x + y = 4$
$y - x = 4$

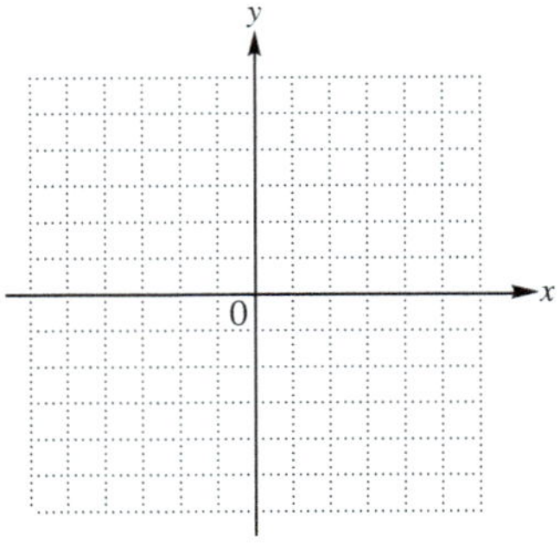

16. $x + y = -5$
$x - y = 5$

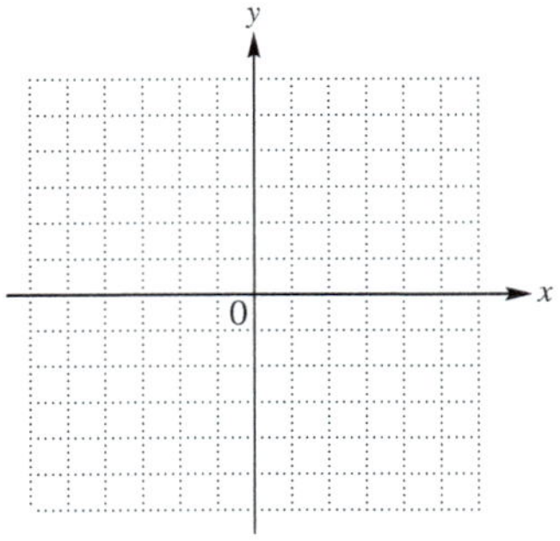

17. $x - 2y = 6$
$x + 2y = 2$

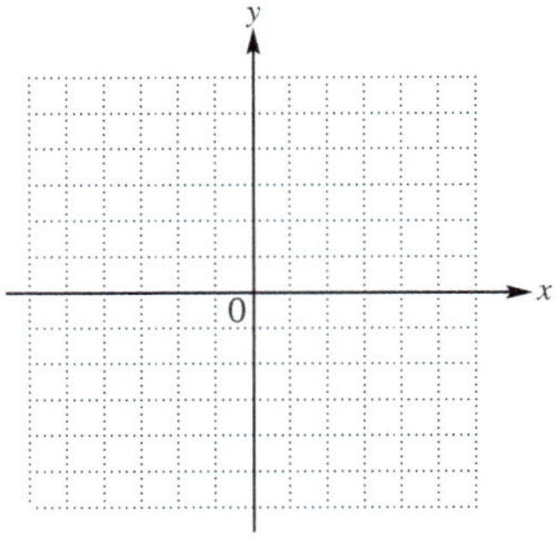

18. $2x - y = 4$
$4x + y = 2$

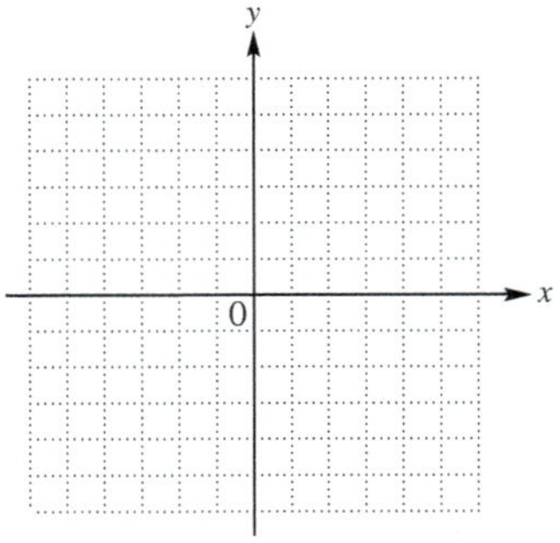

19. $3x - 2y = -3$
$-3x - y = -6$

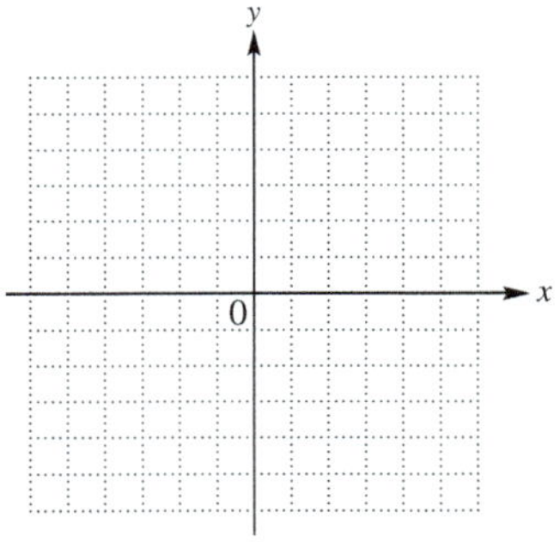

20. $2x - y = 4$
$2x + 3y = 12$

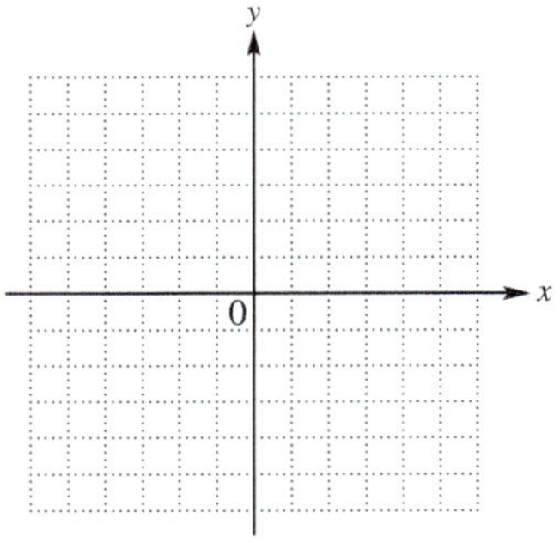

21. $2x - 3y = -6$
$y = -3x + 2$

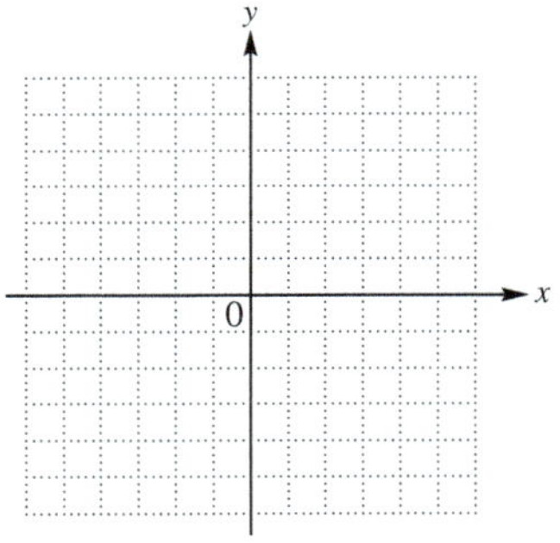

22. $-3x + y = -3$
$y = x - 3$

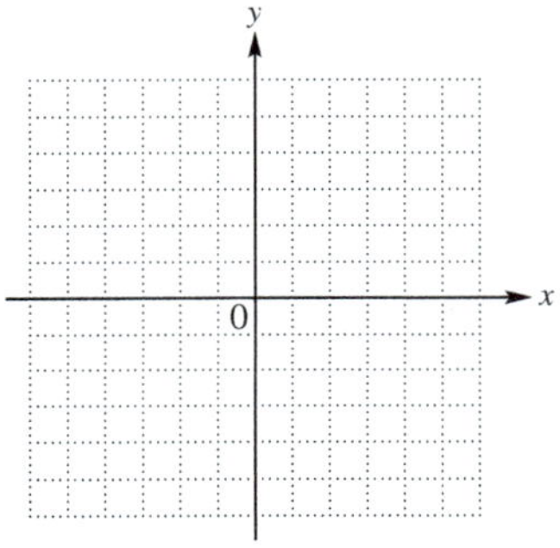

23. $x + 2y = 6$
$2x + 4y = 8$

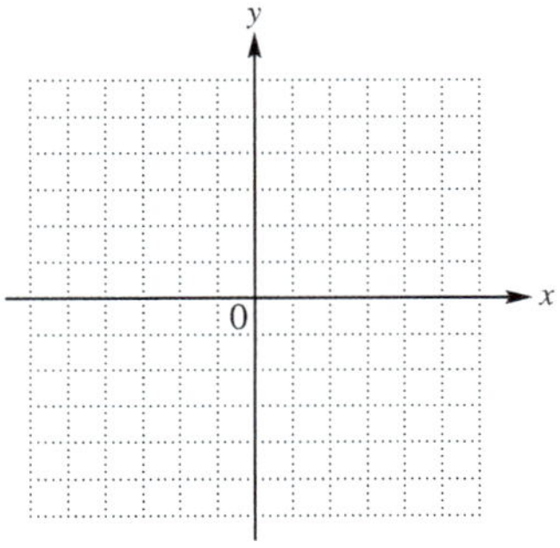

24. $2x - y = 6$
$6x - 3y = 12$

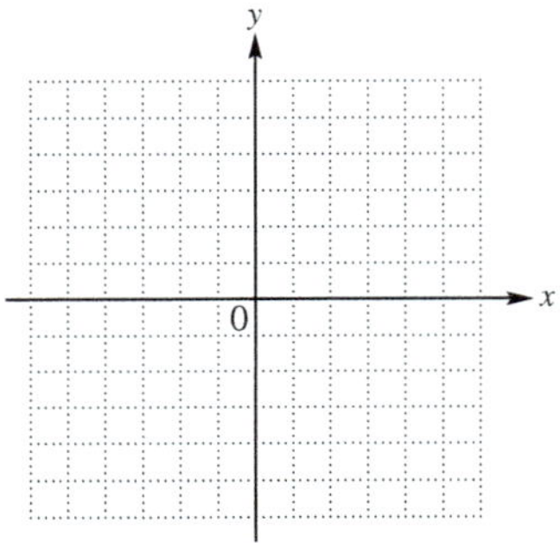

25. $4x - 2y = 8$
$2x = y + 4$

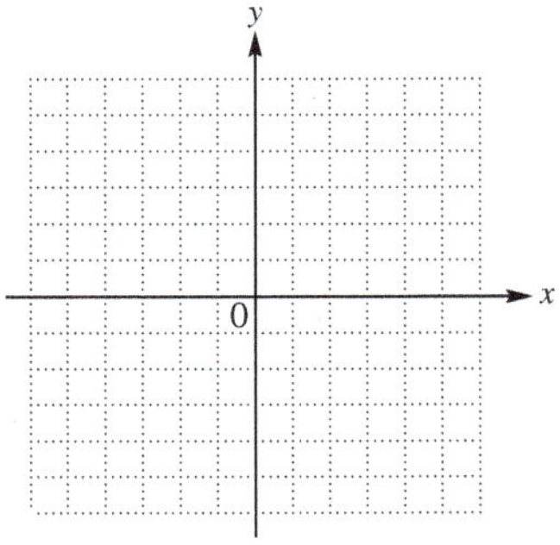

26. $3x = 5 - y$
$6x + 2y = 10$

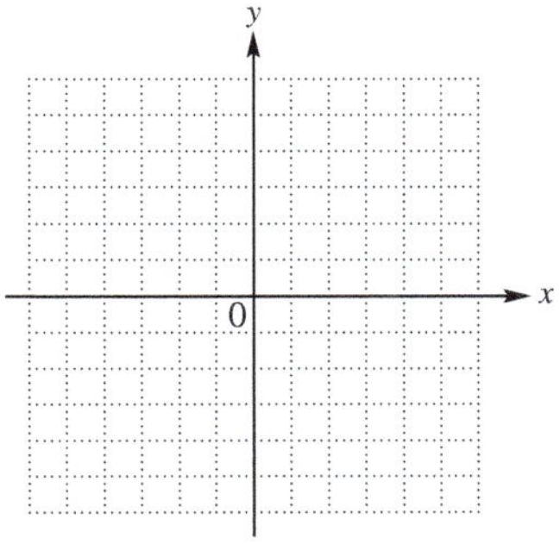

27. $3x - 4y = 24$
$y = -\frac{3}{2}x + 3$

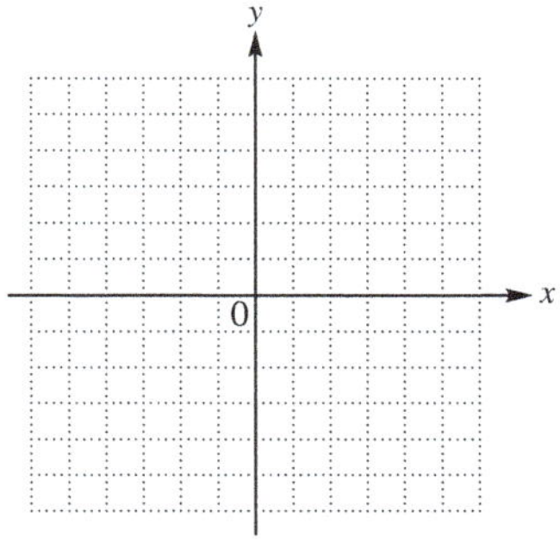

28. $3x - 2y = 12$
$y = -4x + 5$

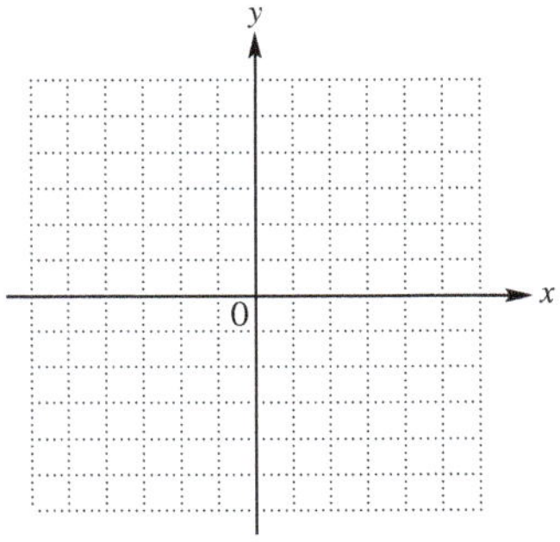

29. $3x = y + 5$
$6x - 5 = 2y$

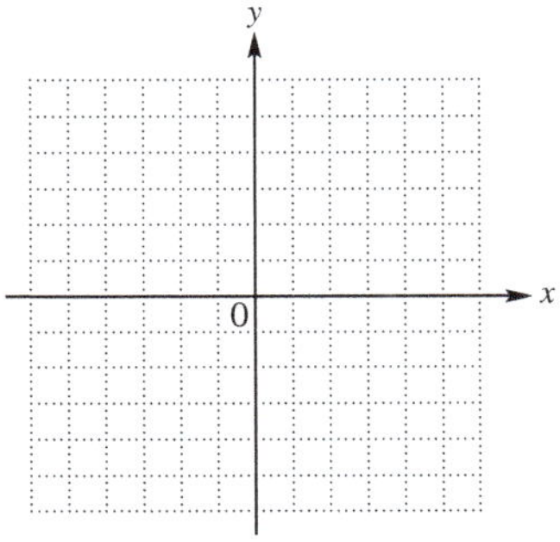

30. $2x = y - 4$
$4x - 2y = -4$

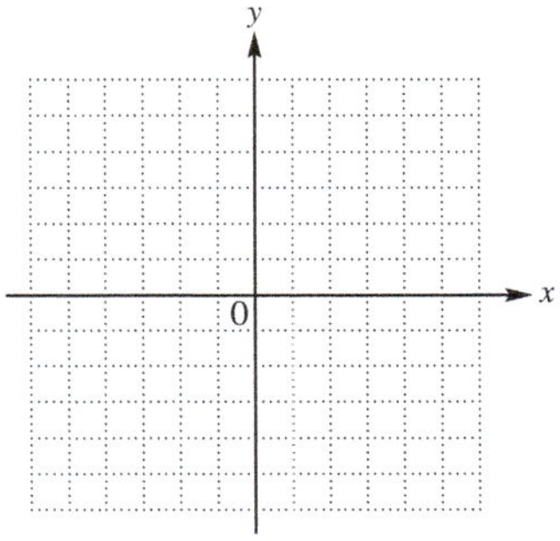

Without graphing, answer the following questions for each linear system. See Example 4.

(a) *Is the system inconsistent, are the equations dependent, or neither?*
(b) *Is the graph a pair of intersecting lines, a pair of parallel lines, or one line?*
(c) *Does the system have one solution, no solution, or an infinite number of solutions?*

31. $y - x = -5$
$x + y = 1$

32. $2x + y = 6$
$x - 3y = -4$

33. $x + 2y = 0$
$4y = -2x$

34. $y = 3x$
$y + 3 = 3x$

35. $5x + 4y = 7$
$10x + 8y = 4$

36. $4x - 6y = 10$
$-6x + 9y = -15$

The numbers of daily morning and evening newspapers in the United States in selected years are shown in the graph. Use the graph to work Exercises 37 and 38.

37. For which years were there more evening dailies than morning dailies?

38. Estimate the year in which the number of evening and morning dailies was closest to the same. About how many newspapers of each type were there in that year?

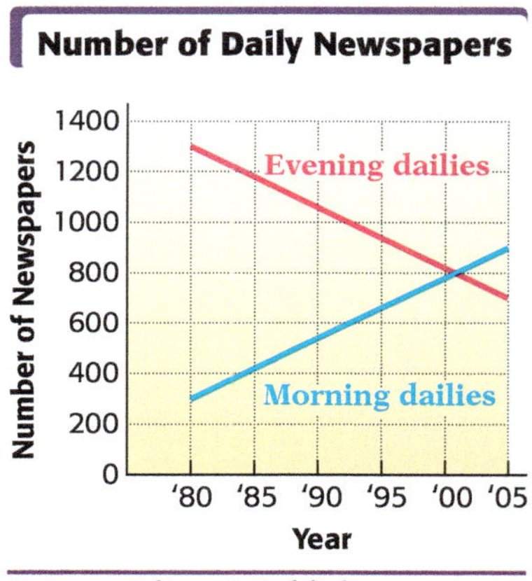

Source: Editor & Publisher International Year Book.

Work Exercises 39 and 40 using the graphs provided.

39. The graph shows how college students managed their money during the period 1997–2004.

(a) During what period did ATM use dominate both credit card *and* debit card use?

(b) In what year did debit card use overtake credit card use?

(c) In what year did debit card use overtake ATM use?

(d) Write an ordered pair for the debit card use data in the year 2004.

(e) Describe the trend in debit card use over this period.

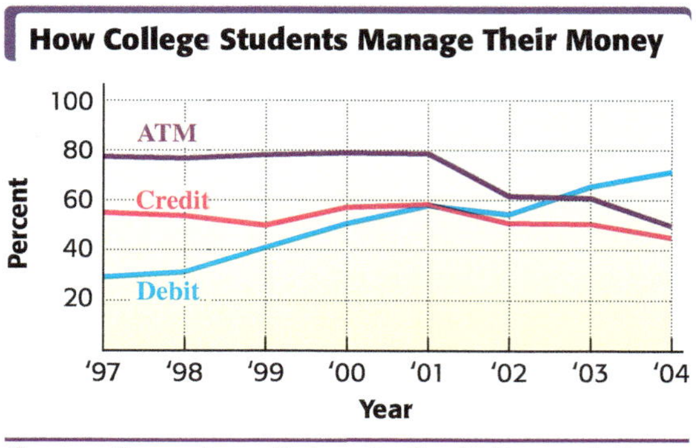

Source: Georgetown University Credit Research Center.

40. The graph shows how the average viewing hours for broadcast TV and cable/satellite TV in the United States has changed during the period 1998–2004.

(a) In approximately what year did Americans spend almost the same number of hours watching broadcast and cable/satellite TV? How many hours per year was this?

(b) Express the point of intersection of the two graphs as an ordered pair of the form (year, hours).

(c) During what period was the time spent watching broadcast TV almost constant?

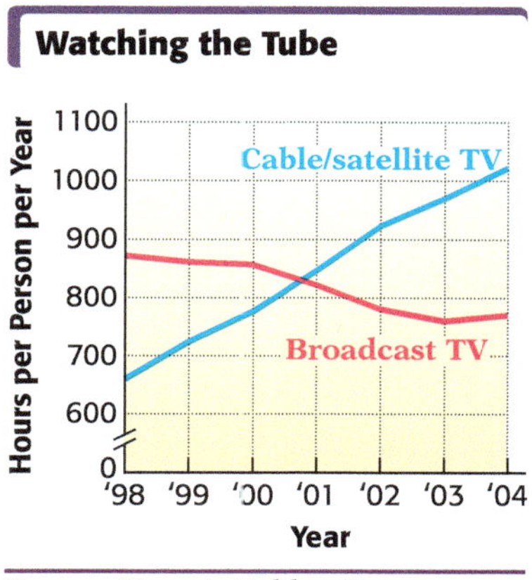

Source: Veronis Suhler Stevenson.

14.2 Solving Systems of Linear Equations by Substitution

OBJECTIVE 1 Solve linear systems by substitution.

Work Problem 1 at the Side. ▶

As we saw in Problem 1 at the side, graphing to solve a system of equations has a serious drawback: It is difficult to accurately find a solution such as $\left(\frac{11}{3}, -\frac{4}{9}\right)$ from a graph. One algebraic method for solving a system of equations is the **substitution method.**

EXAMPLE 1 Using the Substitution Method

Solve the system by the substitution method.

$$3x + 5y = 26$$
$$y = 2x$$

The second equation is already solved for y. This equation says that $y = 2x$. Substituting $2x$ for y in the first equation gives

$$3x + 5y = 26$$
$$3x + 5(2x) = 26 \quad \text{Let } y = 2x.$$
$$3x + 10x = 26 \quad \text{Multiply.}$$
$$13x = 26 \quad \text{Combine like terms.}$$
$$x = 2. \quad \text{Divide by 13.}$$

Because $x = 2$, we find y from the equation $y = 2x$ by substituting 2 for x.

$$y = 2(2) = 4 \quad \text{Let } x = 2.$$

Check that the solution set of the given system is $\{(2, 4)\}$ by substituting 2 for x and 4 for y in *both* equations.

Work Problem 2 at the Side. ▶

EXAMPLE 2 Using the Substitution Method

Solve the system by the substitution method.

$$2x + 5y = 7$$
$$x = -1 - y$$

The second equation gives x in terms of y. Substitute $-1 - y$ for x in the first equation.

$$2x + 5y = 7$$
$$2(-1 - y) + 5y = 7 \quad \text{Let } x = -1 - y.$$
$$-2 - 2y + 5y = 7 \quad \text{Distributive property}$$
$$-2 + 3y = 7 \quad \text{Combine like terms.}$$
$$3y = 9 \quad \text{Add 2.}$$
$$y = 3 \quad \text{Divide by 3.}$$

Distribute 2 to *both* -1 and $-y$.

To find x, substitute 3 for y in the equation $x = -1 - y$ to get

$$x = -1 - 3 = -4.$$

Write the x-coordinate first.

Check that the solution set of the given system is $\{(-4, 3)\}$.

Work Problem 3 at the Side. ▶

OBJECTIVES

1. Solve linear systems by substitution.
2. Solve special systems by substitution.
3. Solve linear systems with fractions and decimals by substitution.

1 Solve the system by graphing.

$$2x + 3y = 6$$
$$x - 3y = 5$$

Can you determine the answer? Why or why not?

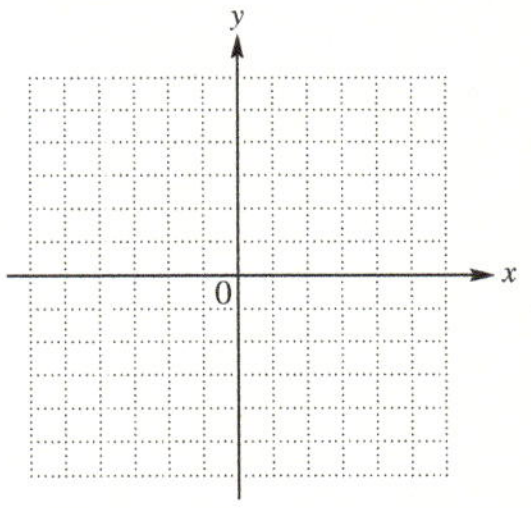

2 Fill in the blanks to solve by the substitution method. Check your solution.

$$3x + 5y = 69$$
$$y = 4x$$

$$3x + 5(\_\_\_) = 69$$
$$\_\_\_ = 69$$
$$x = \_\_\_$$
$$y = 4(\_\_\_) = \_\_\_$$

The solution set is ____.

3 Solve by the substitution method. Check your solution.

$$2x + 7y = -12$$
$$x = 3 - 2y$$

ANSWERS

1. The answer cannot be determined from the graph because it is too difficult to read the exact coordinates.
2. $4x$; $23x$; 3; 3; 12; $\{(3, 12)\}$
3. $\{(15, -6)\}$

CAUTION
Even though we found y first in Example 2, ***the x-coordinate is always written first in the ordered-pair solution of a system.***

To solve a system by substitution, follow these steps.

Solving a Linear System by Substitution

Step 1 **Solve one equation for either variable.** If one of the variables has coefficient 1 or -1, choose it, since it usually makes the substitution easier.

Step 2 **Substitute** for that variable in the other equation. The result should be an equation with just one variable.

Step 3 **Solve** the equation from Step 2.

Step 4 **Substitute** the result from Step 3 into the equation from Step 1 to find the value of the other variable.

Step 5 **Check** the solution in *both* of the original equations. Then write the solution set.

EXAMPLE 3 **Using the Substitution Method**

Use substitution to solve the system.

$$2x = 4 - y \quad (1)$$
$$5x + 3y = 10 \quad (2)$$

Step 1 For the substitution method, we must solve one of the equations for either x or y. Because the coefficient of y in equation (1) is -1, we choose equation (1) and solve for y.

$$2x = 4 - y \quad (1)$$
$$y + 2x = 4 \quad \text{Add } y.$$
$$y = -2x + 4 \quad \text{Subtract } 2x.$$

Step 2 Now substitute $-2x + 4$ for y in equation (2).

$$5x + 3y = 10 \quad (2)$$
$$5x + 3(-2x + 4) = 10 \quad \text{Let } y = -2x + 4.$$

Step 3 Now solve the equation from Step 2.

$$5x - 6x + 12 = 10 \quad \text{Distributive property}$$
$$-x + 12 = 10 \quad \text{Combine like terms.}$$
$$-x = -2 \quad \text{Subtract 12.}$$
$$x = 2 \quad \text{Multiply by } -1.$$

Distribute 3 to *both* $-2x$ and 4.

Step 4 Since $y = -2x + 4$ and $x = 2$, $y = -2(2) + 4 = 0$.

Step 5 Check that $(2, 0)$ is the solution.

Check

$$2x = 4 - y \quad (1) \qquad\qquad 5x + 3y = 10 \quad (2)$$
$$2(2) \stackrel{?}{=} 4 - 0 \qquad\qquad 5(2) + 3(0) \stackrel{?}{=} 10$$
$$4 = 4 \quad \text{True} \qquad\qquad 10 = 10 \quad \text{True}$$

Since both results are true, the solution set of the system is $\{(2, 0)\}$.

Work Problem 4 at the Side.

4 Solve each system by substitution. Check each solution.

(a) Fill in the blanks to solve

$$x + 4y = -1$$
$$2x - 5y = 11.$$

Solve the first equation for x.

$$x = -1 - \_\_\_\_$$

Substitute into the second equation to find y.

$$2(\_\_\_\_) - 5y = 11$$
$$-2 - 8y - 5y = 11$$
$$-2 - \_\_\_\_y = 11$$
$$\_\_\_\_y = 13$$
$$y = \_\_\_\_$$

Find x.

$$x = -1 - \_\_\_\_$$
$$x = \_\_\_\_$$

The solution set is ____.

(b) $2x + 5y = 4$
$x + y = -1$

ANSWERS

4. (a) $4y$; $-1 - 4y$; 13; -13; -1; -4; 3; $\{(3, -1)\}$ **(b)** $\{(-3, 2)\}$

EXAMPLE 4 Using the Substitution Method

Use substitution to solve the system.

$$2x + 3y = 10 \quad (1)$$
$$-3x - 2y = 0 \quad (2)$$

Step 1 To use the substitution method, we must solve one of the equations for one of the variables. We choose equation (1) and solve for x.

$$2x + 3y = 10 \quad (1)$$
$$2x = 10 - 3y \quad \text{Subtract } 3y.$$
$$x = 5 - \frac{3}{2}y \quad \text{Divide by 2.}$$

Step 2 Substitute this expression for x in equation (2).

$$-3x - 2y = 0 \quad (2)$$
$$-3\left(5 - \frac{3}{2}y\right) - 2y = 0 \quad \text{Let } x = 5 - \tfrac{3}{2}y.$$

Step 3

$$-15 + \frac{9}{2}y - 2y = 0 \quad \text{Distributive property}$$

$-3(-\frac{3}{2}) = (-\frac{3}{1})(-\frac{3}{2}) = \frac{9}{2}$

$$-15 + \frac{5}{2}y = 0 \quad \text{Combine like terms.}$$
$$\frac{5}{2}y = 15 \quad \text{Add 15.}$$

$\frac{2}{5}(\frac{5}{2}y) = (\frac{2}{5}\cdot\frac{5}{2})y = 1 \cdot y = y$

$$y = \frac{30}{5} = 6 \quad \text{Multiply by } \tfrac{2}{5}.$$

Step 4 Find x by substituting 6 for y in $x = 5 - \frac{3}{2}y$.

$$x = 5 - \frac{3}{2}(6) = -4$$

Step 5 Check that $(-4, 6)$ is the solution.

Check

$2x + 3y = 10$ (1)	$-3x - 2y = 0$ (2)
$2(-4) + 3(6) \stackrel{?}{=} 10$	$-3(-4) - 2(6) \stackrel{?}{=} 0$
$-8 + 18 \stackrel{?}{=} 10$	$12 - 12 \stackrel{?}{=} 0$
$10 = 10$ True	$0 = 0$ True

Both results are true, so the solution set of the system is $\{(-4, 6)\}$.

Note

In Example 4, we could have started the solution by solving the second equation for either x or y and then substituting the result into the first equation. The solution would be the same.

Work Problem 5 at the Side. ▶

5 Solve the system by substitution. Check your solution.

$$3x + 2y = 1$$
$$3x - 4y = -11$$

ANSWER

5. $\{(-1, 2)\}$

OBJECTIVE 2 Solve special systems by substitution. We can solve inconsistent systems with graphs that are parallel lines and systems of dependent equations with graphs that are the same line using the substitution method.

EXAMPLE 5 Solving an Inconsistent System by Substitution

Use substitution to solve the system.

$$x = 5 - 2y \quad (1)$$
$$2x + 4y = 6 \quad (2)$$

Substitute $5 - 2y$ for x in equation (2).

$$2x + 4y = 6 \quad (2)$$
$$2(5 - 2y) + 4y = 6 \quad \text{Let } x = 5 - 2y.$$
$$10 - 4y + 4y = 6 \quad \text{Distributive property}$$
$$10 = 6 \quad \text{False}$$

This false result means that the equations in the system have graphs that are parallel lines. The system is inconsistent, and the solution set is $\emptyset$. See Figure 6.

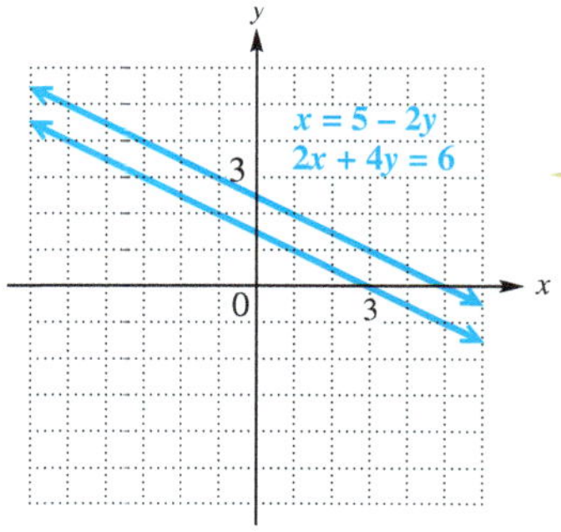

Figure 6

The symbol for an empty solution set is $\emptyset$. Do **not** use set braces. Write $\emptyset$, **not** $\{\emptyset\}$.

CAUTION
It is a common error to give "false" as the answer to an inconsistent system. The correct response is $\emptyset$.

EXAMPLE 6 Solving a System with Dependent Equations by Substitution

Solve the system by the substitution method.

$$3x - y = 4 \quad (1)$$
$$-9x + 3y = -12 \quad (2)$$

Begin by solving equation (1) for y to get $y = 3x - 4$. Substitute $3x - 4$ for y in equation (2) and solve the resulting equation.

$$-9x + 3y = -12 \quad (2)$$
$$-9x + 3(3x - 4) = -12 \quad \text{Let } y = 3x - 4.$$
$$-9x + 9x - 12 = -12 \quad \text{Distributive property}$$
$$0 = 0 \quad \text{Add 12; combine like terms.}$$

Continued on Next Page

This true result means that every solution of one equation is also a solution of the other, so the system has an infinite number of solutions—all the ordered pairs corresponding to points that lie on the common graph. The solution set is $\{(x, y) | 3x - y = 4\}$. A graph of the equations of this system is shown in Figure 7.

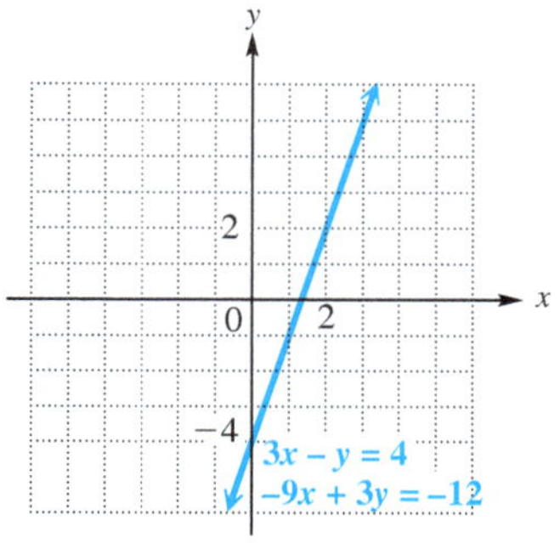

Figure 7

CAUTION
It is a common error to give "true" as the solution of a system of dependent equations. Remember that we give the solution set in set-builder notation using an equation that is in standard form, with integer coefficients having greatest common factor 1.

Work Problem **6** *at the Side.* ▶

OBJECTIVE 3 Solve linear systems with fractions and decimals by substitution. When a system includes an equation with fractions as coefficients, eliminate the fractions by multiplying each side of the equation by a common denominator. Then solve the resulting system.

EXAMPLE 7 Using the Substitution Method with Fractions as Coefficients

Solve the system by the substitution method.

$$3x + \frac{1}{4}y = 2 \quad (1)$$

$$\frac{1}{2}x + \frac{3}{4}y = -\frac{5}{2} \quad (2)$$

Clear equation (1) of fractions by multiplying each side by 4.

$$4\left(3x + \frac{1}{4}y\right) = 4(2) \quad \text{Multiply by 4.}$$

$$4(3x) + 4\left(\frac{1}{4}y\right) = 4(2) \quad \text{Distributive property}$$

$$12x + y = 8 \quad (3)$$

Continued on Next Page

6 Solve each system by substitution.

(a) $8x - y = 4$
$y = 8x + 4$

(b) $7x - 6y = 10$
$-14x + 20 = -12y$

ANSWERS

6. **(a)** $\emptyset$ Do **not** write $\{\emptyset\}$.
(b) $\{(x, y) | 7x - 6y = 10\}$

7 Solve the system by substitution. First clear all fractions.

$$\frac{2}{3}x + \frac{1}{2}y = 6$$
$$\frac{1}{2}x - \frac{3}{4}y = 0$$

8 Complete the process of solving the system given in Example 8 by substitution. (*Hint:* Solve equation (4) for x. Then substitute this result for x in equation (3) to find y.)

Now clear equation (2) of fractions by multiplying each side by the common denominator 4.

$$\frac{1}{2}x + \frac{3}{4}y = -\frac{5}{2} \qquad (2)$$
$$4\left(\frac{1}{2}x + \frac{3}{4}y\right) = 4\left(-\frac{5}{2}\right) \qquad \text{Multiply by 4.}$$
$$4\left(\frac{1}{2}x\right) + 4\left(\frac{3}{4}y\right) = 4\left(-\frac{5}{2}\right) \qquad \text{Distributive property}$$
$$2x + 3y = -10 \qquad (4)$$

The given system of equations has been simplified to the equivalent system

$$12x + y = 8 \qquad (3)$$
$$2x + 3y = -10. \qquad (4)$$

To solve this system by substitution, equation (3) can be solved for y.

$$12x + y = 8 \qquad (3)$$
$$y = -12x + 8 \qquad \text{Subtract } 12x.$$

Now substitute this result for y in equation (4).

$$2x + 3y = -10 \qquad (4)$$
$$2x + 3(-12x + 8) = -10 \qquad \text{Let } y = -12x + 8.$$
$$2x - 36x + 24 = -10 \qquad \text{Distributive property}$$
$$-34x = -34 \qquad \text{Combine like terms; subtract 24.}$$
$$x = 1 \qquad \text{Divide by } -34.$$

Distribute 3 to both $-12x$ and 8.

Substitute 1 for x in $y = -12x + 8$ to get

$$y = -12(1) + 8 = -4.$$

Check by substituting 1 for x and -4 for y in both of the original equations. The solution set is $\{(1, -4)\}$.

◀ Work Problem **7** at the Side.

If any of the coefficients in the equations of a system are decimals, we can eliminate the decimals by multiplying by a power of 10, as we did when solving linear equations with decimal coefficients in **Section 10.2.**

EXAMPLE 8 **Using the Substitution Method with Decimals as Coefficients**

Solve the system by the substitution method.

$$0.5x + 2.4y = 4.2 \qquad (1)$$
$$-0.1x + 1.5y = 5.1 \qquad (2)$$

To eliminate (or "clear") decimals, multiply each equation by 10.

$$5x + 24y = 42 \qquad (3)$$
$$-x + 15y = 51 \qquad (4) \qquad (-0.1x) \cdot 10 = -1x = -x$$

Now we can solve this equivalent system by substitution.

◀ Work problem **8** at the Side.

ANSWERS

7. $\{(6, 4)\}$

8. $\{(-6, 3)\}$

14.2 Exercises

FOR EXTRA HELP

MyMathLab | MathXL PRACTICE | WATCH | DOWNLOAD | READ | REVIEW

1. A student solves the system

$$\begin{aligned} 5x - y &= 15 \\ 7x + y &= 21 \end{aligned}$$

and finds that $x = 3$, which is the correct value for x. The student gives the solution set as $\{3\}$. ***WHAT WENT WRONG?***

2. When you use the substitution method, how can you tell that a system has

(a) no solution? **(b)** an infinite number of solutions?

Solve each system by the substitution method. Check each solution. See Examples 1–8.

3. $x + y = 12$, $y = 3x$

4. $x + 3y = -28$, $y = -5x$

5. $3x + 2y = 27$, $x = y + 4$

6. $4x + 3y = -5$, $x = y - 3$

7. $3x + 5y = 14$, $x - 2y = -10$

8. $5x + 2y = -1$, $2x - y = -13$

9. $3x + 4 = -y$, $2x + y = 0$

10. $2x - 5 = -y$, $x + 3y = 0$

11. $7x + 4y = 13$, $x + y = 1$

12. $3x - 2y = 19$, $x + y = 8$

13. $3x - y = 5$, $y = 3x - 5$

14. $4x - y = -3$, $y = 4x + 3$

15. $6x - 8y = 6$, $2y = -2 + 3x$

16. $3x + 2y = 6$, $6x = 8 + 4y$

17. $2x + 8y = 3$, $x = 8 - 4y$

18. $2x + 10y = 3$
$x = 1 - 5y$

19. $12x - 16y = 8$
$3x = 4y + 2$

20. $6x + 9y = 6$
$2x = 2 - 3y$

21. $\frac{1}{5}x + \frac{2}{3}y = -\frac{8}{5}$
$3x - y = 9$

22. $\frac{1}{3}x - \frac{1}{2}y = \frac{1}{6}$
$3x - 2y = 9$

23. $\frac{x}{2} - \frac{y}{3} = 9$
$\frac{x}{5} - \frac{y}{4} = 5$

24. $\frac{1}{6}x + \frac{1}{6}y = 2$
$-\frac{1}{2}x - \frac{1}{3}y = -8$

25. $\frac{x}{5} + 2y = \frac{16}{5}$
$\frac{3x}{5} + \frac{y}{2} = -\frac{7}{5}$

26. $\frac{x}{3} - \frac{3y}{4} = -\frac{1}{2}$
$\frac{x}{6} + \frac{y}{8} = \frac{3}{4}$

27. $0.1x + 0.9y = -2$
$0.5x - 0.2y = 4.1$

28. $0.2x - 1.3y = -3.2$
$-0.1x + 2.7y = 9.8$

29. $0.08x - 0.01y = 1.3$
$0.22x + 0.15y = 8.9$

Relating Concepts (Exercises 30–33) For Individual or Group Work

A system of linear equations can be used to model the cost and the revenue of a business. ***Work Exercises 30–33 in order.***

30. Suppose that you start a business manufacturing and selling bicycles, and it costs you \$5000 to get started. You determine that each bicycle will cost \$400 to manufacture. Explain why the linear equation $y_1 = 400x + 5000$ gives your *total* cost to manufacture x bicycles (y_1 in dollars).

31. You decide to sell each bike for \$600. What expression in x represents the revenue you will take in if you sell x bikes? Write an equation using y_2 to express your revenue when you sell x bikes (y_2 in dollars).

32. Form a system from the two equations in Exercises 30 and 31, and then solve the system, assuming $y_1 = y_2$, that is, cost = revenue.

33. The value of x from Exercise 32 is the number of bikes it takes to *break even.* Fill in the blanks: When ______ bikes are sold, the break-even point is reached. At that point, you have spent ______ dollars and taken in ______ dollars.

14.3 Solving Systems of Linear Equations by Elimination

OBJECTIVE 1 Solve linear systems by elimination. An algebraic method that depends on the addition property of equality can be used to solve systems. As mentioned earlier, adding the same quantity to each side of an equation results in equal sums.

$$\text{If} \quad A = B, \quad \text{then} \quad A + C = B + C.$$

This addition can be taken a step further. Adding *equal* quantities, rather than the *same* quantity, to both sides of an equation also results in equal sums.

$$\text{If} \quad A = B \quad \text{and} \quad C = D, \quad \text{then} \quad A + C = B + D.$$

Using the addition property to solve systems is called the **elimination method.** When using this method, the idea is to *eliminate* one of the variables. ***To do this, one of the variables in the two equations must have coefficients that are opposites.***

OBJECTIVES

1. **Solve linear systems by elimination.**
2. **Multiply when using the elimination method.**
3. **Use an alternative method to find the second value in a solution.**
4. **Use the elimination method to solve special systems.**

EXAMPLE 1 Using the Elimination Method

Use the elimination method to solve the system.

$$x + y = 5$$
$$x - y = 3$$

Each equation in this system is a statement of equality, so the sum of the left sides equals the sum of the right sides. Adding in this way gives

$$(x + y) + (x - y) = 5 + 3.$$

Combine like terms and simplify to get

$$2x = 8$$
$$x = 4. \qquad \text{Divide by 2.}$$

Notice that y has been eliminated. The result, $x = 4$, gives the x-value of the solution of the given system. To find the y-value of the solution, substitute 4 for x in either of the two equations of the system.

Work Problem 1 at the Side. ▶

Check the solution set found at the side, $\{(4, 1)\}$, by substituting 4 for x and 1 for y in both equations of the given system.

Check

$x + y = 5$		$x - y = 3$	
$4 + 1 \stackrel{?}{=} 5$		$4 - 1 \stackrel{?}{=} 3$	
$5 = 5$	True	$3 = 3$	True

Since both results are true, the solution set of the system is $\{(4, 1)\}$.

CAUTION
A system is not completely solved until values for both x and y are found. Do not stop after finding the value of only one variable. Remember to write the solution set as a set containing an ordered pair.

1 **(a)** Substitute 4 for x in the equation $x + y = 5$ to find the value of y.

(b) Give the solution set of the system.

ANSWERS

1. (a) $y = 1$ **(b)** $\{(4, 1)\}$

◀ Work Problem 2 at the Side.

2 Solve each system by the elimination method. Check each solution.

(a) Fill in the blanks to solve the following system.

$$x + y = 8$$
$$x - y = 2$$

Add.

$$(x + y) + (x - y) = 8 + \_\_\_$$
$$2\,\_\_\_ = \_\_\_$$
$$x = \_\_\_$$

Find y.

$$x - y = 2$$
$$\_\_\_ - y = 2$$
$$-y = \_\_\_$$
$$y = \_\_\_$$

The solution set is ___.

(b) $3x - y = 7$
$2x + y = 3$

Answers

2. (a) 2; x; 10; 5; 5; -3; 3; $\{(5, 3)\}$
(b) $\{(2, -1)\}$

In general, to solve a system by elimination, follow these steps.

Solving a Linear System by Elimination

Step 1 **Write both equations in standard form** $Ax + By = C$.

Step 2 **Transform so that the coefficients of one pair of variable terms are opposites.** Multiply one or both equations by appropriate numbers so that the sum of the coefficients of either the x- or y-terms is 0.

Step 3 **Add** the new equations to eliminate a variable. The sum should be an equation with just one variable.

Step 4 **Solve** the equation from Step 3 for the remaining variable.

Step 5 **Substitute** the result from Step 4 into *either* of the original equations and solve for the other variable.

Step 6 **Check** the solution in *both* of the original equations. Then write the solution set.

It does not matter which variable is eliminated first. Usually we choose the one that is more convenient to work with.

EXAMPLE 2 **Using the Elimination Method**

Solve the system.

$$y + 11 = 2x$$
$$5x = y + 26$$

Step 1 Rewrite both equations in the form $Ax + By = C$ to get the system

$$-2x + y = -11 \qquad \text{Subtract } 2x \text{ and } 11.$$
$$5x - y = 26. \qquad \text{Subtract } y.$$

Step 2 Because the coefficients of y are 1 and -1, adding will eliminate y. It is not necessary to multiply either equation by a number.

Step 3 Add the two equations. This time we use vertical addition.

$$\begin{array}{rcr} -2x + y &=& -11 \\ 5x - y &=& 26 \\ \hline 3x \quad &=& 15 \end{array} \qquad \text{Add in columns.}$$

Step 4 Solve the equation.

$$3x = 15$$

Don't stop here. → $x = 5$ Divide by 3.

Step 5 Find the value of y by substituting 5 for x in either of the original equations. Choosing the first equation gives

$$y + 11 = 2x$$
$$y + 11 = 2(5) \qquad \text{Let } x = 5.$$
$$y + 11 = 10$$
$$y = -1. \qquad \text{Subtract 11.}$$

Continued on Next Page

Step 6 Check the solution by substituting $x = 5$ and $y = -1$ into both of the original equations.

Check

$y + 11 = 2x$	$5x = y + 26$
$-1 + 11 \stackrel{?}{=} 2(5)$	$5(5) \stackrel{?}{=} -1 + 26$
$10 = 10$ True	$25 = 25$ True

Since $(5, -1)$ is a solution of *both* equations, the solution set is $\{(5, -1)\}$.

Work Problem 3 at the Side. ▶

3 Solve each system by the elimination method. Check each solution.

(a) $2x - y = 2$
$4x + y = 10$

(b) $8x - 5y = 32$
$4x + 5y = 4$

OBJECTIVE 2 Multiply when using the elimination method. Sometimes we need to multiply each side of one or both equations in a system by some number before adding the equations will eliminate a variable.

EXAMPLE 3 Multiplying Both Equations When Using the Elimination Method

Solve the system.

$$2x + 3y = -15 \quad (1)$$
$$5x + 2y = 1 \quad (2)$$

Adding the two equations gives $7x + 5y = -14$, which does not eliminate either variable. However, we can multiply each equation by a suitable number so that the coefficients of one of the two variables are opposites. For example, to eliminate x, multiply each side of equation (1) by 5, and each side of equation (2) by -2.

$$\begin{aligned} 10x + 15y &= -75 && \text{Multiply equation (1) by 5.} \\ -10x - 4y &= -2 && \text{Multiply equation (2) by } -2. \\ \hline 11y &= -77 && \text{Add.} \\ y &= -7 && \text{Divide by 11.} \end{aligned}$$

Substituting -7 for y in either equation (1) or (2) gives $x = 3$. Check that the solution set of the system is $\{(3, -7)\}$.

Work Problem 4 at the Side. ▶

4 **(a)** Solve the system in Example 3 by first eliminating the variable y. Check your solution.

(b) Solve

$$6x + 7y = 4$$
$$5x + 8y = -1,$$

and check your solution.

OBJECTIVE 3 Use an alternative method to find the second value in a solution. Sometimes it is easier to find the value of the second variable in a solution by using the elimination method twice.

EXAMPLE 4 Finding the Second Value Using an Alternative Method

Solve the system.

$$4x = 9 - 3y \quad (1)$$
$$5x - 2y = 8 \quad (2)$$

Rearrange the terms in equation (1) so that like terms are aligned in columns. To do this, add $3y$ to each side to get the following system.

$$4x + 3y = 9 \quad (3)$$
$$5x - 2y = 8 \quad (2)$$

One way to proceed is to eliminate y by multiplying each side of equation (3) by 2 and each side of equation (2) by 3, and then adding.

Continued on Next Page

Answers

3. (a) $\{(2, 2)\}$ **(b)** $\left\{\left(3, -\frac{8}{5}\right)\right\}$

4. (a) $\{(3, -7)\}$ **(b)** $\{(3, -2)\}$

5 Solve each system of equations.

(a) $5x = 7 + 2y$
$5y = 5 - 3x$

(b) $3y = 8 + 4x$
$6x = 9 - 2y$

$$\begin{aligned} 8x + \mathbf{6y} &= 18 && \text{Multiply equation (3) by 2.}\\ 15x - \mathbf{6y} &= 24 && \text{Multiply equation (2) by 3.}\\ \hline 23x &= 42 && \text{Add.}\\ x &= \frac{42}{23} && \text{Divide by 23.}\end{aligned}$$

Substituting $\frac{42}{23}$ for x in one of the given equations would give y, but the arithmetic involved would be messy. Instead, solve for y by starting again with the original equations and eliminating x. Multiply each side of equation (3) by 5 and each side of equation (2) by -4, and then add.

$$\begin{aligned} \mathbf{20x} + 15y &= 45 && \text{Multiply equation (3) by 5.}\\ \mathbf{-20x} + 8y &= -32 && \text{Multiply equation (2) by } -4.\\ \hline 23y &= 13 && \text{Add.}\\ y &= \frac{13}{23} && \text{Divide by 23.}\end{aligned}$$

Check that the solution set is $\{(\frac{42}{23}, \frac{13}{23})\}$.

◀ Work Problem **5** at the Side.

When the value of the first variable is a fraction, the method used in Example 4 helps avoid arithmetic errors. Of course, this method could be used to solve any system of equations.

OBJECTIVE 4 **Use the elimination method to solve special systems.**

6 Solve each system by the elimination method.

(a) $4x + 3y = 10$
$2x + \frac{3}{2}y = 12$

(b) $4x - 6y = 10$
$-10x + 15y = -25$

EXAMPLE 5 **Using the Elimination Method for an Inconsistent System or Dependent Equations**

Solve each system by the elimination method.

(a) $2x + 4y = 5$
$4x + 8y = -9$

Multiply each side of $2x + 4y = 5$ by -2; then add to $4x + 8y = -9$.

$$\begin{aligned} -4x - 8y &= -10\\ 4x + 8y &= -9\\ \hline \mathbf{0} &= \mathbf{-19} && \text{False}\end{aligned}$$

Write $\emptyset$, **not** $\{\emptyset\}$.

The false statement $0 = -19$ indicates that the solution set is $\emptyset$.

(b) $3x - y = 4$
$-9x + 3y = -12$

Multiply each side of the first equation by 3; then add the two equations.

$$\begin{aligned} 9x - 3y &= 12\\ -9x + 3y &= -12\\ \hline \mathbf{0} &= \mathbf{0} && \text{True}\end{aligned}$$

A true statement occurs when the equations are equivalent. As before, this indicates that every solution of one equation is also a solution of the other. The solution set is $\{(x, y) \mid 3x - y = 4\}$. (See **Section 14.2,** Example 6, where the same system was solved using substitution.)

◀ Work Problem **6** at the Side.

Answers

5. (a) $\left\{\left(\frac{45}{31}, \frac{4}{31}\right)\right\}$ **(b)** $\left\{\left(\frac{11}{26}, \frac{42}{13}\right)\right\}$

6. (a) $\emptyset$ **(b)** $\{(x, y) \mid 2x - 3y = 5\}$

14.3 Exercises

FOR EXTRA HELP

MyMathLab | MathXL PRACTICE | WATCH | DOWNLOAD | READ | REVIEW

In Exercises 1–4, answer true *or* false *for each statement. If* false, *tell why.*

1. The ordered pair (0, 0) *must* be a solution of a system of the form

$$\begin{aligned} Ax + By &= 0 \\ Cx + Dy &= 0. \end{aligned}$$

2. To eliminate the y-terms in the system

$$\begin{aligned} 2x + 12y &= 7 \\ 3x + 4y &= 1, \end{aligned}$$

we should multiply the bottom equation by 3 and then add.

3. The system

$$\begin{aligned} x + y &= 1 \\ x + y &= 2 \end{aligned}$$

has $\emptyset$ as its solution set.

4. The ordered pair $(4, -5)$ cannot be a solution of a system that contains the equation $5x - 4y = 0$.

Solve each system by the elimination method. Check each solution. See Examples 1 and 2.

5. $\begin{aligned} x + y &= 2 \\ 2x - y &= -5 \end{aligned}$

6. $\begin{aligned} 3x - y &= -12 \\ x + y &= 4 \end{aligned}$

7. $\begin{aligned} 2x + y &= -5 \\ x - y &= 2 \end{aligned}$

8. $\begin{aligned} 2x + y &= -15 \\ -x - y &= 10 \end{aligned}$

9. $\begin{aligned} 3x + 2y &= 0 \\ -3x - y &= 3 \end{aligned}$

10. $\begin{aligned} 5x - y &= 5 \\ -5x + 2y &= 0 \end{aligned}$

11. $\begin{aligned} 6x - y &= -1 \\ 5y &= 17 + 6x \end{aligned}$

12. $\begin{aligned} y &= 9 - 6x \\ -6x + 3y &= 15 \end{aligned}$

Solve each system by the elimination method. Check each solution. See Examples 3–5.

13. $2x - y = 12$
$3x + 2y = -3$

14. $x + y = 3$
$-3x + 2y = -19$

15. $x + 3y = 19$
$2x - y = 10$

16. $4x - 3y = -19$
$2x + y = 13$

17. $x + 4y = 16$
$3x + 5y = 20$

18. $2x + y = 8$
$5x - 2y = -16$

19. $5x - 3y = -20$
$-3x + 6y = 12$

20. $4x + 3y = -28$
$5x - 6y = -35$

21. $2x - 8y = 0$
$4x + 5y = 0$

22. $3x - 15y = 0$
$6x + 10y = 0$

23. $x + y = 7$
$x + y = -3$

24. $x - y = 4$
$x - y = -3$

25. $-x + 3y = 4$
$-2x + 6y = 8$

26. $6x - 2y = 24$
$-3x + y = -12$

27. $4x - 3y = -19$
$3x + 2y = 24$

28. $5x + 4y = 12$
$3x + 5y = 15$

29. $3x - 7 = -5y$
$5x + 4y = -10$

30. $2x + 3y = 13$
$6 + 2y = -5x$

31. $2x + 3y = 0$
$4x + 12 = 9y$

32. $-4x + 3y = 2$
$5x + 3 = -2y$

33. $24x + 12y = -7$
$16x - 17 = 18y$

34. $9x + 4y = -3$
$6x + 7 = -6y$

35. $3x = 3 + 2y$
$-\frac{4}{3}x + y = \frac{1}{3}$

36. $3x = 27 + 2y$
$x - \frac{7}{2}y = -25$

37. $5x - 2y = 3$
$10x - 4y = 5$

38. $3x - 5y = 1$
$6x - 10y = 4$

39. $6x + 3y = 0$
$-18x - 9y = 0$

40. $3x - 5y = 0$
$9x - 15y = 0$

Relating Concepts (Exercises 41–46) For Individual or Group Work

Attending the movies is one of America's favorite forms of entertainment. The graph shows U.S. movie attendance from 1996 through 2004. In 1996, attendance was 1339 *million, as represented by the point P*(1996, 1339)*. In 2004, attendance was* 1536 *million, as represented by the point Q*(2004, 1536)*. We can find an equation of line segment PQ by using a system of equations. Then we use the equation we found to approximate the attendance in any of the years between 1996 and 2004.* ***Work Exercises 41–46 in order.***

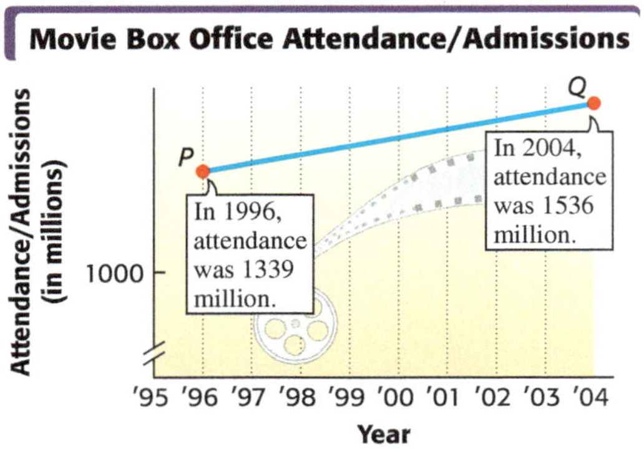

Source: Motion Picture Association of America.

41. The line segment has an equation that can be written in the form $y = ax + b$. Using the coordinates of point P with $x = 1996$ and $y = 1339$, write an equation in the variables a and b.

42. Using the coordinates of point Q with $x = 2004$ and $y = 1536$, write a second equation in the variables a and b.

43. Write the system of equations formed from the two equations in Exercises 41 and 42, and solve the system using the elimination method.

44. What is the equation of the segment PQ?

45. Let $x = 2002$ in the equation of Exercise 44, and solve for y to the nearest tenth. How does the result compare with the actual figure of 1639 million?

46. The actual data points for the years 1996 through 2004 do not lie in a perfectly straight line. Explain the pitfalls of relying too heavily on using the equation in Exercise 44 to predict attendance.

Summary Exercises on Solving Systems of Linear Equations

The exercises in this summary include a variety of problems on solving systems of linear equations. Since we do not usually specify the method of solution, use the following guidelines to help you decide whether to use substitution or elimination.

Guidelines for Choosing a Method to Solve a System of Linear Equations

1. If one of the equations of the system is already solved for one of the variables, as in the systems

$$\begin{aligned} 3x + 4y &= 9 \\ y &= 2x - 6 \end{aligned} \quad \text{or} \quad \begin{aligned} -5x + 3y &= 9 \\ x &= 3y - 7, \end{aligned}$$

the substitution method is the better choice.

2. If both equations are in standard $Ax + By = C$ form, as in

$$\begin{aligned} 4x - 11y &= 3 \\ -2x + 3y &= 4, \end{aligned}$$

and none of the variables has coefficient -1 or 1, the elimination method is the better choice.

3. If one or both of the equations are in standard form and the coefficient of one of the variables is -1 or 1, as in the systems

$$\begin{aligned} 3x + y &= -2 \\ -5x + 2y &= 4 \end{aligned} \quad \text{or} \quad \begin{aligned} -x + 3y &= -4 \\ 3x - 2y &= 8, \end{aligned}$$

either method is appropriate.

Use the preceding guidelines to solve each problem.

1. Assuming you want to minimize the amount of work required, tell whether you would use the substitution or elimination method to solve each system. Explain your answers. *Do not actually solve.*

(a) $\begin{aligned} 3x + 2y &= 18 \\ y &= 3x \end{aligned}$

(b) $\begin{aligned} 3x + y &= -7 \\ x - y &= -5 \end{aligned}$

(c) $\begin{aligned} 3x - 2y &= 0 \\ 9x + 8y &= 7 \end{aligned}$

2. Which one of the following systems would be easier to solve using the substitution method? Why?

$$\begin{aligned} 5x - 3y &= 7 \\ 2x + 8y &= 3 \end{aligned} \qquad \begin{aligned} 7x + 2y &= 4 \\ y &= -3x + 1 \end{aligned}$$

In Exercises 3 and 4, **(a)** *solve the system by the elimination method,* **(b)** *solve the system by the substitution method, and* **(c)** *tell which method you prefer for that particular system and why.*

3. $4x - 3y = -8$
$x + 3y = 13$

4. $2x + 5y = 0$
$x = -3y + 1$

Solve each system by the method of your choice. (For Exercises 5–7, see your answers for Exercise 1.)

5. $3x + 2y = 18$
$y = 3x$

6. $3x + y = -7$
$x - y = -5$

7. $3x - 2y = 0$
$9x + 8y = 7$

8. $x + y = 7$
$x = -3 - y$

9. $5x - 4y = 15$
$-3x + 6y = -9$

10. $4x + 2y = 3$
$y = -x$

11. $3x = 7 - y$
$2y = 14 - 6x$

12. $3x - 5y = 7$
$2x + 3y = 30$

13. $3y = 4x + 2$
$5x - 2y = -3$

14. $4x + 3y = 1$
$3x + 2y = 2$

15. $2x - 3y = 7$
$-4x + 6y = 14$

16. $0.2x + 0.3y = 1.0$
$-0.3x + 0.1y = 1.8$

17. $6x + 5y = 13$
$3x + 3y = 4$

18. $x - 3y = 7$
$4x + y = 5$

19. $\frac{1}{4}x - \frac{1}{5}y = 9$
$y = 5x$

20. $\frac{1}{2}x + \frac{1}{3}y = -\frac{1}{3}$
$\frac{1}{2}x + 2y = -7$

21. $-\frac{1}{2}x - \frac{1}{3}y = -5$
$-\frac{1}{2}x - \frac{1}{3}y = -5$

22. $\frac{x}{5} + 2y = \frac{8}{5}$
$\frac{3x}{5} + \frac{y}{2} = -\frac{7}{10}$

23. $\frac{x}{5} + y = \frac{6}{5}$
$\frac{x}{10} + \frac{y}{3} = \frac{5}{6}$

24. $\frac{2}{5}x + \frac{4}{3}y = -8$
$\frac{7}{10}x - \frac{2}{9}y = 9$

25. $0.5x + 0.2y = 0.2$
$x - 0.6y = -0.5$

14.4 Systems of Linear Equations in Three Variables

OBJECTIVES

1. Understand the geometry of systems of three equations in three variables.
2. Solve linear systems (with three equations and three variables) by elimination.
3. Solve linear systems (with three equations and three variables) in which some of the equations have missing terms.
4. Solve special systems.

A solution of an equation in three variables, such as

$$2x + 3y - z = 4, \quad \text{Linear equation in three variables}$$

is called an **ordered triple** and is written (x, y, z). For example, the ordered triple $(0, 1, -1)$ is a solution of the equation, because

$$2(0) + 3(1) - (-1) = 4$$

is a true statement. Verify that another solution of this equation is $(10, -3, 7)$.

We now extend the term *linear equation* to equations of the form

$$Ax + By + Cz + \dots + Dw = K,$$

where not all the coefficients $A, B, C, \dots, D$ equal 0. For example,

$$2x + 3y - 5z = 7 \quad \text{and} \quad x - 2y - z + 3u - 2w = 8$$

are linear equations, the first with three variables and the second with five.

OBJECTIVE 1 Understand the geometry of systems of three equations in three variables. Consider the solution of a system such as

$$\begin{aligned} 4x + 8y + z &= 2 \\ x + 7y - 3z &= -14 \\ 2x - 3y + 2z &= 3. \end{aligned} \quad \text{System of linear equations in three variables}$$

Theoretically, a system of this type can be solved by graphing. However, the graph of a linear equation with three variables is a *plane,* not a line. Since the graph of each equation of the system is a plane, which requires three-dimensional graphing, the graphing method is not practical for solving such systems. However, it does illustrate the number of solutions possible for these systems, as shown in Figure 6.

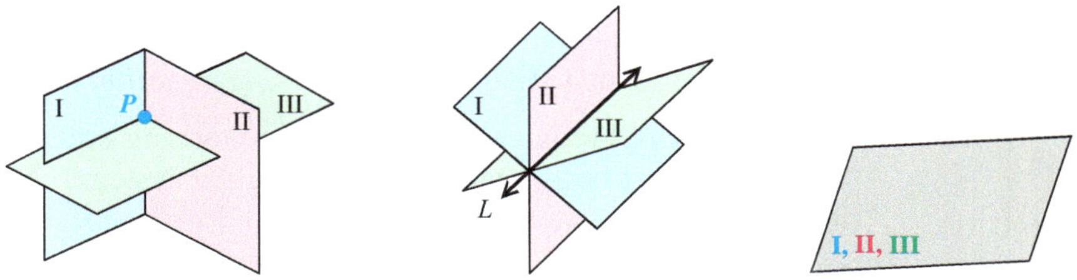

A single solution (a) | Points of a line in common (b) | All points in common (c)

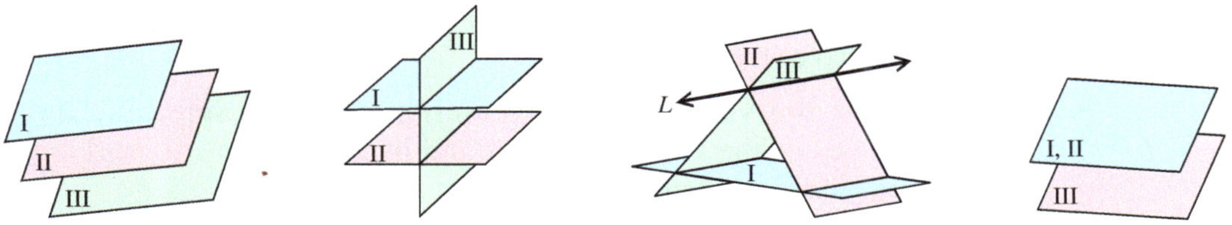

No points in common (d) | No points in common (e) | No points in common (f) | No points in common (g)

Figure 6

Figure 6 on the preceding page illustrates the following cases.

Graphs of Linear Systems in Three Variables

Case 1 **The three planes may meet at a single, common point** that forms the solution set of the system. See Figure 6(a).

Case 2 **The three planes may have the points of a line in common** so that the infinite set of points that satisfy the equation of the line forms the solution set of the system. See Figure 6(b).

Case 3 **The three planes may coincide** so that the solution set of the system is the set of all points on a plane. See Figure 6(c).

Case 4 **The planes may have no points common to all three** so that there is no solution of the system; the solution set is $\emptyset$. See Figures 6(d)–(g).

OBJECTIVE 2 **Solve linear systems (with three equations and three variables) by elimination.** Is it possible to solve a system of three equations in three variables such as the one that follows?

$$\begin{aligned} 4x + 8y + z &= 2 \\ x + 7y - 3z &= -14 \\ 2x - 3y + 2z &= 3 \end{aligned}$$

As mentioned, graphing to find the solution set of such a system is impractical. We solve these systems with an extension of the elimination method from **Section 5.1,** summarized as follows.

Solving a Linear System in Three Variables*

Step 1 **Select a variable and an equation.** A good choice for the variable, which we call the *focus variable,* is one that has coefficient 1 or -1. Then select an equation, usually the one that contains the focus variable, as the *working equation.*

Step 2 **Eliminate the focus variable.** Use the working equation and one of the other two equations of the original system. The result is an equation in two variables.

Step 3 **Eliminate the focus variable again.** Use the working equation and the remaining equation of the original system. The result is another equation in two variables.

Step 4 **Write the equations in two variables that result from Steps 2 and 3 as a system, and solve it.** Doing this gives the values of two of the variables.

Step 5 **Find the value of the remaining variable.** Substitute the values of the two variables found in Step 4 into the working equation to obtain the value of the focus variable.

Step 6 **Check** the ordered-triple solution in *each* of the *original* equations of the system. Then write the solution set.

*The authors wish to thank Christine Heinecke Lehmann of Purdue University North Central for her suggestions here.

Example 1 **Solving a System in Three Variables**

Solve the system.

$$\begin{aligned} 4x + 8y + z &= 2 & (1)\\ x + 7y - 3z &= -14 & (2)\\ 2x - 3y + 2z &= 3 & (3) \end{aligned}$$

Step 1 Since z in equation (1) has coefficient 1, we choose z as the focus variable and (1) as the working equation. (Another option would be to choose x as the focus variable, since it also has coefficient 1, and use (2) as the working equation.)

$$4x + 8y + \underset{\text{Focus variable}}{z} = 2 \qquad (1) \leftarrow \text{Working equation}$$

Step 2 Multiply working equation (1) by 3 and add the result to equation (2) to eliminate focus variable z.

$$\begin{aligned} 12x + 24y + 3z &= 6 && \text{Multiply each side of (1) by 3.}\\ x + 7y - 3z &= -14 && (2)\\ \hline 13x + 31y &= -8 && \text{Add.} \quad (4) \end{aligned}$$

Step 3 Multiply working equation (1) by -2 and add the result to remaining equation (3) to again eliminate focus variable z.

$$\begin{aligned} -8x - 16y - 2z &= -4 && \text{Multiply each side of (1) by } -2.\\ 2x - 3y + 2z &= 3 && (3)\\ \hline -6x - 19y &= -1 && \text{Add.} \quad (5) \end{aligned}$$

Step 4 Write the equations in two variables that result in Steps 2 and 3 as a system.

Make sure these equations have the same variables.

$$\begin{aligned} 13x + 31y &= -8 && (4) \quad \text{The result from Step 2}\\ -6x - 19y &= -1 && (5) \quad \text{The result from Step 3} \end{aligned}$$

Now solve this system. We choose to eliminate x.

$$\begin{aligned} 78x + 186y &= -48 && \text{Multiply each side of (4) by 6.}\\ -78x - 247y &= -13 && \text{Multiply each side of (5) by 13.}\\ \hline -61y &= -61 && \text{Add.}\\ y &= 1 && \text{Divide by } -61. \end{aligned}$$

Substitute 1 for y in either equation (4) or (5) to find x.

$$\begin{aligned} -6x - 19y &= -1 && (5)\\ -6x - 19(1) &= -1 && \text{Let } y = 1.\\ -6x - 19 &= -1 && \text{Multiply.}\\ -6x &= 18 && \text{Add 19.}\\ x &= -3 && \text{Divide by } -6. \end{aligned}$$

Step 5 Now substitute the two values we found in Step 4 in working equation (1) to find the value of the remaining variable, focus variable z.

$$\begin{aligned} 4x + 8y + z &= 2 && (1)\\ 4(-3) + 8(1) + z &= 2 && \text{Let } x = -3 \text{ and } y = 1.\\ -4 + z &= 2 && \text{Multiply; add.}\\ z &= 6 && \text{Add 4.} \end{aligned}$$

Continued on Next Page

1 Check that the solution $(-3, 1, 6)$ satisfies equations (2) and (3) of Example 1.

(a) $x + 7y - 3z = -14$ (2)

Does the solution satisfy equation (2)?

(b) $2x - 3y + 2z = 3$ (3)

Does the solution satisfy equation (3)?

2 Solve each system.

(a)
$$\begin{aligned} x + y + z &= 2 \\ x - y + 2z &= 2 \\ -x + 2y - z &= 1 \end{aligned}$$

(b)
$$\begin{aligned} 2x + y + z &= 9 \\ -x - y + z &= 1 \\ 3x - y + z &= 9 \end{aligned}$$

Answers

1. **(a)** yes **(b)** yes
2. **(a)** $\{(-1, 1, 2)\}$ **(b)** $\{(2, 1, 4)\}$

Step 6 It appears that the ordered triple $(-3, 1, 6)$ is the only solution of the system. We must check that the solution satisfies all three original equations of the system. For equation (1),

Check
$$\begin{aligned} 4x + 8y + z &= 2 \quad (1) \\ 4(-3) + 8(1) + 6 &\stackrel{?}{=} 2 \\ -12 + 8 + 6 &\stackrel{?}{=} 2 \\ 2 &= 2. \quad \text{True} \end{aligned}$$

◀ *Work Problem* **1** *at the Side.*

Because $(-3, 1, 6)$ also satisfies equations (2) and (3), the solution set is $\{(-3, 1, 6)\}$.

◀ *Work Problem* **2** *at the Side.*

OBJECTIVE 3 Solve linear systems (with three equations and three variables) in which some of the equations have missing terms. If a linear system has an equation missing a term or terms, one elimination step can be omitted.

Example 2 Solving a System of Equations with Missing Terms

Solve the system.

$$\begin{aligned} 6x - 12y &= -5 \quad (1) \quad \text{Missing } z \\ 8y + z &= 0 \quad (2) \quad \text{Missing } x \\ 9x - z &= 12 \quad (3) \quad \text{Missing } y \end{aligned}$$

Since equation (3) is missing the variable y, one way to begin is to eliminate y again using equations (1) and (2).

Leave space for the missing terms.

$$\begin{aligned} 12x - 24y &= -10 \quad \text{Multiply each side of (1) by 2.} \\ 24y + 3z &= 0 \quad \text{Multiply each side of (2) by 3.} \\ \hline 12x + 3z &= -10 \quad \text{Add.} \quad (4) \end{aligned}$$

Use the resulting equation (4) in x and z, together with equation (3), $9x - z = 12$, to eliminate z. Multiply equation (3) by 3.

$$\begin{aligned} 27x - 3z &= 36 \quad \text{Multiply each side of (3) by 3.} \\ 12x + 3z &= -10 \quad (4) \\ \hline 39x &= 26 \quad \text{Add.} \\ x = \frac{26}{39} &= \frac{2}{3} \quad \text{Divide by 39; lowest terms} \end{aligned}$$

We can find z by substituting this value for x into equation (3).

$$\begin{aligned} 9x - z &= 12 \quad (3) \\ 9\left(\frac{2}{3}\right) - z &= 12 \quad \text{Let } x = \tfrac{2}{3}. \\ 6 - z &= 12 \quad \text{Multiply.} \\ z &= -6 \quad \text{Subtract 6; multiply by } -1. \end{aligned}$$

Continued on Next Page

We can find y by substituting -6 for z in equation (2).

$$8y + z = 0 \quad (2)$$
$$8y - 6 = 0 \quad \text{Let } z = -6.$$
$$8y = 6 \quad \text{Add 6.}$$
$$y = \frac{6}{8} = \frac{3}{4} \quad \text{Divide by 8; lowest terms}$$

Thus, $x = \frac{2}{3}$, $y = \frac{3}{4}$, and $z = -6$. Check these values in each of the original equations of the system to verify that the solution set of the system is $\left\{\left(\frac{2}{3}, \frac{3}{4}, -6\right)\right\}$.

Note

Another way to solve the system in Example 2 is to begin by eliminating the variable z from equations (2) and (3). The resulting equation together with equation (1) forms a system of two equations in the variables x and y. Try working Example 2 this way to see that the same solution results.

There are often multiple ways to solve a system of equations. Some ways may involve more work than others.

Work Problem 3 at the Side. ▶

OBJECTIVE 4 Solve special systems. Linear systems with three variables may be inconsistent or may include dependent equations.

Example 3 Solving an Inconsistent System with Three Variables

Solve the system.

$$2x - 4y + 6z = 5 \quad (1)$$
$$-x + 3y - 2z = -1 \quad (2)$$
$$x - 2y + 3z = 1 \quad (3)$$

Use as the working equation, with focus variable x.

Eliminate x by adding equations (2) and (3) to get the equation

$$y + z = 0.$$

Eliminate x again, using equations (1) and (3).

$$-2x + 4y - 6z = -2 \quad \text{Multiply each side of (3) by } -2.$$
$$2x - 4y + 6z = 5 \quad (1)$$
$$0 = 3 \quad \text{False}$$

This false statement indicates that equations (1) and (3) have no common solution. Thus, the system is inconsistent and the solution set is $\emptyset$. The graph of this system would show these two planes parallel to one another.

Note

If a false statement results when adding as in Example 3, it is not necessary to go any further with the solution. Since two of the three planes are parallel, it is not possible for the three planes to have any common points.

Work Problem 4 at the Side. ▶

3 Solve each system.

(a) $x - y = 6$
$2y + 5z = 1$
$3x - 4z = 8$

(b) $5x - y = 26$
$4y + 3z = -4$
$x + z = 5$

4 Solve each system.

(a) $3x - 5y + 2z = 1$
$5x + 8y - z = 4$
$-6x + 10y - 4z = 5$

(b) $7x - 9y + 2z = 0$
$y + z = 0$
$8x - z = 0$

Answers

3. **(a)** $\{(4, -2, 1)\}$ **(b)** $\{(5, -1, 0)\}$
4. **(a)** $\emptyset$ **(b)** $\{(0, 0, 0)\}$

5 Solve the system.

$$\begin{aligned} x - y + z &= 4 \\ -3x + 3y - 3z &= -12 \\ 2x - 2y + 2z &= 8 \end{aligned}$$

Example 4 **Solving a System of Dependent Equations with Three Variables**

Solve the system.

$$\begin{aligned} 2x - 3y + 4z &= 8 &\quad (1) \\ -x + \frac{3}{2}y - 2z &= -4 &\quad (2) \\ 6x - 9y + 12z &= 24 &\quad (3) \end{aligned}$$

Multiplying each side of equation (1) by 3 gives equation (3). Multiplying each side of equation (2) by -6 also gives equation (3). Because of this, the equations are dependent. All three equations have the same graph, as illustrated in Figure 6(c). The solution set is written

$$\{(x, y, z) \mid 2x - 3y + 4z = 8\}. \quad \text{Set-builder notation}$$

Although any one of the three equations could be used to write the solution set, we use the equation with coefficients that are integers with greatest common factor 1, as we did in **Section 5.1.**

◀ *Work Problem* **5** *at the Side.*

6 Solve the system.

$$\begin{aligned} 2x + 3y - z &= 8 \\ \frac{1}{2}x + \frac{3}{4}y - \frac{1}{4}z &= 2 \\ x + \frac{3}{2}y - \frac{1}{2}z &= -6 \end{aligned}$$

Example 5 **Solving Another Special System**

Solve the system.

$$\begin{aligned} 2x - y + 3z &= 6 &\quad (1) \\ x - \frac{1}{2}y + \frac{3}{2}z &= 3 &\quad (2) \\ 4x - 2y + 6z &= 1 &\quad (3) \end{aligned}$$

Multiplying each side of equation (2) by 2 gives equation (1). Thus, these two equations are dependent.

Equations (1) and (3) are not equivalent, however. Multiplying equation (3) by $\frac{1}{2}$ gives

$$2x - y + 3z = \frac{1}{2}, \quad \text{(3) multiplied by } \tfrac{1}{2}$$

which is *not* equivalent to equation (1). Instead, we obtain two equations with the same coefficients, but with different constant terms. The graphs of equations (1) and (3) have no points in common (that is, the planes are parallel). Thus, the system is inconsistent and the solution set is $\emptyset$, as illustrated in Figure 6(g).

◀ *Work Problem* **6** *at the Side.*

Answers

5. $\{(x, y, z) \mid x - y + z = 4\}$
6. $\emptyset$

14.4 Exercises

FOR EXTRA HELP MyMathLab

MathXL PRACTICE · WATCH · DOWNLOAD · READ · REVIEW

1. Explain what the following statement means: The solution set of the system

$$\begin{aligned} 2x + y + z &= 3 \\ 3x - y + z &= -2 \\ 4x - y + 2z &= 0 \end{aligned}$$

is $\{(-1, 2, 3)\}$.

2. The two equations

$$\begin{aligned} x + y + z &= 6 \\ 2x - y + z &= 3 \end{aligned}$$

have a common solution of $(1, 2, 3)$. Which equation would complete a system of three linear equations in three variables having solution set $\{(1, 2, 3)\}$?

A. $3x + 2y - z = 1$ **B.** $3x + 2y - z = 4$

C. $3x + 2y - z = 5$ **D.** $3x + 2y - z = 6$

Solve each system of equations. See Example 1.

3. $$\begin{aligned} 2x - 5y + 3z &= -1 \\ x + 4y - 2z &= 9 \\ x - 2y - 4z &= -5 \end{aligned}$$

4. $$\begin{aligned} x + 3y - 6z &= 7 \\ 2x - y + z &= 1 \\ x + 2y + 2z &= -1 \end{aligned}$$

5. $$\begin{aligned} 3x + 2y + z &= 8 \\ 2x - 3y + 2z &= -16 \\ x + 4y - z &= 20 \end{aligned}$$

6. $$\begin{aligned} -3x + y - z &= -10 \\ -4x + 2y + 3z &= -1 \\ 2x + 3y - 2z &= -5 \end{aligned}$$

7. $$\begin{aligned} x + 2y + z &= 4 \\ 2x + y - z &= -1 \\ x - y - z &= -2 \end{aligned}$$

8. $$\begin{aligned} x - 2y + 5z &= -7 \\ -2x - 3y + 4z &= -14 \\ -3x + 5y - z &= -7 \end{aligned}$$

9. $$\begin{aligned} -x + 2y + 6z &= 2 \\ 3x + 2y + 6z &= 6 \\ x + 4y - 3z &= 1 \end{aligned}$$

10. $$\begin{aligned} 2x + y + 2z &= 1 \\ x + 2y + z &= 2 \\ x - y - z &= 0 \end{aligned}$$

11.
$$\begin{aligned} 2x + 5y + 2z &= 0 \\ 4x - 7y - 3z &= 1 \\ 3x - 8y - 2z &= -6 \end{aligned}$$

12.
$$\begin{aligned} 5x - 2y + 3z &= -9 \\ 4x + 3y + 5z &= 4 \\ 2x + 4y - 2z &= 14 \end{aligned}$$

13.
$$\begin{aligned} x + 2y + 3z &= 1 \\ -x - y + 3z &= 2 \\ -6x + y + z &= -2 \end{aligned}$$

14.
$$\begin{aligned} x + y - z &= -2 \\ 2x - y + z &= -5 \\ -x + 2y - 3z &= -4 \end{aligned}$$

Solve each system of equations. See Example 2.

15.
$$\begin{aligned} 2x - 3y + 2z &= -1 \\ x + 2y + z &= 17 \\ 2y - z &= 7 \end{aligned}$$

16.
$$\begin{aligned} 2x - y + 3z &= 6 \\ x + 2y - z &= 8 \\ 2y + z &= 1 \end{aligned}$$

17.
$$\begin{aligned} 4x + 2y - 3z &= 6 \\ x - 4y + z &= -4 \\ -x + 2z &= 2 \end{aligned}$$

18.
$$\begin{aligned} 2x + 3y - 4z &= 4 \\ x - 6y + z &= -16 \\ -x + 3z &= 8 \end{aligned}$$

19.
$$\begin{aligned} -5x + 2y + z &= 5 \\ -3x - 2y - z &= 3 \\ -x + 6y &= 1 \end{aligned}$$

20.
$$\begin{aligned} x + y - z &= 0 \\ 2y - z &= 1 \\ 2x + 3y - 4z &= -4 \end{aligned}$$

21.
$$\begin{aligned} 2x + y &= 6 \\ 3y - 2z &= -4 \\ 3x - 5z &= -7 \end{aligned}$$

22.
$$\begin{aligned} 4x - 8y &= -7 \\ 4y + z &= 7 \\ -8x + z &= -4 \end{aligned}$$

23. Using your immediate surroundings, give an example of three planes that

(a) intersect in a single point;

(b) do not intersect;

(c) intersect in infinitely many points.

24. Suppose that a system has infinitely many ordered triple solutions of the form (x, y, z) such that

$$x + y + 2z = 1.$$

Give three specific ordered triples that are solutions of the system.

Solve each system of equations. If the system is inconsistent or has dependent equations, say so. See Examples 1–5.

25. $$\begin{aligned} 2x + 2y - 6z &= 5 \\ -3x + y - z &= -2 \\ -x - y + 3z &= 4 \end{aligned}$$

26. $$\begin{aligned} -2x + 5y + z &= -3 \\ 5x + 14y - z &= -11 \\ 7x + 9y - 2z &= -5 \end{aligned}$$

27. $$\begin{aligned} -5x + 5y - 20z &= -40 \\ x - y + 4z &= 8 \\ 3x - 3y + 12z &= 24 \end{aligned}$$

28. $$\begin{aligned} x + 4y - z &= 3 \\ -2x - 8y + 2z &= -6 \\ 3x + 12y - 3z &= 9 \end{aligned}$$

29. $$\begin{aligned} 2x + y - z &= 6 \\ 4x + 2y - 2z &= 12 \\ -x - \frac{1}{2}y + \frac{1}{2}z &= -3 \end{aligned}$$

30. $$\begin{aligned} 2x - 8y + 2z &= -10 \\ -x + 4y - z &= 5 \\ \frac{1}{8}x - \frac{1}{2}y + \frac{1}{8}z &= -\frac{5}{8} \end{aligned}$$

31. $$\begin{aligned} x + y - 2z &= 0 \\ 3x - y + z &= 0 \\ 4x + 2y - z &= 0 \end{aligned}$$

32. $$\begin{aligned} 2x + 3y - z &= 0 \\ x - 4y + 2z &= 0 \\ 3x - 5y - z &= 0 \end{aligned}$$

33. $$\begin{aligned} x - 2y + \frac{1}{3}z &= 4 \\ 3x - 6y + z &= 12 \\ -6x + 12y - 2z &= -3 \end{aligned}$$

34. $$\begin{aligned} 4x + y - 2z &= 3 \\ x + \frac{1}{4}y - \frac{1}{2}z &= \frac{3}{4} \\ 2x + \frac{1}{2}y - z &= 1 \end{aligned}$$

35.
$$\begin{aligned} x + 5y - 2z &= -1 \\ -2x + 8y + z &= -4 \\ 3x - y + 5z &= 19 \end{aligned}$$

36.
$$\begin{aligned} x + 3y + z &= 2 \\ 4x + y + 2z &= -4 \\ 5x + 2y + 3z &= -2 \end{aligned}$$

Relating Concepts (Exercises 37–44) For Individual or Group Work

Suppose that on a distant planet a function of the form

$$f(x) = ax^2 + bx + c \quad (a \neq 0)$$

describes the height in feet of a projectile x seconds after it has been projected upward. ***Work Exercises 37–44 in order,*** *to see how this can be related to a system of three equations in three variables a, b, and c.*

37. After 1 sec, the height of a certain projectile is 128 ft. Thus, $f(1) = 128$. Use this information to write one equation in the variables a, b, and c. (*Hint:* Substitute 1 for x and 128 for $f(x)$.)

38. After 1.5 sec, the height is 140 ft. Write a second equation in a, b, and c.

39. After 3 sec, the height is 80 ft. Write a third equation in a, b, and c.

40. Write a system of three equations in a, b, and c, based on your answers in Exercises 37–39. Solve the system.

41. What is the function f for this particular projectile?

42. In the function f written in Exercise 41, the ________ of the projectile is a function of the ________ elapsed after it was projected.

43. What was the initial height of the projectile? (*Hint:* Find $f(0)$.)

44. The projectile reaches its maximum height in 1.625 sec. Find its maximum height.

14.5 Applications of Systems of Linear Equations

OBJECTIVES

1. Solve geometry problems using two variables.
2. Solve money problems using two variables.
3. Solve mixture problems using two variables.
4. Solve distance-rate-time problems using two variables.
5. Solve problems with three variables using a system of three equations.

Many applied problems involve more than one unknown quantity. Although some problems with two unknowns can be solved using just one variable (as in **Chapter 2**), an alternative method of solution uses two variables. To solve a problem in this way using two unknowns, we must write two equations that relate the unknown quantities. The system formed by the pair of equations can then be solved using the methods of this chapter.

Problems that can be solved by writing a system of equations have been of interest historically. The following problem, which is given in the exercises for this section, first appeared in a Hindu work that dates back to about A.D. 850.

> *The mixed price of 9 citrons [a lemonlike fruit shown in the photo] and 7 fragrant wood apples is 107; again, the mixed price of 7 citrons and 9 fragrant wood apples is 101. O you arithmetician, tell me quickly the price of a citron and the price of a wood apple here, having distinctly separated those prices well.*

The following steps, based on the six-step problem-solving method first introduced in **Section 2.3**, give a strategy for solving applied problems using more than one variable.

Solving an Applied Problem by Writing a System of Equations

Step 1 **Read** the problem, several times if necessary, until you understand what is given and what is to be found.

Step 2 **Assign variables** to represent the unknown values, using diagrams or tables as needed. *Write down* what each variable represents.

Step 3 **Write a system of equations** that relates the unknowns.

Step 4 **Solve** the system of equations.

Step 5 **State the answer** to the problem. Does it seem reasonable?

Step 6 **Check** the answer in the words of the original problem.

OBJECTIVE 1 Solve geometry problems using two variables. Problems about the perimeter of a geometric figure often involve two unknowns and can be solved using systems of equations.

1 Solve the problem.

The length of the foundation of a rectangular house is to be 6 m more than its width. Find the length and width of the house if the perimeter must be 48 m.

Example 1 Finding the Dimensions of a Soccer Field

Unlike football, where the dimensions of a playing field cannot vary, a rectangular soccer field may have a width between 50 and 100 yd and a length between 100 and 130 yd. Suppose that one particular field has a perimeter of 320 yd. Its length measures 40 yd more than its width. What are the dimensions of this field? (*Source: Microsoft Encarta Encyclopedia.*)

Step 1 **Read** the problem again. We must find the dimensions of the field.

Step 2 **Assign variables.** Let $L =$ the length and $W =$ the width. Figure 7 shows a soccer field with these variables as labels.

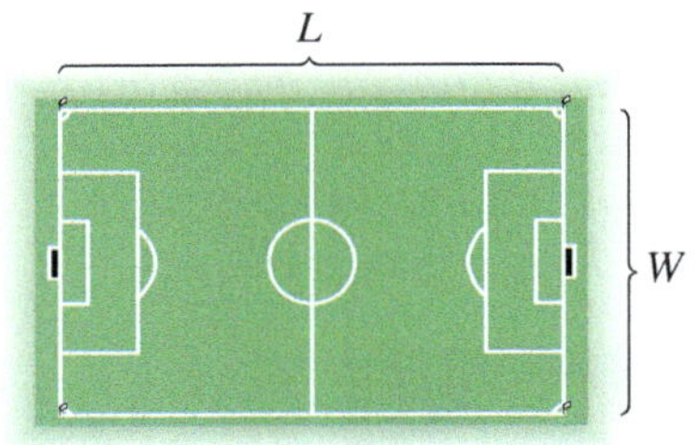

Figure 7

Step 3 **Write a system of equations.** Because the perimeter is 320 yd, we find one equation by using the perimeter formula:

$$2L + 2W = 320.$$

Because the length is 40 yd more than the width, we have

$$L = W + 40.$$

The system is, therefore,

$$\begin{aligned} 2L + 2W &= 320 && (1) \\ L &= W + 40. && (2) \end{aligned}$$

Step 4 **Solve** the system of equations. Since equation (2) is solved for L, we can use the substitution method. We substitute $W + 40$ for L in equation (1), and solve for W.

$$\begin{aligned} 2L + 2W &= 320 && (1) \\ 2(W + 40) + 2W &= 320 && \text{Let } L = W + 40. \\ 2W + 80 + 2W &= 320 && \text{Distributive property} \\ 4W + 80 &= 320 && \text{Combine like terms.} \\ 4W &= 240 && \text{Subtract 80.} \\ W &= 60 && \text{Divide by 4.} \end{aligned}$$

Be sure to use parentheses around $W + 40$.

Let $W = 60$ in the equation $L = W + 40$ to find L.

$$L = 60 + 40 = 100$$

Step 5 **State the answer.** The length is 100 yd, and the width is 60 yd. Both dimensions are within the ranges given in the problem.

Step 6 **Check.** The perimeter is $2(100) + 2(60) = 320$ yd, and the length, 100 yd, is indeed 40 yd more than the width, since $100 - 40 = 60$. The answer is correct.

◀ *Work Problem* **1** *at the Side.*

Answer

1. length: 15 m; width: 9 m

OBJECTIVE 2 Solve money problems using two variables.

Example 2 Solving a Problem about Ticket Prices

For the 2005–2006 National Hockey League and National Basketball Association seasons, two hockey tickets and one basketball ticket purchased at their average prices would have cost \$128.30. One hockey ticket and two basketball tickets would have cost \$133.03. What were the average ticket prices for the two sports? (*Source:* Team Marketing Report, Chicago.)

Step 1 **Read** the problem again. There are two unknowns.

Step 2 **Assign variables.**

Let h = the average price for a hockey ticket

and b = the average price for a basketball ticket.

Step 3 **Write a system of equations.** Because two hockey tickets and one basketball ticket cost a total of \$128.30, one equation for the system is

$$2h + b = 128.30.$$

By similar reasoning, the second equation is

$$h + 2b = 133.03.$$

Therefore, the system is

$$\begin{aligned} 2h + b &= 128.30 && (1) \\ h + 2b &= 133.03. && (2) \end{aligned}$$

Step 4 **Solve** the system of equations. To eliminate h, multiply equation (2) by -2 and add.

$$\begin{aligned} 2h + b &= 128.30 && (1) \\ -2h - 4b &= -266.06 && \text{Multiply each side of (2) by } -2. \\ \hline -3b &= -137.76 && \text{Add.} \\ b &= 45.92 && \text{Divide by } -3. \end{aligned}$$

To find the value of h, let $b = 45.92$ in equation (2).

$$\begin{aligned} h + 2b &= 133.03 && (2) \\ h + 2(45.92) &= 133.03 && \text{Let } b = 45.92. \\ h + 91.84 &= 133.03 && \text{Multiply.} \\ h &= 41.19 && \text{Subtract } 91.84. \end{aligned}$$

Step 5 **State the answer.** The average price for one basketball ticket was \$45.92. For one hockey ticket, the average price was \$41.19.

Step 6 **Check** that these values satisfy the conditions stated in the problem.

Work Problem 2 at the Side. ▶

2 Solve the problem.

For recent Major League Baseball and National Football League seasons, based on average ticket prices, three baseball tickets and two football tickets would have cost \$181.41, while two baseball tickets and one football ticket would have cost \$101.29. What were the average ticket prices for the two sports? (*Source:* Team Marketing Report, Chicago.)

ANSWER

2. baseball: \$21.17; football: \$58.95

OBJECTIVE 3 Solve mixture problems using two variables. We solved mixture problems in **Section 2.3** using one variable. For many mixture problems we can use more than one variable and a system of equations.

Example 3 Solving a Mixture Problem

How many ounces each of 5% hydrochloric acid and 20% hydrochloric acid must be combined to get 10 oz of solution that is 12.5% hydrochloric acid?

Step 1 **Read** the problem. Two solutions of different strengths are being mixed together to get a specific amount of a solution with an "in-between" strength.

Step 2 **Assign variables.**

Let x = the number of ounces of 5% solution

and y = the number of ounces of 20% solution.

Use a table to summarize the information from the problem. We multiply the amount of each solution (given in the first column) by its concentration of acid (given in the second column) to get the amount of acid in that solution (given in the third column).

Ounces of Solution	Percent (as a decimal)	Ounces of Pure Acid
x	$5\% = 0.05$	$0.05x$
y	$20\% = 0.20$	$0.20y$
10	$12.5\% = 0.125$	$(0.125)10$

Gives equation (1) (first column) — Gives equation (2) (third column)

Figure 8 also illustrates what is happening in the problem.

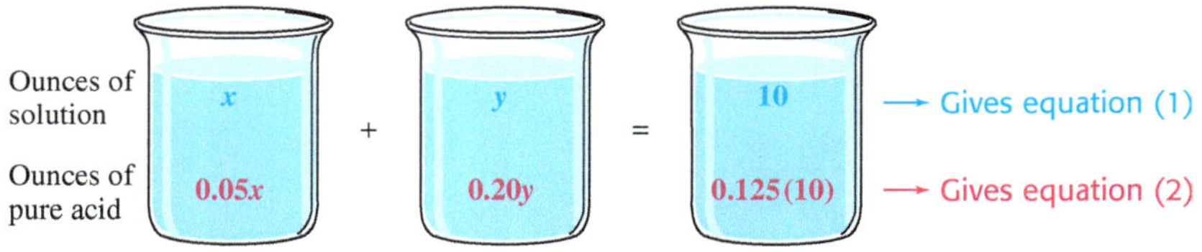

Figure 8

Step 3 **Write a system of equations.** When the x ounces of 5% solution and the y ounces of 20% solution are combined, the total number of ounces is 10, so

$$x + y = 10. \qquad (1)$$

The ounces of acid in the 5% solution ($0.05x$) plus the ounces of acid in the 20% solution ($0.20y$) should equal the total ounces of acid in the mixture, which is $(0.125)10$, or 1.25. That is,

$$0.05x + 0.20y = 1.25. \qquad (2)$$

Notice that these equations can be quickly determined by reading down in the table or across in Figure 8.

Continued on Next Page

Step 4 **Solve** the system of equations (1) and (2). Eliminate x by first multiplying equation (2) by 100 to clear it of decimals and then multiplying equation (1) by -5.

$$\begin{aligned} 5x + 20y &= 125 && \text{Multiply each side of (2) by 100.} \\ -5x - 5y &= -50 && \text{Multiply each side of (1) by } -5. \\ \hline 15y &= 75 && \text{Add.} \\ y &= 5 && \text{Divide by 15.} \end{aligned}$$

Because $y = 5$ and $x + y = 10$, the value of x is also 5.

Step 5 **State the answer.** The desired mixture will require 5 oz of the 5% solution and 5 oz of the 20% solution.

Step 6 **Check.**

Total amount of solution: $x + y = 5 \text{ oz} + 5 \text{ oz}$
$= 10$ oz, as required.

Total amount of acid: 5% of 5 oz + 20% of 5 oz
$= 0.05(5) + 0.20(5)$
$= 1.25$ oz

Percent of acid in solution:

$$\frac{\text{Total acid} \rightarrow 1.25}{\text{Total solution} \rightarrow 10} = 0.125, \quad \text{or} \quad 12.5\%, \quad \text{as required.}$$

Work Problem 3 *at the Side.* ▶

OBJECTIVE 4 Solve distance-rate-time problems using two variables. Motion problems require the distance formula, $d = rt$, where d is distance, r is rate (or speed), and t is time. These applications often lead to systems of equations.

Example 4 Solving a Motion Problem

A car travels 250 km in the same time that a truck travels 225 km. If the speed of the car is 8 km per hr faster than the speed of the truck, find both speeds.

Step 1 **Read** the problem again. Given the distances traveled, we need to find the speed of each vehicle.

Step 2 **Assign variables.**

Let x = the speed of the car
and y = the speed of the truck.

Fill in the given information for each vehicle (distances) and the assigned variables for the unknown speeds (rates) in a table.

	d	*r*	*t*
Car	250	x	$\frac{250}{x}$
Truck	225	y	$\frac{225}{y}$

The times must be equal.

To get the expressions for time, solve the distance formula, $d = rt$, for t. Since $\frac{d}{r} = t$, the two times can be written as $\frac{250}{x}$ and $\frac{225}{y}$.

Continued on Next Page

3 Solve each problem.

(a) A grocer has some \$4 per lb coffee and some \$8 per lb coffee, which he will mix to make 50 lb of \$5.60 per lb coffee. How many pounds of each should be used?

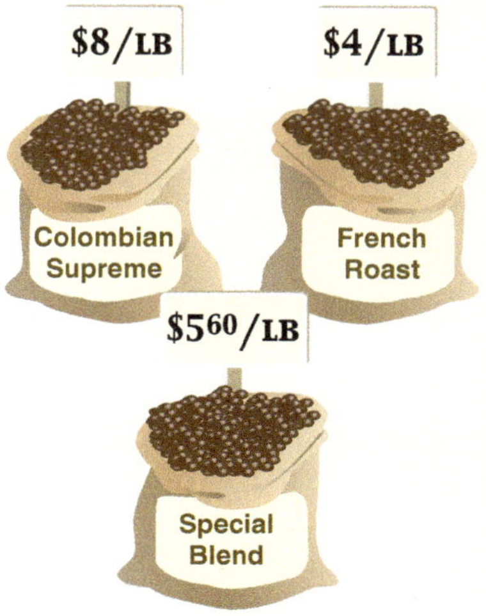

(b) Some 40% ethyl alcohol solution is to be mixed with some 80% solution to get 200 L of a 50% solution. How many liters of each should be used?

Answers

3. (a) 30 lb of \$4; 20 lb of \$8
(b) 150 L of 40%; 50 L of 80%

4 Solve the problem.

A train travels 600 mi in the same time that a truck travels 520 mi. Find the speed of each vehicle if the train's average speed is 8 mph faster than the truck's.

Step 3 **Write a system of equations.** The problem states that the car travels 8 km per hr faster than the truck. Since the two speeds are x and y,

$$x = y + 8. \qquad (1)$$

Both vehicles travel for the same time, so from the table,

$$\text{Time for car} \rightarrow \frac{250}{x} = \frac{225}{y}. \leftarrow \text{Time for truck}$$

This is not a linear equation. Multiplying each side by xy gives

$$xy \cdot \frac{250}{x} = \frac{225}{y} \cdot xy \qquad \text{Multiply by the LCD, } xy.$$

$$\frac{250xy}{x} = \frac{225xy}{y}$$

$$250y = 225x, \qquad (2)$$

which is linear. The system is

$$x = y + 8 \qquad (1)$$
$$250y = 225x. \qquad (2)$$

Step 4 **Solve** the system of equations by substitution. Replace x with $y + 8$ in equation (2).

$$250y = 225x \qquad (2)$$

$$250y = 225(y + 8) \qquad \text{Let } x = y + 8.$$

Be sure to use parentheses around $y + 8$.

$$250y = 225y + 1800 \qquad \text{Distributive property}$$

$$25y = 1800 \qquad \text{Subtract } 225y.$$

$$y = 72 \qquad \text{Divide by 25.}$$

Because $x = y + 8$, the value of x is

$$72 + 8 = 80.$$

Step 5 **State the answer.** The car's speed is 80 km per hr, and the truck's speed is 72 km per hr.

Step 6 **Check.** This is especially important since one of the equations had variable denominators.

$$\text{Car: } t = \frac{d}{r} = \frac{250}{80} = 3.125$$

$$\text{Truck: } t = \frac{d}{r} = \frac{225}{72} = 3.125$$

Times are equal.

Since $80 - 72 = 8$, the conditions of the problem are satisfied.

Work Problem **4** *at the Side.*

OBJECTIVE 5 Solve problems with three variables using a system of three equations.

Problem-Solving Hint

If an application requires finding *three* unknown quantities, we can use a system of *three* equations to solve it. We extend the method used for two unknowns.

Answer

4. train: 60 mph; truck: 52 mph

Example 5 **Solving a Problem Involving Prices**

At Panera Bread, a loaf of honey wheat bread costs \$2.95, a loaf of sunflower bread costs \$2.99, and a loaf of French bread costs \$5.79. On a recent day, three times as many loaves of honey wheat bread were sold as sunflower bread. The number of loaves of French bread sold was 5 less than the number of loaves of honey wheat bread sold. Total receipts for these breads were \$87.89. How many loaves of each type of bread were sold? (*Source:* Panera Bread menu.)

Step 1 **Read** the problem again. There are three unknowns in this problem.

Step 2 **Assign variables** to represent the three unknowns.

Let x = the number of loaves of honey wheat bread,
y = the number of loaves of sunflower bread,
and z = the number of loaves of French bread.

Step 3 **Write a system of three equations.** Since three times as many loaves of honey wheat bread were sold as sunflower bread,

$x = 3y$, or $x - 3y = 0$. Subtract $3y$. (1)

Also,

Number of loaves of French	equals	5 less than the number of loaves of honey wheat.
↓	↓	↓
z	$=$	$x - 5$,

$-x + z = -5$ Subtract x.
$x - z = 5$. Multiply by -1. (2)

Multiplying the cost of a loaf of each kind of bread by the number of loaves of that kind sold and adding gives the total receipts.

$$2.95x + 2.99y + 5.79z = 87.89$$

Multiply each side of this equation by 100 to clear it of decimals.

$$295x + 299y + 579z = 8789 \quad (3)$$

Step 4 **Solve** the system of three equations,

$$x - 3y = 0 \quad (1)$$
$$x - z = 5 \quad (2)$$
$$295x + 299y + 579z = 8789, \quad (3)$$

using the method shown in **Section 5.2.**

Work Problem **5** *at the Side.* ▶

Thus, we find that $x = 12$, $y = 4$, and $z = 7$.

Step 5 **State the answer.** The solution set is $\{(12, 4, 7)\}$, meaning that 12 loaves of honey wheat bread, 4 loaves of sunflower bread, and 7 loaves of French bread were sold.

Step 6 **Check.** Since $12 = 3 \cdot 4$, the number of loaves of honey wheat bread is three times the number of loaves of sunflower bread. Also, $12 - 7 = 5$, so the number of loaves of French bread is 5 less than the number of loaves of honey wheat bread. Multiply the appropriate cost per loaf by the number of loaves sold and add the results to check that total receipts were \$87.89.

Work Problem **6** *at the Side.* ▶

5 Solve the system of equations from Example 5.

$$x - 3y = 0 \quad (1)$$
$$x - z = 5 \quad (2)$$
$$295x + 299y + 579z = 8789 \quad (3)$$

6 Solve the problem.

A department store display features three kinds of perfume: Felice, Vivid, and Joy. There are 10 more bottles of Felice than Vivid, and 3 fewer bottles of Joy than Vivid. Each bottle of Felice costs \$8, Vivid costs \$15, and Joy costs \$32. The total value of all the perfume is \$589. How many bottles of each are there?

ANSWERS

5. $\{(12, 4, 7)\}$
6. 21 bottles of Felice; 11 of Vivid; 8 of Joy

7 Solve the problem.

A paper mill makes newsprint, bond, and copy machine paper. Each ton of newsprint requires 3 tons of recycled paper and 1 ton of wood pulp. Each ton of bond requires 2 tons of recycled paper, 4 tons of wood pulp, and 3 tons of rags. A ton of copy machine paper requires 2 tons of recycled paper, 3 tons of wood pulp, and 2 tons of rags. The mill has 4200 tons of recycled paper, 5800 tons of wood pulp, and 3900 tons of rags. How much of each kind of paper can be made from these supplies?

Example 6 Solving a Business Production Problem

A company produces three color television sets, models X, Y, and Z. Each model X set requires 2 hr of electronics work, 2 hr of assembly time, and 1 hr of finishing time. Each model Y requires 1, 3, and 1 hr of electronics, assembly, and finishing time, respectively. Each model Z requires 3, 2, and 2 hr of the same work, respectively. There are 100 hr available for electronics, 100 hr available for assembly, and 65 hr available for finishing per week. How many of each model should be produced each week if all available time must be used?

Step 1 **Read** the problem again. There are three unknowns.

Step 2 **Assign variables.**

Let x = the number of model X produced per week,

y = the number of model Y produced per week,

and z = the number of model Z produced per week.

We organize the information in a table.

	Each Model X	Each Model Y	Each Model Z	Totals	
Hours of Electronics Work	2	1	3	100	→ Gives equation (1)
Hours of Assembly Time	2	3	2	100	→ Gives equation (2)
Hours of Finishing Time	1	1	2	65	→ Gives equation (3)

Step 3 **Write a system of three equations.** The x model X sets require $2x$ hours of electronics, the y model Y sets require $1y$ (or y) hours of electronics, and the z model Z sets require $3z$ hours of electronics. Since 100 hr are available for electronics,

$$2x + y + 3z = 100. \qquad (1)$$

Similarly, from the fact that 100 hr are available for assembly,

$$2x + 3y + 2z = 100, \qquad (2)$$

and the fact that 65 hr are available for finishing leads to the equation

$$x + y + 2z = 65. \qquad (3)$$

Notice that by reading *across* the table, we can quickly determine the coefficients and constants in the equations of the system.

Step 4 **Solve** the system of equations (1), (2), and (3), namely,

$$\begin{aligned} 2x + y + 3z &= 100 \qquad (1) \\ 2x + 3y + 2z &= 100 \qquad (2) \\ x + y + 2z &= 65 \qquad (3) \end{aligned}$$

to find $x = 15$, $y = 10$, and $z = 20$.

Step 5 **State the answer.** The company should produce 15 model X, 10 model Y, and 20 model Z sets per week.

Step 6 **Check** that these values satisfy the conditions of the problem.

◀ Work Problem 7 at the Side.

ANSWER

7. 400 tons of newsprint; 900 tons of bond; 600 tons of copy machine paper

14.5 Exercises

FOR EXTRA HELP MyMathLab MathXL PRACTICE WATCH DOWNLOAD READ REVIEW

Solve each problem. See Example 1.

1. During the 2007 Major League Baseball regular season, the Cleveland Indians played 162 games. They won 30 more games than they lost. What was their win-loss record that year?

2. Refer to Exercise 1. During the same 162-game season, the Chicago White Sox lost 18 more games than they won. What was the team's win-loss record?

2007 MLB Final Standings
American League Central

Team	W	L
Cleveland	___	___
Detroit	88	74
Minnesota	79	83
Chicago	___	___
Kansas City	69	93

Source: www.mlb.com

3. Venus and Serena measured a tennis court and found that it was 42 ft longer than it was wide and had a perimeter of 228 ft. What were the length and the width of the tennis court?

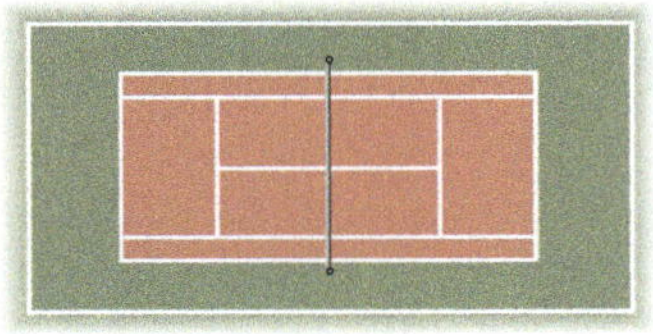

4. Wilt and Oscar found that the width of their basketball court was 44 ft less than the length. If the perimeter was 288 ft, what were the length and the width of their court?

5. The two biggest Fortune 500 companies in 2005 were Wal-Mart and ExxonMobil. ExxonMobil's revenue was $24 billion more than that of Wal-Mart. Total revenue for the two companies was $656 billion. What was the revenue for each company? (*Source:* Fortune 500.)

6. In 2007, U.S. exports to Canada were $112 billion more than exports to Mexico. Together, exports to these two countries totaled $386 billion. How much were exports to each country? (*Source:* U.S. Census Bureau.)

In Exercises 7 and 8, find the measures of the angles marked x and y. Remember that (1) *the sum of the measures of the angles of a triangle is* 180°, (2) *supplementary angles have a sum of* 180°, *and* (3) *vertical angles have equal measures.*

7.

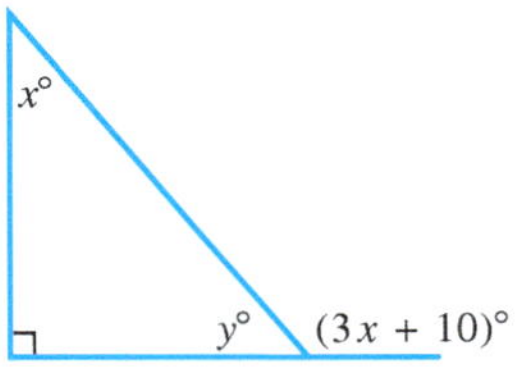

8.

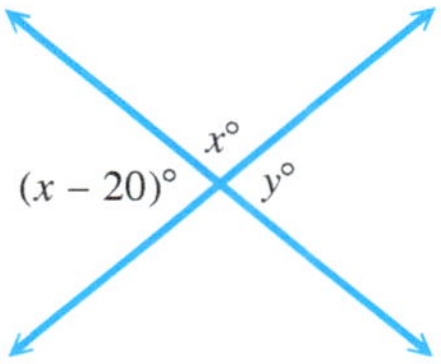

The Fan Cost Index (FCI) represents the cost of four average-price tickets, four small soft drinks, two small beers, four hot dogs, parking for one car, two game programs, and two souvenir caps to a sporting event. (*Source:* www.teammarketing.com)

Use the concept of FCI in Exercises 9 and 10. See Example 2.

9. For the 2005–2006 season, the FCI prices for the National Hockey League (NHL) and the National Basketball Association (NBA) totaled \$514.69. The hockey FCI was \$20.05 less than that of basketball. What were the FCIs for these sports?

10. In 2005, the FCI prices for Major League Baseball (MLB) and the National Football League (NFL) totaled \$501.01. The football FCI was \$158.63 more than that of baseball. What were the FCIs for these sports?

Solve each problem. See Example 2.

11. Andrew McGinnis works at Arby's. During one particular day he sold 15 Junior Roast Beef sandwiches and 10 Big Montana sandwiches, totaling \$75.25. Another day he sold 30 Junior Roast Beef sandwiches and 5 Big Montana sandwiches, totaling \$84.65. How much did each type of sandwich cost? (*Source:* Arby's menu.)

12. London and New York are among the most expensive cities worldwide for business travelers. On the basis of average costs per day for each city (which include the costs of business-class lodging and three meals), 2 days in London and 3 days in New York costs \$2099, while 4 days in London and 2 days in New York costs \$2586. What is the average cost per day for each city? (*Source:* Runzheimer International.)

The formulas $p = br$ *(percentage = base × rate) and* $I = prt$ *(simple interest = principal × rate × time) are used in the applications in Exercises 17–24. To prepare to use these formulas, answer the questions in Exercises 13 and 14.*

13. If a container of liquid contains 60 oz of solution, what is the number of ounces of pure acid if the given solution contains the following acid concentrations?

(a) 10% **(b)** 25% **(c)** 40% **(d)** 50%

14. If \$5000 is invested in an account paying simple annual interest, how much interest will be earned during the first year at the following rates?

(a) 2% **(b)** 3% **(c)** 4% **(d)** 3.5%

15. If a pound of turkey costs \$1.29, how much will x pounds cost?

16. If a ticket to the movie *The Final Season* costs \$8.50 and y tickets are sold, how much is collected from the sale?

Solve each problem. See Example 3.

17. How many gallons each of 25% alcohol and 35% alcohol should be mixed to get 20 gal of 32% alcohol?

Gallons of Solution	Percent (as a Decimal)	Gallons of Pure Alcohol
x	25% = 0.25	
y	35% = 0.35	
20	32% =	

18. How many liters each of 15% acid and 33% acid should be mixed to get 120 L of 21% acid?

Liters of Solution	Percent (as a Decimal)	Liters of Pure Acid
x	15% = 0.15	
y	33% =	
120	21% =	

19. Pure acid is to be added to a 10% acid solution to obtain 54 L of a 20% acid solution. What amounts of each should be used? (*Hint:* Pure acid is 100% acid.)

20. A truck radiator holds 36 L of fluid. How much pure antifreeze must be added to a mixture that is 4% antifreeze to fill the radiator with a mixture that is 20% antifreeze?

21. A party mix is made by adding nuts that sell for \$2.50 per kg to a cereal mixture that sells for \$1 per kg. How much of each should be added to get 30 kg of a mix that will sell for \$1.70 per kg?

	Number of Kilograms	Price per Kilogram	Value
Nuts	x	2.50	
Cereal	y	1.00	
Mixture		1.70	

22. A popular fruit drink is made by mixing fruit juices. Such a drink with 50% juice is to be mixed with another drink that is 30% juice to get 200 L of a drink that is 45% juice. How much of each should be used?

	Liters of Drink	Percent (as a Decimal)	Liters of Pure Juice
50% Juice	x	0.50	
30% Juice	y	0.30	
Mixture		0.45	

23. A total of \$3000 is invested, part at 2% simple interest and part at 4%. If the total annual return from the two investments is \$100, how much is invested at each rate?

Principal	Rate (as a Decimal)	Interest
x	0.02	$0.02x$
y	0.04	$0.04y$
3000	XXXXX	100

24. An investor must invest a total of \$15,000 in two accounts, one paying 4% annual simple interest, and the other 3%. If he wants to earn \$550 annual interest, how much should he invest at each rate?

Principal	Rate (as a Decimal)	Interest
x	0.04	
y	0.03	
15,000	XXXXX	

The formula $d = rt$ (distance = rate × time) is used in the applications in Exercises 27–30. To prepare to use this formula, answer the questions in Exercises 25 and 26.

25. If the speed of a boat in still water is 10 mph, and the speed of the current of a river is x mph, what is the speed of the boat

(a) going upstream (that is, against the current, which slows the boat down);

(b) going downstream (that is, with the current, which speeds the boat up)?

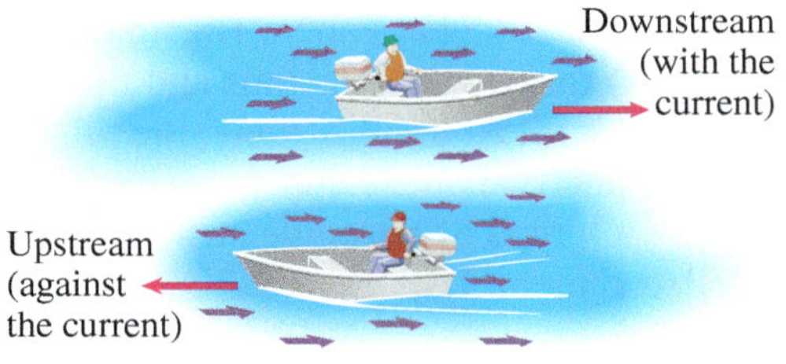

26. If the speed of a killer whale is 25 mph and the whale swims for y hours, how many miles does the whale travel?

Solve each problem. See Example 4.

27. A motor scooter travels 20 mi in the same time that a bicycle covers 8 mi. If the speed of the scooter is 5 mph more than twice the speed of the bicycle, find both speeds.

28. A train travels 150 km in the same time that a plane covers 400 km. If the speed of the plane is 20 km per hr less than 3 times the speed of the train, find both speeds.

29. In his motorboat, Bill Ruhberg travels upstream at top speed to his favorite fishing spot, a distance of 36 mi, in 2 hr. Returning, he finds that the trip downstream, still at top speed, takes only 1.5 hr. Find the speed of Bill's boat and the speed of the current. Let $x =$ the speed of the boat in still water and $y =$ the speed of the current.

	r	t	d
Upstream	$x - y$	2	
Downstream	$x + y$		

30. Traveling for 3 hr into a steady headwind, a plane flies 1650 mi. The pilot determines that flying *with* the same wind for 2 hr, he could make a trip of 1300 mi. Find the speed of the plane and the speed of the wind.

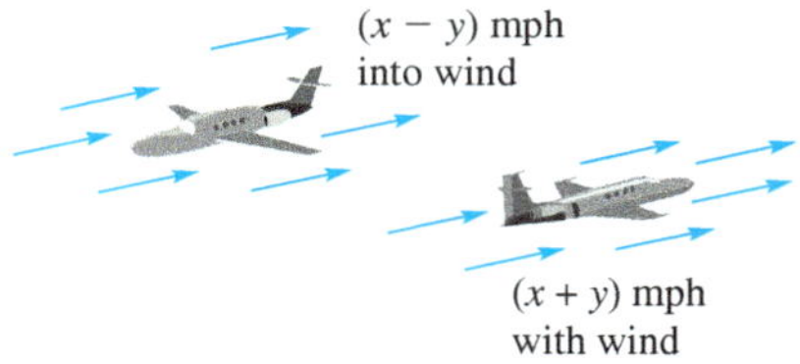

Solve each problem. See Examples 1–6.

31. (See the Chapter Introduction.) How many pounds of candy that sells for \$0.75 per lb must be mixed with candy that sells for \$1.25 per lb to obtain 9 lb of a mixture that should sell for \$0.96 per lb?

32. The average cost of tuition and fees at a 4-yr public college or university during the 2007–2008 school year was \$3824 more than at a 2-yr community college. Suppose a student plans to attend a local community college for 2 yr and then transfer to a public 4-yr university for 2 yr. The student expected to pay \$17,092 tuition and fees for the 4 yr, assuming the 2007–2008 rates were locked in. What was the cost of tuition and fees during the 2007–2008 year at each type of school? (*Source:* The College Board.)

33. Tickets to a production of *Cats* at Shelton State Community College cost \$5 for general admission or \$4 with a student ID. If 184 people paid to see a performance and \$812 was collected, how many of each type of ticket were sold?

34. At a business meeting at Panera Bread, the bill (without tax) for two cappuccinos and three house lattes was \$14.55. At another table, the bill for one cappuccino and two house lattes was \$8.77. How much did each type of beverage cost? (*Source:* Panera Bread menu.)

35. The mixed price of 9 citrons and 7 fragrant wood apples is 107; again, the mixed price of 7 citrons and 9 fragrant wood apples is 101. O you arithmetician, tell me quickly the price of a citron and the price of a wood apple here, having distinctly separated those prices well. (*Source:* Hindu work, A.D. 850.) (*Hint:* "Mixed price" refers to the price of a mixture of the two fruits.)

36. Braving blizzard conditions on the planet Hoth, Luke Skywalker sets out at top speed in his snow speeder for a rebel base 4800 mi away. He travels into a steady headwind and makes the trip in 3 hr. Returning, he finds that the trip back, still at top speed but now with a tailwind, takes only 2 hr. Find the top speed of Luke's snow speeder and the speed of the wind.

	r	*t*	*d*
Into Headwind			
With Tailwind			

Solve each problem involving three unknowns. See Examples 5 and 6. (In Exercises 37–40, remember that the sum of the measures of the angles of a triangle is 180°.)

37. In the figure, $z = x + 10$ and $x + y = 100$. Determine a third equation involving x, y, and z, and then find the measures of the three angles.

38. In the figure, x is 10 less than y and x is 20 less than z. Write a system of equations and find the measures of the three angles.

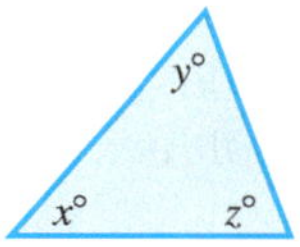

39. In a certain triangle, the measure of the second angle is 10° more than three times the first. The third angle measure is equal to the sum of the measures of the other two. Find the measures of the three angles.

40. The measure of the largest angle of a triangle is 12° less than the sum of the measures of the other two. The smallest angle measures 58° less than the largest. Find the measures of the angles.

41. The perimeter of a triangle is 70 cm. The longest side is 4 cm less than the sum of the other two sides. Twice the shortest side is 9 cm less than the longest side. Find the length of each side of the triangle.

42. The perimeter of a triangle is 56 in. The longest side measures 4 in. less than the sum of the other two sides. Three times the shortest side is 4 in. more than the longest side. Find the lengths of the three sides.

43. In a random sample of 100 Americans of voting age, 10 more Americans identify themselves as Independents than Republicans. Six fewer Americans identify themselves as Republicans than Democrats. Assuming that all of those sampled are Republican, Democrat, or Independent, how many of those in the sample identify themselves with each political affiliation? (*Source:* The Gallup Organization.)

44. In the 2004 Summer Olympics in Athens, Greece, the United States earned 6 more gold medals than bronze. The number of silver medals earned was 19 less than twice the number of bronze medals. The United Stated earned a total of 103 medals. How many of each kind of medal did the United States earn? (*Source: World Almanac and Book of Facts.*)

45. Tickets for one show on the Harlem Globetrotters' 2006 "Unstoppable" Tour cost \$14, \$20, or, for VIP seats, \$50. Five times as many \$14 tickets were sold as VIP tickets. The number of \$14 tickets was 15 more than the sum of the number of \$20 tickets and the number of VIP tickets. Sales of all three kinds of tickets totaled \$11,700. How many of each kind of ticket were sold? (*Source*: www.ticketmaster.com)

46. Three kinds of tickets are available for a *Third Day* concert: "up close," "in the middle," and "far out." "Up close" tickets cost \$10 more than "in the middle" tickets, while "in the middle" tickets cost \$10 more than "far out" tickets. Twice the cost of an "up close" ticket is \$20 more than 3 times the cost of a "far out" ticket. Find the price of each kind of ticket.

47. A wholesaler supplies college T-shirts to three college bookstores: A, B, and C. The wholesaler recently shipped a total of 800 T-shirts to the three bookstores. In order to meet student demand at the three colleges, twice as many T-shirts were shipped to bookstore B as to bookstore A, and the number shipped to bookstore C was 40 less than the sum of the numbers shipped to the other two bookstores. How many T-shirts were shipped to each bookstore?

48. An office supply store sells three models of computer desks: A, B, and C. In January, the store sold a total of 85 computer desks. The number of model B desks was five more than the number of model C desks, and the number of model A desks was four more than twice the number of model C desks. How many of each model did the store sell in January?

49. During the 2005–2006 National Hockey League regular season, the Calgary Flames played 82 games. Their wins and losses totaled 71. They tied 14 fewer games than they lost. How many wins, losses, and ties did they have?

50. (Refer to Exercise 49.) During the same 82-game season, the Minnesota Wild had a total of 44 losses and ties. They had two more wins than losses. How many wins, losses, and ties did they have?

2005-2006 NHL Final Standings

Team	W	L	T	Pts
Calgary	___	___	___	103
Colorado	43	30	9	95
Edmonton	41	28	13	95
Vancouver	42	32	8	92
Minnesota	___	___	___	84

Source: www.sportzdomain.com

14.6 Solving Systems of Linear Equations by Matrix Methods

OBJECTIVES

1. **Define a matrix.**
2. **Write the augmented matrix for a system.**
3. **Use row operations to solve a system with two equations.**
4. **Use row operations to solve a system with three equations.**
5. **Use row operations to solve special systems.**

OBJECTIVE 1 Define a matrix. An ordered array of numbers such as

$$\text{Rows}\quad \begin{bmatrix} 2 & 3 & 5 \\ 7 & 1 & 2 \end{bmatrix} \quad \text{Matrix}$$

(Columns)

is called a **matrix.** The numbers are called **elements** of the matrix. *Matrices* (the plural of *matrix*) are named according to the number of **rows** and **columns** they contain. The rows are read horizontally, and the columns are read vertically. For example, the first row in the preceding matrix is 2 3 5 and the first column is $\begin{matrix} 2 \\ 7 \end{matrix}$. This matrix is a 2×3 (read "two by three") matrix because it has 2 rows and 3 columns. The number of rows followed by the number of columns gives the **dimensions** of the matrix.

$$\begin{bmatrix} -1 & 0 \\ 1 & -2 \end{bmatrix} \quad 2 \times 2 \text{ matrix} \qquad \begin{bmatrix} 8 & -1 & -3 \\ 2 & 1 & 6 \\ 0 & 5 & -3 \\ 5 & 9 & 7 \end{bmatrix} \quad 4 \times 3 \text{ matrix}$$

A **square matrix** is one that has the same number of rows as columns. The 2×2 matrix is a square matrix.

Calculator Tip Figure 9 shows how a graphing calculator displays the preceding two matrices. Work with matrices is made much easier by using technology when available. Consult your owner's manual for details.

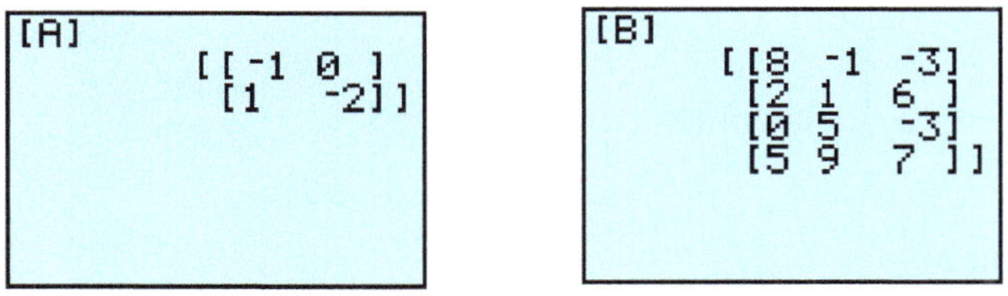

Figure 9

In this section, we discuss a method of solving linear systems that uses matrices. The advantage of this new method is that it can be done by a graphing calculator or a computer, allowing large systems of equations to be solved easily.

OBJECTIVE 2 Write the augmented matrix for a system. To solve a linear system using matrices, we begin by writing an *augmented matrix* for the system. An **augmented matrix** has a vertical bar that separates the columns of the matrix into two groups. For example, to solve the system

$$\begin{aligned} x - 3y &= 1 \\ 2x + y &= -5, \end{aligned}$$

we start with the augmented matrix

$$\left[\begin{array}{cc|c} 1 & -3 & 1 \\ 2 & 1 & -5 \end{array}\right]. \quad \text{Augmented matrix}$$

System of equations:

$$\begin{aligned} x - 3y &= 1 \\ 2x + y &= -5 \end{aligned}$$

Augmented matrix:

$$\left[\begin{array}{cc|c} 1 & -3 & 1 \\ 2 & 1 & -5 \end{array}\right]$$

Coefficients of the variables — The bar separates the coefficients from the constants. — Constants

Notice that we place the coefficients of the variables to the left of the bar, and the constants to the right. ***The matrix is just a shorthand way of writing the system of equations, so the rows of the augmented matrix can be treated the same as the equations of a system of equations.***

We know that exchanging the position of two equations in a system does not change the system. Also, multiplying any equation in a system by a nonzero number does not change the system. Comparable changes to the augmented matrix of a system of equations produce new matrices that correspond to systems with the same solutions as the original system.

The following **row operations** produce new matrices that lead to systems having the same solutions as the original system.

Matrix Row Operations

1. Any two rows of the matrix may be interchanged.
2. The numbers in any row may be multiplied by any nonzero real number.
3. Any row may be transformed by adding to the numbers of the row the product of a real number and the corresponding numbers of another row.

Examples of these row operations follow.

Row operation 1:

$$\begin{bmatrix} 2 & 3 & 9 \\ 4 & 8 & -3 \\ 1 & 0 & 7 \end{bmatrix} \text{ becomes } \begin{bmatrix} 1 & 0 & 7 \\ 4 & 8 & -3 \\ 2 & 3 & 9 \end{bmatrix}.$$

Interchange row 1 and row 3.

Row operation 2:

$$\begin{bmatrix} 2 & 3 & 9 \\ 4 & 8 & -3 \\ 1 & 0 & 7 \end{bmatrix} \text{ becomes } \begin{bmatrix} 6 & 9 & 27 \\ 4 & 8 & -3 \\ 1 & 0 & 7 \end{bmatrix}.$$

Multiply the numbers in row 1 by 3.

Row operation 3:

$$\begin{bmatrix} 2 & 3 & 9 \\ 4 & 8 & -3 \\ 1 & 0 & 7 \end{bmatrix} \text{ becomes } \begin{bmatrix} 0 & 3 & -5 \\ 4 & 8 & -3 \\ 1 & 0 & 7 \end{bmatrix}.$$

Multiply the numbers in row 3 by −2; add them to the corresponding numbers in row 1.

The third row operation corresponds to the way we eliminated a variable from a pair of equations in the previous sections.

OBJECTIVE 3 Use row operations to solve a system with two equations. Row operations can be used to rewrite a matrix. The goal is a matrix in the form

$$\left[\begin{array}{cc|c} 1 & a & b \\ 0 & 1 & c \end{array}\right] \quad \text{or} \quad \left[\begin{array}{ccc|c} 1 & a & b & c \\ 0 & 1 & d & e \\ 0 & 0 & 1 & f \end{array}\right]$$

for systems with two or three equations, respectively. Notice that there are 1s down the diagonal from upper left to lower right and 0s below the 1s. A matrix written this way is said to be in **row echelon form.** When these matrices are rewritten as systems of equations, the value of one variable is known, and the rest can be found by substitution. The following examples illustrate this method.

Example 1 **Using Row Operations to Solve a System with Two Variables**

Use row operations to solve the system.

$$\begin{aligned} x - 3y &= 1 \\ 2x + y &= -5 \end{aligned}$$

We start with the augmented matrix of the system.

$$\left[\begin{array}{cc|c} 1 & -3 & 1 \\ 2 & 1 & -5 \end{array}\right] \quad \text{Augmented matrix}$$

Now we use the various row operations to change this matrix into one that leads to a system that is easier to solve.

It is best to work by columns. We start with the first column and make sure that there is a 1 in the first row, first column position. There is already a 1 in this position. Next, we get 0 in every position below the first. To get a 0 in row two, column one, we use the third row operation and add to the numbers in row two the result of multiplying each number in row one by -2. (We abbreviate this as $-2R_1 + R_2$.) Row one remains unchanged.

$$\left[\begin{array}{cc|c} 1 & -3 & 1 \\ 2 + 1(-2) & 1 + -3(-2) & -5 + 1(-2) \end{array}\right]$$

Original number from row two ↑   ↑ -2 times number from row one

$$\left[\begin{array}{cc|c} 1 & -3 & 1 \\ 0 & 7 & -7 \end{array}\right] \quad -2R_1 + R_2$$

The matrix now has a 1 in the first position of column one, with 0 in every position below the first.

Now we go to column two. An entry of 1 is needed in row two, column two. We get this 1 by using the second row operation, multiplying each number of row two by $\frac{1}{7}$.

$$\left[\begin{array}{cc|c} 1 & -3 & 1 \\ 0 & 1 & -1 \end{array}\right] \quad \tfrac{1}{7}R_2$$

Stop here—this matrix is in row echelon form.

This augmented matrix leads to the system of equations

$$\begin{aligned} 1x - 3y &= 1 \\ 0x + 1y &= -1, \end{aligned} \quad \text{or} \quad \begin{aligned} x - 3y &= 1 \\ y &= -1. \end{aligned}$$

From the second equation, $y = -1$. We substitute -1 for y in the first equation to get

$$\begin{aligned} x - 3y &= 1 \\ x - 3(-1) &= 1 && \text{Let } y = -1. \\ x + 3 &= 1 && \text{Multiply.} \\ x &= -2. && \text{Subtract 3.} \end{aligned}$$

Write the values of x and y in the correct order.

The solution set of the system is $\{(-2, -1)\}$. Check this solution by substitution in both equations of the system.

Work Problem **1** at the Side. ▶

1 Use row operations to solve the system.

$$\begin{aligned} x - 2y &= 9 \\ 3x + y &= 13 \end{aligned}$$

Answer

1. $\{(5, -2)\}$

Calculator Tip If the augmented matrix of the system in Example 1 is entered as matrix A in a graphing calculator (Figure 10(a)) and the row echelon form of the matrix is found (Figure 10(b)), the system becomes

$$x + \frac{1}{2}y = -\frac{5}{2}$$
$$y = -1.$$

While this system looks different from the one we obtained in Example 1, it is equivalent, since its solution set is also $\{(-2, -1)\}$.

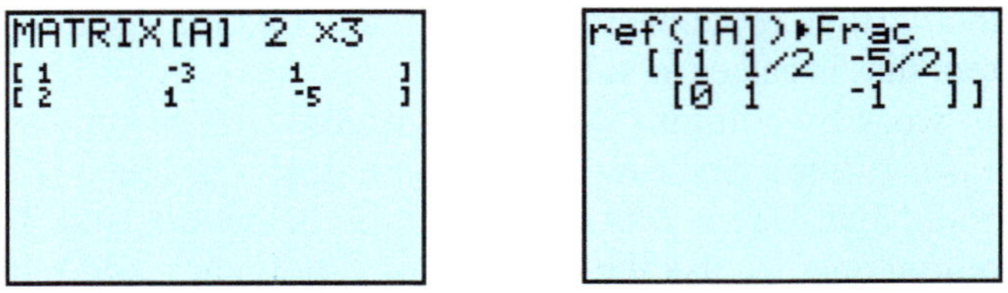

(a) (b)

Figure 10

OBJECTIVE 4 Use row operations to solve a system with three equations. As before, we use row operations to get 1s down the diagonal from left to right and all 0s below each 1.

Example 2 Using Row Operations to Solve a System with Three Variables

Use row operations to solve the system.

$$\begin{aligned} x - y + 5z &= -6 \\ 3x + 3y - z &= 10 \\ x + 3y + 2z &= 5 \end{aligned}$$

Start by writing the augmented matrix of the system.

$$\left[\begin{array}{ccc|c} 1 & -1 & 5 & -6 \\ 3 & 3 & -1 & 10 \\ 1 & 3 & 2 & 5 \end{array}\right] \qquad \text{Augmented matrix}$$

This matrix already has 1 in row one, column one. Next get 0s in the rest of column one. First, add to row two the results of multiplying each number of row one by -3. This gives the matrix

$$\left[\begin{array}{ccc|c} 1 & -1 & 5 & -6 \\ 0 & 6 & -16 & 28 \\ 1 & 3 & 2 & 5 \end{array}\right]. \qquad -3R_1 + R_2$$

Now add to the numbers in row three the results of multiplying each number of row one by -1.

$$\left[\begin{array}{ccc|c} 1 & -1 & 5 & -6 \\ 0 & 6 & -16 & 28 \\ 0 & 4 & -3 & 11 \end{array}\right] \qquad -1R_1 + R_3$$

Continued on Next Page

Get 1 in row two, column two by multiplying each number in row two by $\frac{1}{6}$.

$$\left[\begin{array}{rrr|r} 1 & -1 & 5 & -6 \\ 0 & 1 & -\frac{8}{3} & \frac{14}{3} \\ 0 & 4 & -3 & 11 \end{array}\right] \quad \frac{1}{6}R_2$$

Introduce 0 in row three, column two by adding to row three the results of multiplying each number in row two by -4.

$$\left[\begin{array}{rrr|r} 1 & -1 & 5 & -6 \\ 0 & 1 & -\frac{8}{3} & \frac{14}{3} \\ 0 & 0 & \frac{23}{3} & -\frac{23}{3} \end{array}\right] \quad -4R_2 + R_3$$

Finally, obtain 1 in row three, column three by multiplying each number in row three by $\frac{3}{23}$.

$$\left[\begin{array}{rrr|r} 1 & -1 & 5 & -6 \\ 0 & 1 & -\frac{8}{3} & \frac{14}{3} \\ 0 & 0 & 1 & -1 \end{array}\right] \quad \frac{3}{23}R_3$$

This final matrix gives the system of equations

$$\begin{aligned} x - y + 5z &= -6 \\ y - \frac{8}{3}z &= \frac{14}{3} \\ z &= -1. \end{aligned}$$

Substitute -1 for z in the second equation, $y - \frac{8}{3}z = \frac{14}{3}$, to get $y = 2$. Substitute 2 for y and -1 for z in the first equation, $x - y + 5z = -6$, to get $x = 1$. The solution set is $\{(1, 2, -1)\}$. Check by substitution in the original system.

Work Problem **2** *at the Side.* ▶

2 Use row operations to solve the system.

$$\begin{aligned} 2x - y + z &= 7 \\ x - 3y - z &= 7 \\ -x + y - 5z &= -9 \end{aligned}$$

OBJECTIVE 5 Use row operations to solve special systems.

Example 3 Recognizing Inconsistent Systems or Dependent Equations

Use row operations to solve each system.

(a) $2x - 3y = 8$
$-6x + 9y = 4$

$$\left[\begin{array}{rr|r} 2 & -3 & 8 \\ -6 & 9 & 4 \end{array}\right] \quad \text{Write the augmented matrix.}$$

$$\left[\begin{array}{rr|r} 1 & -\frac{3}{2} & 4 \\ -6 & 9 & 4 \end{array}\right] \quad \frac{1}{2}R_1$$

$$\left[\begin{array}{rr|r} 1 & -\frac{3}{2} & 4 \\ \mathbf{0} & \mathbf{0} & \mathbf{28} \end{array}\right] \quad 6R_1 + R_2$$

The corresponding system of equations is

$$\begin{aligned} x - \frac{3}{2}y &= 4 \\ 0 &= 28, \quad \text{False} \end{aligned}$$

which has no solution and is inconsistent. The solution set is $\emptyset$.

Continued on Next Page

Answer

2. $\{(2, -2, 1)\}$

(b) $-10x + 12y = 30$
$5x - 6y = -15$

$$\left[\begin{array}{cc|c} -10 & 12 & 30 \\ 5 & -6 & -15 \end{array}\right]$$ Write the augmented matrix.

$$\left[\begin{array}{cc|c} 1 & -\frac{6}{5} & -3 \\ 5 & -6 & -15 \end{array}\right]$$ $-\frac{1}{10}R_1$

$$\left[\begin{array}{cc|c} 1 & -\frac{6}{5} & -3 \\ 0 & 0 & 0 \end{array}\right]$$ $-5R_1 + R_2$

The corresponding system is

$$x - \frac{6}{5}y = -3$$

$$0 = 0, \quad \text{True}$$

which has dependent equations. Using the second equation of the original system, we write the solution set as

$$\{(x, y) \mid 5x - 6y = -15\}.$$

◀ *Work Problem* **3** *at the Side.*

3 Use row operations to solve each system.

(a) $x - y = 2$
$-2x + 2y = 2$

(b) $x - y = 2$
$-2x + 2y = -4$

ANSWERS

3. (a) $\emptyset$ **(b)** $\{(x, y) \mid x - y = 2\}$

14.6 Exercises

FOR EXTRA HELP — MyMathLab — Math XL PRACTICE — WATCH — DOWNLOAD — READ — REVIEW

1. Consider the matrix $\begin{bmatrix} -2 & 3 & 1 \\ 0 & 5 & -3 \\ 1 & 4 & 8 \end{bmatrix}$, and answer the following.

(a) What are the elements of the second row?

(b) What are the elements of the third column?

(c) Is this a square matrix? Explain.

(d) Give the matrix obtained by interchanging the first and third rows.

(e) Give the matrix obtained by multiplying the first row by $-\frac{1}{2}$.

(f) Give the matrix obtained by multiplying the third row by 3 and adding it to the first row.

2. Give the dimensions of each matrix.

(a) $\begin{bmatrix} 3 & -7 \\ 4 & 5 \\ -1 & 0 \end{bmatrix}$ **(b)** $\begin{bmatrix} 4 & 9 & 0 \\ -1 & 2 & -4 \end{bmatrix}$ **(c)** $\begin{bmatrix} 6 & 3 \\ -2 & 5 \\ 4 & 10 \\ 1 & -11 \end{bmatrix}$

Complete the steps in the matrix solution of each system by filling in the blanks. Give the final system and the solution set. See Example 1.

3. $4x + 8y = 44$
$2x - y = -3$

$\left[\begin{array}{cc|c} 4 & 8 & 44 \\ 2 & -1 & -3 \end{array}\right]$

$\left[\begin{array}{cc|c} 1 & \underline{\quad} & \underline{\quad} \\ 2 & -1 & -3 \end{array}\right]$ $\frac{1}{4}R_1$

$\left[\begin{array}{cc|c} 1 & 2 & 11 \\ 0 & \underline{\quad} & \underline{\quad} \end{array}\right]$ $-2R_1 + R_2$

$\left[\begin{array}{cc|c} 1 & 2 & 11 \\ 0 & 1 & \underline{\quad} \end{array}\right]$ $-\frac{1}{5}R_2$

4. $2x - 5y = -1$
$3x + y = 7$

$\left[\begin{array}{cc|c} 2 & -5 & -1 \\ 3 & 1 & 7 \end{array}\right]$

$\left[\begin{array}{cc|c} 1 & -\frac{5}{2} & \underline{\quad} \\ 3 & 1 & 7 \end{array}\right]$ $\frac{1}{2}R_1$

$\left[\begin{array}{cc|c} 1 & -\frac{5}{2} & -\frac{1}{2} \\ 0 & \underline{\quad} & \underline{\quad} \end{array}\right]$ $-3R_1 + R_2$

$\left[\begin{array}{cc|c} 1 & -\frac{5}{2} & -\frac{1}{2} \\ 0 & 1 & \underline{\quad} \end{array}\right]$ $\frac{2}{17}R_2$

Use row operations to solve each system. See Examples 1 and 3.

5. $x + y = 5$
$x - y = 3$

6. $x + 2y = 7$
$x - y = -2$

7. $2x + 4y = 6$
$3x - y = 2$

8. $4x + 5y = -7$
$x - y = 5$

9. $3x + 4y = 13$
$2x - 3y = -14$

10. $5x + 2y = 8$
$3x - y = 7$

11. $-4x + 12y = 36$
$x - 3y = 9$

12. $2x - 4y = 8$
$-3x + 6y = 5$

13. $\begin{aligned} 2x + y &= 4 \\ 4x + 2y &= 8 \end{aligned}$

14. $\begin{aligned} -3x - 4y &= 1 \\ 6x + 8y &= -2 \end{aligned}$

15. $\begin{aligned} \frac{1}{2}x + \frac{1}{3}y &= 0 \\ \frac{2}{3}x + \frac{3}{4}y &= 0 \end{aligned}$

16. $\begin{aligned} 1.2x + 0.3y &= 0 \\ 2.9x - 0.6y &= 0 \end{aligned}$

Complete the steps in the matrix solution of each system by filling in the blanks. Give the final system and the solution set. See Example 2.

17. $\begin{aligned} x + y - z &= -3 \\ 2x + y + z &= 4 \\ 5x - y + 2z &= 23 \end{aligned}$

$$\left[\begin{array}{ccc|c} 1 & 1 & -1 & -3 \\ 2 & 1 & 1 & 4 \\ 5 & -1 & 2 & 23 \end{array}\right]$$

$$\left[\begin{array}{ccc|c} 1 & 1 & -1 & -3 \\ 0 & \_\_ & \_\_ & \_\_ \\ 0 & \_\_ & \_\_ & \_\_ \end{array}\right] \quad \begin{array}{l} -2R_1 + R_2 \\ -5R_1 + R_3 \end{array}$$

$$\left[\begin{array}{ccc|c} 1 & 1 & -1 & -3 \\ 0 & 1 & \_\_ & \_\_ \\ 0 & -6 & 7 & 38 \end{array}\right] \quad -1R_2$$

$$\left[\begin{array}{ccc|c} 1 & 1 & -1 & -3 \\ 0 & 1 & -3 & -10 \\ 0 & 0 & \_\_ & \_\_ \end{array}\right] \quad 6R_2 + R_3$$

$$\left[\begin{array}{ccc|c} 1 & 1 & -1 & -3 \\ 0 & 1 & -3 & -10 \\ 0 & 0 & 1 & \_\_ \end{array}\right] \quad -\tfrac{1}{11}R_3$$

18. $\begin{aligned} 2x + y + 2z &= 11 \\ 2x - y - z &= -3 \\ 3x + 2y + z &= 9 \end{aligned}$

$$\left[\begin{array}{ccc|c} 2 & 1 & 2 & 11 \\ 2 & -1 & -1 & -3 \\ 3 & 2 & 1 & 9 \end{array}\right]$$

$$\left[\begin{array}{ccc|c} 1 & \_\_ & \_\_ & \_\_ \\ 2 & -1 & -1 & -3 \\ 3 & 2 & 1 & 9 \end{array}\right] \quad \tfrac{1}{2}R_1$$

$$\left[\begin{array}{ccc|c} 1 & \frac{1}{2} & 1 & \frac{11}{2} \\ 0 & \_\_ & \_\_ & \_\_ \\ 0 & \_\_ & \_\_ & \_\_ \end{array}\right] \quad \begin{array}{l} -2R_1 + R_2 \\ -3R_1 + R_3 \end{array}$$

$$\left[\begin{array}{ccc|c} 1 & \frac{1}{2} & 1 & \frac{11}{2} \\ 0 & 1 & \_\_ & \_\_ \\ 0 & \frac{1}{2} & -2 & -\frac{15}{2} \end{array}\right] \quad -\tfrac{1}{2}R_2$$

$$\left[\begin{array}{ccc|c} 1 & \frac{1}{2} & 1 & \frac{11}{2} \\ 0 & 1 & \frac{3}{2} & 7 \\ 0 & 0 & \_\_ & \_\_ \end{array}\right] \quad -\tfrac{1}{2}R_2 + R_3$$

$$\left[\begin{array}{ccc|c} 1 & \frac{1}{2} & 1 & \frac{11}{2} \\ 0 & 1 & \frac{3}{2} & 7 \\ 0 & 0 & 1 & \_\_ \end{array}\right] \quad -\tfrac{4}{11}R_3$$

Use row operations to solve each system. See Examples 2 and 3.

19. $\begin{aligned} x + y - 3z &= 1 \\ 2x - y + z &= 9 \\ 3x + y - 4z &= 8 \end{aligned}$

20. $\begin{aligned} 2x + 4y - 3z &= -18 \\ 3x + y - z &= -5 \\ x - 2y + 4z &= 14 \end{aligned}$

21. $\begin{aligned} x + y - z &= 6 \\ 2x - y + z &= -9 \\ x - 2y + 3z &= 1 \end{aligned}$

22. $\begin{aligned} x + 3y - 6z &= 7 \\ 2x - y + 2z &= 0 \\ x + y + 2z &= -1 \end{aligned}$

23. $\begin{aligned} x - y &= 1 \\ y - z &= 6 \\ x + z &= -1 \end{aligned}$

24. $\begin{aligned} x + y &= 1 \\ 2x - z &= 0 \\ y + 2z &= -2 \end{aligned}$

25. $\begin{aligned} 4x + 8y + 4z &= 9 \\ x + 3y + 4z &= 10 \\ 5x + 10y + 5z &= 12 \end{aligned}$

26. $\begin{aligned} x + 2y + 3z &= -2 \\ 2x + 4y + 6z &= -5 \\ x - y + 2z &= 6 \end{aligned}$

27. $\begin{aligned} x - 2y + z &= 4 \\ 3x - 6y + 3z &= 12 \\ -2x + 4y - 2z &= -8 \end{aligned}$

28. $\begin{aligned} x + 3y + z &= 1 \\ 2x + 6y + 2z &= 2 \\ 3x + 9y + 3z &= 3 \end{aligned}$

29. $\begin{aligned} 5x + 3y - z &= 0 \\ 2x - 3y + z &= 0 \\ x + 4y - 2z &= 0 \end{aligned}$

30. $\begin{aligned} 4x + 5y - z &= 0 \\ 7x - 5y + z &= 0 \\ x + 3y - 2z &= 0 \end{aligned}$

15

Factoring

15.1 Factors: The Greatest Common Factor

OBJECTIVES

1. Find the greatest common factor of a list of numbers.
2. Find the greatest common factor of a list of variable terms.
3. Factor out the greatest common factor.
4. Factor by grouping.

Recall from **Section 4.2** that to **factor** a number means to write it as the product of two or more numbers. The product is called the **factored form** of the number. Here is an example.

$$12 = \overbrace{\underbrace{6 \cdot 2}_{\text{Factored form}}}^{\text{Factors}}$$

Factoring is a process that "undoes" multiplying. We multiply $6 \cdot 2$ to get 12, but we factor 12 by writing it as $6 \cdot 2$.

OBJECTIVE 1 Find the greatest common factor of a list of numbers. An integer that is a factor of two or more integers is a **common factor** of those integers. For example, 6 is a common factor of 18 and 24 because 6 is a factor of both 18 and 24. Other common factors of 18 and 24 are 1, 2, and 3. The **greatest common factor (GCF)** of a list of integers is the largest common factor of those integers. This means 6 is the greatest common factor of 18 and 24, since it is the largest of their common factors.

> **Note**
> Factors of a number are also divisors of the number. The greatest common factor is the same as the greatest common divisor.

EXAMPLE 1 Finding the Greatest Common Factor for Numbers

Find the greatest common factor for each list of numbers.

(a) 30, 45

First write each number in prime factored form.

$$30 = 2 \cdot 3 \cdot 5$$
$$45 = 3 \cdot 3 \cdot 5$$

Use each prime the *least* number of times it appears in *all* the factored forms. There is no 2 in the prime factored form of 45, so there will be no 2 in the greatest common factor. The least number of times 3 appears in all the factored forms is 1. The least number of times 5 appears is also 1. From this, the

$$\text{GCF} = 3^1 \cdot 5^1 = 3 \cdot 5 = 15.$$

(b) 72, 120, 432

Find the prime factored form of each number.

$$72 = 2 \cdot 2 \cdot 2 \cdot 3 \cdot 3$$
$$120 = 2 \cdot 2 \cdot 2 \cdot 3 \cdot 5$$
$$432 = 2 \cdot 2 \cdot 2 \cdot 2 \cdot 3 \cdot 3 \cdot 3$$

The least number of times 2 appears in all the factored forms is 3, and the least number of times 3 appears is 1. There is no 5 in the prime factored form of either 72 or 432, so the

$$\text{GCF} = 2^3 \cdot 3^1 = 24.$$

Continued on Next Page

(c) 10, 11, 14

Write the prime factored form of each number.

$$10 = 2 \cdot 5$$
$$11 = 11$$
$$14 = 2 \cdot 7$$

There are no primes common to all three numbers, so the GCF is 1.

Work Problem **1** *at the Side.* ▶

OBJECTIVE 2 Find the greatest common factor of a list of variable terms. The terms x^4, x^5, x^6, and x^7 have x^4 as the greatest common factor because the least exponent on the variable x is 4.

$$x^4 = 1 \cdot x^4, \quad x^5 = x \cdot x^4, \quad x^6 = x^2 \cdot x^4, \quad x^7 = x^3 \cdot x^4$$

Note

The exponent on a variable in the GCF is the *least* exponent that appears on that variable in *all* the terms.

EXAMPLE 2 Finding the Greatest Common Factor for Variable Terms

Find the greatest common factor for each list of terms.

(a) $21m^7, -18m^6, 45m^8$

$$21m^7 = 3 \cdot 7 \cdot m^7$$
$$-18m^6 = -1 \cdot 2 \cdot 3 \cdot 3 \cdot m^6$$
$$45m^8 = 3 \cdot 3 \cdot 5 \cdot m^8$$

First, 3 is the greatest common factor of the coefficients 21, -18, and 45. The least exponent on m is 6, so the

$$\text{GCF} = 3m^6.$$

(b) $x^4y^2, x^7y^5, x^3y^7, y^{15}$

$$x^4y^2, \quad x^7y^5, \quad x^3y^7, \quad y^{15}$$

There is no x in the last term, y^{15}, so x will not appear in the greatest common factor. There is a y in each term, however, and 2 is the least exponent on y. The GCF is y^2.

(c) $-a^2b, -ab^2$

$$-a^2b = -1a^2b = -1 \cdot 1 \cdot a^2b$$
$$-ab^2 = -1ab^2 = -1 \cdot 1 \cdot ab^2$$

The factors of -1 are -1 and 1. Since $1 > -1$, the GCF is $1ab$, or ab.

Note

In a list of negative terms, sometimes a negative common factor is preferable (even though it is not the greatest common factor). In Example 2(c), for instance, we might prefer $-ab$ as the common factor. In factoring exercises like this, either answer will be acceptable.

1 Find the greatest common factor for each list of numbers.

(a) 30, 20, 15

$30 = 2 \cdot 3 \cdot 5$

$20 = 2 \cdot$ ____ $\cdot$ ____

$15 = 3 \cdot$ ____

GCF = ____

(b) 42, 28, 35

(c) 12, 18, 26, 32

(d) 10, 15, 21

ANSWERS

1. **(a)** 2; 5; 5; 5 **(b)** 7 **(c)** 2 **(d)** 1

2 Find the greatest common factor for each list of terms.

(a) $6m^4, 9m^2, 12m^5$

$6m^4 = 2 \cdot \_\_\_\_ \cdot m^4$

$9m^2 = 3 \cdot \_\_\_\_ \cdot \_\_\_\_$

$12m^5 = 2 \cdot 2 \cdot \_\_\_\_ \cdot \_\_\_\_$

GCF = ____

(b) $-12p^5, -18q^4$

(c) y^4z^2, y^6z^8, z^9

(d) $12p^{11}, 17q^5$

Answers

2. (a) 3; 3; m^2; 3; m^5; $3m^2$
(b) 6 **(c)** z^2 **(d)** 1

Finding the Greatest Common Factor (GCF)

Step 1 **Factor.** Write each number in prime factored form.

Step 2 **List common factors.** List each prime number or each variable that is a factor of every term in the list. (If a prime does not appear in one of the prime factored forms, it cannot appear in the greatest common factor.)

Step 3 **Choose least exponents.** Use as exponents on the common prime factors the *least* exponents from the prime factored forms.

Step 4 **Multiply.** Multiply the primes from Step 3. If there are no primes left after Step 3, the greatest common factor is 1.

◀ Work Problem **2** at the Side.

OBJECTIVE 3 Factor out the greatest common factor. The polynomial

$$3m + 12$$

has two terms, $3m$ and 12. The greatest common factor of these two terms is 3. We can write $3m + 12$ so that each term is a product with 3 as one factor.

$$3m + 12$$
$$= 3 \cdot m + 3 \cdot 4$$
$$= 3(m + 4) \qquad \text{Distributive property}$$

The factored form of $3m + 12$ is $3(m + 4)$. This process is called **factoring out the greatest common factor.**

CAUTION

The polynomial $3m + 12$ is *not* in factored form when written as the *sum*

$$3 \cdot m + 3 \cdot 4. \qquad \text{Not in factored form}$$

The terms are factored, but the polynomial is not. The factored form of $3m + 12$ is the *product*

$$3(m + 4). \qquad \text{In factored form}$$

Writing a polynomial as a product, that is, in factored form, is called **factoring** the polynomial.

EXAMPLE 3 Factoring Out the Greatest Common Factor

Factor out the greatest common factor.

(a) $5y^2 + 10y$

$$= 5y(y) + 5y(2) \qquad \text{GCF} = 5y$$
$$= 5y(y + 2) \qquad \text{Distributive property}$$

Check Multiply the factored form.

$$5y(y + 2)$$
$$= 5y(y) + 5y(2) \qquad \text{Distributive property}$$
$$= 5y^2 + 10y \qquad \text{Original polynomial}$$

Continued on Next Page

(b) $20m^5 + 10m^4 - 15m^3$

$= 5m^3(4m^2) + 5m^3(2m) - 5m^3(3)$ GCF $= 5m^3$

$= 5m^3(4m^2 + 2m - 3)$ Factor out $5m^3$.

Check $5m^3(4m^2 + 2m - 3)$

$= 20m^5 + 10m^4 - 15m^3$ Original polynomial

(c) $x^5 + x^3$

$= x^3(x^2) + x^3(1)$

$= x^3(x^2 + 1)$ Don't forget the 1.

(d) $20m^7p^2 - 36m^3p^4$

$= 4m^3p^2(5m^4) - 4m^3p^2(9p^2)$ GCF $= 4m^3p^2$

$= 4m^3p^2(5m^4 - 9p^2)$ Factor out $4m^3p^2$.

(e) $\frac{1}{6}n^2 + \frac{5}{6}n$

$= \frac{1}{6}n(n) + \frac{1}{6}n(5)$ GCF $= \frac{1}{6}n$

$= \frac{1}{6}n(n + 5)$

CAUTION

Be sure to include the 1 in a problem like Example 3(c). ***Check that the factored form can be multiplied out to give the original polynomial.***

Work Problem 3 at the Side. ▶

3 Factor out the greatest common factor.

(a) $4x^2 + 6x$

(b) $10y^5 - 8y^4 + 6y^2$

(c) $m^7 + m^9$

(d) $8p^5q^2 + 16p^6q^3 - 12p^4q^7$

(e) $\frac{1}{3}b^2 - \frac{2}{3}b$

EXAMPLE 4 Factoring Out a Negative Common Factor

Factor $-8x^4 + 16x^3 - 4x^2$.

We can factor out either $4x^2$ or $-4x^2$ here. We factor out $-4x^2$ so that the coefficient of the first term in the trinomial factor will be positive.

$-8x^4 + 16x^3 - 4x^2$ Be careful with signs.

$= -4x^2(2x^2) - 4x^2(-4x) - 4x^2(1)$ $-4x^2$ is a common factor.

$= -4x^2(2x^2 - 4x + 1)$ Factor out $-4x^2$.

Check $-4x^2(2x^2 - 4x + 1)$

$= -4x^2(2x^2) - 4x^2(-4x) - 4x^2(1)$ Distributive property

$= -8x^4 + 16x^3 - 4x^2$ Original polynomial

Work Problem 4 at the Side. ▶

4 Factor

$-14a^3b^2 - 21a^2b^3 + 7ab$

by factoring out a negative common factor.

Note

Whenever we factor a polynomial in which the coefficient of the first term of a polynomial is negative, we will factor out the negative common factor, even if it is just -1. However, it would also be correct to factor out $4x^2$ in Example 4 to obtain $4x^2(-2x^2 + 4x - 1)$.

ANSWERS

3. **(a)** $2x(2x + 3)$
(b) $2y^2(5y^3 - 4y^2 + 3)$
(c) $m^7(1 + m^2)$
(d) $4p^4q^2(2p + 4p^2q - 3q^5)$
(e) $\frac{1}{3}b(b - 2)$

4. $-7ab(2a^2b + 3ab^2 - 1)$

5 Factor out the greatest common factor.

(a) $r(t - 4) + 5(t - 4)$

(b) $y^2(y + 2) - 3(y + 2)$

(c) $x(x - 1) - 5(x - 1)$

EXAMPLE 5 Factoring Out a Common Binomial Factor

Factor out the greatest common factor.

(a) $a(a + 3) + 4(a + 3)$

Sometimes the GCF has a factor with more than one term. The binomial $a + 3$ is the greatest common factor here.

Same

$$a(a + 3) + 4(a + 3)$$
$$= (a + 3)(a + 4)$$

(b) $x^2(x + 1) - 5(x + 1)$

$= (x + 1)(x^2 - 5)$ Factor out $x + 1$.

◀ Work Problem **5** at the Side.

OBJECTIVE 4 Factor by grouping. ***When a polynomial has four terms, common factors can sometimes be used to factor by grouping.***

EXAMPLE 6 Factoring by Grouping

Factor by grouping.

(a) $2x + 6 + ax + 3a$

Group the first two terms and the last two terms, since the first two terms have a common factor of 2 and the last two terms have a common factor of a.

$$2x + 6 + ax + 3a$$
$$= (2x + 6) + (ax + 3a)$$
$$= 2(x + 3) + a(x + 3)$$

The expression is still not in factored form because it is the ***sum*** of two terms. Now, however, $x + 3$ is a common factor and can be factored out.

$2x + 6 + ax + 3a$

$= (2x + 6) + (ax + 3a)$ Group the terms.

$= 2(x + 3) + a(x + 3)$ Factor each group.

$= (x + 3)(2 + a)$ Factor out $x + 3$.

The final result is in factored form because it is a ***product.*** Note that the goal in factoring by grouping is to get a common factor, $x + 3$ here, so that the last step is possible. Check by multiplying the binomials using the FOIL method from **Section 13.3.**

Check $(x + 3)(2 + a)$

$= 2x + ax + 6 + 3a$ FOIL

$= 2x + 6 + ax + 3a,$ Rearrange terms.

which is the original polynomial.

Continued on Next Page

ANSWERS

5. **(a)** $(t - 4)(r + 5)$
(b) $(y + 2)(y^2 - 3)$
(c) $(x - 1)(x - 5)$

(b) $6ax + 24x + a + 4$

$= (6ax + 24x) + (a + 4)$ Group the terms.

$= 6x(a + 4) + 1(a + 4)$ Factor each group.

Remember the 1.

$= (a + 4)(6x + 1)$ Factor out $a + 4$.

Check $(a + 4)(6x + 1)$

$= 6ax + a + 24x + 4$ FOIL

$= 6ax + 24x + a + 4,$ Rearrange terms.

which is the original polynomial.

(c) $2x^2 - 10x + 3xy - 15y$

$= (2x^2 - 10x) + (3xy - 15y)$ Group the terms.

$= 2x(x - 5) + 3y(x - 5)$ Factor each group.

$= (x - 5)(2x + 3y)$ Factor out the common factor, $x - 5$.

Check $(x - 5)(2x + 3y)$

$= 2x^2 + 3xy - 10x - 15y$ FOIL

$= 2x^2 - 10x + 3xy - 15y$ Original polynomial

(d) $t^3 + 2t^2 - 3t - 6$

Be sure to write a + sign between the groups.

$= (t^3 + 2t^2) + (-3t - 6)$ Group the terms.

$= t^2(t + 2) - 3(t + 2)$ Factor out -3 so there is a common factor, $t + 2$; $-3(t + 2) = -3t - 6$.

Be careful with signs.

$= (t + 2)(t^2 - 3)$ Factor out $t + 2$.

Check by multiplying.

CAUTION

Be careful with signs when grouping in a problem like Example 6(d). It is wise to check the factoring in the second step, as shown in the example side comment, before continuing.

Work Problem 6 at the Side. ▶

Factoring a Polynomial with Four Terms by Grouping

Step 1 **Group terms.** Collect the terms into two groups so that each group has a common factor.

Step 2 **Factor within groups.** Factor out the greatest common factor from each group.

Step 3 **Factor the entire polynomial.** Factor a common binomial factor from the results of Step 2.

Step 4 **If necessary, rearrange terms.** If Step 2 does not result in a common binomial factor, try a different grouping.

6 Factor by grouping.

(a) $pq + 5q + 2p + 10$

(b) $2xy + 3y + 2x + 3$

(c) $2a^2 - 4a + 3ab - 6b$

(d) $x^3 + 3x^2 - 5x - 15$

Answers

6. **(a)** $(p + 5)(q + 2)$
(b) $(2x + 3)(y + 1)$
(c) $(a - 2)(2a + 3b)$
(d) $(x + 3)(x^2 - 5)$

7 Factor by grouping.

(a) $6y^2 - 20w + 15y - 8yw$

(b) $9mn - 4 + 12m - 3n$

EXAMPLE 7 **Rearranging Terms Before Factoring by Grouping**

Factor by grouping.

(a) $10x^2 - 12y + 15x - 8xy$

Factoring out the common factor of 2 from the first two terms and the common factor of x from the last two terms gives

$$10x^2 - 12y + 15x - 8xy$$
$$= 2(5x^2 - 6y) + x(15 - 8y).$$

This does not lead to a common factor, so we try rearranging the terms. There is usually more than one way to do this. We try the following.

$$10x^2 - 12y + 15x - 8xy$$
$$= 10x^2 - 8xy - 12y + 15x \qquad \text{Commutative property}$$
$$= (10x^2 - 8xy) + (-12y + 15x) \qquad \text{Group the terms.}$$
$$= 2x(5x - 4y) + 3(-4y + 5x) \qquad \text{Factor each group.}$$
$$= 2x(5x - 4y) + 3(5x - 4y) \qquad \text{Rewrite } -4y + 5x.$$
$$= (5x - 4y)(2x + 3) \qquad \text{Factor out } 5x - 4y.$$

Check $(5x - 4y)(2x + 3)$

$$= 10x^2 + 15x - 8xy - 12y \qquad \text{FOIL}$$
$$= 10x^2 - 12y + 15x - 8xy \qquad \text{Original polynomial}$$

(b) $2xy + 12 - 3y - 8x$

We need to rearrange these terms to get two groups that each have a common factor. Trial and error suggests the following grouping.

$$2xy + 12 - 3y - 8x$$

Always write a + sign between the two groups.

$$= (2xy - 3y) + (-8x + 12) \qquad \text{Group the terms.}$$
$$= y(2x - 3) - 4(2x - 3) \qquad \text{Factor each group.}$$

Be careful with signs.

$$= (2x - 3)(y - 4) \qquad \text{Factor out } 2x - 3.$$

Since the quantities in parentheses in the second step must be the same, we factored out -4 rather than 4. *Check* by multiplying.

CAUTION

Use negative signs carefully when grouping, as in Example 7(b), or a sign error will occur. ***Always check by multiplying.***

◀ *Work Problem* **7** *at the Side.*

ANSWERS

7. (a) $(2y + 5)(3y - 4w)$
(b) $(3m - 1)(3n + 4)$

15.1 Exercises

FOR EXTRA HELP

MyMathLab

MathXL PRACTICE WATCH DOWNLOAD READ REVIEW

Find the greatest common factor for each list of numbers. See Example 1.

1. 12, 16 **2.** 18, 24 **3.** 40, 20, 4 **4.** 50, 30, 5

5. 18, 24, 36, 48 **6.** 15, 30, 45, 75 **7.** 4, 9, 12 **8.** 9, 16, 24

Find the greatest common factor for each list of terms. See Example 2.

9. $16y, 24$ **10.** $18w, 27$ **11.** $30x^3, 40x^6, 50x^7$

12. $60z^4, 70z^8, 90z^9$ **13.** $-x^4y^3, -xy^2$ **14.** $-a^4b^5, -a^3b$

15. $42ab^3, -36a, 90b, -48ab$ **16.** $45c^3d, 75c, 90d, -105cd$

Complete each factoring.

17. $9m^4 = 3m^2(\quad)$ **18.** $12p^5 = 6p^3(\quad)$ **19.** $-8z^9 = -4z^5(\quad)$

20. $-15k^{11} = -5k^8(\quad)$ **21.** $6m^4n^5 = 3m^3n(\quad)$ **22.** $27a^3b^2 = 9a^2b(\quad)$

23. $12y + 24 = 12(\qquad)$ **24.** $18p + 36 = 18(\qquad)$

25. $10a^2 - 20a = 10a(\qquad)$ **26.** $15x^2 - 30x = 15x(\qquad)$

27. $8x^2y + 12x^3y^2 = 4x^2y(\qquad)$ **28.** $18s^3t^2 + 10st = 2st(\qquad)$

Factor out the greatest common factor, or a negative common factor if the coefficient of the term of greatest degree is negative. See Examples 3–5.

29. $x^2 - 4x$ **30.** $m^2 - 7m$ **31.** $6t^2 + 15t$ **32.** $8x^2 + 6x$

33. $\frac{1}{4}d^2 - \frac{3}{4}d$ **34.** $\frac{1}{5}z^2 + \frac{3}{5}z$ **35.** $-12x^3 - 6x^2$ **36.** $-21b^3 + 7b^2$

37. $65y^{10} + 35y^6$ **38.** $100a^5 + 16a^3$ **39.** $11w^3 - 100$

40. $13z^5 - 80$ **41.** $8m^2n^3 + 24m^2n^2$ **42.** $19p^2y - 38p^2y^3$

43. $-4x^3 + 10x^2 - 6x$ **44.** $-9z^3 + 6z^2 - 12z$ **45.** $13y^8 + 26y^4 - 39y^2$

46. $5x^5 + 25x^4 - 20x^3$

47. $45q^4p^5 + 36qp^6 + 81q^2p^3$

48. $125a^3z^5 + 60a^4z^4 - 85a^5z^2$

49. $c(x + 2) + d(x + 2)$

50. $r(5 - x) + t(5 - x)$

51. $a^2(2a + b) - b(2a + b)$

52. $3x(x^2 + 5) - y(x^2 + 5)$

53. $q(p + 4) - 1(p + 4)$

54. $y^2(x - 4) + 1(x - 4)$

Factor by grouping. See Examples 6 and 7.

55. $5m + mn + 20 + 4n$

56. $ts + 5t + 2s + 10$

57. $6xy - 21x + 8y - 28$

58. $2mn - 8n + 3m - 12$

59. $3xy + 9x + y + 3$

60. $6n + 4mn + 3 + 2m$

61. $7z^2 + 14z - az - 2a$

62. $2b^2 + 3b - 8ab - 12a$

63. $18r^2 + 12ry - 3xr - 2xy$

64. $5m^2 + 15mp - 2mr - 6pr$

65. $w^3 + w^2 + 9w + 9$

66. $y^3 + y^2 + 6y + 6$

67. $3a^3 + 6a^2 - 2a - 4$

68. $10x^3 + 15x^2 - 8x - 12$

69. $16m^3 - 4m^2p^2 - 4mp + p^3$

70. $10t^3 - 2t^2s^2 - 5ts + s^3$

71. $y^2 + 3x + 3y + xy$

72. $m^2 + 14p + 7m + 2mp$

73. $2z^2 + 6w - 4z - 3wz$

74. $2a^2 + 20b - 8a - 5ab$

Relating Concepts (Exercises 75–78) For Individual or Group Work

In many cases, the choice of which pairs of terms to group when factoring by grouping can be made in different ways. To see this for Example 7(b), ***work Exercises 75–78 in order.***

75. Start with the polynomial from Example 7(b), $2xy + 12 - 3y - 8x$, and rearrange the terms as follows: $2xy - 8x - 3y + 12$. What property from **Section 1.3** allows this?

76. Group the first two terms and the last two terms of the rearranged polynomial in Exercise 75. Then factor each group.

77. Is your result from Exercise 76 in factored form? Explain your answer.

78. If your answer to Exercise 77 is *no*, factor the polynomial. Is the result the same as the one shown for Example 7(b)?

15.2 Factoring Trinomials

OBJECTIVES

1. **Factor trinomials with a coefficient of 1 for the squared term.**
2. **Factor trinomials after factoring out the greatest common factor.**

Using FOIL, the product of the binomials $k - 3$ and $k + 1$ is

$$(k - 3)(k + 1) = k^2 - 2k - 3. \quad \text{Multiplying}$$

Suppose instead that we are given the polynomial $k^2 - 2k - 3$ and want to rewrite it as the product $(k - 3)(k + 1)$. That is,

$$k^2 - 2k - 3 = (k - 3)(k + 1). \quad \text{Factoring}$$

Recall from **Section 15.1** that this process is called *factoring* the polynomial. Factoring reverses or, "undoes," multiplying.

OBJECTIVE 1 Factor trinomials with a coefficient of 1 for the squared term. When factoring polynomials with integer coefficients, we use only integers in the factors. For example, we can factor $x^2 + 5x + 6$ by finding integers m and n such that

$$x^2 + 5x + 6 \quad \text{is written as} \quad (x + m)(x + n).$$

To find these integers m and n, we first use FOIL to multiply the two binomials on the right above:

$$\begin{aligned} &(x + m)(x + n) \\ &= x^2 + nx + mx + mn \\ &= x^2 + (n + m)x + mn. \quad \text{Distributive property} \end{aligned}$$

Comparing this result with $x^2 + 5x + 6$ shows that we must find integers m and n having a sum of 5 and a product of 6.

Product of m and n is 6.

$$x^2 + 5x + 6 = x^2 + (n + m)x + mn$$

Sum of m and n is 5.

Because many pairs of integers have a sum of 5, it is best to begin by listing those pairs of integers whose product is 6. Both 5 and 6 are positive, so we consider only pairs in which both integers are positive.

Work Problem 1 at the Side. ▶

From Problem 1 at the side, we see that the numbers 1 and 6 and the numbers 2 and 3 both have a product of 6, but only the pair 2 and 3 has a sum of 5. So 2 and 3 are the required integers, and

$$x^2 + 5x + 6 \quad \text{is factored as} \quad (x + 2)(x + 3).$$

Check by multiplying the binomials using FOIL. ***Make sure that the sum of the outer and inner products produces the correct middle term.***

Check $(x + 2)(x + 3) = x^2 + 5x + 6$ Correct

$2x$
$3x$
$5x$ Add.

This method of factoring can be used only for trinomials that have 1 as the coefficient of the squared term.

1 **(a)** List all pairs of positive integers whose product is 6.

(b) Find the pair from part (a) whose sum is 5.

ANSWERS

1. **(a)** 1, 6; 2, 3 **(b)** 2, 3

2 Factor each trinomial.

(a) $y^2 + 12y + 20$

First complete the given list of numbers.

Factors of 20	Sums of Factors
20, 1	$20 + 1 \quad = 21$
10, ___	$10 +$ ___ $=$ ___
5, ___	$5 +$ ___ $=$ ___

(b) $x^2 + 9x + 18$

3 Factor each trinomial.

(a) $t^2 - 12t + 32$

First complete the given list of numbers.

Factors of 32	Sums of Factors
$-32, -1$	$-32 + (-1) = -33$
-16, ___	$-16 + ($___$) =$ ___
-8, ___	$-8 + ($___$) =$ ___

(b) $y^2 - 10y + 24$

ANSWERS

2. **(a)** 2; 2; 12; 4; 4; 9; $(y + 10)(y + 2)$
(b) $(x + 3)(x + 6)$

3. **(a)** -2; -2; -18; -4; -4; -12; $(t - 8)(t - 4)$
(b) $(y - 6)(y - 4)$

EXAMPLE 1 **Factoring a Trinomial with All Positive Terms**

Factor $m^2 + 9m + 14$.

Look for two integers whose product is 14 and whose sum is 9. List the pairs of integers whose products are 14. Then examine the sums. Only positive integers are needed since all signs in $m^2 + 9m + 14$ are positive.

Factors of 14	Sums of Factors	
14, 1	$14 + 1 = 15$	
7, 2	$7 + 2 = 9$	Sum is 9.

From the list, 7 and 2 are the required integers, since $7 \cdot 2 = 14$ and $7 + 2 = 9$. Thus,

$$m^2 + 9m + 14 \quad \text{factors as} \quad (m + 2)(m + 7).$$

Check $(m + 2)(m + 7)$

$= m^2 + 7m + 2m + 14$ FOIL

$= m^2 + 9m + 14$ Original polynomial

Note

In Example 1, the answer $(m + 2)(m + 7)$ also could have been written

$$(m + 7)(m + 2).$$

Because of the commutative property of multiplication, the order of the factors does not matter. ***Always check by multiplying.***

◀ *Work Problem* **2** *at the Side.*

EXAMPLE 2 **Factoring a Trinomial with a Negative Middle Term**

Factor $x^2 - 9x + 20$.

Find two integers whose product is 20 and whose sum is -9. Since the numbers we are looking for have a *positive product* and a *negative sum*, we consider only pairs of negative integers.

Factors of 20	Sums of Factors	
$-20, -1$	$-20 + (-1) = -21$	
$-10, -2$	$-10 + (-2) = -12$	
$-5, -4$	$-5 + (-4) = -9$	Sum is -9.

The required integers are -5 and -4, so

$$x^2 - 9x + 20 \quad \text{factors as} \quad (x - 5)(x - 4).$$

Check $(x - 5)(x - 4)$

$= x^2 - 4x - 5x + 20$ FOIL

$= x^2 - 9x + 20$ Original polynomial

◀ *Work Problem* **3** *at the Side.*

EXAMPLE 3 **Factoring a Trinomial with Two Negative Terms**

Factor $p^2 - 2p - 15$.

Find two integers whose product is -15 and whose sum is -2. If these numbers do not come to mind right away, find them (if they exist) by listing all the pairs of integers whose product is -15. Because the last term, -15, is negative, we need pairs of integers with different signs.

Factors of -15	Sums of Factors	
$15, -1$	$15 + (-1) = 14$	
$-15, 1$	$-15 + 1 = -14$	
$5, -3$	$5 + (-3) = 2$	
$\mathbf{-5, 3}$	$-5 + 3 = \mathbf{-2}$	Sum is -2.

The required integers are -5 and 3, so

$$p^2 - 2p - 15 \quad \text{factors as} \quad (p - 5)(p + 3).$$

Check Multiply $(p - 5)(p + 3)$ to obtain $p^2 - 2p - 15$.

Note

In Examples 1–3, notice that we listed factors in descending order (disregarding sign) when we were looking for the required pair of integers. This helps avoid skipping the correct combination.

Work Problem **4** *at the Side.* ▶

As shown in the next example, some trinomials cannot be factored using only integers. We call such trinomials **prime polynomials.**

EXAMPLE 4 **Deciding Whether Polynomials Are Prime**

Factor each trinomial.

(a) $x^2 - 5x + 12$

As in Example 2, both factors must be negative to give a positive product and a negative sum. First, list all the pairs of negative integers whose product is 12. Then examine the sums.

Factors of 12	Sums of Factors
$-12, -1$	$-12 + (-1) = -13$
$-6, -2$	$-6 + (-2) = -8$
$-4, -3$	$-4 + (-3) = -7$

None of the pairs of integers has a sum of -5. Therefore, the trinomial $x^2 - 5x + 12$ *cannot be factored using only integers;* it is a *prime polynomial.*

(b) $k^2 - 8k + 11$

There is no pair of integers whose product is 11 and whose sum is -8, so $k^2 - 8k + 11$ is a prime polynomial.

Work Problem **5** *at the Side.* ▶

4 Factor each trinomial.

(a) $a^2 - 9a - 22$

(b) $r^2 - 6r - 16$

5 Factor each trinomial, if possible.

(a) $r^2 - 3r - 4$

(b) $m^2 - 2m + 5$

ANSWERS

4. (a) $(a - 11)(a + 2)$ **(b)** $(r - 8)(r + 2)$
5. (a) $(r - 4)(r + 1)$ **(b)** prime

6 Factor each trinomial.

(a) $b^2 - 3ab - 4a^2$

(b) $r^2 - 6rs + 8s^2$

7 Factor each trinomial completely.

(a) $2p^3 + 6p^2 - 8p$

(b) $-3x^4 + 15x^3 - 18x^2$

Answers

6. (a) $(b - 4a)(b + a)$
(b) $(r - 4s)(r - 2s)$

7. (a) $2p(p + 4)(p - 1)$
(b) $-3x^2(x - 3)(x - 2)$

Guidelines for factoring a trinomial of the form $x^2 + bx + c$ are summarized here.

Factoring $x^2 + bx + c$

Find two integers whose product is c and whose sum is b.

1. Both integers must be positive if b and c are positive.
2. Both integers must be negative if c is positive and b is negative.
3. One integer must be positive and one must be negative if c is negative.

EXAMPLE 5 Factoring a Trinomial with Two Variables

Factor $z^2 - 2bz - 3b^2$.

Here, the coefficient of z in the middle term is $-2b$, so we need to find two expressions whose product is $-3b^2$ and whose sum is $-2b$. The expressions are $-3b$ and b, so

$$z^2 - 2bz - 3b^2 \quad \text{factors as} \quad (z - 3b)(z + b).$$

Check $(z - 3b)(z + b)$

$= z^2 + zb - 3bz - 3b^2$ FOIL

$= z^2 + 1bz - 3bz - 3b^2$ Identity and commutative properties

$= z^2 - 2bz - 3b^2$ Combine like terms.

◀ Work Problem 6 at the Side.

OBJECTIVE 2 Factor trinomials after factoring out the greatest common factor. The trinomial in the next example does not have a coefficient of 1 for the squared term. (In fact, there is no squared term.) However, there may be a common factor.

EXAMPLE 6 Factoring a Trinomial with a Common Factor

Factor $4x^5 - 28x^4 + 40x^3$.

First, factor out the greatest common factor, $4x^3$.

$$4x^5 - 28x^4 + 40x^3$$
$$= 4x^3(x^2 - 7x + 10)$$

Now factor $x^2 - 7x + 10$. The integers -5 and -2 have a product of 10 and a sum of -7. The complete factored form is

Include $4x^3$. $\quad 4x^3(x - 5)(x - 2)$.

Check $4x^3(x - 5)(x - 2)$

$= 4x^3(x^2 - 7x + 10)$ FOIL

$= 4x^5 - 28x^4 + 40x^3$ Distributive property

◀ Work Problem 7 at the Side.

CAUTION

When factoring, ***always look for a common factor first.*** Remember to include the common factor as part of the answer. As a check, multiplying out the complete factored form should give the original polynomial.

15.2 Exercises

FOR EXTRA HELP — MyMathLab — Math XL PRACTICE — WATCH — DOWNLOAD — READ — REVIEW

1. When factoring a trinomial in x as $(x + a)(x + b)$, what must be true of a and b, if the last term of the trinomial is negative?

2. In Exercise 1, what must be true of a and b if the last term is positive?

3. What is meant by a *prime polynomial?*

4. How can you check your work when factoring a trinomial? Does the check ensure that the trinomial is *completely* factored?

In Exercises 5–8, list all pairs of integers with the given product. Then find the pair whose sum is given. See the tables in Examples 1–4.

5. Product: 12; Sum: 7

6. Product: 18; Sum: 9

7. Product: -24; Sum: -5

8. Product: -36; Sum: -16

9. Which one of the following is the correct factored form of $x^2 - 12x + 32$?

A. $(x - 8)(x + 4)$ **B.** $(x + 8)(x - 4)$

C. $(x - 8)(x - 4)$ **D.** $(x + 8)(x + 4)$

10. What would be the first step in factoring

$$2x^3 + 8x^2 - 10x?$$

Complete each factoring.

11. $x^2 + 15x + 44 = (x + 4)(\quad)$

12. $r^2 + 15r + 56 = (r + 7)(\quad)$

13. $x^2 - 9x + 8 = (x - 1)(\quad)$

14. $t^2 - 14t + 24 = (t - 2)(\quad)$

15. $y^2 - 2y - 15 = (y + 3)(\quad)$

16. $t^2 - t - 42 = (t + 6)(\quad)$

17. $x^2 + 9x - 22 = (x - 2)(\quad)$

18. $x^2 + 6x - 27 = (x - 3)(\quad)$

19. $y^2 - 7y - 18 = (y + 2)(\quad)$

20. $y^2 - 2y - 24 = (y + 4)(\quad)$

Factor completely. If a polynomial cannot be factored, write prime. See Examples 1–4.

21. $y^2 + 9y + 8$

22. $a^2 + 9a + 20$

23. $b^2 + 8b + 15$

24. $x^2 + 6x + 8$

25. $m^2 + m - 20$

26. $p^2 + 4p - 5$

27. $x^2 + 3x - 40$

28. $d^2 + 4d - 45$

29. $y^2 - 8y + 15$

30. $y^2 - 6y + 8$

31. $z^2 - 15z + 56$

32. $x^2 - 13x + 36$

33. $r^2 - r - 30$

34. $q^2 - q - 42$

35. $a^2 - 8a - 48$

36. $m^2 - 10m - 24$

37. $x^2 + 4x + 5$

38. $t^2 + 11t + 12$

Factor completely. See Examples 5 and 6.

39. $r^2 + 3ra + 2a^2$

40. $x^2 + 5xa + 4a^2$

41. $x^2 + 4xy + 3y^2$

42. $p^2 + 9pq + 8q^2$

43. $t^2 - tz - 6z^2$

44. $a^2 - ab - 12b^2$

45. $v^2 - 11vw + 30w^2$

46. $v^2 - 11vx + 24x^2$

47. $4x^2 + 12x - 40$

48. $5y^2 - 5y - 30$

49. $2t^3 + 8t^2 + 6t$

50. $3t^3 + 27t^2 + 24t$

51. $-2x^6 - 8x^5 + 42x^4$

52. $-4y^5 - 12y^4 + 40y^3$

53. $a^5 + 3a^4b - 4a^3b^2$

54. $z^{10} - 4z^9y - 21z^8y^2$

55. $m^3n - 10m^2n^2 + 24mn^3$

56. $y^3z + 3y^2z^2 - 54yz^3$

57. Use the FOIL method from **Section 13.3** to show that $(2x + 4)(x - 3) = 2x^2 - 2x - 12$. Why, then, is it incorrect to completely factor $2x^2 - 2x - 12$ as $(2x + 4)(x - 3)$?

58. Why is it incorrect to completely factor $3x^2 + 9x - 12$ as the product $(x - 1)(3x + 12)$?

15.3 Factoring Trinomials by Grouping

Trinomials like $2x^2 + 7x + 6$, in which the coefficient of the squared term is *not* 1, are factored with extensions of the methods from the previous sections. One such method uses factoring by grouping from **Section 15.1.**

OBJECTIVE 1 Factor trinomials by grouping when the coefficient of the squared term is not 1. Recall that a trinomial such as $m^2 + 3m + 2$ is factored by finding two integers whose product is 2 and whose sum is 3. To factor $2x^2 + 7x + 6$, we look for two integers whose product is $2 \cdot 6 = 12$ and whose sum is 7.

$$2x^2 + 7x + 6$$

Sum is 7. (coefficient 7 of the middle term)
Product is $2 \cdot 6 = 12$.

By considering pairs of positive integers whose product is 12, the necessary integers are found to be 3 and 4. We use these integers to write the middle term, $7x$, as $7x = 3x + 4x$. The trinomial $2x^2 + 7x + 6$ becomes

$$\begin{aligned} &2x^2 + 7x + 6 \\ &= 2x^2 + \underbrace{3x + 4x}_{7x} + 6 \\ &= (2x^2 + 3x) + (4x + 6) && \text{Group the terms.} \\ &= x(2x + 3) + 2(2x + 3) && \text{Factor each group.} \end{aligned}$$

Must be the same

$$= (2x + 3)(x + 2). \qquad \text{Factor out } 2x + 3.$$

Check Multiply $(2x + 3)(x + 2)$ to obtain $2x^2 + 7x + 6$.

In the preceding example, we could have written $7x$ as $4x + 3x$. Factoring by grouping this way would give the same answer.

Work Problem 1 at the Side. ▶

EXAMPLE 1 Factoring Trinomials by Grouping

Factor each trinomial.

(a) $6r^2 + r - 1$

We must find two integers with a product of $6(-1) = -6$ and a sum of 1. The integers are -2 and 3. We write the middle term, r, as $-2r + 3r$.

$$\begin{aligned} &6r^2 + r - 1 \\ &= 6r^2 - 2r + 3r - 1 && r = -2r + 3r \\ &= (6r^2 - 2r) + (3r - 1) && \text{Group the terms.} \\ &= 2r(3r - 1) + 1(3r - 1) && \text{The binomials must be the same.} \end{aligned}$$

Remember the 1.

$$= (3r - 1)(2r + 1) \qquad \text{Factor out } 3r - 1.$$

Check Multiply $(3r - 1)(2r + 1)$ to obtain $6r^2 + r - 1$.

Continued on Next Page

OBJECTIVE

1 **Factor trinomials by grouping when the coefficient of the squared term is not 1.**

1 **(a)** Factor $2x^2 + 7x + 6$ by writing $7x$ as $4x + 3x$. Complete the following.

$2x^2 + 7x + 6$
$= 2x^2 + 4x + 3x + 6$
$= (2x^2 + \_\_\_) + (3x + \_\_\_)$
$= 2x(x + \_\_\_) + 3(x + \_\_\_)$
$= (\_\_\_\_)(2x + 3)$

(b) Is the answer in part (a) the same as in the example? (Remember that the order of the factors does not matter.)

ANSWERS

1. **(a)** $4x$; 6; 2; 2; $x + 2$ **(b)** yes

(b) $12z^2 - 5z - 2$

Look for two integers whose product is $12(-2) = -24$ and whose sum is -5. The required integers are 3 and -8, so

$$12z^2 - 5z - 2$$
$$= 12z^2 + 3z - 8z - 2 \qquad -5z = 3z - 8z$$
$$= (12z^2 + 3z) + (-8z - 2) \qquad \text{Group the terms.}$$
$$= 3z(4z + 1) - 2(4z + 1) \qquad \text{Factor each group.}$$

Be careful with signs.

$$= (4z + 1)(3z - 2). \qquad \text{Factor out } 4z + 1.$$

Check Multiply $(4z + 1)(3z - 2)$ to obtain $12z^2 - 5z - 2$.

(c) $10m^2 + mn - 3n^2$

Two integers whose product is $10(-3) = -30$ and whose sum is 1 are -5 and 6. Rewrite the trinomial with four terms.

$$10m^2 + mn - 3n^2$$
$$= 10m^2 - 5mn + 6mn - 3n^2 \qquad mn = -5mn + 6mn$$
$$= 5m(2m - n) + 3n(2m - n) \qquad \text{Group the terms; factor each group.}$$
$$= (2m - n)(5m + 3n) \qquad \text{Factor out } 2m - n.$$

Check by multiplying.

◀ *Work Problem* **2** *at the Side.*

2 Factor each trinomial by grouping.

(a) $2m^2 + 7m + 3$

(b) $5p^2 - 2p - 3$

(c) $15k^2 - km - 2m^2$

EXAMPLE 2 **Factoring a Trinomial with a Common Factor by Grouping**

Factor $28x^5 - 58x^4 - 30x^3$.

First factor out the greatest common factor, $2x^3$.

$$28x^5 - 58x^4 - 30x^3$$
$$= 2x^3(14x^2 - 29x - 15)$$

To factor $14x^2 - 29x - 15$, find two integers whose product is $14(-15) = -210$ and whose sum is -29. Factoring 210 into prime factors gives

$$210 = 2 \cdot 3 \cdot 5 \cdot 7.$$

Combine these prime factors in pairs in different ways, using one positive factor and one negative factor to get -210. The factors 6 and -35 have the correct sum, -29. Now rewrite the given trinomial and factor it.

$$28x^5 - 58x^4 - 30x^3$$

Remember the common factor.

$$= 2x^3(14x^2 - 29x - 15)$$
$$= 2x^3(14x^2 + 6x - 35x - 15)$$
$$= 2x^3[(14x^2 + 6x) + (-35x - 15)]$$
$$= 2x^3[2x(7x + 3) - 5(7x + 3)]$$
$$= 2x^3[(7x + 3)(2x - 5)]$$
$$= 2x^3(7x + 3)(2x - 5)$$

◀ *Work Problem* **3** *at the Side.*

3 Factor each trinomial completely.

(a) $-4x^2 + 2x + 30$

(b) $18p^4 + 63p^3 + 27p^2$

(c) $6a^2 + 3ab - 18b^2$

Answers

2. **(a)** $(2m + 1)(m + 3)$
(b) $(5p + 3)(p - 1)$
(c) $(5k - 2m)(3k + m)$

3. **(a)** $-2(2x + 5)(x - 3)$
(b) $9p^2(2p + 1)(p + 3)$
(c) $3(2a - 3b)(a + 2b)$

15.3 Exercises

The middle term of each trinomial has been rewritten. Now factor by grouping. See Example 1.

1. $m^2 + 8m + 12$
$= m^2 + 6m + 2m + 12$

2. $x^2 + 9x + 14$
$= x^2 + 7x + 2x + 14$

3. $a^2 + 3a - 10$
$= a^2 + 5a - 2a - 10$

4. $y^2 - 2y - 24$
$= y^2 + 4y - 6y - 24$

5. $10t^2 + 9t + 2$
$= 10t^2 + 5t + 4t + 2$

6. $6x^2 + 13x + 6$
$= 6x^2 + 9x + 4x + 6$

7. $15z^2 - 19z + 6$
$= 15z^2 - 10z - 9z + 6$

8. $12p^2 - 17p + 6$
$= 12p^2 - 9p - 8p + 6$

9. $8s^2 + 2st - 3t^2$
$= 8s^2 - 4st + 6st - 3t^2$

10. $3x^2 - xy - 14y^2$
$= 3x^2 - 7xy + 6xy - 14y^2$

11. $15a^2 + 22ab + 8b^2$
$= 15a^2 + 10ab + 12ab + 8b^2$

12. $25m^2 + 25mn + 6n^2$
$= 25m^2 + 15mn + 10mn + 6n^2$

13. Which pair of integers would be used to rewrite the middle term when factoring $12y^2 + 5y - 2$ by grouping?
A. $-8, 3$ **B.** $8, -3$ **C.** $-6, 4$ **D.** $6, -4$

14. Which pair of integers would be used to rewrite the middle term when factoring $20b^2 - 13b + 2$ by grouping?
A. $10, 3$ **B.** $-10, -3$ **C.** $8, 5$ **D.** $-8, -5$

Complete the steps to factor each trinomial by grouping.

15. $2m^2 + 11m + 12$

(a) Find two integers whose product is ________ · ________ = ________ and whose sum is ________.

(b) The required integers are ________ and ________.

(c) Write the middle term $11m$ as ________ + ________.

(d) Rewrite the given trinomial using four terms.

(e) Factor the polynomial in part (d) by grouping.

(f) Check by multiplying.

16. $6y^2 - 19y + 10$

(a) Find two integers whose product is ________ · ________ = ________ and whose sum is ________.

(b) The required integers are ________ and ________.

(c) Write the middle term $-19y$ as ________ + ________.

(d) Rewrite the given trinomial using four terms.

(e) Factor the polynomial in part (d) by grouping.

(f) Check by multiplying.

Factor each trinomial by grouping. See Examples 1 and 2.

17. $2x^2 + 7x + 3$

18. $3y^2 + 13y + 4$

19. $4r^2 + r - 3$

20. $4r^2 + 3r - 10$

21. $8m^2 - 10m - 3$

22. $20x^2 - 28x - 3$

23. $21m^2 + 13m + 2$

24. $38x^2 + 23x + 2$

25. $6b^2 + 7b + 2$

26. $6w^2 + 19w + 10$

27. $12y^2 - 13y + 3$

28. $15a^2 - 16a + 4$

29. $24x^2 - 42x + 9$

30. $48b^2 - 74b - 10$

31. $2m^3 + 2m^2 - 40m$

32. $3x^3 + 12x^2 - 36x$

33. $-32z^5 + 20z^4 + 12z^3$

34. $-18x^5 - 15x^4 + 75x^3$

35. $12p^2 + 7pq - 12q^2$

36. $6m^2 - 5mn - 6n^2$

37. $6a^2 - 7ab - 5b^2$

38. $25g^2 - 5gh - 2h^2$

39. $5 - 6x + x^2$

40. $7 + 8x + x^2$

41. On a quiz, a student factored $16x^2 - 24x + 5$ by grouping as follows.

$$16x^2 - 24x + 5$$
$$= 16x^2 - 4x - 20x + 5$$
$$= 4x(4x - 1) - 5(4x - 1) \quad \text{His answer}$$

He thought his answer was correct since it checked by multiplying. Why was his answer marked wrong? What is the correct factored form?

42. On the same quiz, another student factored $3k^3 - 12k^2 - 15k$ by first factoring out the common factor $3k$ to get $3k(k^2 - 4k - 5)$. Then she wrote

$$k^2 - 4k - 5$$
$$= k^2 - 5k + k - 5$$
$$= k(k - 5) + 1(k - 5)$$
$$= (k - 5)(k + 1). \quad \text{Her answer}$$

Why was her answer marked wrong? What is the correct factored form?

15.4 Factoring Trinomials Using FOIL

OBJECTIVE

1 **Factor trinomials using FOIL.**

OBJECTIVE 1 Factor trinomials using FOIL. This section shows an alternative method of factoring trinomials in which the coefficient of the squared term is not 1. This method uses trial and error.

To factor $2x^2 + 7x + 6$ (the same trinomial factored at the beginning of **Section 15.3**) by trial and error, we use FOIL backwards. We want to write $2x^2 + 7x + 6$ as the product of two binomials.

$$(\qquad)(\qquad)$$

The product of the two first terms of the binomials is $2x^2$. The possible factors of $2x^2$ are $2x$ and x or $-2x$ and $-x$. Since all terms of the trinomial are positive, we consider only positive factors. Thus, we have

$$(2x \qquad)(x \qquad).$$

The product of the two last terms, 6, can be factored as $1 \cdot 6$, $6 \cdot 1$, $2 \cdot 3$, or $3 \cdot 2$. Try each pair to find the pair that gives the correct middle term, $7x$.

Work Problem **1** *at the Side.* ▶

1 Multiply to decide whether each factored form is correct or incorrect for

$$2x^2 + 7x + 6.$$

(a) $(2x + 1)(x + 6)$

(b) $(2x + 6)(x + 1)$

(c) $(2x + 2)(x + 3)$

In part (b) at the side, since $2x + 6 = 2(x + 3)$, the binomial $2x + 6$ has a common factor of 2, while $2x^2 + 7x + 6$ has no common factor other than 1. The product $(2x + 6)(x + 1)$ cannot be correct. (Part (c) also has one binomial factor with a common factor.)

> **Note**
> If the original polynomial has no common factor, then none of its binomial factors will either.

Now try the remaining numbers 3 and 2 as factors of 6.

$$(2x + 3)(x + 2) = 2x^2 + 7x + 6 \qquad \text{Correct}$$

$3x$
$4x$
$7x$ Add.

Finally, we see that $2x^2 + 7x + 6$ factors as $(2x + 3)(x + 2)$.

Check Multiply $(2x + 3)(x + 2)$ to obtain $2x^2 + 7x + 6$.

EXAMPLE 1 Factoring a Trinomial with All Positive Terms Using FOIL

Factor $8p^2 + 14p + 5$.

The number 8 has several possible pairs of factors, but 5 has only 1 and 5 or -1 and -5. For this reason, it is easier to begin by considering the factors of 5. Ignore the negative factors, since all coefficients in the trinomial are positive. If $8p^2 + 14p + 5$ can be factored, the factors will have the form

$$(\quad + 5)(\quad + 1).$$

Continued on Next Page

ANSWERS

1. **(a)** incorrect **(b)** incorrect **(c)** incorrect

2 Factor each trinomial.

(a) $2p^2 + 9p + 9$

(b) $6p^2 + 19p + 10$

(c) $8x^2 + 14x + 3$

When factoring $8p^2 + 14p + 5$, the possible pairs of factors of $8p^2$ are $8p$ and p, or $4p$ and $2p$. Try various combinations, checking to see if the middle term is $14p$ in each case.

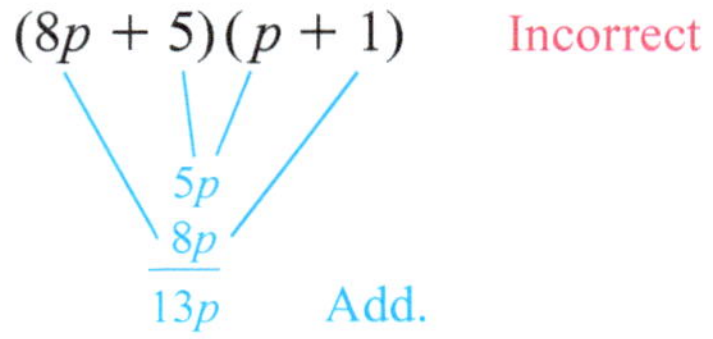

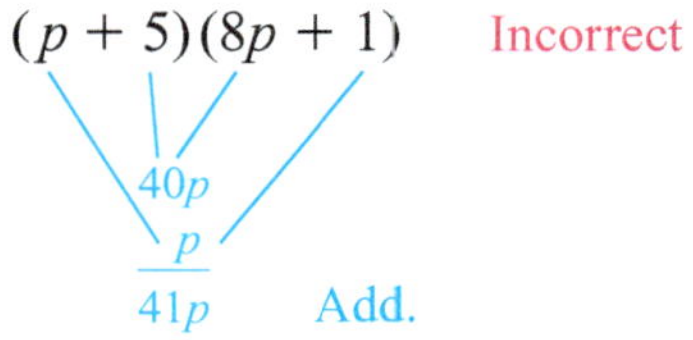

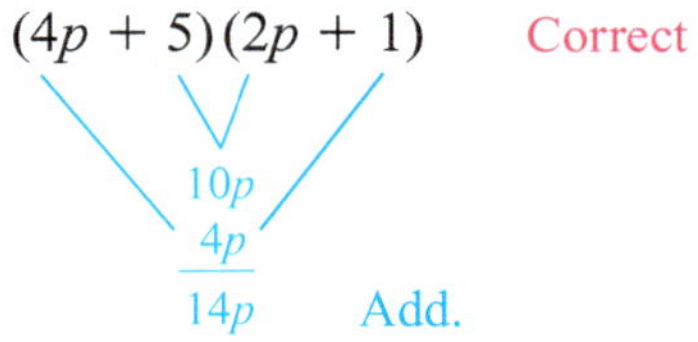

Since $14p$ is the correct middle term,

$$8p^2 + 14p + 5 \quad \text{factors as} \quad (4p + 5)(2p + 1).$$

Check Multiply $(4p + 5)(2p + 1)$ to obtain $8p^2 + 14p + 5$.

◀ Work Problem **2** at the Side.

3 Factor each trinomial.

(a) $4y^2 - 11y + 6$

(b) $9x^2 - 21x + 10$

EXAMPLE 2 **Factoring a Trinomial with a Negative Middle Term Using FOIL**

Factor $6x^2 - 11x + 3$.

Since 3 has only 1 and 3 or -1 and -3 as factors, it is better here to begin by factoring 3. The last term of the trinomial $6x^2 - 11x + 3$ is positive and the middle term has a negative coefficient, so we consider only negative factors. We need two negative factors because the *product* of two negative factors is positive and their *sum* is negative, as required.

Try -3 and -1 as factors of 3:

$$(\quad - 3)(\quad - 1).$$

The factors of $6x^2$ may be either $6x$ and x, or $2x$ and $3x$.

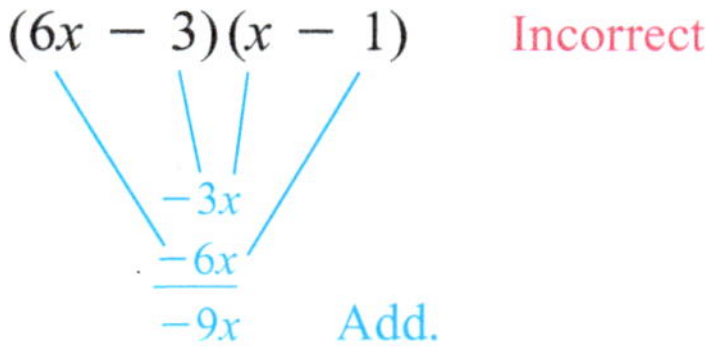

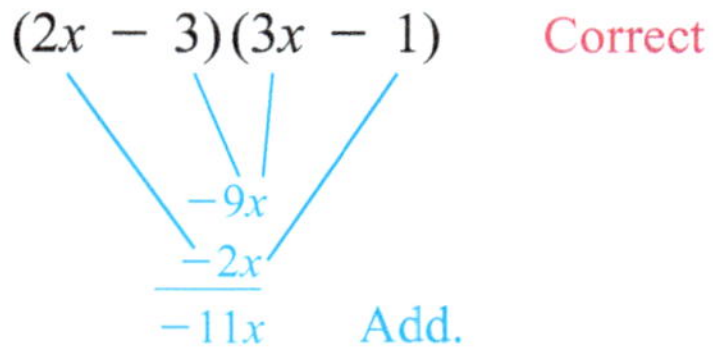

The factors $2x$ and $3x$ produce $-11x$, the correct middle term, so

$$6x^2 - 11x + 3 \quad \text{factors as} \quad (2x - 3)(3x - 1).$$

Check Multiply $(2x - 3)(3x - 1)$ to obtain $6x^2 - 11x + 3$.

Note

In Example 2, we might also realize that our initial attempt to factor $6x^2 - 11x + 3$ as $(6x - 3)(x - 1)$ *cannot* be correct since $6x - 3$ has a common factor of 3 and the original polynomial does not.

◀ Work Problem **3** at the Side.

Answers

2. (a) $(2p + 3)(p + 3)$
   (b) $(3p + 2)(2p + 5)$
   (c) $(4x + 1)(2x + 3)$
3. (a) $(4y - 3)(y - 2)$
   (b) $(3x - 5)(3x - 2)$

EXAMPLE 3 **Factoring a Trinomial with a Negative Last Term Using FOIL**

Factor $8x^2 + 6x - 9$.

The integer 8 has several possible pairs of factors, as does -9. Since the last term is negative, one positive factor and one negative factor of -9 are needed. Since the coefficient of the middle term is small, it is wise to avoid large factors such as 8 or 9. We try 4 and 2 as factors of 8, and 3 and -3 as factors of -9, and check the middle term.

$$(4x + 3)(2x - 3) \quad \text{Incorrect}$$

$$\begin{array}{r} 6x \\ -12x \\ \hline -6x \end{array} \quad \text{Add.}$$

Now we try interchanging 3 and -3, since only the sign of the middle term is incorrect.

$$(4x - 3)(2x + 3) \quad \text{Correct}$$

$$\begin{array}{r} -6x \\ 12x \\ \hline 6x \end{array} \quad \text{Add.}$$

This combination produces $6x$, the correct middle term, so

$$8x^2 + 6x - 9 \quad \text{factors as} \quad (4x - 3)(2x + 3).$$

Work Problem **4** at the Side. ▶

4 Factor each trinomial, if possible.

(a) $6x^2 + 5x - 4$

(b) $6m^2 - 11m - 10$

(c) $4x^2 - 3x - 7$

(d) $3y^2 + 8y - 6$

EXAMPLE 4 **Factoring a Trinomial with Two Variables**

Factor $12a^2 - ab - 20b^2$.

There are several pairs of factors of $12a^2$, including

$$12a \text{ and } a, \quad 6a \text{ and } 2a, \quad \text{and} \quad 4a \text{ and } 3a,$$

just as there are many possible pairs of factors of $-20b^2$, including

$$20b \text{ and } -b, \quad -20b \text{ and } b, \quad 10b \text{ and } -2b,$$
$$-10b \text{ and } 2b, \quad 4b \text{ and } -5b, \quad \text{and} \quad -4b \text{ and } 5b.$$

Once again, since the coefficient of the middle term is small, avoid the larger factors. Try the factors $6a$ and $2a$ and $4b$ and $-5b$.

$$(6a + 4b)(2a - 5b)$$

This cannot be correct, as mentioned before, since $6a + 4b$ has 2 as a common factor, while the given trinomial does not. Try $3a$ and $4a$ with $4b$ and $-5b$.

$$(3a + 4b)(4a - 5b)$$
$$= 12a^2 + ab - 20b^2 \quad \text{Incorrect}$$

Here the middle term is ab, rather than $-ab$. Interchange the signs of the last two terms in the factors.

$$(3a - 4b)(4a + 5b)$$
$$= 12a^2 - ab - 20b^2 \quad \text{Correct}$$

Work Problem **5** at the Side. ▶

5 Factor each trinomial.

(a) $2x^2 - 5xy - 3y^2$

(b) $8a^2 + 2ab - 3b^2$

ANSWERS

4. **(a)** $(3x + 4)(2x - 1)$
(b) $(2m - 5)(3m + 2)$
(c) $(4x - 7)(x + 1)$
(d) prime

5. **(a)** $(2x + y)(x - 3y)$
(b) $(4a + 3b)(2a - b)$

6 Factor each trinomial.

(a) $36z^3 - 6z^2 - 72z$

(b) $-24x^3 + 32x^2y + 6xy^2$

EXAMPLE 5 Factoring Trinomials with Common Factors

Factor each trinomial.

(a) $15y^3 + 55y^2 + 30y$

First factor out the greatest common factor, $5y$.

$$15y^3 + 55y^2 + 30y$$
$$= 5y(3y^2 + 11y + 6)$$

Now factor $3y^2 + 11y + 6$. Try $3y$ and y as factors of $3y^2$ and 2 and 3 as factors of 6.

$$(3y + 2)(y + 3)$$
$$= 3y^2 + 11y + 6 \quad \text{Correct}$$

The complete factored form of $15y^3 + 55y^2 + 30y$ is

$$5y(3y + 2)(y + 3).$$

Remember the common factor.

Check $5y(3y + 2)(y + 3)$

$$= 5y(3y^2 + 11y + 6) \quad \text{FOIL}$$
$$= 15y^3 + 55y^2 + 30y \quad \text{Distributive property}$$

(b) $-24a^3 - 42a^2 + 45a$

The common factor could be $3a$ or $-3a$. If we factor out $-3a$, the first term of the trinomial will be positive, which makes it easier to factor.

$$-24a^3 - 42a^2 + 45a$$
$$= -3a(8a^2 + 14a - 15) \quad \text{Factor out } -3a.$$
$$= -3a(4a - 3)(2a + 5) \quad \text{Use trial and error.}$$

Check $-3a(4a - 3)(2a + 5)$

$$= -3a(8a^2 + 14a - 15)$$
$$= -24a^3 - 42a^2 + 45a$$

CAUTION

This caution bears repeating: ***Remember to include the common factor in the final factored form.***

Work Problem 6 at the Side.

ANSWERS

6. **(a)** $6z(2z - 3)(3z + 4)$
(b) $-2x(6x + y)(2x - 3y)$

15.4 Exercises

FOR EXTRA HELP MyMathLab MathXL PRACTICE WATCH DOWNLOAD READ REVIEW

Decide which is the correct factored form of the given polynomial.

1. $2x^2 - x - 1$
A. $(2x - 1)(x + 1)$ **B.** $(2x + 1)(x - 1)$

2. $3a^2 - 5a - 2$
A. $(3a + 1)(a - 2)$ **B.** $(3a - 1)(a + 2)$

3. $4y^2 + 17y - 15$
A. $(y + 5)(4y - 3)$ **B.** $(2y - 5)(2y + 3)$

4. $12c^2 - 7c - 12$
A. $(6c - 2)(2c + 6)$ **B.** $(4c + 3)(3c - 4)$

5. $4k^2 + 13mk + 3m^2$
A. $(4k + m)(k + 3m)$ **B.** $(4k + 3m)(k + m)$

6. $2x^2 + 11x + 12$
A. $(2x + 3)(x + 4)$ **B.** $(2x + 4)(x + 3)$

Complete each factoring.

7. $6a^2 + 7ab - 20b^2$
$= (3a - 4b)(\qquad)$

8. $9m^2 - 3mn - 2n^2$
$= (3m + n)(\qquad)$

9. $2x^2 + 6x - 8$
$= 2(\qquad)$
$= 2(\qquad)(\qquad)$

10. $3x^2 - 9x - 30$
$= 3(\qquad)$
$= 3(\qquad)(\qquad)$

11. $4z^3 - 10z^2 - 6z$
$= 2z(\qquad)$
$= 2z(\qquad)(\qquad)$

12. $15r^3 - 39r^2 - 18r$
$= 3r(\qquad)$
$= 3r(\qquad)(\qquad)$

13. For the polynomial $12x^2 + 7x - 12$, 2 is not a common factor. Explain why the binomial $2x - 6$, then, cannot be a factor of the polynomial.

14. How are the signs of the last terms of the two binomial factors of a trinomial determined?

Factor each trinomial completely. See Examples 1–5.

15. $3a^2 + 10a + 7$

16. $7r^2 + 8r + 1$

17. $2y^2 + 7y + 6$

18. $5z^2 + 12z + 4$

19. $15m^2 + m - 2$

20. $6x^2 + x - 1$

21. $12s^2 + 11s - 5$

22. $20x^2 + 11x - 3$

23. $10m^2 - 23m + 12$

24. $6x^2 - 17x + 12$

25. $8w^2 - 14w + 3$

26. $9p^2 - 18p + 8$

27. $20y^2 - 39y - 11$

28. $10x^2 - 11x - 6$

29. $3x^2 - 15x + 16$

30. $2t^2 + 13t - 18$

31. $20x^2 + 22x + 6$

32. $36y^2 + 81y + 45$

33. $-40m^2q - mq + 6q$

34. $-15a^2b - 22ab - 8b$

35. $15n^4 - 39n^3 + 18n^2$

36. $24a^4 + 10a^3 - 4a^2$

37. $-15x^2y^2 + 7xy^2 + 4y^2$

38. $-14a^2b^3 - 15ab^3 + 9b^3$

39. $5a^2 - 7ab - 6b^2$

40. $6x^2 - 5xy - y^2$

41. $12s^2 + 11st - 5t^2$

42. $25a^2 + 25ab + 6b^2$

43. $6m^6n + 7m^5n^2 + 2m^4n^3$

44. $12k^3q^4 - 4k^2q^5 - kq^6$

If a trinomial has a negative coefficient for the squared term, such as $-2x^2 + 11x - 12$, *it may be easier to factor by first factoring out the common factor* -1:

$$\begin{aligned} &-2x^2 + 11x - 12 \\ &= -1(2x^2 - 11x + 12) \\ &= -1(2x - 3)(x - 4). \end{aligned}$$

Use this method to factor the trinomials in Exercises 45–50.

45. $-x^2 - 4x + 21$

46. $-x^2 + x + 72$

47. $-3x^2 - x + 4$

48. $-5x^2 + 2x + 16$

49. $-2a^2 - 5ab - 2b^2$

50. $-3p^2 + 13pq - 4q^2$

Relating Concepts (Exercises 51–56) For Individual or Group Work

One of the most common problems that beginning algebra students face is this: If an answer obtained doesn't look exactly like the one given in the back of the book, is it necessarily incorrect? Often there are several different equivalent forms of an answer that are all correct. ***Work Exercises 51–56 in order,*** *to see how and why this is possible for factoring problems.*

51. Factor the integer 35 as the product of two prime numbers.

52. Factor the integer 35 as the product of the negatives of two prime numbers.

53. Verify that $6x^2 - 11x + 4$ factors as $(3x - 4)(2x - 1)$.

54. Verify that $6x^2 - 11x + 4$ factors as $(4 - 3x)(1 - 2x)$.

55. Compare the two valid factored forms in Exercises 53 and 54. How do the factors in each case compare?

56. Suppose you know that the correct factored form of a particular trinomial is $(7t - 3)(2t - 5)$. Based on your observations in Exercises 51–55, what is another valid factored form?

15.5 Special Factoring Techniques

OBJECTIVES

1. Factor a difference of squares.
2. Factor a perfect square trinomial.

By reversing the rules for multiplication of binomials from **Section 13.4,** we obtain rules for factoring polynomials in certain forms.

OBJECTIVE 1 Factor a difference of squares. The formula for the product of the sum and difference of the same two terms is

$$(a + b)(a - b) = a^2 - b^2.$$

Reversing this rule leads to the following special factoring rule.

Factoring a Difference of Squares

$$\mathbf{a^2 - b^2 = (a + b)(a - b)}$$

For example,

$$m^2 - 16$$
$$= m^2 - 4^2$$
$$= (m + 4)(m - 4).$$

As the next examples show, the following conditions must be true for a binomial to be a difference of squares.

1. Both terms of the binomial must be squares, such as

$$x^2, \quad 9y^2, \quad 25, \quad 1, \quad m^4.$$

2. The terms of the binomial must have different signs (one positive and one negative).

EXAMPLE 1 Factoring Differences of Squares

Factor each binomial, if possible. (In part (c), use fractions.)

$$a^2 - b^2 = (a + b)(a - b)$$

$$\downarrow \quad \downarrow \quad \downarrow \quad \downarrow \quad \downarrow \quad \downarrow$$

(a) $x^2 - 49 = x^2 - 7^2 = (x + 7)(x - 7)$

(b) $y^2 - m^2$

$= (y + m)(y - m)$

(c) $z^2 - \dfrac{9}{16}$

$= \left(z + \dfrac{3}{4}\right)\left(z - \dfrac{3}{4}\right) \qquad \frac{9}{16} = \left(\frac{3}{4}\right)^2$

(d) $x^2 - 8$

Because 8 is not the square of an integer, this binomial is not a difference of squares. It is a prime polynomial.

(e) $p^2 + 16$

Since $p^2 + 16$ is a *sum* of squares, it is not equal to $(p + 4)(p - 4)$. Also, using FOIL,

$$(p - 4)(p - 4)$$
$$= p^2 - 8p + 16, \quad \text{not} \quad p^2 + 16,$$

and

$$(p + 4)(p + 4)$$
$$= p^2 + 8p + 16, \quad \text{not} \quad p^2 + 16,$$

so $p^2 + 16$ is a prime polynomial.

1 Factor, if possible. (In part (b), use fractions.)

(a) $p^2 - 100$

(b) $x^2 - \frac{25}{36}$

(c) $x^2 + y^2$

(d) $9m^2 - 49$

(e) $64a^2 - 25$

2 Factor completely.

(a) $50r^2 - 32$

(b) $27y^2 - 75$

(c) $25a^2 - 64b^2$

(d) $k^4 - 49$

(e) $81r^4 - 16$

CAUTION

As Example 1(e) suggests, ***after any common factor is removed, a sum of squares cannot be factored.***

EXAMPLE 2 Factoring Differences of Squares

Factor each difference of squares.

$$a^2 - b^2 = (a+b)(a-b)$$

(a) $25m^2 - 16 = (5m)^2 - 4^2 = (5m+4)(5m-4)$

(b) $49z^2 - 64$

$= (7z)^2 - 8^2$

$= (7z+8)(7z-8)$

◀ Work Problem **1** at the Side.

Note

Always check a factored form by multiplying.

EXAMPLE 3 Factoring More Complex Differences of Squares

Factor completely.

(a) $81y^2 - 36$

$= 9(9y^2 - 4)$ Factor out the GCF, 9.

$= 9[(3y)^2 - 2^2]$

$= 9(3y+2)(3y-2)$ Difference of squares

(b) $9x^2 - 4z^2$

$= (3x)^2 - (2z)^2$

$= (3x+2z)(3x-2z)$

(c) $p^4 - 36$

$= (p^2)^2 - 6^2$

$= (p^2+6)(p^2-6)$

(d) $m^4 - 16$

$= (m^2)^2 - 4^2$

$= (m^2+4)(m^2-4)$ Difference of squares

Don't stop here.

$= (m^2+4)(m+2)(m-2)$ Difference of squares again

CAUTION

Factor again when any of the factors is a difference of squares, as in Example 3(d). Check by multiplying.

◀ Work Problem **2** at the Side.

OBJECTIVE 2 Factor a perfect square trinomial. The expressions 144, $4x^2$, and $81m^6$ are called *perfect squares* because

$$144 = 12^2, \quad 4x^2 = (2x)^2, \quad \text{and} \quad 81m^6 = (9m^3)^2.$$

Answers

1. (a) $(p+10)(p-10)$
   (b) $\left(x+\frac{5}{6}\right)\left(x-\frac{5}{6}\right)$
   (c) prime
   (d) $(3m+7)(3m-7)$
   (e) $(8a+5)(8a-5)$
2. (a) $2(5r+4)(5r-4)$
   (b) $3(3y+5)(3y-5)$
   (c) $(5a+8b)(5a-8b)$
   (d) $(k^2+7)(k^2-7)$
   (e) $(9r^2+4)(3r+2)(3r-2)$

A **perfect square trinomial** is a trinomial that is the square of a binomial. For example, $x^2 + 8x + 16$ is a perfect square trinomial because it is the square of the binomial $x + 4$:

$$\begin{aligned} x^2 + 8x + 16 & \\ &= (x + 4)(x + 4) \\ &= (x + 4)^2. \end{aligned}$$

On the one hand, a necessary condition for a trinomial to be a perfect square is that *two of its terms must be perfect squares.* For this reason, $16x^2 + 4x + 15$ is not a perfect square trinomial because only the term $16x^2$ is a perfect square.

On the other hand, even if two of the terms are perfect squares, the trinomial may not be a perfect square trinomial. For example, $x^2 + 6x + 36$ has two perfect square terms, but it is not a perfect square trinomial. (Try to find a binomial that can be squared to give $x^2 + 6x + 36$.)

We can multiply to see that the square of a binomial gives one of the following perfect square trinomials.

Factoring Perfect Square Trinomials

$$a^2 + 2ab + b^2 = (a + b)^2$$

$$a^2 - 2ab + b^2 = (a - b)^2$$

The middle term of a perfect square trinomial is always twice the product of the two terms in the squared binomial. (See **Section 13.4.**) Use this to check any attempt to factor a trinomial that appears to be a perfect square.

EXAMPLE 4 **Factoring a Perfect Square Trinomial**

Factor $x^2 + 10x + 25$.

The term x^2 is a perfect square, and so is 25. Try to factor the trinomial

$$x^2 + 10x + 25 \quad \text{as} \quad (x + 5)^2.$$

To check, take twice the product of the two terms in the squared binomial.

$$2 \cdot x \cdot 5 = 10x$$

Twice (2); First term of binomial (x); Last term of binomial (5)

Since $10x$ is the middle term of the trinomial, the trinomial is a perfect square and can be factored as $(x + 5)^2$. Thus,

$$x^2 + 10x + 25 \quad \text{factors as} \quad (x + 5)^2.$$

Work Problem 3 at the Side. ▶

EXAMPLE 5 **Factoring Perfect Square Trinomials**

Factor each trinomial.

(a) $x^2 - 22x + 121$

The first and last terms are perfect squares ($121 = 11^2$ or $(-11)^2$). Check to see whether the middle term of $x^2 - 22x + 121$ is twice the product of the first and last terms of the binomial $x - 11$.

Continued on Next Page

3 Factor each trinomial.

(a) $p^2 + 14p + 49$

(b) $m^2 + 8m + 16$

(c) $x^2 + 2x + 1$

Answers

3. **(a)** $(p + 7)^2$ **(b)** $(m + 4)^2$ **(c)** $(x + 1)^2$

4 Factor each trinomial.

(a) $p^2 - 18p + 81$

(b) $16a^2 + 56a + 49$

(c) $121p^2 + 110p + 100$

(d) $64x^2 - 48x + 9$

(e) $27y^3 + 72y^2 + 48y$

$$2 \cdot x \cdot (-11) = -22x$$

(Twice; First term; Last term)

Since twice the product of the first and last terms of the binomial is the middle term, $x^2 - 22x + 121$ is a perfect square trinomial and

$$x^2 - 22x + 121 \quad \text{factors as} \quad (x - 11)^2.$$

(Same sign)

Notice that the sign of the second term in the squared binomial is the same as the sign of the middle term in the trinomial.

(b) $9m^2 - 24m + 16 = (3m)^2 + 2(3m)(-4) + (-4)^2 = (3m - 4)^2$

(Twice; First term; Last Term)

(c) $25y^2 + 20y + 16$

The first and last terms are perfect squares.

$$25y^2 = (5y)^2 \quad \text{and} \quad 16 = 4^2$$

However, twice the product of the first and last terms of the binomial $5y + 4$ is $2 \cdot 5y \cdot 4 = 40y$, which is not the middle term of $25y^2 + 20y + 16$. This trinomial is not a perfect square. In fact, the trinomial cannot be factored, even with the methods of the previous sections. It is a prime polynomial.

(d) $12z^3 + 60z^2 + 75z$

$= 3z(4z^2 + 20z + 25)$ — Factor out $3z$.

$= 3z[(2z)^2 + 2(2z)(5) + 5^2]$ — $4z^2 + 20z + 25$ is a perfect square trinomial.

$= 3z(2z + 5)^2$ — Factor.

Note

1. The sign of the second term in the squared binomial is always the same as the sign of the middle term in the trinomial.
2. The first and last terms of a perfect square trinomial must be *positive,* because they are squares. For example, the polynomial $x^2 - 2x - 1$ cannot be a perfect square because the last term is negative.
3. Perfect square trinomials can also be factored using grouping or FOIL, although using the method of this section is often easier.

◀ *Work Problem* **4** *at the Side.*

The methods of factoring discussed in this section are summarized here.

Special Factoring Rules

Difference of squares	$a^2 - b^2 = (a + b)(a - b)$
Perfect square trinomials	$a^2 + 2ab + b^2 = (a + b)^2$
	$a^2 - 2ab + b^2 = (a - b)^2$

Answers

4. **(a)** $(p - 9)^2$ **(b)** $(4a + 7)^2$ **(c)** prime **(d)** $(8x - 3)^2$ **(e)** $3y(3y + 4)^2$

15.5 Exercises

FOR EXTRA HELP MyMathLab MathXL PRACTICE WATCH DOWNLOAD READ REVIEW

1. To help you factor a difference of squares, complete the following list of squares.

$1^2 =$ ______	$2^2 =$ ______	$3^2 =$ ______	$4^2 =$ ______	$5^2 =$ ______
$6^2 =$ ______	$7^2 =$ ______	$8^2 =$ ______	$9^2 =$ ______	$10^2 =$ ______
$11^2 =$ ______	$12^2 =$ ______	$13^2 =$ ______	$14^2 =$ ______	$15^2 =$ ______
$16^2 =$ ______	$17^2 =$ ______	$18^2 =$ ______	$19^2 =$ ______	$20^2 =$ ______

2. To use the factoring techniques described in this section, you will sometimes need to recognize fourth powers of integers. Complete the following list of fourth powers.

$1^4 =$ ______ $2^4 =$ ______ $3^4 =$ ______ $4^4 =$ ______ $5^4 =$ ______

3. The following powers of x are all perfect squares: $x^2, x^4, x^6, x^8, x^{10}$. Based on this observation, we may make a conjecture (an educated guess) that if the power of a variable is divisible by ______ (with 0 remainder), then it is a perfect square.

4. Which of the following are differences of squares?

A. $x^2 - 4$ **B.** $y^2 + 9$ **C.** $2a^2 - 25$ **D.** $9m^2 - 1$

Factor each binomial completely. In Exercises 7, 8, 13, and 14, use fractions. See Examples 1–3.

5. $y^2 - 25$ **6.** $t^2 - 16$ **7.** $p^2 - \frac{1}{9}$ **8.** $q^2 - \frac{1}{4}$

9. $m^2 - 12$ **10.** $k^2 - 18$ **11.** $9r^2 - 4$ **12.** $4x^2 - 9$

13. $4m^2 - \frac{9}{25}$ **14.** $100b^2 - \frac{49}{81}$ **15.** $36x^2 - 16$ **16.** $32a^2 - 8$

17. $196p^2 - 225$ **18.** $361q^2 - 400$ **19.** $16r^2 - 25a^2$ **20.** $49m^2 - 100p^2$

21. $100x^2 + 49$ **22.** $81w^2 + 16$ **23.** $p^4 - 49$ **24.** $r^4 - 25$

25. $x^4 - 1$ **26.** $y^4 - 10{,}000$ **27.** $p^4 - 256$ **28.** $16k^4 - 1$

29. When a student was directed to factor $x^4 - 81$ completely, his teacher did not give him full credit for the answer $(x^2 + 9)(x^2 - 9)$. The student argued that because his answer does indeed give $x^4 - 81$ when multiplied out, he should be given full credit. ***WHAT WENT WRONG?*** Give the correct factored form.

30. The binomial $4x^2 + 16$ is a sum of squares that *can* be factored. How is this binomial factored? When can a sum of squares be factored?

31. In the polynomial $9y^2 + 14y + 25$, the first and last terms are perfect squares. Can the polynomial be factored? If it can, factor it. If it cannot, explain why it is not a perfect square trinomial.

32. Which of the following are perfect square trinomials?

A. $y^2 - 13y + 36$ **B.** $x^2 + 6x + 9$ **C.** $4z^2 - 4z + 1$ **D.** $16m^2 + 10m + 1$

Factor each trinomial completely. It may be necessary to factor out the greatest common factor first. In Exercises 37–40, use fractions or decimals, as appropriate. See Examples 4 and 5.

33. $w^2 + 2w + 1$

34. $p^2 + 4p + 4$

35. $x^2 - 8x + 16$

36. $x^2 - 10x + 25$

37. $t^2 + t + \frac{1}{4}$

38. $m^2 + \frac{2}{3}m + \frac{1}{9}$

39. $x^2 - 1.0x + 0.25$

40. $y^2 - 1.4y + 0.49$

41. $2x^2 + 24x + 72$

42. $3y^2 - 48y + 192$

43. $16x^2 - 40x + 25$

44. $36y^2 - 60y + 25$

45. $49x^2 - 28xy + 4y^2$

46. $4z^2 - 12zw + 9w^2$

47. $64x^2 + 48xy + 9y^2$

48. $9t^2 + 24tr + 16r^2$

49. $-50h^3 + 40h^2y - 8hy^2$

50. $-18x^3 - 48x^2y - 32xy^2$

Relating Concepts (Exercises 51–54) For Individual or Group Work

We have seen that multiplication and factoring are reverse processes. We know that multiplication and division are also related. To check a division problem, we multiply the quotient by the divisor to get the dividend. To see how factoring and division are related, ***work Exercises 51–54 in order.***

51. Factor $10x^2 + 11x - 6$.

52. Use long division from **Section 13.7** to divide $10x^2 + 11x - 6$ by $2x + 3$.

53. Could we have predicted the result in Exercise 52 from the result in Exercise 51? Explain.

54. Divide $x^3 - 1$ by $x - 1$. Use your answer to factor $x^3 - 1$.

Summary Exercises on Factoring

As you factor a polynomial, we suggest asking yourself these questions to decide on a suitable factoring technique.

Factoring a Polynomial

1. **Is there a common factor other than 1?** If so, factor out the greatest common factor (GCF) of all terms of the given polynomial.
2. **How many terms are in the polynomial?**

   Two terms: Check to see whether it is a difference of squares.

   Three terms: Is it a perfect square trinomial? If the trinomial is not a perfect square, check to see whether the coefficient of the second-degree term is 1. If so, use the method of **Section 15.2.** If the coefficient of the squared term of the trinomial is not 1, use the general factoring methods of **Sections 15.3** and **15.4.**

   Four terms: Try to factor the polynomial by grouping.
3. **Can any factors be factored further?** If so, factor them.

CAUTION

Be careful when checking your answer to a factoring problem.

1. *Check* that the product of all the factors does indeed yield the original polynomial.
2. *Check* that the original polynomial has been factored *completely*. (See Question 3 above.)

Checking by multiplication alone will not ensure that you have factored the original polynomial completely.

Suppose we are asked to completely factor the trinomial

$$6x^2 + 24xy + 24y^2$$

and give

$$(2x + 4y)(3x + 6y)$$

as the answer. If we only check by multiplying, we might conclude that our answer is correct because, by FOIL,

$$(2x + 4y)(3x + 6y)$$
$$= 6x^2 + 24xy + 24y^2.$$

However, we would not have factored the given polynomial *completely* because the binomial $2x + 4y$ has a common factor of 2 and the binomial $3x + 6y$ has a common factor of 3. Rather than factoring out these common factors at the *end* of our work, it is more efficient to factor out the GCF, 6 (which is $2 \cdot 3$), at the beginning, as follows.

$$6x^2 + 24xy + 24y^2$$
$$= 6(x^2 + 4xy + 4y^2) \quad \text{Factor out the GCF, 6.}$$
$$= 6(x + 2y)^2 \quad \text{Factor the perfect square trinomial.}$$

To avoid leaving out common factors in your final answer, ***remember to always factor out the GCF first.***

We now use the factoring guidelines given on the preceding page to factor $12x^2 + 26xy + 12y^2$ completely.

Question 1: ***Is there a common factor?*** Yes, there is a common factor. The GCF is 2, so factor it out.

$$12x^2 + 26xy + 12y^2$$
$$= 2(6x^2 + 13xy + 6y^2)$$

Question 2: ***How many terms are in the polynomial?*** The polynomial $6x^2 + 13xy + 6y^2$ has three terms, but it is not a perfect square. To factor the trinomial by grouping, as in **Section 15.3,** begin by finding two integers with a product of $6 \cdot 6$, or 36, and a sum of 13. These integers are 4 and 9.

$12x^2 + 26xy + 12y^2$

$= 2(6x^2 + 13xy + 6y^2)$	Factor out the GCF, 2.
$= 2(6x^2 + 4xy + 9xy + 6y^2)$	$4 \cdot 9 = 36$; $4 + 9 = 13$
$= 2[(6x^2 + 4xy) + (9xy + 6y^2)]$	Group the terms.
$= 2[2x(3x + 2y) + 3y(3x + 2y)]$	Factor each group.
$= 2(3x + 2y)(2x + 3y)$	Factor out the common factor, $3x + 2y$.

(The trinomial $6x^2 + 13xy + 6y^2$ could also be factored by trial and error, using FOIL backwards, as in **Section 15.4.**)

Question 3: ***Can any factors be factored further?*** None of the factors can be factored further, so the original polynomial has been factored completely.

Match each polynomial in Column I with the method you would use to factor it in Column II. The choices in Column II may be used once, more than once, or not at all.

I

1. $12x^2 + 20x + 8$

2. $x^2 - 17x + 72$

3. $-16m^2n + 24mn - 40mn^2$

4. $64a^2 - 121b^2$

5. $36p^2 - 60pq + 25q^2$

6. $z^2 - 4z + 6$

7. $625 - r^4$

8. $x^6 + 4x^4 - 3x^2 - 12$

9. $4w^2 + 49$

10. $144 - 24z + z^2$

II

A. Factor out the GCF; no further factoring is possible.

B. Factor a difference of squares once.

C. Factor a difference of squares twice.

D. Factor a perfect square trinomial.

E. Factor by grouping.

F. Factor out the GCF; then factor a trinomial by grouping or trial and error.

G. Factor into two binomials by finding two integers whose product is the constant in the trinomial and whose sum is the coefficient of the middle term.

H. The polynomial is prime.

Factor each polynomial completely. In Exercises 26, 32, 64, and 81, use fractions or decimals, as appropriate. Remember to check by multiplying.

11. $32m^9 + 16m^5 + 24m^3$

12. $2m^2 - 10m - 48$

13. $14k^3 + 7k^2 - 70k$

14. $9z^2 + 64$

15. $6z^2 + 31z + 5$

16. $m^2 - 3mn - 4n^2$

17. $49z^2 - 16y^2$

18. $100n^2r^2 + 30nr^3 - 50n^2r$

19. $16x^2 + 20x$

20. $20 + 5m + 12n + 3mn$

21. $10y^2 - 7yz - 6z^2$

22. $y^4 - 81$

23. $m^2 + 2m - 15$

24. $6y^2 - 5y - 4$

25. $32z^3 + 56z^2 - 16z$

26. $p^2 - 2.4p + 1.44$

27. $z^2 - 12z + 36$

28. $9m^2 - 64$

29. $y^2 - 4yk - 12k^2$

30. $16z^2 - 8z + 1$

31. $6y^2 - 6y - 12$

32. $x^2 + \frac{1}{2}x + \frac{1}{16}$

33. $p^2 - 17p + 66$

34. $a^2 + 17a + 72$

35. $k^2 + 100$

36. $108m^2 - 36m + 3$

37. $z^2 - 3za - 10a^2$

38. $2a^3 + a^2 - 14a - 7$

39. $4k^2 - 12k + 9$

40. $a^2 - 3ab - 28b^2$

41. $16r^2 + 24rm + 9m^2$

42. $3k^2 + 4k - 4$

43. $n^2 - 12n - 35$

44. $a^4 - 625$

45. $16k^2 - 48k + 36$

46. $8k^2 - 10k - 3$

47. $36y^6 - 42y^5 - 120y^4$

48. $5z^3 - 45z^2 + 70z$

49. $8p^2 + 23p - 3$

50. $8k^2 - 2kh - 3h^2$

51. $54m^2 - 24z^2$

52. $4k^2 - 20kz + 25z^2$

53. $6a^2 + 10a - 4$

54. $15h^2 + 11hg - 14g^2$

55. $28a^2 - 63b^2$

56. $10z^2 - 7z - 6$

57. $125m^4 - 400m^3n + 195m^2n^2$

58. $9y^2 + 12y - 5$

59. $9u^2 + 66uv + 121v^2$

60. $36x^2 + 32x + 9$

61. $27p^{10} - 45p^9 - 252p^8$

62. $10m^2 + 25m - 60$

63. $4 - 2q - 6p + 3pq$

64. $k^2 - \frac{64}{121}$

65. $64p^2 - 100m^2$

66. $m^3 + 4m^2 - 6m - 24$

67. $100a^2 - 81y^2$

68. $8a^2 + 23ab - 3b^2$

69. $a^2 + 8a + 16$

70. $4y^2 - 25$

71. $2x^2 + 5x + 6$

72. $-3x^3 + 12xy^2$

73. $25a^2 - 70ab + 49b^2$

74. $8t^4 - 8$

75. $-4x^2 + 24xy - 36y^2$

76. $100a^2 - 25b^2$

77. $-2x^2 + 26x - 72$

78. $2m^2 - 15n - 5mn + 6m$

79. $12x^2 + 22x - 20$

80. $y^6 + 5y^4 - 3y^2 - 15$

81. $y^2 - 0.64$

82. $12p^3 - 54p^2 - 30p$

15.6 A General Approach to Factoring

A polynomial is completely factored when **(1)** it is written as a *product* of prime polynomials with integer coefficients, and **(2)** none of the polynomial factors can be factored further.

OBJECTIVES

1. Factor out any common factor.
2. Factor binomials.
3. Factor trinomials.
4. Factor polynomials with more than three terms.

Factoring a Polynomial

Step 1 **Factor out any common factor.**

Step 2 **If the polynomial is a binomial,** check to see if it is the difference of squares, the difference of cubes, or the sum of cubes.

If the polynomial is a trinomial, check to see if it is a perfect square trinomial. If it is not, factor as in **Section 15.2.**

If the polynomial has more than three terms, try to factor by grouping.

Step 3 ***Check the factored form by multiplying.***

OBJECTIVE 1 Factor out any common factor. ***This step is always the same, regardless of the number of terms in the polynomial.***

EXAMPLE 1 Factoring Out a Common Factor

Factor each polynomial.

(a) $9p + 45$

$= 9(p + 5)$ GCF = 9

(b) $8m^2p^2 + 4mp$

$= 4mp(2mp + 1)$

(c) $5x(a + b) - y(a + b)$

$= (a + b)(5x - y)$ Factor out $a + b$.

Work Problem 1 at the Side. ▶

1 Factor each polynomial.

(a) $8x - 80$

(b) $2x^3 + 10x^2 - 2x$

(c) $12m(p - q) - 7n(p - q)$

OBJECTIVE 2 Factor binomials. Use one of the following rules.

Factoring a Binomial

For a **binomial** (two terms), check for the following patterns.

Difference of squares	$x^2 - y^2 = (x + y)(x - y)$
Difference of cubes	$x^3 - y^3 = (x - y)(x^2 + xy + y^2)$
Sum of cubes	$x^3 + y^3 = (x + y)(x^2 - xy + y^2)$

EXAMPLE 2 Factoring Binomials

Factor each binomial if possible.

(a) $64m^2 - 9n^2$

$= (8m)^2 - (3n)^2$ Difference of squares

$= (8m + 3n)(8m - 3n)$

(b) $8p^3 - 27$ $8p^3 = (2p)^3$; $27 = 3^3$

$= (2p - 3)[(2p)^2 + (2p)(3) + 3^2]$

$= (2p - 3)(4p^2 + 6p + 9)$

(c) $1000m^3 + 1$

$= (10m)^3 + 1^3$ Sum of cubes

$= (10m + 1)[(10m)^2 - (10m)(1) + 1^2]$

$= (10m + 1)(100m^2 - 10m + 1)$

(d) $25m^2 + 121$ is prime. It is the *sum* of squares.

Work Problem 2 at the Side. ▶

2 Factor each binomial if possible.

(a) $36x^2 - y^2$

(b) $4t^2 + 1$

(c) $125x^3 - 27y^3$

(d) $x^3 + 343y^3$

ANSWERS

1. (a) $8(x - 10)$
   (b) $2x(x^2 + 5x - 1)$
   (c) $(p - q)(12m - 7n)$
2. (a) $(6x + y)(6x - y)$
   (b) prime
   (c) $(5x - 3y)(25x^2 + 15xy + 9y^2)$
   (d) $(x + 7y)(x^2 - 7xy + 49y^2)$

3 Factor each trinomial.

(a) $16m^2 + 56m + 49$

(b) $r^2 + 18r + 72$

(c) $8t^2 - 13t + 5$

(d) $6x^2 - 3x - 63$

4 Factor each polynomial.

(a) $p^3 - 2pq^2 + p^2q - 2q^3$

(b) $9x^2 + 24x + 16 - y^2$

(c) $64a^3 + 16a^2 + b^3 - b^2$

Answers

3. (a) $(4m + 7)^2$
   (b) $(r + 6)(r + 12)$
   (c) $(8t - 5)(t - 1)$
   (d) $3(2x - 7)(x + 3)$
4. (a) $(p + q)(p^2 - 2q^2)$
   (b) $(3x + 4 + y)(3x + 4 - y)$
   (c) $(4a + b)(16a^2 - 4ab + b^2 + 4a - b)$

Note

The binomial $25m^2 + 625$ is a sum of squares. It *can* be factored, however, as $25(m^2 + 25)$ because it has a common factor, 25.

OBJECTIVE 3 Factor trinomials. Consider the following.

Factoring a Trinomial

For a **trinomial** (three terms), decide whether it is a perfect square trinomial of the form

$$x^2 + 2xy + y^2 = (x + y)^2 \quad \text{or} \quad x^2 - 2xy + y^2 = (x - y)^2.$$

If not, use the general factoring methods of **Section 7.2.**

EXAMPLE 3 Factoring Trinomials

Factor each trinomial.

(a) $p^2 + 10p + 25$
$= (p + 5)^2$ $\quad 2(p)(5) = 10p$

(b) $49z^2 - 42z + 9$
$= (7z - 3)^2$ $\quad 2(7z)(3) = 42z$

(c) $y^2 - 5y - 6$
$= (y - 6)(y + 1)$ $\quad$ The numbers -6 and 1 have a product of -6 and a sum of -5.

(d) $2k^2 - k - 6$
$= (2k + 3)(k - 2)$

(e) $28z^2 + 6z - 10$
$= 2(14z^2 + 3z - 5)$
$= 2(7z + 5)(2z - 1)$

◀ Work Problem 3 at the Side.

OBJECTIVE 4 Factor polynomials with more than three terms. Consider factoring by grouping.

EXAMPLE 4 Factoring Polynomials with More than Three Terms

Factor each polynomial.

(a) $20k^3 + 4k^2 - 45k - 9$
$= (20k^3 + 4k^2) - (45k + 9)$ $\quad$ Group the terms.
$= 4k^2(5k + 1) - 9(5k + 1)$ $\quad$ Factor each group.
$= (5k + 1)(4k^2 - 9)$ $\quad$ $5k + 1$ is a common factor.
$= (5k + 1)(2k + 3)(2k - 3)$ $\quad$ Difference of squares

(b) $4a^2 + 4a + 1 - b^2$
$= (4a^2 + 4a + 1) - b^2$ $\quad$ Group the first three terms.
$= (2a + 1)^2 - b^2$ $\quad$ Perfect square trinomial
$= (2a + 1 + b)(2a + 1 - b)$ $\quad$ Difference of squares

(c) $8m^3 + 4m^2 - n^3 - n^2$
$= (8m^3 - n^3) + (4m^2 - n^2)$ $\quad$ Rearrange and group the terms.
$= (2m - n)(4m^2 + 2mn + n^2) + (2m - n)(2m + n)$ $\quad$ Factor each group.
$= (2m - n)(4m^2 + 2mn + n^2 + 2m + n)$ $\quad$ Factor out $2m - n$.

◀ Work Problem 4 at the Side.

15.6 Exercises

FOR EXTRA HELP MyMathLab Math XL PRACTICE WATCH DOWNLOAD READ REVIEW

Factor each polynomial completely. We have randomly included all the different types of factoring exercises here to give you practice in applying factoring strategies. See Examples 1–4.

1. $100a^2 - 9b^2$

2. $10r^2 + 13r - 3$

3. $18p^5 - 24p^3 + 12p^6$

4. $15x^2 - 20x$

5. $x^2 + 2x - 35$

6. $9 - a^2 + 2ab - b^2$

7. $225p^2 + 256$

8. $x^3 + 1000$

9. $6b^2 - 17b - 3$

10. $k^2 - 6k + 16$

11. $18m^3n + 3m^2n^2 - 6mn^3$

12. $6t^2 + 19tu - 77u^2$

13. $2p^2 + 11pq + 15q^2$

14. $9m^2 - 45m + 18m^3$

15. $4k^2 + 28kr + 49r^2$

16. $54m^3 - 2000$

17. $mn - 2n + 5m - 10$

18. $9m^2 - 30mn + 25n^2 - p^2$

19. $x^3 + 3x^2 - 9x - 27$

20. $56k^3 - 875$

21. $9r^2 + 100$

22. $8p^3 - 125$

23. $6k^2 - k - 1$

24. $27m^2 + 144mn + 192n^2$

25. $x^4 - 625$

26. $125m^6 + 216$

27. $ab + 6b + ac + 6c$

28. $p^3 + 64$

29. $4y^2 - 8y$

30. $6a^4 - 11a^2 - 10$

31. $14z^2 - 3zk - 2k^2$

32. $12z^3 - 6z^2 + 18z$

33. $256b^2 - 400c^2$

34. $z^2 - zp + 20p^2$

35. $1000z^3 + 512$

36. $64m^2 - 25n^2$

37. $10r^2 + 23rs - 5s^2$

38. $12k^2 - 17kq - 5q^2$

39. $32x^2 + 16x^3 - 24x^5$

40. $48k^4 - 243$

41. $14x^2 - 25xq - 25q^2$

42. $5p^2 - 10p$

43. $y^2 + 3y - 10$

44. $b^2 - 7ba - 18a^2$

45. $2a^3 + 6a^2 - 4a$

46. $12m^2rx + 4mnrx + 40n^2rx$

47. $18p^2 + 53pr - 35r^2$

48. $21a^2 - 5ab - 4b^2$

49. $(x - 2y)^2 - 4$

50. $(3m - n)^2 - 25$

51. $(5r + 2s)^2 - 6(5r + 2s) + 9$

52. $(p + 8q)^2 - 10(p + 8q) + 25$

53. $z^4 - 9z^2 + 20$

54. $21m^4 - 32m^2 - 5$

55. $4(p + 2) + m(p + 2)$

56. $kq - 9q + kr - 9r$

57. $50p^2 - 162$

58. $25x^2 - 20xy + 4y^2$

59. $16a^2 + 8ab + b^2$

60. $40p - 32r$

15.7 Solving Quadratic Equations by Factoring

OBJECTIVES

1. Solve quadratic equations by factoring.
2. Solve other equations by factoring.

Galileo Galilei developed theories to explain physical phenomena and set up experiments to test his ideas. According to legend, Galileo dropped objects of different weights from the Leaning Tower of Pisa to disprove the belief that heavier objects fall faster than lighter objects. He developed a formula for freely falling objects described by $\mathbf{d = 16t^2}$, where d is the distance in feet that an object falls (disregarding air resistance) in t seconds, regardless of weight.

The equation $d = 16t^2$ is a *quadratic equation.* A quadratic equation contains a squared term and no terms of higher degree.

Galileo Galilei (1564–1642)

Quadratic Equation

A **quadratic equation** is an equation that can be written in the form

$$ax^2 + bx + c = 0,$$

where a, b, and c are real numbers, with $a \neq 0$. The given form is called **standard form**.

$$x^2 + 5x + 6 = 0, \quad 2t^2 - 5t = 3, \quad y^2 = 4 \qquad \text{Quadratic equations}$$

In these examples, only $x^2 + 5x + 6 = 0$ is in standard form.

Work Problems 1 and 2 at the Side. ▶

Up to now, we have factored *expressions,* including many quadratic expressions of the form $ax^2 + bx + c$. In this section, we use factored quadratic expressions to solve quadratic *equations*.

OBJECTIVE 1 Solve quadratic equations by factoring. We use the **zero-factor property** to solve a quadratic equation by factoring.

Zero-Factor Property

If a and b are real numbers and $ab = 0$, then $a = 0$ or $b = 0$.

In words, if the product of two numbers is 0, then at least one of the numbers must be 0. One number *must* be 0, but both *may* be 0.

EXAMPLE 1 Using the Zero-Factor Property

Solve each equation.

(a) $(x + 3)(2x - 1) = 0$

The product $(x + 3)(2x - 1)$ is equal to 0. By the zero-factor property, the only way that the product of these two factors can be 0 is if at least one of the factors equals 0. Therefore, either $x + 3 = 0$ or $2x - 1 = 0$.

$$x + 3 = 0 \quad \text{or} \quad 2x - 1 = 0 \qquad \text{Zero-factor property}$$

$$x = -3 \quad \text{or} \quad 2x = 1 \qquad \text{Solve each equation.}$$

$$x = \frac{1}{2}$$

Continued on Next Page

1 Which of the following equations are quadratic equations?

A. $y^2 - 4y - 5 = 0$

B. $x^3 - x^2 + 16 = 0$

C. $2z^2 + 7z = -3$

D. $x + 2y = -4$

2 Write each quadratic equation in standard form.

(a) $x^2 - 3x = 4$

(b) $y^2 = 9y - 8$

Answers

1. A, C
2. (a) $x^2 - 3x - 4 = 0$
(b) $y^2 - 9y + 8 = 0$

3 Solve each equation. Check your solutions.

(a) $(x - 5)(x + 2) = 0$

(b) $(3x - 2)(x + 6) = 0$

(c) $z(2z + 5) = 0$

The given equation, $(x + 3)(2x - 1) = 0$, has two solutions, -3 and $\frac{1}{2}$. *Check* these solutions by substituting -3 for x in the original equation, $(x + 3)(2x - 1) = 0$. Then start over and substitute $\frac{1}{2}$ for x.

Check If $x = -3$, then

$$(x + 3)(2x - 1) = 0$$
$$(-3 + 3)[2(-3) - 1] \stackrel{?}{=} 0$$
$$0(-7) = 0. \quad \text{True}$$

If $x = \frac{1}{2}$, then

$$(x + 3)(2x - 1) = 0$$
$$\left(\frac{1}{2} + 3\right)\left(2 \cdot \frac{1}{2} - 1\right) \stackrel{?}{=} 0$$
$$\frac{7}{2}(1 - 1) \stackrel{?}{=} 0$$
$$\frac{7}{2} \cdot 0 = 0. \quad \text{True}$$

Both -3 and $\frac{1}{2}$ result in true equations, so the solution set is $\{-3, \frac{1}{2}\}$.

(b)
$$y(3y - 4) = 0$$
$$y = 0 \quad \text{or} \quad 3y - 4 = 0 \qquad \text{Zero-factor property}$$
$$3y = 4$$
$$y = \frac{4}{3}$$

Don't forget that 0 is a solution.

Check these solutions by substituting each one in the original equation. The solution set is $\{0, \frac{4}{3}\}$.

◀ Work Problem **3** at the Side.

Note

The word *or* as used in Example 1 means "one or the other or both."

In Example 1, each equation to be solved was given with the polynomial in factored form. If the polynomial in an equation is not already factored, first make sure that the equation is in standard form. Then factor and solve.

EXAMPLE 2 Solving Quadratic Equations

Solve each equation.

(a) $x^2 - 5x = -6$

First, write the equation in standard form by adding 6 to each side.

Don't factor x out at this step.

$$x^2 - 5x = -6$$
$$x^2 - 5x + 6 = 0 \qquad \text{Add 6.}$$

Now factor $x^2 - 5x + 6$. Find two numbers whose product is 6 and whose sum is -5. These two numbers are -2 and -3, so the equation becomes

$$(x - 2)(x - 3) = 0. \qquad \text{Factor the trinomial.}$$
$$x - 2 = 0 \quad \text{or} \quad x - 3 = 0 \qquad \text{Zero-factor property}$$
$$x = 2 \quad \text{or} \quad x = 3 \qquad \text{Solve each equation.}$$

Continued on Next Page

Answers

2. (a) $\{-2, 5\}$ (b) $\left\{-6, \frac{2}{3}\right\}$ (c) $\left\{-\frac{5}{2}, 0\right\}$

Check If $x = 2$, then

$$2^2 - 5(2) \stackrel{?}{=} -6$$
$$4 - 10 \stackrel{?}{=} -6$$
$$-6 = -6. \quad \text{True}$$

If $x = 3$, then

$$3^2 - 5(3) \stackrel{?}{=} -6$$
$$9 - 15 \stackrel{?}{=} -6$$
$$-6 = -6. \quad \text{True}$$

Both solutions check, so the solution set is $\{2, 3\}$.

(b)

$y^2 = y + 20$	
$y^2 - y - 20 = 0$	Write in standard form.
$(y - 5)(y + 4) = 0$	Factor the trinomial.
$y - 5 = 0$ or $y + 4 = 0$	Zero-factor property
$y = 5$ or $y = -4$	Solve each equation.

Check by substituting in the original equation. The solution set is $\{-4, 5\}$.

Work Problem 4 at the Side. ▶

Solving a Quadratic Equation by Factoring

Step 1 **Write the equation in standard form,** that is, with all terms on one side of the equals sign in descending powers of the variable and 0 on the other side.

Step 2 **Factor** completely.

Step 3 **Use the zero-factor property** to set each factor with a variable equal to 0.

Step 4 **Solve** the resulting equations.

Step 5 **Check** each solution in the original equation.

EXAMPLE 3 Solving a Quadratic Equation (Common Factor)

Solve $4p^2 + 40 = 26p$.

$4p^2 - 26p + 40 = 0$	Standard form
$2(2p^2 - 13p + 20) = 0$	Factor out 2.
$2p^2 - 13p + 20 = 0$	Divide each side by 2.
$(2p - 5)(p - 4) = 0$	Factor.
$2p - 5 = 0$ or $p - 4 = 0$	Zero-factor property
$2p = 5$ or $p = 4$	Solve each equation.
$p = \frac{5}{2}$	

Check that the solution set is $\{\frac{5}{2}, 4\}$ by substituting in the original equation.

CAUTION
A common error is to include the common factor 2 as a solution in Example 3. ***Only factors containing variables lead to solutions.***

Work Problem 5 at the Side. ▶

4 Solve each equation. Check your solutions.

(a) $m^2 - 3m - 10 = 0$

(b) $r^2 + 2r = 8$

5 Solve each equation. Check your solutions.

(a) $10a^2 - 5a - 15 = 0$

(b) $4x^2 - 2x = 42$

Answers

4. (a) $\{-2, 5\}$ **(b)** $\{-4, 2\}$

5. (a) $\left\{-1, \frac{3}{2}\right\}$ **(b)** $\left\{-3, \frac{7}{2}\right\}$

6 Solve each equation. Check your solutions.

(a) $49m^2 - 9 = 0$

(b) $p(4p + 7) = 2$

(c) $m^2 = 3m$

EXAMPLE 4 Solving Quadratic Equations

Solve each equation.

(a) $16m^2 - 25 = 0$

We can factor the left side of the equation as the difference of squares (**Section 14.5**).

$$16m^2 - 25 = 0$$

$$(4m + 5)(4m - 5) = 0 \quad \text{Factor the difference of squares.}$$

$$4m + 5 = 0 \quad \text{or} \quad 4m - 5 = 0 \quad \text{Zero-factor property}$$

$$4m = -5 \quad \text{or} \quad 4m = 5 \quad \text{Solve each equation.}$$

$$m = -\frac{5}{4} \quad \text{or} \quad m = \frac{5}{4}$$

Check the solutions $-\frac{5}{4}$ and $\frac{5}{4}$ in the original equation. The solution set is $\{-\frac{5}{4}, \frac{5}{4}\}$.

(b) $k(2k + 5) = 3$

We need to write this equation in standard form.

$$k(2k + 5) = 3$$

To be in standard form, 0 must be on one side.

$$2k^2 + 5k = 3 \quad \text{Multiply.}$$

$$2k^2 + 5k - 3 = 0 \quad \text{Subtract 3.}$$

$$(2k - 1)(k + 3) = 0 \quad \text{Factor.}$$

$$2k - 1 = 0 \quad \text{or} \quad k + 3 = 0 \quad \text{Zero-factor property}$$

$$2k = 1 \quad \text{or} \quad k = -3$$

$$k = \frac{1}{2}$$

The solution set is $\{-3, \frac{1}{2}\}$.

(c) $y^2 = 2y$

$$y^2 - 2y = 0 \quad \text{Standard form}$$

Don't forget to set the variable factor y equal to 0.

$$y(y - 2) = 0 \quad \text{Factor.}$$

$$y = 0 \quad \text{or} \quad y - 2 = 0 \quad \text{Zero-factor property}$$

$$y = 2$$

The solution set is $\{0, 2\}$.

CAUTION

In Example 4(b), the zero-factor property could not be used to solve the equation $k(2k + 5) = 3$ in its given form because of the 3 on the right. ***The zero-factor property applies only to a product that equals 0.***

In Example 4(c), it is tempting to begin by dividing each side of the equation $y^2 = 2y$ by y to get $y = 2$. Note that we do not get the other solution, 0, if we divide by a variable. (We *may* divide each side of an equation by a *nonzero* real number, however. For instance, in Example 3 we divided each side by 2.)

◀ Work Problem 6 at the Side.

ANSWERS

6. (a) $\left\{-\frac{3}{7}, \frac{3}{7}\right\}$ (b) $\left\{-2, \frac{1}{4}\right\}$ (c) $\{0, 3\}$

EXAMPLE 5 **Solving a Quadratic Equation with a Double Solution**

Solve each equation.

(a)

$$z^2 + 121 = 22z$$

$$z^2 - 22z + 121 = 0 \qquad \text{Standard form}$$

Because $121 = 11^2$ and $22z = 2 \cdot z \cdot 11$, the trinomial on the left is a perfect square.

$$(z - 11)^2 = 0 \qquad \text{Factor.}$$

To apply the zero-product property, write $(z - 11)^2$ as two separate factors.

$$(z - 11)(z - 11) = 0 \qquad a^2 = a \cdot a$$

$$z - 11 = 0 \quad \text{or} \quad z - 11 = 0 \qquad \text{Zero-factor property}$$

Because the two factors are identical, they both lead to the same solution. (This is called a **double solution.**) Thus,

$$z - 11 = 0$$

$$z = 11. \qquad \text{Add 11.}$$

Check

$$z^2 + 121 = 22z$$

$$11^2 + 121 \stackrel{?}{=} 22(11) \qquad \text{Let } z = 11.$$

$$121 + 121 \stackrel{?}{=} 242$$

$$242 = 242 \qquad \text{True}$$

The solution set is $\{11\}$.

(b)

$$9t^2 - 30t = -25$$

$$9t^2 - 30t + 25 = 0 \qquad \text{Standard form}$$

$$(3t - 5)^2 = 0 \qquad \text{Factor the perfect square trinomial.}$$

$$3t - 5 = 0 \quad \text{or} \quad 3t - 5 = 0 \qquad \text{Zero-factor property}$$

$$3t = 5 \qquad \text{Solve the equation.}$$

$$t = \frac{5}{3}$$

Check the double solution by substituting $\frac{5}{3}$ in the original equation. The solution set is $\{\frac{5}{3}\}$.

Work Problem 7 at the Side. ▶

7 Solve each equation. Check your solutions.

(a) $x^2 + 16x = -64$

(b) $4x^2 - 4x + 1 = 0$

CAUTION

When a trinomial has two identical factors (a perfect square trinomial), as in Examples 5(a) and (b), it is common for students to write the solution of the corresponding quadratic equation twice in the solution set. Each of these equations has only *one* distinct solution. ***There is no need to write the same number more than once in a solution set.***

Note

Not all quadratic equations can be solved by factoring. A more general method for solving such equations is given in **Chapter 17.**

ANSWERS

7. (a) $\{-8\}$ **(b)** $\left\{\frac{1}{2}\right\}$

OBJECTIVE 2 Solve other equations by factoring. We can extend the zero-factor property to solve equations that involve more than two factors with variables, as shown in Examples 6 and 7. (These equations are *not* quadratic equations. Why not?)

8 Solve each equation. Check your solutions.

(a) $r^3 - 16r = 0$

(b) $x^3 - 3x^2 - 18x = 0$

EXAMPLE 6 Solving an Equation with More Than Two Variable Factors

Solve $6z^3 - 6z = 0$.

$$6z^3 - 6z = 0$$

$$6z(z^2 - 1) = 0 \quad \text{Factor out } 6z.$$

$$6z(z + 1)(z - 1) = 0 \quad \text{Factor } z^2 - 1.$$

By an extension of the zero-factor property, this product can equal 0 only if at least one of the factors equals 0. Write and solve three equations, one for each factor with a variable.

$$6z = 0 \quad \text{or} \quad z + 1 = 0 \quad \text{or} \quad z - 1 = 0$$

$$z = 0 \quad \text{or} \quad z = -1 \quad \text{or} \quad z = 1$$

Check by substituting, in turn, 0, −1, and 1 in the original equation. The solution set is $\{-1, 0, 1\}$.

◀ Work Problem 8 at the Side.

9 Solve each equation. Check your solutions.

(a) $(m + 3)(m^2 - 11m + 10) = 0$

(b) $(2x + 5)(4x^2 - 9) = 0$

EXAMPLE 7 Solving an Equation with a Quadratic Factor

Solve $(2x - 1)(x^2 - 9x + 20) = 0$.

$$(2x - 1)(x^2 - 9x + 20) = 0$$

$$(2x - 1)(x - 5)(x - 4) = 0 \quad \text{Factor } x^2 - 9x + 20.$$

$$2x - 1 = 0 \quad \text{or} \quad x - 5 = 0 \quad \text{or} \quad x - 4 = 0 \quad \text{Zero-factor property}$$

$$x = \frac{1}{2} \quad \text{or} \quad x = 5 \quad \text{or} \quad x = 4$$

Check to verify that the solution set is $\{\frac{1}{2}, 4, 5\}$.

◀ Work Problem 9 at the Side.

CAUTION
In Example 7, it would be unproductive to begin by multiplying the two factors together. Keep in mind that the zero-factor property and its extension requires the product of two or more factors to equal 0. ***Always consider first whether an equation is given in the appropriate form to apply the zero-factor property.***

ANSWERS

8. (a) $\{-4, 0, 4\}$ **(b)** $\{-3, 0, 6\}$

9. (a) $\{-3, 1, 10\}$ **(b)** $\left\{-\frac{5}{2}, -\frac{3}{2}, \frac{3}{2}\right\}$

15.7 Exercises

FOR EXTRA HELP

MyMathLab

MathXL PRACTICE WATCH DOWNLOAD READ REVIEW

Solve each equation, and check your solutions. See Example 1.

1. $(x + 5)(x - 2) = 0$

2. $(x - 1)(x + 8) = 0$

3. $(2m - 7)(m - 3) = 0$

4. $(6k + 5)(k + 4) = 0$

5. $t(6t + 5) = 0$

6. $w(4w + 1) = 0$

7. $2x(3x - 4) = 0$

8. $6x(4x + 9) = 0$

9. $\left(x + \frac{1}{2}\right)\left(2x - \frac{1}{3}\right) = 0$

10. $\left(a + \frac{2}{3}\right)\left(5a - \frac{1}{2}\right) = 0$

11. $(x - 9)(x - 9) = 0$

12. $(2x + 1)(2x + 1) = 0$

13. Look at this "solution." ***WHAT WENT WRONG?***

$$2x(3x - 4) = 0$$

$$x = 2 \quad \text{or} \quad x = 0 \quad \text{or} \quad 3x - 4 = 0$$

$$x = \frac{4}{3}$$

The solution set is $\{2, 0, \frac{4}{3}\}$.

14. Look at this "solution." ***WHAT WENT WRONG?***

$$x(7x - 1) = 0$$

$$7x - 1 = 0 \quad \text{Zero-factor property}$$

$$x = \frac{1}{7}$$

The solution set is $\{\frac{1}{7}\}$.

Solve each equation, and check your solutions. See Examples 2–7.

15. $y^2 + 3y + 2 = 0$

16. $p^2 + 8p + 7 = 0$

17. $y^2 - 3y + 2 = 0$

18. $r^2 - 4r + 3 = 0$

19. $x^2 = 24 - 5x$

20. $t^2 = 2t + 15$

21. $x^2 = 3 + 2x$

22. $m^2 = 4 + 3m$

23. $z^2 + 3z = -2$

24. $p^2 - 2p = 3$

25. $m^2 + 8m + 16 = 0$

26. $b^2 - 6b + 9 = 0$

27. $3x^2 + 5x - 2 = 0$

28. $6r^2 - r - 2 = 0$

29. $6p^2 = 4 - 5p$

30. $6x^2 = 4 + 5x$

31. $9s^2 + 12s = -4$

32. $36x^2 + 60x = -25$

33. $y^2 - 9 = 0$

34. $m^2 - 100 = 0$

35. $16k^2 - 49 = 0$

36. $4w^2 - 9 = 0$

37. $n^2 = 121$

38. $x^2 = 400$

39. $x^2 = 7x$

40. $t^2 = 9t$

41. $6r^2 = 3r$

42. $10y^2 = -5y$

43. $g(g - 7) = -10$

44. $r(r - 5) = -6$

45. $z(2z + 7) = 4$

46. $b(2b + 3) = 9$

47. $2(y^2 - 66) = -13y$

48. $3(t^2 + 4) = 20t$

49. $5x^3 - 20x = 0$

50. $3x^3 - 48x = 0$

51. $9y^3 - 49y = 0$

52. $16r^3 - 9r = 0$

53. $(2r + 5)(3r^2 - 16r + 5) = 0$

54. $(3m + 4)(6m^2 + m - 2) = 0$

55. $(2x + 7)(x^2 + 2x - 3) = 0$

56. $(x + 1)(6x^2 + x - 12) = 0$

57. Galileo's formula for freely falling objects, $d = 16t^2$, was given at the beginning of this section. The distance d in feet an object falls depends on the time elapsed t in seconds. (This is an example of an important mathematical concept, a **function.**)

(a) Use Galileo's formula and complete the following table. (*Hint:* Substitute each given value into the formula and solve for the unknown value.)

t in seconds	0	1	2	3		
d in feet	0	16			256	576

(b) When $t = 0$, $d = 0$. Explain this in the context of the problem.

58. In Exercise 57, when you substituted 256 for d and solved for t, you should have found two solutions: 4 and -4. Why doesn't -4 make sense as an answer?

15.8 Applications of Quadratic Equations

OBJECTIVES

1. Solve problems about geometric figures.
2. Solve problems about consecutive integers.
3. Solve problems using the Pythagorean formula.
4. Solve problems using given quadratic models.

We can use factoring to solve quadratic equations that arise in applications. We follow the same basic problem-solving steps used throughout this text.

Solving an Applied Problem

Step 1 **Read** the problem, several times if necessary, until you *understand* what is given and what is to be found.

Step 2 **Assign a variable** to represent the unknown value, using diagrams or tables as needed. Write a statement that tells what the variable represents. Express any other unknown values in terms of the variable.

Step 3 **Write an equation** using the variable expression(s).

Step 4 **Solve** the equation.

Step 5 **State the answer.** Does it seem reasonable?

Step 6 **Check** the answer in the words of the original problem.

OBJECTIVE 1 Solve problems about geometric figures. Refer to the formulas given on the inside back cover of this text, if necessary.

EXAMPLE 1 Solving an Area Problem

The Monroes want to plant a rectangular garden in their yard. The width of the garden will be 4 ft less than its length, and they want it to have an area of 96 ft^2. (ft^2 means square feet.) Find the length and width of the garden.

Step 1 **Read** the problem carefully. We need to find the dimensions of a garden with area 96 ft^2.

Step 2 **Assign a variable.**

Let x = the length of the garden.

Then $x - 4$ = the width. (The width is 4 ft less than the length.)

See Figure 1.

Figure 1

Step 3 **Write an equation.** The area of a rectangle is given by

$$\text{Area} = lw = \text{length} \times \text{width}. \quad \text{Area formula}$$

Substitute 96 for area, x for length, and $x - 4$ for width.

$$A = lw$$

$$96 = x(x - 4) \quad \text{Let } A = 96,\ l = x,\ w = x - 4.$$

Continued on Next Page

1 Solve each problem.

(a) The length of a rectangular room is 2 m more than the width. The area of the floor is 48 m^2. Find the length and width of the room.

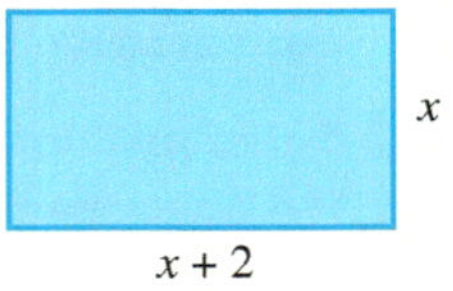

Step 4 **Solve.**

$$96 = x(x - 4)$$

$$96 = x^2 - 4x \quad \text{Distributive property}$$

$$x^2 - 4x - 96 = 0 \quad \text{Standard form}$$

$$(x - 12)(x + 8) = 0 \quad \text{Factor.}$$

$$x - 12 = 0 \quad \text{or} \quad x + 8 = 0 \quad \text{Zero-factor property}$$

$$x = 12 \quad \text{or} \quad x = -8$$

Step 5 **State the answer.** The solutions are 12 and -8. A rectangle cannot have a side of negative length, so discard -8. The length of the garden will be 12 ft. The width will be $12 - 4 = 8$ ft.

Step 6 **Check.** The width of the garden is 4 ft less than the length; the area is $12 \cdot 8 = 96$ ft^2.

Problem-Solving Hint

When solving applied problems, ***always check solutions against physical facts*** and discard any answers that are not appropriate.

◀ *Work Problem* **1** *at the Side.*

OBJECTIVE 2 Solve problems about consecutive integers. Two **consecutive integers** are integers that are next to each other on a number line, such as 5 and 6, or -11 and -10. **Consecutive odd integers** are *odd* integers that are next to each other, such as 5 and 7, or -13 and -11. **Consecutive even integers** are defined similarly; 4 and 6 are consecutive even integers, as are -10 and -8. (In this book, we will list consecutive integers in increasing order from left to right.)

(b) The length of each side of a square is increased by 4 in. The sum of the areas of the original square and the larger square is 106 in^2. What is the length of a side of the original square?

Problem-Solving Hint

In consecutive integer problems, if x represents the first integer, then for

two consecutive integers, use	$x,\ x + 1;$
three consecutive integers, use	$x,\ x + 1,\ x + 2;$
two consecutive even or odd integers, use	$x,\ x + 2;$
three consecutive even or odd integers, use	$x,\ x + 2,\ x + 4.$

EXAMPLE 2 Solving a Consecutive Integer Problem

The product of the numbers on two consecutive post-office boxes is 210. Find the box numbers.

Step 1 **Read** the problem. Note that the boxes are numbered consecutively.

Step 2 **Assign a variable.**

Let x = the first box number.

Then $x + 1$ = the next consecutive box number.

See Figure 2.

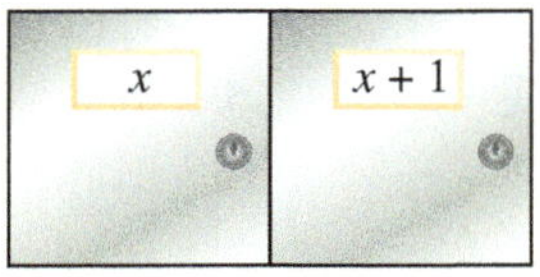

Figure 2

Step 3 **Write an equation.** The product of the box numbers is 210, so

$$x(x + 1) = 210.$$

Continued on Next Page

ANSWERS

1. **(a)** length: 8 m; width: 6 m **(b)** 5 in.

Step 4 **Solve.**

$$x(x + 1) = 210$$

$$x^2 + x = 210 \quad \text{Distributive property}$$

$$x^2 + x - 210 = 0 \quad \text{Standard form}$$

$$(x + 15)(x - 14) = 0 \quad \text{Factor.}$$

$$x + 15 = 0 \quad \text{or} \quad x - 14 = 0 \quad \text{Zero-factor property}$$

$$x = -15 \quad \text{or} \quad x = 14$$

Step 5 **State the answer**. The solutions are -15 and 14. Discard the solution -15 since a box number cannot be negative. When $x = 14$, then $x + 1 = 15$, so the post-office boxes have the numbers 14 and 15.

Step 6 **Check.** The numbers 14 and 15 are consecutive and their product is $14 \cdot 15 = 210$, as required.

Work Problem 2 at the Side. ▶

2 Solve the problem.

The product of the numbers on two consecutive lockers at a health club is 132. Find the locker numbers.

EXAMPLE 3 **Solving a Consecutive Integer Problem**

The product of two consecutive odd integers is 1 less than five times their sum. Find the integers.

Step 1 **Read** carefully. This problem is a little more complicated.

Step 2 **Assign a variable.** We must find two consecutive *odd* integers.

Let x = the lesser integer.

Then $x + 2$ = the next greater odd integer.

Step 3 **Write an equation.** According to the problem, the product is 1 less than five times the sum.

The product	is	five times the sum	less 1.
↓	↓	↓	↓
$x(x + 2)$	$=$	$5(x + x + 2)$	$-\ 1$

Step 4 **Solve.**

$$x^2 + 2x = 5x + 5x + 10 - 1 \quad \text{Distributive property}$$

$$x^2 + 2x = 10x + 9 \quad \text{Combine like terms.}$$

$$x^2 - 8x - 9 = 0 \quad \text{Standard form}$$

$$(x - 9)(x + 1) = 0 \quad \text{Factor.}$$

$$x - 9 = 0 \quad \text{or} \quad x + 1 = 0 \quad \text{Zero-factor property}$$

$$x = 9 \quad \text{or} \quad x = -1$$

Step 5 **State the answer.** We need to find two consecutive odd integers.

If $x = 9$ is the lesser, then $x + 2 = 9 + 2 = 11$ is the greater.

If $x = -1$ is the lesser, then $x + 2 = -1 + 2 = 1$ is the greater.

There are two sets of answers here since integers can be positive or negative.

Step 6 **Check.** The product of the first pair of integers is $9 \cdot 11 = 99$. One less than five times their sum is $5(9 + 11) - 1 = 99$. Thus 9 and 11 satisfy the problem. Repeat the check with -1 and 1.

Work Problem 3 at the Side. ▶

3 Solve each problem.

(a) The product of two consecutive even integers is 4 more than two times their sum. Find the integers.

(b) Find three consecutive odd integers such that the product of the least and greatest is 16 more than the middle integer.

Answers

2. 11 and 12

3. **(a)** 4 and 6 or -2 and 0 **(b)** 3, 5, 7

CAUTION

Do *not* use $x, x + 1, x + 3$, and so on to represent consecutive odd integers. To see why, let $x = 3$. Then $x + 1 = 3 + 1 = 4$ and $x + 3 = 3 + 3 = 6$, and 3, 4, and 6 are not consecutive odd integers.

OBJECTIVE 3 Solve problems using the Pythagorean formula. The next example uses the Pythagorean formula from **Section 5.8.**

Pythagorean Formula

If a right triangle (a triangle with a 90° angle) has longest side of length c and two other sides of lengths a and b, then

$$a^2 + b^2 = c^2.$$

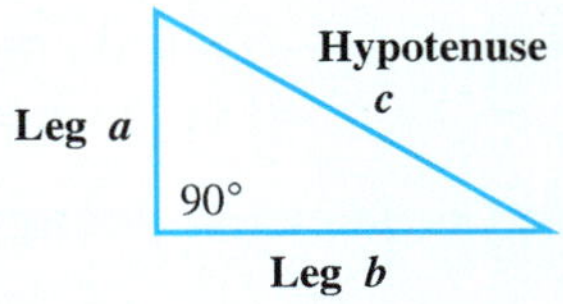

Recall that the longest side, the **hypotenuse,** is opposite the right angle. The two shorter sides are the **legs** of the triangle.

EXAMPLE 4 Using the Pythagorean Formula

Amy and Kevin leave their office, with Amy traveling north and Kevin traveling east. When Kevin is 1 mi farther than Amy from the office, the distance between them is 2 mi more than Amy's distance from the office. Find their distances from the office and the distance between them.

Step 1 **Read** the problem again. We must find three distances.

Step 2 **Assign a variable.** Let x represent Amy's distance from the office, $x + 1$ represent Kevin's distance from the office, and $x + 2$ represent the distance between them. Place these on a right triangle, as in Figure 3.

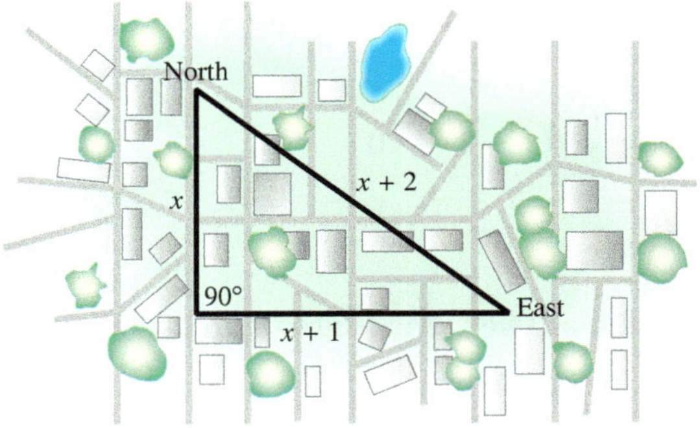

Figure 3

Step 3 **Write an equation.** Substitute into the Pythagorean formula.

$$a^2 + b^2 = c^2$$

$$x^2 + (x + 1)^2 = (x + 2)^2$$

Be careful to substitute properly.

Continued on Next Page

Step 4 **Solve.**

$$x^2 + x^2 + 2x + 1 = x^2 + 4x + 4$$
$$x^2 - 2x - 3 = 0 \quad \text{Standard form}$$
$$(x - 3)(x + 1) = 0 \quad \text{Factor.}$$
$$x - 3 = 0 \quad \text{or} \quad x + 1 = 0 \quad \text{Zero-factor property}$$
$$x = 3 \quad \text{or} \quad x = -1$$

Step 5 **State the answer.** Since -1 cannot represent a distance, 3 is the only possible answer. Amy's distance is 3 mi, Kevin's distance is $3 + 1 = 4$ mi, and the distance between them is $3 + 2 = 5$ mi.

Step 6 **Check.** Since $3^2 + 4^2 = 5^2$, the answer is correct.

CAUTION

When solving a problem involving the Pythagorean formula, be sure that the expressions for the sides are properly placed.

$$\textbf{leg}^2 + \textbf{leg}^2 = \textbf{hypotenuse}^2$$

Work Problem **4** *at the Side.* ▶

OBJECTIVE 4 **Solve problems using given quadratic models.** In Examples 1–4, we wrote quadratic equations to model, or mathematically describe, various situations and then solved the equations. Now we are given the quadratic models and must use them to determine data.

EXAMPLE 5 **Finding the Height of a Ball**

A tennis player can hit a ball 180 ft per sec. If she hits a ball directly upward, the height h of the ball in feet at time t in seconds is modeled by the quadratic equation

$$h = -16t^2 + 180t + 6.$$

When will the ball be 206 ft above the ground?

A height of 206 ft means $h = 206$, so we substitute 206 for h in the equation and then solve for t.

$$\mathbf{206} = -16t^2 + 180t + 6 \quad \text{Let } h = 206.$$
$$-16t^2 + 180t + 6 = 206 \quad \text{Interchange sides.}$$
$$-16t^2 + 180t - 200 = 0 \quad \text{Standard form}$$
$$4t^2 - 45t + 50 = 0 \quad \text{Divide by } -4.$$
$$(4t - 5)(t - 10) = 0 \quad \text{Factor.}$$
$$4t - 5 = 0 \quad \text{or} \quad t - 10 = 0 \quad \text{Zero-factor property}$$
$$t = \frac{5}{4} \quad \text{or} \quad t = 10$$

Since we found two acceptable answers, the ball will be 206 ft above the ground twice (once on its way up and once on its way down)—at $\frac{5}{4}$ sec and at 10 sec after it is hit. See Figure 4.

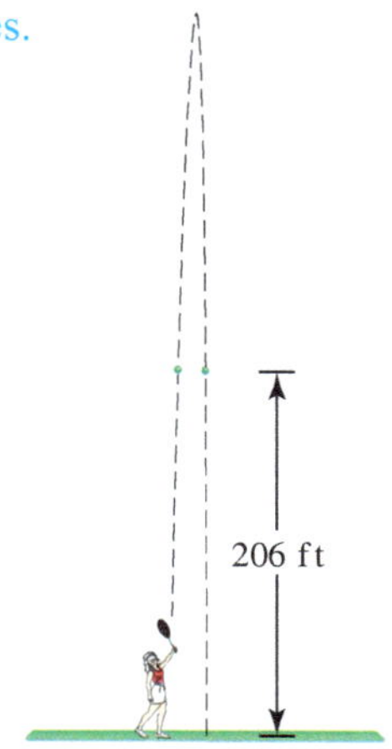

Figure 4

Work Problem **5** *at the Side.* ▶

4 Solve the problem.

The hypotenuse of a right triangle is 3 in. longer than the longer leg. The shorter leg is 3 in. shorter than the longer leg. Find the lengths of the sides of the triangle.

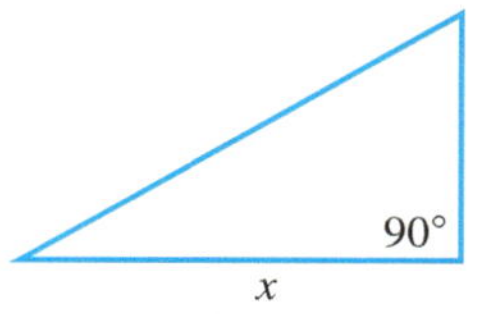

5 Solve the problem.

The number of impulses fired after a nerve has been stimulated is modeled by

$$I = -x^2 + 2x + 60,$$

where x is in milliseconds (ms) after the stimulation. When will 45 impulses occur? Do you get two solutions? Why is only one answer given?

ANSWERS

4. 9 in., 12 in., 15 in.

5. After 5 ms; There are two solutions, -3 and 5; Only one answer makes sense here because a negative answer is not appropriate.

6 Use the model in Example 6 to find the foreign-born population of the United States in 1990. Give your answer to the nearest tenth of a million. How does it compare to the actual value from the table?

EXAMPLE 6 Modeling the Foreign-Born Population of the United States

After decreasing in the middle of the 20th century, the foreign-born population of the United States started to increase in the later part of the century and is now increasing rapidly. The foreign-born population over the years 1930–2004 can be modeled by the quadratic equation

$$y = 0.01036x^2 - 0.5316x + 15.36,$$

where $x = 0$ represents 1930, $x = 10$ represents 1940, and so on, and y is the number of people in millions. (*Source:* U.S. Census Bureau.)

(a) Use the model to find the foreign-born population in 1980 to the nearest tenth of a million.

Since $x = 0$ represents 1930, $x = 50$ represents 1980. Substitute 50 for x in the equation.

$y = 0.01036(50)^2 - 0.5316(50) + 15.36$ Let $x = 50$.

$y = 14.7$ Round to the nearest tenth.

In 1980, the foreign-born population of the United States was about 14.7 million.

(b) Repeat part (a) for 2004.

$y = 0.01036(74)^2 - 0.5316(74) + 15.36$ For 2004, let $x = 74$.

$y = 32.8$ Round to the nearest tenth.

In 2004, the foreign-born population of the United States was about 32.8 million.

(c) The model used in parts (a) and (b) was developed using the data in the table below. How do the results in parts (a) and (b) compare to the actual data from the table?

Year	Foreign-Born Population (millions)
1930	14.2
1940	11.6
1950	10.3
1960	9.7
1970	9.6
1980	14.1
1990	19.8
2000	28.4
2004	34.2

From the table, the actual value for 1980 is 14.1 million. Our answer in part (a), 14.7 million, is slightly high. For 2004, the actual value is 34.2 million, so our answer of 32.8 million in part (b) is somewhat low.

◀ *Work Problem* **6** *at the Side.*

ANSWER

6. 20.8 million; The actual value is 19.8 million, so our answer using the model is somewhat high.

15.8 Exercises

1. To review the six problem-solving steps, complete each statement.

Step 1: ________ the problem, several times if necessary, until you understand what is given and what must be found.

Step 2: Assign a ________ to represent the unknown value.

Step 3: Write a(n) ________ using the variable expression(s).

Step 4: ________ the equation.

Step 5: State the ________.

Step 6: ________ the answer in the words of the ________ problem.

2. A student solves an applied problem and gets 6 or -3 for the length of the side of a square. Which of these answers is reasonable? Explain.

In Exercises 3–6, a figure and a corresponding geometric formula are given. Using x as the variable, complete Steps 3–6 for each problem. (Refer to the steps in Exercise 1 as needed.)

3.

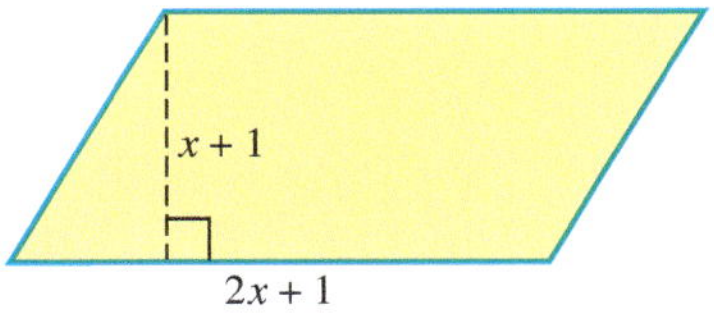

Area of a parallelogram: $A = bh$

The area of this parallelogram is 45 sq. units. Find its base and height.

4.

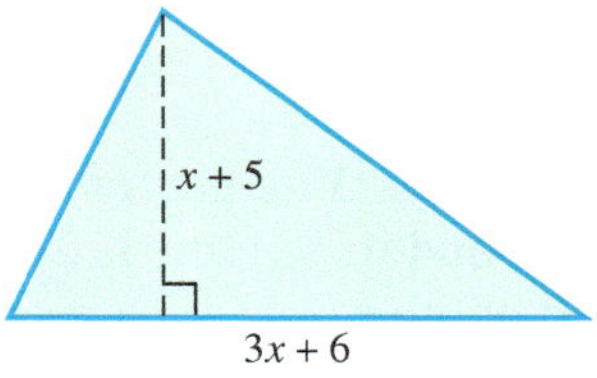

Area of a triangle: $A = \frac{1}{2}bh$

The area of this triangle is 60 sq. units. Find its base and height.

5.

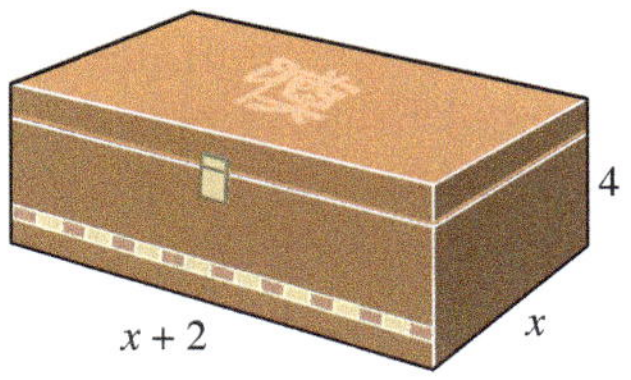

Volume of a rectangular Chinese box: $V = lwh$

The volume of this box is 192 cu. units. Find its length and width.

6.

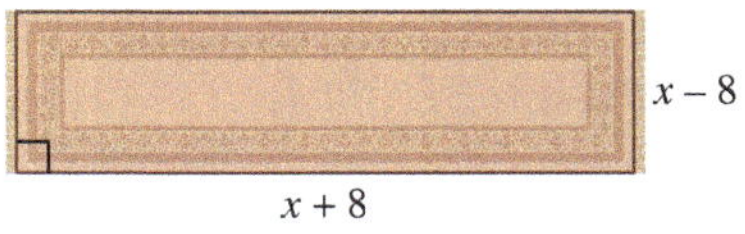

Area of a rectangular rug: $A = lw$

The area of this rug is 80 sq. units. Find its length and width.

Solve each problem. Check your answers to be sure they are reasonable. Refer to the formulas on the inside back cover. See Example 1.

7. The length of a standard jewel case is 2 cm more than its width. The area of the rectangular top of the case is 168 cm^2. Find the length and width of the jewel case.

8. A standard DVD case is 6 cm longer than it is wide. The area of the rectangular top of the case is 247 cm^2. Find the length and width of the case.

9. The dimensions of an HPf1905 flat-panel monitor are such that its length is 3 in. more than its width. If the length were doubled and if the width were decreased by 1 in., the area would be increased by 150 in.^2. What are the length and width of the flat panel?

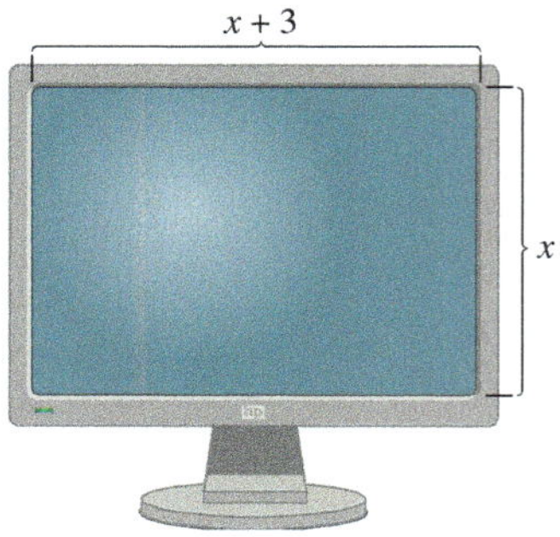

10. The keyboard of the computer in Exercise 9 is 11 in. longer than it is wide. If both its length and width are increased by 2 in., the area of the top of the keyboard is increased by 54 in.^2. Find the length and width of the keyboard. (*Source:* Author's computer.)

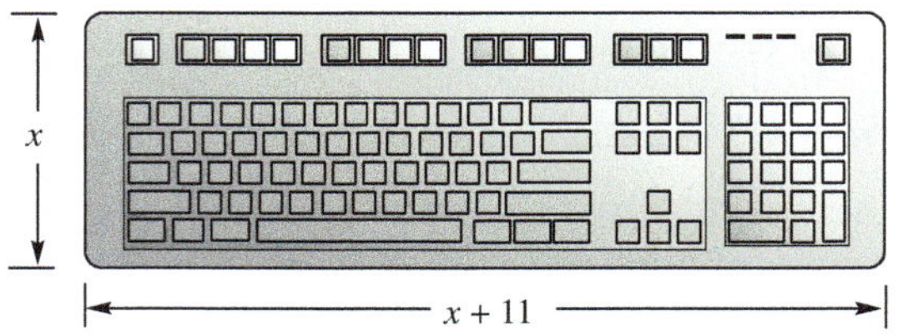

11. A 10-gal aquarium is 3 in. higher than it is wide. Its length is 21 in., and its volume is 2730 in.^3. What are the height and width of the aquarium?

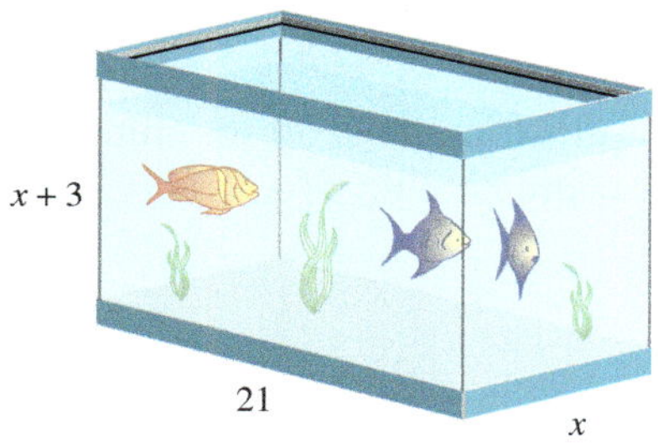

12. A toolbox is 2 ft high, and its width is 3 ft less than its length. If its volume is 80 ft^3, find the length and width of the box.

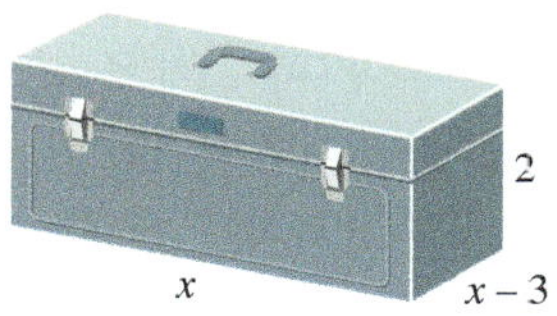

13. A square mirror has sides measuring 2 ft less than the sides of a square painting. If the difference between their areas is 32 ft^2, find the lengths of the sides of the mirror and the painting.

14. The sides of one square have length 3 m more than the sides of a second square. If the area of the larger square is subtracted from 4 times the area of the smaller square, the result is 36 m^2. What are the lengths of the sides of each square?

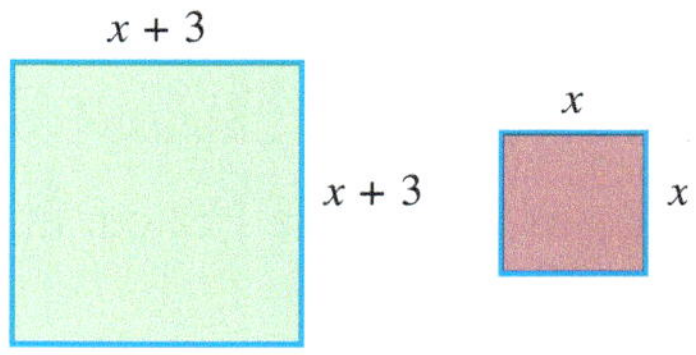

Solve each problem about consecutive integers. See Examples 2 and 3.

15. The product of the numbers on two consecutive volumes of research data is 420. Find the volume numbers.

16. The product of the page numbers on two facing pages of a book is 600. Find the page numbers.

17. The product of two consecutive integers is 11 more than their sum. Find the integers.

18. The product of two consecutive integers is 4 less than four times their sum. Find the integers.

19. Find two consecutive odd integers such that their product is 15 more than three times their sum.

20. Find two consecutive odd integers such that five times their sum is 23 less than their product.

21. Find three consecutive even integers such that the sum of the squares of the lesser two is equal to the square of the greatest.

22. Find three consecutive even integers such that the square of the sum of the lesser two is equal to twice the greatest.

Use the Pythagorean formula to solve each problem. See Example 4.

23. The hypotenuse of a right triangle is 1 cm longer than the longer leg. The shorter leg is 7 cm shorter than the longer leg. Find the length of the longer leg of the triangle.

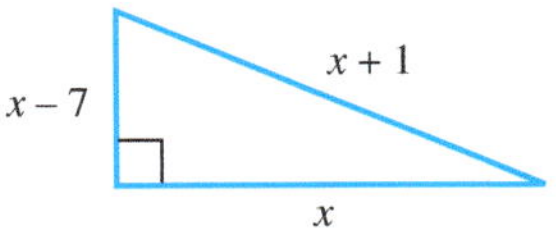

24. The longer leg of a right triangle is 1 m longer than the shorter leg. The hypotenuse is 1 m shorter than twice the shorter leg. Find the length of the shorter leg of the triangle.

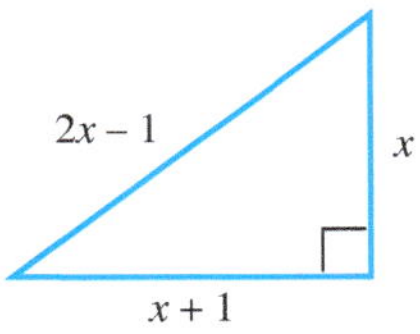

25. Terri works due north of home. Her husband Denny works due east. They leave for work at the same time. By the time Terri is 5 mi from home, the distance between them is 1 mi more than Denny's distance from home. How far from home is Denny?

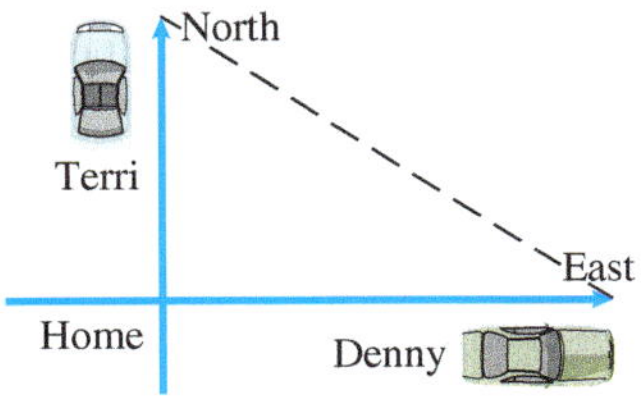

26. Two cars left an intersection at the same time. One traveled north. The other traveled 14 mi farther, but to the east. How far apart were they then, if the distance between them was 4 mi more than the distance traveled east?

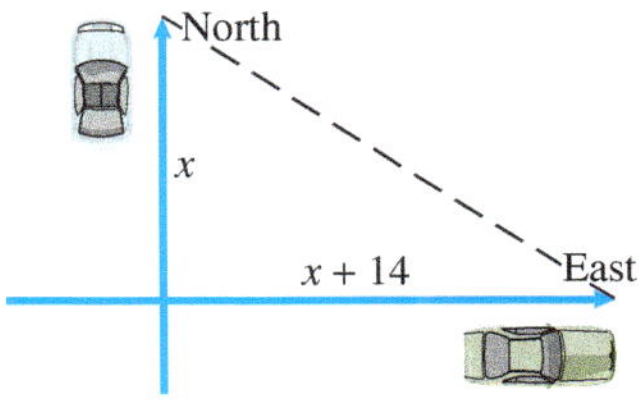

27. A ladder is leaning against a building. The distance from the bottom of the ladder to the building is 4 ft less than the length of the ladder. How high up the side of the building is the top of the ladder if that distance is 2 ft less than the length of the ladder?

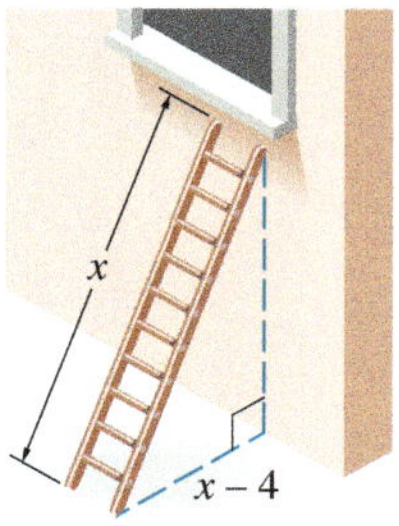

28. A lot has the shape of a right triangle with one leg 2 m longer than the other. The hypotenuse is 2 m less than twice the length of the shorter leg. Find the length of the shorter leg.

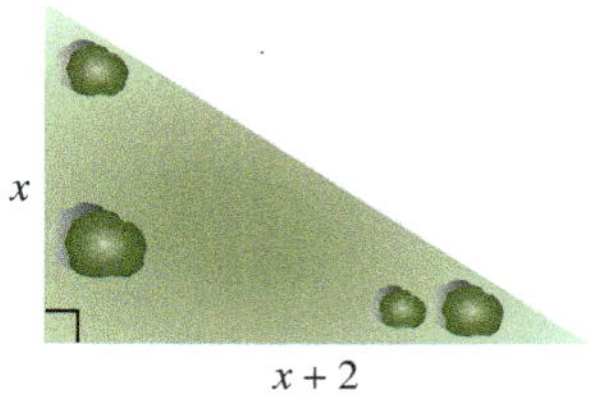

Solve each problem. See Examples 5 and 6.

29. An object projected from a height of 48 ft with an initial velocity of 32 ft per sec after t seconds has height

$$h = -16t^2 + 32t + 48.$$

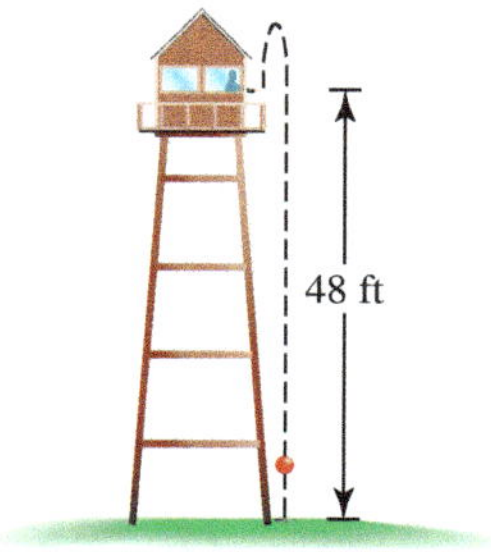

(a) After how many seconds is the height 64 ft? (*Hint:* Let $h = 64$ and solve.)

(b) After how many seconds is the height 60 ft?

(c) After how many seconds does the object hit the ground? (*Hint:* When the object hits the ground, $h = 0$.)

(d) The quadratic equation from part (c) has two solutions, yet only one of them is appropriate for answering the question. Why is this so?

30. If an object is projected upward from ground level with an initial velocity of 64 ft per sec, its height h in feet t seconds later is

$$h = -16t^2 + 64t.$$

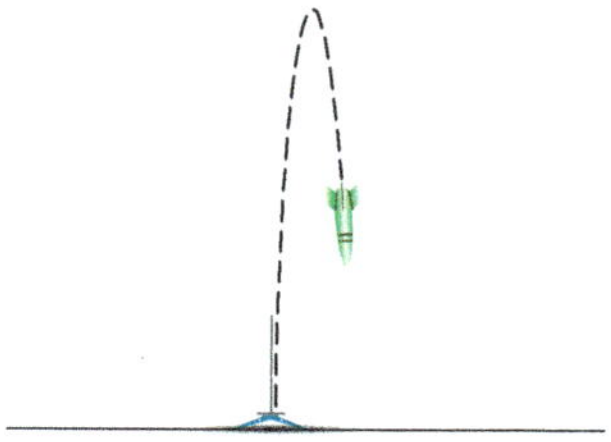

(a) After how many seconds is the height 48 ft?

(b) The object reaches its maximum height 2 sec after it is projected. What is this maximum height?

(c) After how many seconds does the object hit the ground?

(d) The quadratic equation from part (c) has two solutions, yet only one of them is appropriate for answering the question. Why is this so?

31. The table shows the number of cellular phone subscribers (in millions) in the United States.

Year	Subscribers (in millions)
1990	5
1992	11
1994	24
1996	44
1998	69
2000	109
2002	141
2004	182
2006	233

Source: CTIA-The Wireless Association.

We used the data to develop the quadratic equation

$$y = 0.734x^2 + 2.62x + 3.37,$$

which models the number of cellular phone subscribers y (in millions) in the year x, where $x = 0$ represents 1990, $x = 2$ represents 1992, and so on.

(a) Use the model to find the number of cellular phones in 1996 to the nearest million. How does the result compare to the actual data in the table?

(b) What value of x corresponds to 2004?

(c) Use the model to find the number of cellular phones in 2004 to the nearest million. How does the result compare to the actual data in the table?

(d) Assuming that the trend in the data continues, use the quadratic equation to estimate the number of cellular phones in 2009 to the nearest million.

Relating Concepts (Exercises 32–40) For Individual or Group Work

The U.S. trade deficit represents the amount by which exports are less than imports. It provides not only a sign of economic prosperity but also a warning of potential decline. The data in the table shows the U.S. trade deficit in goods and services for 2001 through 2005.

Year	Deficit (in billions of dollars)
2001	365.1
2002	423.7
2003	496.9
2004	612.1
2005	714.4

Source: U.S. Census Bureau.

Use the data to ***work Exercises 32–40 in order.***

32. How much did the trade deficit in goods and services increase from 2001 to 2002? What percent increase is this (to the nearest percent)?

33. The U.S. trade deficit might be approximated by the linear equation

$$y = 88.7x + 256,$$

where y is the deficit in billions of dollars. Here $x = 1$ represents 2001, $x = 2$ represents 2002, and so on. Use this equation to approximate the trade deficits in 2003, 2004, and 2005.

34. How do your answers from Exercise 33 compare to the actual data in the table?

35. The trade deficit y (in billions of dollars) might also be approximated by the quadratic equation

$$y = 9.24x^2 + 33.24x + 321,$$

where $x = 1$ again represents 2001, $x = 2$ represents 2002, and so on. Use this equation to approximate the trade deficits in 2003, 2004, and 2005.

36. Compare your answers from Exercise 35 to the actual data in the table. Which equation, the linear one in Exercise 33 or the quadratic one in Exercise 35, models the data better?

37. Write the data from the table as a set of ordered pairs (x, y), where x represents the years starting with 2001, such that $x = 1$ for 2001, $x = 2$ for 2002, and so on, and y represents the trade deficit in billions of dollars.

Year	Deficit (in billions of dollars)
2001	365.1
2002	423.7
2003	496.9
2004	612.1
2005	714.4

Source: U.S. Census Bureau.

38. Plot the ordered pairs from Exercise 37 on the graph.

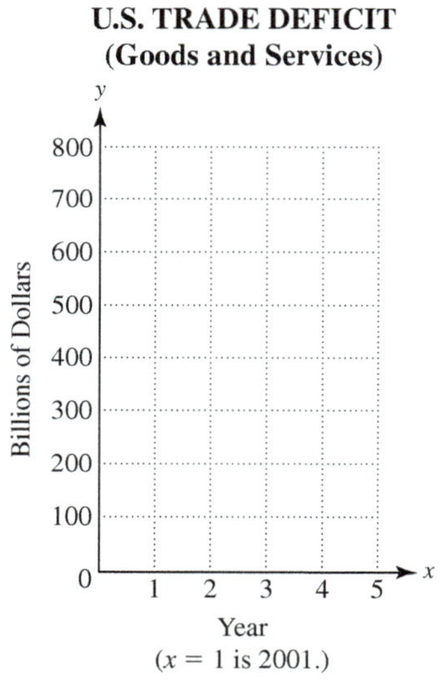

39. Assuming that the trend in the data continues, use the quadratic equation from Exercise 35 to estimate the trade deficit for the year 2006.

40. The actual trade deficit for 2006 was 758.2 billion dollars.

(a) How does the actual deficit for 2006 compare to your estimate from Exercise 39?

(b) Should the quadratic equation be used to estimate the U.S. trade deficit for years after 2005? Explain.

16

Rational Expressions

16.1 The Fundamental Property of Rational Expressions

OBJECTIVES

1. Find the values of the variable for which a rational expression is undefined.
2. Find the numerical value of a rational expression.
3. Write rational expressions in lowest terms.
4. Recognize equivalent forms of rational expressions.

The quotient of two integers (with denominator not 0), such as $\frac{2}{3}$ or $-\frac{3}{4}$, is called a *rational number*. In the same way, the quotient of two polynomials with denominator not equal to 0 is called a *rational expression.*

Rational Expression

A **rational expression** is an expression of the form $\frac{P}{Q}$, where P and Q are polynomials, with $Q \neq 0$.

Examples of rational expressions include

$$\frac{-6x}{x^3+8}, \quad \frac{9x}{y+3}, \quad \text{and} \quad \frac{2m^3}{8}. \qquad \text{Rational expressions}$$

Our work with rational expressions will require much of what we learned in **Chapters 5 and 6** on polynomials and factoring, as well as the rules for fractions from **Chapter R.**

OBJECTIVE 1 Find the values of the variable for which a rational expression is undefined. In the definition of a rational expression $\frac{P}{Q}$, Q cannot equal 0. ***The denominator of a rational expression cannot equal 0 because division by 0 is undefined.***

For instance, in the rational expression

$$\frac{8x^2}{x-3}, \leftarrow \text{Denominator cannot equal 0.}$$

the variable x can take on any value except 3. When x is 3, the denominator becomes $3 - 3 = 0$, making the expression undefined. Thus, x cannot equal 3. We can indicate this restriction by writing $x \neq 3$.

To determine the values for which a rational expression is undefined, use the following procedure.

Determining When a Rational Expression Is Undefined

Step 1 Set the denominator of the rational expression equal to 0.

Step 2 Solve this equation.

Step 3 The solutions of the equation are the values that make the rational expression undefined.

EXAMPLE 1 Finding Values That Make Rational Expressions Undefined

Find any values of the variable for which each rational expression is undefined.

(a) $\frac{p+5}{3p+2}$

Remember that the *numerator* may be any number; we must find any value of p that makes the *denominator* equal to 0 since division by 0 is undefined.

Continued on Next Page

Step 1 Set the denominator equal to 0.

$$3p + 2 = 0$$

Step 2 Solve this equation.

$$3p = -2$$

$$p = -\frac{2}{3}$$

Step 3 The given expression is undefined for $-\frac{2}{3}$, since substituting $-\frac{2}{3}$ for p makes the denominator 0. Thus, $p \neq -\frac{2}{3}$.

(b) $\dfrac{9m^2}{m^2 - 5m + 6}$

$$m^2 - 5m + 6 = 0 \quad \text{Set the denominator equal to 0.}$$

$$(m - 2)(m - 3) = 0 \quad \text{Factor.}$$

$$m - 2 = 0 \quad \text{or} \quad m - 3 = 0 \quad \text{Zero-factor property}$$

$$m = 2 \quad \text{or} \quad m = 3 \quad \text{Solve.}$$

The original expression is undefined for 2 and 3, so $m \neq 2$ and $m \neq 3$.

(c) $\dfrac{2r}{r^2 + 1}$

This denominator will not equal 0 for any value of r because r^2 is always greater than or equal to 0, and adding 1 makes the sum greater than 0. Thus, there are no values for which this rational expression is undefined.

Work Problem **1** *at the Side.* ▶

1 Find any values of the variable for which each rational expression is undefined. Write answers with $\neq$.

(a) $\dfrac{x + 2}{x - 5}$

(b) $\dfrac{3r}{r^2 + 6r + 8}$

(c) $\dfrac{-5m}{m^2 + 4}$

OBJECTIVE 2 Find the numerical value of a rational expression. We use substitution to evaluate a rational expression for a given value of the variable.

EXAMPLE 2 Evaluating Rational Expressions

Find the numerical value of $\dfrac{3x + 6}{2x - 4}$ for each value of x.

(a) $x = 1$

$$\frac{3x + 6}{2x - 4}$$

$$= \frac{3(1) + 6}{2(1) - 4} \quad \text{Let } x = 1.$$

$$= \frac{9}{-2}$$

$$= -\frac{9}{2} \quad \frac{a}{-b} = -\frac{a}{b}$$

(b) $x = 0$

$$\frac{3x + 6}{2x - 4}$$

$$= \frac{3(0) + 6}{2(0) - 4} \quad \text{Let } x = 0.$$

$$= \frac{6}{-4}$$

$$= -\frac{3}{2} \quad \text{Lowest terms}$$

Continued on Next Page

ANSWERS

1. **(a)** $x \neq 5$ **(b)** $r \neq -4, r \neq -2$ **(c)** never undefined

2 Find the numerical value of each rational expression when $x = -3$, $x = 0$, and $x = 3$.

(a) $\dfrac{x}{2x + 1}$

(b) $\dfrac{2x + 6}{x - 3}$

(c) $x = 2$

$$\frac{3x + 6}{2x - 4}$$

$$= \frac{3(2) + 6}{2(2) - 4} \quad \text{Let } x = 2.$$

$$= \frac{12}{0}$$

Substituting 2 for x makes the denominator 0, so the expression is undefined when $x = 2$.

(d) $x = -2$

$$\frac{3x + 6}{2x - 4}$$

$$= \frac{3(-2) + 6}{2(-2) - 4} \quad \text{Let } x = -2.$$

$$= \frac{0}{-8}$$

$$= 0 \quad \tfrac{0}{b} = 0$$

Note

The numerator of a rational expression may be any real number. If the numerator equals 0 and the denominator does *not* equal 0, then the rational expression equals 0. See Example 2(d).

◀ Work Problem **2** at the Side.

OBJECTIVE 3 Write rational expressions in lowest terms. A fraction such as $\frac{2}{3}$ is said to be in *lowest terms*. How can "lowest terms" be defined? We use the idea of greatest common factor for this definition, which applies to all rational expressions.

Lowest Terms

A rational expression $\frac{P}{Q}$ $(Q \neq 0)$ is in **lowest terms** if the greatest common factor of its numerator and denominator is 1.

The properties of rational numbers also apply to rational expressions. We use the **fundamental property of rational expressions** to write a rational expression in lowest terms.

Fundamental Property of Rational Expressions

If $\frac{P}{Q}$ $(Q \neq 0)$ is a rational expression and if K represents any polynomial, where $K \neq 0$, then

$$\frac{PK}{QK} = \frac{P}{Q}.$$

This property is based on the identity property of multiplication, which says that multiplying by 1 leaves any number unchanged.

$$\frac{PK}{QK} = \frac{P}{Q} \cdot \frac{K}{K} = \frac{P}{Q} \cdot 1 = \frac{P}{Q}$$

The next example shows how to write both a rational number and a rational expression in lowest terms. Notice the similarity in the procedures. In both cases, we factor and then divide out the greatest common factor.

Answers

2. (a) $\frac{3}{5}$; 0; $\frac{3}{7}$ (b) 0; -2; undefined

EXAMPLE 3 **Writing in Lowest Terms**

Write each expression in lowest terms.

(a) $\dfrac{30}{72}$

Begin by factoring.

$$= \frac{2 \cdot 3 \cdot 5}{2 \cdot 2 \cdot 2 \cdot 3 \cdot 3}$$

(b) $\dfrac{14k^2}{2k^3}$

Write k^2 as $k \cdot k$ and k^3 as $k \cdot k \cdot k$.

$$= \frac{2 \cdot 7 \cdot k \cdot k}{2 \cdot k \cdot k \cdot k}$$

Group any factors common to the numerator and denominator.

$$= \frac{5 \cdot (2 \cdot 3)}{2 \cdot 2 \cdot 3 \cdot (2 \cdot 3)} \qquad = \frac{7(2 \cdot k \cdot k)}{k(2 \cdot k \cdot k)}$$

Use the fundamental property.

$$= \frac{5}{2 \cdot 2 \cdot 3}, \text{ or } \frac{5}{12} \qquad = \frac{7}{k}$$

Work Problem **3** at the Side. ▶

3 Write each expression in lowest terms.

(a) $\dfrac{15}{40}$

(b) $\dfrac{5x^4}{15x^2}$

(c) $\dfrac{6p^3}{2p^2}$

Writing a Rational Expression in Lowest Terms

Step 1 **Factor** the numerator and denominator completely.

Step 2 **Use the fundamental property** to divide out any common factors.

EXAMPLE 4 **Writing in Lowest Terms**

Write each rational expression in lowest terms.

(a) $\dfrac{3x - 12}{5x - 20}$

$$= \frac{3(x - 4)}{5(x - 4)} \qquad \text{Factor. (Step 1)}$$

$$= \frac{3}{5} \qquad \text{Fundamental property (Step 2)}$$

The given expression is equal to $\frac{3}{5}$ for all values of x, where $x \neq 4$ (since the denominator of the original rational expression is 0 when x is 4).

(b) $\dfrac{m^2 + 2m - 8}{2m^2 - m - 6}$

$$= \frac{(m + 4)(m - 2)}{(2m + 3)(m - 2)} \qquad \text{Factor. (Step 1)}$$

$$= \frac{m + 4}{2m + 3} \qquad \text{Fundamental property (Step 2)}$$

Here $m \neq -\frac{3}{2}$ and $m \neq 2$, since the denominator of the original expression is 0 for these values of m.

ANSWERS

3. **(a)** $\frac{3}{8}$ **(b)** $\frac{x^2}{3}$ **(c)** $3p$

4 Write each rational expression in lowest terms.

(a) $\dfrac{4y+2}{6y+3}$

(b) $\dfrac{8p+8q}{5p+5q}$

(c) $\dfrac{x^2+4x+4}{4x+8}$

(d) $\dfrac{a^2-b^2}{a^2+2ab+b^2}$

5 Write $\frac{x-y}{y-x}$ from Example 5 in lowest terms by factoring -1 from the numerator. How does the result compare to the result in Example 5?

Answers

4. **(a)** $\frac{2}{3}$ **(b)** $\frac{8}{5}$ **(c)** $\frac{x+2}{4}$ **(d)** $\frac{a-b}{a+b}$

5. The result is -1, the same as in Example 5.

From now on, we write statements of equality of rational expressions with the understanding that they apply only to those real numbers that make neither denominator equal to 0.

CAUTION

Rational expressions cannot be written in lowest terms until after the numerator and denominator have been factored. Only common factors can be divided out, not common terms. For example,

$$\frac{6x+9}{4x+6} = \frac{3(2x+3)}{2(2x+3)} = \frac{3}{2} \qquad \Bigg| \qquad \frac{6+x}{4x} \leftarrow \text{Numerator cannot be factored.}$$

Divide out the common factor. (arrow to $(2x+3)$)

This expression is already in lowest terms.

◀ Work Problem 4 at the Side.

EXAMPLE 5 Writing in Lowest Terms (Factors Are Opposites)

Write $\dfrac{x-y}{y-x}$ in lowest terms.

At first glance, there might not seem to be any way in which $x-y$ and $y-x$ can be factored to get a common factor. However, $y-x$ can be factored as

$$\begin{aligned} & y - x \\ &= -1(-y+x) && \text{Factor out } -1. \\ &= -1(x-y) && \text{Commutative property} \end{aligned}$$

Be careful with signs.

With this result in mind, we simplify as follows.

$$\begin{aligned} & \frac{x-y}{y-x} \\ &= \frac{1(x-y)}{-1(x-y)} && y-x = -1(x-y) \text{ from above.} \\ &= \frac{1}{-1}, \quad \text{or} \quad -1 && \text{Fundamental property} \end{aligned}$$

◀ Work Problem 5 at the Side.

In Example 5, notice that $y-x$ is the **opposite** (or **additive inverse**) of $x-y$. A general rule for this situation follows.

If the numerator and the denominator of a rational expression are opposites, such as in $\frac{x-y}{y-x}$, then the rational expression is equal to -1.

CAUTION

Although x and y appear in both the numerator and denominator in Example 5, we cannot use the fundamental property right away because they are *terms,* not *factors.* ***Terms are added, while factors are multiplied.***

EXAMPLE 6 **Writing in Lowest Terms (Factors Are Opposites)**

Write each rational expression in lowest terms.

(a) $\dfrac{2-m}{m-2}$

Since $2-m$ and $m-2$ (or $-2+m$) are opposites, this expression equals -1.

(b) $\dfrac{4x^2-9}{6-4x}$

$$= \frac{(2x+3)(2x-3)}{2(3-2x)} \quad \text{Factor the numerator and denominator.}$$

$$= \frac{(2x+3)(2x-3)}{2(-1)(2x-3)} \quad \text{Write } 3-2x \text{ as } -1(2x-3).$$

$$= \frac{2x+3}{2(-1)} \quad \text{Fundamental property}$$

$$= \frac{2x+3}{-2}, \text{ or } -\frac{2x+3}{2} \quad \frac{a}{-b} = -\frac{a}{b}$$

(c) $\dfrac{3+r}{3-r}$

The quantity $3-r$ *is not* the opposite of $3+r$. This rational expression is already in lowest terms.

Work Problem **6** at the Side. ▶

6 Write each rational expression in lowest terms.

(a) $\dfrac{5-y}{y-5}$

(b) $\dfrac{m-n}{n-m}$

(c) $\dfrac{25x^2-16}{12-15x}$

(d) $\dfrac{9-k}{9+k}$

OBJECTIVE 4 **Recognize equivalent forms of rational expressions.** When working with rational expressions, it is important to be able to recognize equivalent forms of an expression. For example, the common fraction $-\frac{5}{6}$ can also be written as $\frac{-5}{6}$ and as $\frac{5}{-6}$.

Consider the final rational expression from Example 6(b),

$$-\frac{2x+3}{2}.$$

The $-$ sign representing the factor -1 is in front of the expression, even with the fraction bar. As with the common fraction $-\frac{5}{6}$, the factor -1 may instead be placed in the numerator or in the denominator. Some other equivalent forms of this rational expression are

Use parentheses.

$$\frac{-(2x+3)}{2} \quad \text{and} \quad \frac{2x+3}{-2}.$$

By the distributive property,

$$\frac{-(2x+3)}{2} \quad \text{can also be written} \quad \frac{-2x-3}{2}.$$

Multiply *each* term in the parentheses by -1.

CAUTION
$\frac{-2x+3}{2}$ is *not* an equivalent form of $\frac{-(2x+3)}{2}$. The sign preceding 3 in the numerator of $\frac{-2x+3}{2}$ should be $-$ rather than $+$. ***Be careful to apply the distributive property correctly.***

ANSWERS

6. (a) -1 **(b)** -1 **(c)** $\dfrac{5x+4}{-3}$, or $-\dfrac{5x+4}{3}$
(d) already in lowest terms

7 Decide whether each rational expression is equivalent to

$$-\frac{2x-6}{x+3}.$$

(a) $\dfrac{-(2x-6)}{x+3}$

(b) $\dfrac{-2x+6}{x+3}$

(c) $\dfrac{-2x-6}{x+3}$

(d) $\dfrac{2x-6}{-(x+3)}$

(e) $\dfrac{2x-6}{-x-3}$

(f) $\dfrac{2x-6}{x-3}$

EXAMPLE 7 Writing Equivalent Forms of a Rational Expression

Write four equivalent forms of the rational expression

$$-\frac{3x+2}{x-6}.$$

If we apply the negative sign to the numerator, we obtain the equivalent forms

$$\frac{-(3x+2)}{x-6}, \quad \text{and, by the distributive property,} \quad \frac{-3x-2}{x-6}.$$

If we apply the negative sign to the denominator of the given rational expression, we obtain the equivalent forms

$$\frac{3x+2}{-(x-6)} \quad \text{and, by distributing again,} \quad \frac{3x+2}{-x+6}.$$

CAUTION

Recall that $-\frac{5}{6} \neq \frac{-5}{-6}$. Thus, in Example 7, it would be incorrect to distribute the negative sign in $-\frac{3x+2}{x-6}$ to *both* the numerator *and* the denominator. (Doing this would actually lead to the *opposite* of the original expression.)

Work Problem 7 at the Side.

ANSWERS

7. **(a)** equivalent **(b)** equivalent **(c)** not equivalent **(d)** equivalent **(e)** equivalent **(f)** not equivalent

16.1 Exercises

FOR EXTRA HELP MyMathLab MathXL PRACTICE WATCH DOWNLOAD READ REVIEW

1. Fill in each blank with the correct response.

(a) The rational expression $\frac{x+5}{x-3}$ is undefined when $x =$ _____, and is equal to 0 when $x =$ _____.

(b) The rational expression $\frac{p-q}{q-p}$ is undefined when $p =$ _____, and in all other cases when written in lowest terms is equal to _____.

2. Make the correct choice for each blank.

(a) $\frac{4-r^2}{4+r^2}$ _______ (is/is not) equal to -1.

(b) $\frac{5+2x}{3-x}$ and $\frac{-5-2x}{x-3}$ _______ (are/are not) equivalent rational expressions.

3. Define *rational expression* in your own words, and give an example.

4. Why can't the denominator of a rational expression equal 0?

Find any value(s) of the variable for which each rational expression is undefined. Write answers with ≠. See Example 1.

5. $\frac{2}{5y}$ **6.** $\frac{7}{3z}$ **7.** $\frac{x+1}{x+6}$ **8.** $\frac{m+2}{m+5}$

9. $\frac{4x^2}{3x-5}$ **10.** $\frac{2x^3}{3x-4}$ **11.** $\frac{m+2}{m^2+m-6}$ **12.** $\frac{r-5}{r^2-5r+4}$

13. $\frac{x^2-3x}{4}$ **14.** $\frac{x^2-4x}{6}$ **15.** $\frac{3x}{x^2+2}$ **16.** $\frac{4q}{q^2+9}$

*Find the numerical value of each rational expression when **(a)** $x = 2$ and **(b)** $x = -3$. See Example 2.*

17. $\frac{5x-2}{4x}$ **18.** $\frac{3x+1}{5x}$ **19.** $\frac{x^2-4}{2x+1}$ **20.** $\frac{2x^2-4x}{3x-1}$

21. $\frac{(-3x)^2}{4x+12}$ **22.** $\frac{(-2x)^3}{3x+9}$ **23.** $\frac{5x+2}{2x^2+11x+12}$ **24.** $\frac{7-3x}{3x^2-7x+2}$

Write each rational expression in lowest terms. See Examples 3 and 4.

25. $\frac{18r^3}{6r}$ **26.** $\frac{27p^2}{3p}$ **27.** $\frac{4(y-2)}{10(y-2)}$

28. $\dfrac{15(m-1)}{9(m-1)}$

29. $\dfrac{(x+1)(x-1)}{(x+1)^2}$

30. $\dfrac{(t+5)(t-3)}{(t-1)(t+5)}$

31. $\dfrac{7m+14}{5m+10}$

32. $\dfrac{8z-24}{4z-12}$

33. $\dfrac{m^2-n^2}{m+n}$

34. $\dfrac{a^2-b^2}{a-b}$

35. $\dfrac{12m^2-3}{8m-4}$

36. $\dfrac{20p^2-45}{6p-9}$

37. $\dfrac{3m^2-3m}{5m-5}$

38. $\dfrac{6t^2-6t}{2t-2}$

39. $\dfrac{9r^2-4s^2}{9r+6s}$

40. $\dfrac{16x^2-9y^2}{12x-9y}$

41. $\dfrac{2x^2-3x-5}{2x^2-7x+5}$

42. $\dfrac{3x^2+8x+4}{3x^2-4x-4}$

43. $\dfrac{zw+4z-3w-12}{zw+4z+5w+20}$

44. $\dfrac{km+4k+4m+16}{km+4k+5m+20}$

45. $\dfrac{ac-ad+bc-bd}{ac-ad-bc+bd}$

46. Which rational expression can be simplified?

A. $\dfrac{x^2+2}{x^2}$ **B.** $\dfrac{x^2+2}{2}$ **C.** $\dfrac{x^2+y^2}{y^2}$ **D.** $\dfrac{x^2-5x}{x}$

Write each rational expression in lowest terms. See Examples 5 and 6.

47. $\dfrac{6-t}{t-6}$

48. $\dfrac{2-k}{k-2}$

49. $\dfrac{m^2-1}{1-m}$

50. $\dfrac{a^2-b^2}{b-a}$

51. $\dfrac{q^2-4q}{4q-q^2}$

52. $\dfrac{z^2-5z}{5z-z^2}$

Write four equivalent forms for each rational expression. See Example 7.

53. $-\dfrac{x+4}{x-3}$

54. $-\dfrac{x+6}{x-1}$

55. $-\dfrac{2x-3}{x+3}$

56. $-\dfrac{5x-6}{x+4}$

57. $-\dfrac{3x-1}{5x-6}$

58. $-\dfrac{2x-9}{7x-1}$

59. The area of the rectangle is represented by x^4+10x^2+21. What is the width? (*Hint*: Use $w = \frac{A}{l}$.)

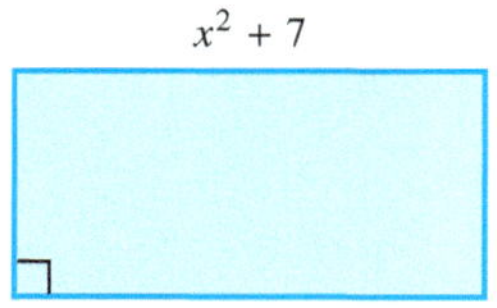

60. The volume of the box is represented by

$$(x^2+8x+15)(x+4).$$

Find the polynomial that represents the area of the bottom of the box.

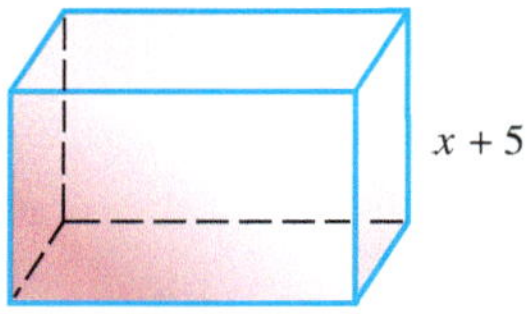

16.2 Multiplying and Dividing Rational Expressions

OBJECTIVE 1 Multiply rational expressions. The product of two fractions is found by multiplying the numerators and multiplying the denominators. Rational expressions are multiplied in the same way.

Multiplying Rational Expressions

The product of the rational expressions $\frac{P}{Q}$ and $\frac{R}{S}$ is shown below.

$$\frac{P}{Q} \cdot \frac{R}{S} = \frac{PR}{QS}$$

In words: To multiply rational expressions, multiply the numerators and multiply the denominators.

EXAMPLE 1 Multiplying Rational Expressions

Multiply. Write each answer in lowest terms.

(a) $\frac{3}{10} \cdot \frac{5}{9}$ **(b)** $\frac{6}{x} \cdot \frac{x^2}{12}$

Indicate the product of the numerators and the product of the denominators.

$= \frac{3 \cdot 5}{10 \cdot 9}$ $= \frac{6 \cdot x^2}{x \cdot 12}$

Leave the products in factored form because common factors are needed to write the product in lowest terms. Factor the numerator and denominator to further identify any common factors. Then use the fundamental property to write each product in lowest terms.

$= \frac{3 \cdot 5}{2 \cdot 5 \cdot 3 \cdot 3}$ $= \frac{6 \cdot x \cdot x}{2 \cdot 6 \cdot x}$

$= \frac{1}{6}$ $= \frac{x}{2}$

Work Problem 1 at the Side. ▶

EXAMPLE 2 Multiplying Rational Expressions

Multiply. Write the answer in lowest terms.

$$\frac{x+y}{2x} \cdot \frac{x^2}{(x+y)^2}$$

$= \frac{(x+y)x^2}{2x(x+y)^2}$ Multiply numerators. Multiply denominators.

$= \frac{(x+y)x \cdot x}{2x(x+y)(x+y)}$ Factor; identify common factors.

$= \frac{x}{2(x+y)}$ $\frac{(x+y)x}{x(x+y)} = 1$; lowest terms

Work Problem 2 at the Side. ▶

OBJECTIVES

1 Multiply rational expressions.

2 Find reciprocals.

3 Divide rational expressions.

1 Multiply. Write each answer in lowest terms.

(a) $\frac{2}{7} \cdot \frac{5}{10}$

(b) $\frac{3m^2}{2} \cdot \frac{10}{m}$

(c) $\frac{8p^2q}{3} \cdot \frac{9}{q^2p}$

2 Multiply. Write each answer in lowest terms.

(a) $\frac{a+b}{5} \cdot \frac{30}{2(a+b)}$

(b) $\frac{3(p-q)}{q^2} \cdot \frac{q}{2(p-q)^2}$

ANSWERS

1. (a) $\frac{1}{7}$ (b) $15m$ (c) $\frac{24p}{q}$

2. (a) 3 (b) $\frac{3}{2q(p-q)}$

3 Multiply. Write each answer in lowest terms.

(a) $\dfrac{x^2 + 7x + 10}{3x + 6} \cdot \dfrac{6x - 6}{x^2 + 2x - 15}$

(b) $\dfrac{m^2 + 4m - 5}{m + 5} \cdot \dfrac{m^2 + 8m + 15}{m - 1}$

4 Find the reciprocal of each rational expression.

(a) $\dfrac{5}{8}$

(b) $\dfrac{6b^5}{3r^2b}$

(c) $\dfrac{t^2 - 4t}{t^2 + 2t - 3}$

ANSWERS

3. (a) $\dfrac{2(x - 1)}{x - 3}$ **(b)** $(m + 5)(m + 3)$

4. (a) $\dfrac{8}{5}$ **(b)** $\dfrac{3r^2b}{6b^5}$ **(c)** $\dfrac{t^2 + 2t - 3}{t^2 - 4t}$

EXAMPLE 3 Multiplying Rational Expressions

Multiply. Write the answer in lowest terms.

$$\frac{x^2 + 3x}{x^2 - 3x - 4} \cdot \frac{x^2 - 5x + 4}{x^2 + 2x - 3}$$

$$= \frac{(x^2 + 3x)(x^2 - 5x + 4)}{(x^2 - 3x - 4)(x^2 + 2x - 3)} \quad \text{Definition of multiplication}$$

$$= \frac{x(x + 3)(x - 4)(x - 1)}{(x - 4)(x + 1)(x + 3)(x - 1)} \quad \text{Factor.}$$

$$= \frac{x}{x + 1} \quad \text{Divide out the common factors.}$$

The quotients $\frac{x + 3}{x + 3}$, $\frac{x - 4}{x - 4}$, and $\frac{x - 1}{x - 1}$ are all equal to 1, justifying the final product $\frac{x}{x + 1}$.

◀ Work Problem **3** at the Side.

OBJECTIVE 2 Find reciprocals. If the product of two rational expressions is 1, the rational expressions are called **reciprocals** (or **multiplicative inverses**) of each other. The reciprocal of a rational expression is found by interchanging the numerator and the denominator. For example,

$$\frac{2x - 1}{x - 5} \text{ has reciprocal } \frac{x - 5}{2x - 1}.$$

EXAMPLE 4 Finding Reciprocals of Rational Expressions

Find the reciprocal of each rational expression.

(a) $\dfrac{4p^3}{9q}$ has reciprocal $\dfrac{9q}{4p^3}$. Interchange the numerator and denominator.

(b) $\dfrac{k^2 - 9}{k^2 - k - 20}$ has reciprocal $\dfrac{k^2 - k - 20}{k^2 - 9}$.

◀ Work Problem **4** at the Side.

OBJECTIVE 3 Divide rational expressions. Suppose we have $\frac{7}{8}$ gal of milk and want to find how many quarts we have. Since 1 qt is $\frac{1}{4}$ gal, we ask, "How many $\frac{1}{4}$s are there in $\frac{7}{8}$?" This would be interpreted as

$$\frac{7}{8} \div \frac{1}{4}, \quad \text{or} \quad \frac{\frac{7}{8}}{\frac{1}{4}} \leftarrow \text{The fraction bar means division.}$$

The fundamental property of rational expressions discussed earlier can be applied to rational number values of P, Q, and K.

$$\frac{P}{Q} = \frac{P \cdot K}{Q \cdot K} = \frac{\frac{7}{8} \cdot 4}{\frac{1}{4} \cdot 4} = \frac{\frac{7}{8} \cdot 4}{1} = \frac{7}{8} \cdot \frac{4}{1}.$$

Let $P = \frac{7}{8}$, $Q = \frac{1}{4}$, and $K = 4$ (the reciprocal of Q).

So, to divide $\frac{7}{8}$ by $\frac{1}{4}$, we multiply $\frac{7}{8}$ by the reciprocal of $\frac{1}{4}$, namely $\frac{4}{1}$, or 4. Since $\frac{7}{8}(4) = \frac{7}{2}$, there are $\frac{7}{2}$ qt, or $3\frac{1}{2}$ qt, in $\frac{7}{8}$ gal.

The preceding discussion illustrates dividing common fractions. Division of rational expressions is defined in the same way.

Dividing Rational Expressions

If $\frac{P}{Q}$ and $\frac{R}{S}$ are any two rational expressions, with $\frac{R}{S} \neq 0$, then

$$\frac{P}{Q} \div \frac{R}{S} = \frac{P}{Q} \cdot \frac{S}{R} = \frac{PS}{QR}$$

In words: To divide one rational expression by another rational expression, multiply the first rational expression (dividend) by the reciprocal of the second rational expression (divisor).

The next example shows the division of two rational numbers and the division of two rational expressions.

EXAMPLE 5 Dividing Rational Expressions

Divide. Write each answer in lowest terms.

(a) $\frac{5}{8} \div \frac{7}{16}$

Multiply the first expression by the reciprocal of the second.

$$= \frac{5}{8} \cdot \frac{16}{7} \quad \leftarrow \text{Reciprocal of } \tfrac{7}{16}$$

$$= \frac{5 \cdot 16}{8 \cdot 7}$$

$$= \frac{5 \cdot 8 \cdot 2}{8 \cdot 7}$$

$$= \frac{10}{7}$$

(b) $\frac{y}{y+3} \div \frac{4y}{y+5}$

$$= \frac{y}{y+3} \cdot \frac{y+5}{4y} \quad \leftarrow \text{Reciprocal of } \tfrac{4y}{y+5}$$

$$= \frac{y(y+5)}{(y+3)(4y)}$$

$$= \frac{y+5}{4(y+3)}$$

Work Problem **5** at the Side. ▶

EXAMPLE 6 Dividing Rational Expressions

Divide. Write the answer in lowest terms.

$$\frac{(3m)^2}{(2p)^3} \div \frac{6m^3}{16p^2}$$

$$= \frac{(3m)^2}{(2p)^3} \cdot \frac{16p^2}{6m^3} \quad \text{Multiply by the reciprocal.}$$

$(3m)^2 = 3^2m^2$; $(2p)^3 = 2^3p^3$

$$= \frac{9m^2}{8p^3} \cdot \frac{16p^2}{6m^3} \quad \text{Power rule for exponents}$$

$$= \frac{9 \cdot 16m^2p^2}{8 \cdot 6p^3m^3} \quad \text{Multiply numerators. Multiply denominators.}$$

$$= \frac{3}{mp} \quad \text{Lowest terms}$$

Work Problem **6** at the Side. ▶

5 Divide. Write each answer in lowest terms.

(a) $\frac{3}{4} \div \frac{5}{16}$

(b) $\frac{r}{r-1} \div \frac{3r}{r+4}$

(c) $\frac{6x-4}{3} \div \frac{15x-10}{9}$

6 Divide. Write each answer in lowest terms.

(a) $\frac{5a^2b}{2} \div \frac{10ab^2}{8}$

(b) $\frac{(3t)^2}{w} \div \frac{3t^2}{5w^4}$

ANSWERS

5. (a) $\frac{12}{5}$ **(b)** $\frac{r+4}{3(r-1)}$ **(c)** $\frac{6}{5}$

6. (a) $\frac{2a}{b}$ **(b)** $15w^3$

EXAMPLE 7 Dividing Rational Expressions

Divide. Write the answer in lowest terms.

$$\frac{x^2-4}{(x+3)(x-2)} \div \frac{(x+2)(x+3)}{-2x}$$

$$= \frac{x^2-4}{(x+3)(x-2)} \cdot \frac{-2x}{(x+2)(x+3)}$$ Multiply by the reciprocal.

$$= \frac{-2x(x^2-4)}{(x+3)(x-2)(x+2)(x+3)}$$ Multiply numerators. Multiply denominators.

$$= \frac{-2x(x+2)(x-2)}{(x+3)(x-2)(x+2)(x+3)}$$ Factor the numerator.

$$= \frac{-2x}{(x+3)^2}, \quad \text{or} \quad -\frac{2x}{(x+3)^2}$$ Lowest terms: $\frac{-a}{b} = -\frac{a}{b}$

◀ Work Problem 7 at the Side.

7 Divide. Write each answer in lowest terms.

(a) $\dfrac{y^2+4y+3}{y+3} \div \dfrac{y^2-4y-5}{y-3}$

(b) $\dfrac{4x(x+3)}{2x+1} \div \dfrac{-x^2(x+3)}{4x^2-1}$

EXAMPLE 8 Dividing Rational Expressions (Factors Are Opposites)

Divide. Write the answer in lowest terms.

$$\frac{m^2-4}{m^2-1} \div \frac{2m^2+4m}{1-m}$$

$$= \frac{m^2-4}{m^2-1} \cdot \frac{1-m}{2m^2+4m}$$ Multiply by the reciprocal.

$$= \frac{(m^2-4)(1-m)}{(m^2-1)(2m^2+4m)}$$ Multiply numerators. Multiply denominators.

$$= \frac{(m+2)(m-2)(1-m)}{(m+1)(m-1)(2m)(m+2)}$$ Factor; $1-m$ and $m-1$ are opposites.

$$= \frac{-1(m-2)}{2m(m+1)}$$ From **Section 15.1**, $\frac{1-m}{m-1} = -1$.

$$= \frac{-m+2}{2m(m+1)}, \quad \text{or} \quad \frac{2-m}{2m(m+1)}$$ Distribute -1 in the numerator.

◀ Work Problem 8 at the Side.

8 Divide. Write each answer in lowest terms.

(a) $\dfrac{ab-a^2}{a^2-1} \div \dfrac{a-b}{a-1}$

(b) $\dfrac{x^2-9}{2x+6} \div \dfrac{9-x^2}{4x-12}$

In summary, follow these steps to multiply or divide rational expressions.

Multiplying or Dividing Rational Expressions

Step 1 **Note the operation.** If the operation is division, use the definition of division to rewrite as multiplication.

Step 2 **Multiply** numerators and multiply denominators.

Step 3 **Factor** all numerators and denominators completely.

Step 4 **Write in lowest terms** using the fundamental property.

Steps 2 and 3 may be interchanged based on personal preference.

Answers

7. (a) $\dfrac{y-3}{y-5}$ (b) $-\dfrac{4(2x-1)}{x}$

8. (a) $\dfrac{-a}{a+1}$ (b) $\dfrac{-2x+6}{3+x}$

16.2 Exercises

FOR EXTRA HELP MyMathLab — Math XL PRACTICE — WATCH — DOWNLOAD — READ — REVIEW

1. Match each multiplication problem in Column I with the correct product in Column II.

I	II
(a) $\frac{5x^3}{10x^4} \cdot \frac{10x^7}{2x}$	A. $\frac{2}{5x^5}$
(b) $\frac{10x^4}{5x^3} \cdot \frac{10x^7}{2x}$	B. $\frac{5x^5}{2}$
(c) $\frac{5x^3}{10x^4} \cdot \frac{2x}{10x^7}$	C. $\frac{1}{10x^7}$
(d) $\frac{10x^4}{5x^3} \cdot \frac{2x}{10x^7}$	D. $10x^7$

2. Match each division problem in Column I with the correct quotient in Column II.

I	II
(a) $\frac{5x^3}{10x^4} \div \frac{10x^7}{2x}$	A. $\frac{5x^5}{2}$
(b) $\frac{10x^4}{5x^3} \div \frac{10x^7}{2x}$	B. $10x^7$
(c) $\frac{5x^3}{10x^4} \div \frac{2x}{10x^7}$	C. $\frac{2}{5x^5}$
(d) $\frac{10x^4}{5x^3} \div \frac{2x}{10x^7}$	D. $\frac{1}{10x^7}$

Multiply. Write each answer in lowest terms. See Examples 1 and 2.

3. $\frac{10m^2}{7} \cdot \frac{14}{15m}$

4. $\frac{36z^3}{6z} \cdot \frac{28}{z^2}$

5. $\frac{16y^4}{18y^5} \cdot \frac{15y^5}{y^2}$

6. $\frac{20x^5}{-2x^2} \cdot \frac{8x^4}{35x^3}$

7. $\frac{2(c+d)}{3} \cdot \frac{18}{6(c+d)^2}$

8. $\frac{4(y-2)}{x} \cdot \frac{3x}{6(y-2)^2}$

Find the reciprocal of each rational expression. See Example 4.

9. $\frac{3p^3}{16q}$

10. $\frac{6x^4}{9y^2}$

11. $\frac{r^2+rp}{7}$

12. $\frac{16}{9a^2+36a}$

13. $\frac{z^2+7z+12}{z^2-9}$

14. $\frac{p^2-4p+3}{p^2-3p}$

Divide. Write each answer in lowest terms. See Examples 5 and 6.

15. $\frac{9z^4}{3z^5} \div \frac{3z^2}{5z^3}$

16. $\frac{35q^8}{9q^5} \div \frac{25q^6}{10q^5}$

17. $\frac{4t^4}{2t^5} \div \frac{(2t)^3}{-6}$

18. $\frac{-12a^6}{3a^2} \div \frac{(2a)^3}{27a}$

19. $\frac{3}{2y-6} \div \frac{6}{y-3}$

20. $\frac{4m+16}{10} \div \frac{3m+12}{18}$

21. Explain in your own words how to multiply rational expressions.

22. Explain in your own words how to divide rational expressions.

Multiply or divide. Write each answer in lowest terms. See Examples 3, 7, and 8.

23. $\frac{5x-15}{3x+9} \cdot \frac{4x+12}{6x-18}$

24. $\frac{8r+16}{24r-24} \cdot \frac{6r-6}{3r+6}$

25. $\frac{2-t}{8} \div \frac{t-2}{6}$

26. $\frac{4}{m-2} \div \frac{16}{2-m}$

27. $\frac{5-4x}{5+4x} \cdot \frac{4x+5}{4x-5}$

28. $\frac{5-x}{5+x} \cdot \frac{x+5}{x-5}$

29. $\frac{6(m-2)^2}{5(m+4)^2} \cdot \frac{15(m+4)}{2(2-m)}$

30. $\frac{7(q-1)}{3(q+1)^2} \cdot \frac{6(q+1)}{3(1-q)^2}$

31. $\frac{p^2+4p-5}{p^2+7p+10} \div \frac{p-1}{p+4}$

32. $\frac{z^2-3z+2}{z^2+4z+3} \div \frac{z-1}{z+1}$

33. $\frac{2k^2-k-1}{2k^2+5k+3} \div \frac{4k^2-1}{2k^2+k-3}$

34. $\frac{2m^2-5m-12}{m^2+m-20} \div \frac{4m^2-9}{m^2+4m-5}$

35. $\frac{2k^2+3k-2}{6k^2-7k+2} \cdot \frac{4k^2-5k+1}{k^2+k-2}$

36. $\frac{2m^2-5m-12}{m^2-10m+24} \div \frac{4m^2-9}{m^2-9m+18}$

37. $\frac{m^2+2mp-3p^2}{m^2-3mp+2p^2} \div \frac{m^2+4mp+3p^2}{m^2+2mp-8p^2}$

38. $\frac{r^2+rs-12s^2}{r^2-rs-20s^2} \div \frac{r^2-2rs-3s^2}{r^2+rs-30s^2}$

39. $\left(\frac{x^2+10x+25}{x^2+10x} \cdot \frac{10x}{x^2+15x+50}\right) \div \frac{x+5}{x+10}$

40. $\left(\frac{m^2-12m+32}{8m} \cdot \frac{m^2-8m}{m^2-8m+16}\right) \div \frac{m-8}{m-4}$

41. If the rational expression $\frac{5x^2y^3}{2pq}$ represents the area of a rectangle and $\frac{2xy}{p}$ represents the length, what rational expression represents the width?

Length = $\frac{2xy}{p}$

The area is $\frac{5x^2y^3}{2pq}$.

42. Consider the division problem $\frac{x-6}{x+4} \div \frac{x+7}{x+5}$.
We know that division by 0 is undefined, so the restrictions on x are $x \neq -4$, $x \neq -5$, and $x \neq -7$. Why is the last restriction needed?

16.3 Least Common Denominators

OBJECTIVE 1 Find the least common denominator for a list of fractions. Just as with common fractions, adding or subtracting rational expressions (to be discussed in the next section) often requires a **least common denominator (LCD),** the simplest expression that is divisible by all denominators. For example, the least common denominator for $\frac{2}{9}$ and $\frac{5}{12}$ is 36, because 36 is the smallest positive number divisible by both 9 and 12.

We can often find least common denominators by inspection. For example, the LCD for $\frac{1}{6}$ and $\frac{2}{3m}$ is $6m$. In other cases, we find the LCD by a procedure similar to that used in **Section 14.1** for finding the greatest common factor.

OBJECTIVES

1 Find the least common denominator for a list of fractions.

2 Write equivalent rational expressions.

Finding the Least Common Denominator (LCD)

Step 1 **Factor** each denominator into prime factors.

Step 2 **List each different denominator factor** the *greatest* number of times it appears in any of the denominators.

Step 3 **Multiply** the denominator factors from Step 2 to get the LCD.

When each denominator is factored into prime factors, every prime factor must be a factor of the least common denominator.

In Example 1, we find the LCD for both numerical and algebraic denominators.

EXAMPLE 1 Finding Least Common Denominators

Find the LCD for each pair of fractions.

(a) $\frac{1}{24}, \frac{7}{15}$ **(b)** $\frac{1}{8x}, \frac{3}{10x}$

Step 1 Write each denominator in factored form with numerical coefficients in prime factored form.

(a)
$$24 = 2 \cdot 2 \cdot 2 \cdot 3 = 2^3 \cdot 3$$
$$15 = 3 \cdot 5$$

(b)
$$8x = 2 \cdot 2 \cdot 2 \cdot x = 2^3 \cdot x$$
$$10x = 2 \cdot 5 \cdot x$$

Step 2 We find the LCD by taking each different factor the *greatest* number of times it appears as a factor in any of the denominators.

(a) The factor 2 appears three times in one product and not at all in the other, so the greatest number of times 2 appears is three. The greatest number of times both 3 and 5 appear is one.

(b) Here, 2 appears three times in one product and once in the other, so the greatest number of times 2 appears is three. The greatest number of times 5 appears is one, and the greatest number of times x appears in either product is one.

Step 3

(a)
$$\text{LCD} = 2 \cdot 2 \cdot 2 \cdot 3 \cdot 5 = 2^3 \cdot 3 \cdot 5 = 120$$

(b)
$$\text{LCD} = 2 \cdot 2 \cdot 2 \cdot 5 \cdot x = 2^3 \cdot 5 \cdot x = 40x$$

Work Problem 1 at the Side. ▶

1 Find the LCD for each pair of fractions.

(a) $\frac{7}{10}, \frac{1}{25}$

(b) $\frac{7}{20p}, \frac{11}{30p}$

(c) $\frac{4}{5x}, \frac{12}{10x}$

ANSWERS

1. (a) 50 **(b)** $60p$ **(c)** $10x$

2 Find the LCD.

(a) $\frac{4}{16m^3n}, \frac{5}{9m^5}$

(b) $\frac{3}{25a^2}, \frac{2}{10a^3b}$

3 Find the LCD for the fractions in each list.

(a) $\frac{7}{3a}, \frac{11}{a^2 - 4a}$

(b) $\frac{2m}{m^2 - 3m + 2}, \frac{5m - 3}{m^2 + 3m - 10}, \frac{4m + 7}{m^2 + 4m - 5}$

(c) $\frac{6}{x - 4}, \frac{3x - 1}{4 - x}$

ANSWERS

2. (a) $144m^5n$ (b) $50a^3b$
3. (a) $3a(a - 4)$
(b) $(m - 1)(m - 2)(m + 5)$
(c) either $x - 4$ or $4 - x$

EXAMPLE 2 Finding the LCD

Find the LCD for $\frac{5}{6r^2}$ and $\frac{3}{4r^3}$.

Step 1 Factor each denominator.

$$6r^2 = 2 \cdot 3 \cdot r^2$$
$$4r^3 = 2^2 \cdot r^3$$

Step 2 The greatest number of times 2 appears is two, the greatest number of times 3 appears is one, and the greatest number of times r appears is three; therefore,

Step 3 $$\text{LCD} = 2^2 \cdot 3 \cdot r^3 = 12r^3.$$

◀ Work Problem **2** at the Side.

CAUTION
When finding the LCD, use each factor the *greatest* number of times it appears in any *single* denominator, not the *total* number of times it appears. For instance, the greatest number of times r appears as a factor in one denominator in Example 2 is 3, *not* 5.

EXAMPLE 3 Finding LCDs

Find the LCD for the fractions in each list.

(a) $\frac{6}{5m}, \frac{4}{m^2 - 3m}$

$$5m = 5 \cdot m$$
$$m^2 - 3m = m(m - 3)$$
} Factor each denominator.

Use each different factor the greatest number of times it appears.

$$\text{LCD} = 5 \cdot m \cdot (m - 3) = 5m(m - 3)$$

Be sure to include m as a factor in the LCD.

Because m is not a *factor* of $m - 3$, both factors, m and $m - 3$, must appear in the LCD.

(b) $\frac{1}{r^2 - 4r - 5}, \frac{3}{r^2 - r - 20}, \frac{1}{r^2 - 10r + 25}$

$$r^2 - 4r - 5 = (r - 5)(r + 1)$$
$$r^2 - r - 20 = (r - 5)(r + 4)$$
$$r^2 - 10r + 25 = (r - 5)^2$$
} Factor each denominator.

Use each different factor the greatest number of times it appears as a factor.

$$\text{LCD} = (r - 5)^2(r + 1)(r + 4)$$

(c) $\frac{1}{q - 5}, \frac{3}{5 - q}$

The expressions $q - 5$ and $5 - q$ are opposites of each other because

$$-(q - 5) = -q + 5 = 5 - q.$$

Therefore, either $q - 5$ or $5 - q$ can be used as the LCD.

◀ Work Problem **3** at the Side.

OBJECTIVE 2 Write equivalent rational expressions. Once the LCD has been found, the next step in preparing to add or subtract two rational expressions is to use the fundamental property to write equivalent rational expressions. We use the following steps.

Writing a Rational Expression with a Specified Denominator

Step 1 **Factor** both denominators.

Step 2 **Decide what factor(s) the denominator must be multiplied by** in order to equal the specified denominator.

Step 3 **Multiply** the rational expression by that factor divided by itself. (That is, multiply by 1.)

EXAMPLE 4 Writing Equivalent Rational Expressions

Write each rational expression as an equivalent expression with the indicated denominator.

(a) $\frac{3}{8} = \frac{?}{40}$

(b) $\frac{9k}{25} = \frac{?}{50k}$

Step 1 For each example, first factor the denominator on the right. Then compare the denominator on the left with the one on the right to decide what factors are missing.

(a) $\frac{3}{8} = \frac{?}{5 \cdot 8}$

Step 2 A factor of 5 is missing.

Step 3 Multiply $\frac{3}{8}$ by $\frac{5}{5}$.

$$\frac{3}{8} = \frac{3}{8} \cdot \frac{5}{5} = \frac{15}{40}$$

$\frac{5}{5} = 1$

(b) $\frac{9k}{25} = \frac{?}{25 \cdot 2k}$

Factors of 2 and k are missing.

Multiply $\frac{9k}{25}$ by $\frac{2k}{2k}$.

$$\frac{9k}{25} = \frac{9k}{25} \cdot \frac{2k}{2k} = \frac{18k^2}{50k}$$

$\frac{2k}{2k} = 1$

Work Problem **4** *at the Side.* ▶

EXAMPLE 5 Writing Equivalent Rational Expressions

Write each rational expression as an equivalent expression with the indicated denominator.

(a)

$$\frac{8}{3x+1} = \frac{?}{12x+4}$$

$$\frac{8}{3x+1} = \frac{?}{4(3x+1)} \quad \text{Factor the denominator on the right.}$$

The missing factor is 4, so multiply the fraction on the left by $\frac{4}{4}$.

$$\frac{8}{3x+1} \cdot \frac{4}{4} = \frac{32}{12x+4} \quad \text{Fundamental property}$$

Continued on Next Page

4 Write each rational expression as an equivalent expression with the indicated denominator.

(a) $\frac{3}{4} = \frac{?}{36}$

(b) $\frac{7k}{5} = \frac{?}{30p}$

Answers

4. (a) $\frac{27}{36}$ **(b)** $\frac{42kp}{30p}$

(b) $\dfrac{12p}{p^2 + 8p} = \dfrac{?}{p^3 + 4p^2 - 32p}$

Factor the denominator in each rational expression.

$$\frac{12p}{p(p+8)} = \frac{?}{p(p+8)(p-4)}$$

$p^3 + 4p^2 - 32p$
$= p(p^2 + 4p - 32)$
$= p(p+8)(p-4)$

The factor $p - 4$ is missing, so multiply $\frac{12p}{p(p+8)}$ by $\frac{p-4}{p-4}$.

$$\frac{12p}{p^2+8p} = \frac{12p}{p(p+8)} \cdot \frac{p-4}{p-4} \quad \text{Fundamental property}$$

$$= \frac{12p(p-4)}{p(p+8)(p-4)} \quad \begin{array}{l}\text{Multiply numerators.}\\ \text{Multiply denominators.}\end{array}$$

$$= \frac{\mathbf{12p^2 - 48p}}{p^3 + 4p^2 - 32p} \quad \text{Multiply the factors.}$$

5 Write each rational expression as an equivalent expression with the indicated denominator.

(a) $\dfrac{9}{2a+5} = \dfrac{?}{6a+15}$

(b) $\dfrac{5k+1}{k^2+2k} = \dfrac{?}{k^3+k^2-2k}$

Note

In the next section we add and subtract rational expressions, which sometimes requires the steps illustrated in Examples 4 and 5. While it may be beneficial to leave the equivalent expression in factored form, we multiplied out the factors in the numerator and the denominator in Example 5(b),

$$\frac{12p(p-4)}{p(p+8)(p-4)},$$

in order to give the answer,

$$\frac{12p^2 - 48p}{p^3 + 4p^2 - 32p},$$

in the same form as the original problem.

◀ *Work Problem* **5** *at the Side.*

Answers

5. (a) $\dfrac{27}{6a+15}$

(b) $\dfrac{(5k+1)(k-1)}{k^3+k^2-2k}$, or $\dfrac{5k^2-4k-1}{k^3+k^2-2k}$

16.3 Exercises

FOR EXTRA HELP MyMathLab — Math XL PRACTICE · WATCH · DOWNLOAD · READ · REVIEW

Choose the correct response in Exercises 1–4.

1. Suppose that the greatest common factor of a and b is 1. Then the least common denominator for $\frac{1}{a}$ and $\frac{1}{b}$ is

A. a **B.** b **C.** ab **D.** 1.

2. If a is a factor of b, then the least common denominator for $\frac{1}{a}$ and $\frac{1}{b}$ is

A. a **B.** b **C.** ab **D.** 1.

3. The least common denominator for $\frac{11}{20}$ and $\frac{1}{2}$ is

A. 40 **B.** 2 **C.** 20 **D.** none of these.

4. Suppose that we wish to write the rational expression $\dfrac{1}{(x-4)^2(y-3)}$ with denominator $(x-4)^3(y-3)^2$. We must multiply both the numerator and the denominator by

A. $(x-4)(y-3)$ **B.** $(x-4)^2$

C. $x-4$ **D.** $(x-4)^2(y-3)$.

Find the least common denominator for the fractions in each list. See Examples 1 and 2.

5. $\dfrac{2}{15}, \dfrac{3}{10}, \dfrac{7}{30}$

6. $\dfrac{5}{24}, \dfrac{7}{12}, \dfrac{9}{28}$

7. $\dfrac{3}{x^4}, \dfrac{5}{x^7}$

8. $\dfrac{2}{y^5}, \dfrac{3}{y^6}$

9. $\dfrac{5}{36q}, \dfrac{17}{24q}$

10. $\dfrac{4}{30p}, \dfrac{9}{50p}$

11. $\dfrac{6}{21r^3}, \dfrac{8}{12r^5}$

12. $\dfrac{9}{35t^2}, \dfrac{5}{49t^6}$

13. If the denominators of two fractions in prime factored form are $2^3 \cdot 3$ and $2^2 \cdot 5$, what is the factored form of their LCD?

14. Suppose two rational expressions have denominators $(t+4)^3(t-3)$ and $(t+4)^2(t+8)$. Find the factored form of their LCD. What is the similarity between the answers for this problem and for Exercise 13?

15. If two denominators have greatest common factor equal to 1, how can you easily find their least common denominator?

16. Suppose two fractions have denominators a^k and a^r, where k and r are natural numbers, with $k > r$. What is their least common denominator?

Find the least common denominator for the fractions in each list. See Examples 1–3.

17. $\dfrac{9}{28m^2}, \dfrac{3}{12m-20}$

18. $\dfrac{15}{27a^3}, \dfrac{8}{9a-45}$

19. $\dfrac{7}{5b-10}, \dfrac{11}{6b-12}$

20. $\dfrac{3}{7x^2+21x}, \dfrac{1}{5x^2+15x}$

21. $\dfrac{5}{c-d}, \dfrac{8}{d-c}$

22. $\dfrac{4}{y-x}, \dfrac{7}{x-y}$

23. $\dfrac{3}{k^2 + 5k}, \dfrac{2}{k^2 + 3k - 10}$

24. $\dfrac{1}{z^2 - 4z}, \dfrac{4}{z^2 - 3z - 4}$

25. $\dfrac{5}{p^2 + 8p + 15}, \dfrac{3}{p^2 - 3p - 18}, \dfrac{2}{p^2 - p - 30}$

26. $\dfrac{10}{y^2 - 10y + 21}, \dfrac{2}{y^2 - 2y - 3}, \dfrac{5}{y^2 - 6y - 7}$

Write each rational expression as an equivalent expression with the indicated denominator. See Examples 4 and 5.

27. $\dfrac{4}{11} = \dfrac{?}{55}$

28. $\dfrac{6}{7} = \dfrac{?}{42}$

29. $\dfrac{-5}{k} = \dfrac{?}{9k}$

30. $\dfrac{-3}{q} = \dfrac{?}{6q}$

31. $\dfrac{13}{40y} = \dfrac{?}{80y^3}$

32. $\dfrac{5}{27p} = \dfrac{?}{108p^4}$

33. $\dfrac{5t^2}{6r} = \dfrac{?}{42r^4}$

34. $\dfrac{8y^2}{3x} = \dfrac{?}{30x^3}$

35. $\dfrac{5}{2(m + 3)} = \dfrac{?}{8(m + 3)}$

36. $\dfrac{7}{4(y - 1)} = \dfrac{?}{16(y - 1)}$

37. $\dfrac{-4t}{3t - 6} = \dfrac{?}{12 - 6t}$

38. $\dfrac{-7k}{5k - 20} = \dfrac{?}{40 - 10k}$

39. $\dfrac{14}{z^2 - 3z} = \dfrac{?}{z(z - 3)(z - 2)}$

40. $\dfrac{12}{x(x + 4)} = \dfrac{?}{x(x + 4)(x - 9)}$

41. $\dfrac{2(b - 1)}{b^2 + b} = \dfrac{?}{b^3 + 3b^2 + 2b}$

42. $\dfrac{3(c + 2)}{c(c - 1)} = \dfrac{?}{c^3 - 5c^2 + 4c}$

16.4 Adding and Subtracting Rational Expressions

To add and subtract rational expressions, we find least common denominators and write equivalent fractions with the LCD.

OBJECTIVES

1. **Add rational expressions having the same denominator.**
2. **Add rational expressions having different denominators.**
3. **Subtract rational expressions.**

OBJECTIVE 1 Add rational expressions having the same denominator. We find the sum of two such rational expressions with the same procedure that we used in **Section 4.4** for adding two fractions having the same denominator.

Adding Rational Expressions (Same Denominator)

If $\frac{P}{Q}$ and $\frac{R}{Q}$ $(Q \neq 0)$ are rational expressions, then

$$\frac{P}{Q} + \frac{R}{Q} = \frac{P+R}{Q}.$$

In words: To add rational expressions with the same denominator, add the numerators and keep the same denominator.

EXAMPLE 1 Adding Rational Expressions (Same Denominator)

Add. Write each answer in lowest terms.

(a) $\frac{4}{9} + \frac{2}{9}$

(b) $\frac{3x}{x+1} + \frac{3}{x+1}$

The denominators are the same, so the sum is found by adding the two numerators and keeping the same (common) denominator.

(a)

$= \frac{4+2}{9}$ Add.

$= \frac{6}{9}$

$= \frac{2 \cdot 3}{3 \cdot 3}$ Factor.

$= \frac{2}{3}$ Lowest terms

(b)

$= \frac{3x+3}{x+1}$ Add.

$= \frac{3(x+1)}{x+1}$ Factor.

$= 3$ Lowest terms

Work Problem 1 at the Side. ▶

1 Add. Write each answer in lowest terms.

(a) $\frac{7}{15} + \frac{3}{15}$

(b) $\frac{3}{y+4} + \frac{2}{y+4}$

(c) $\frac{x}{x+y} + \frac{1}{x+y}$

(d) $\frac{a}{a+b} + \frac{b}{a+b}$

(e) $\frac{x^2}{x+1} + \frac{x}{x+1}$

OBJECTIVE 2 Add rational expressions having different denominators. As in **Section 4.4,** we use the following steps to add two rational expressions having different denominators.

Adding Rational Expressions (Different Denominators)

Step 1 **Find the least common denominator (LCD).**

Step 2 **Write each rational expression** as an equivalent rational expression with the LCD as the denominator.

Step 3 **Add** the numerators to get the numerator of the sum. The LCD is the denominator of the sum.

Step 4 **Write in lowest terms** using the fundamental property.

ANSWERS

1. (a) $\frac{2}{3}$ (b) $\frac{5}{y+4}$ (c) $\frac{x+1}{x+y}$ (d) 1 (e) x

2 Add. Write each answer in lowest terms.

(a) $\frac{1}{10} + \frac{1}{15}$

(b) $\frac{6}{5x} + \frac{9}{2x}$

(c) $\frac{m}{3n} + \frac{2}{7n}$

EXAMPLE 2 Adding Rational Expressions (Different Denominators)

Add. Write each answer in lowest terms.

(a) $\frac{1}{12} + \frac{7}{15}$ **(b)** $\frac{2}{3y} + \frac{1}{4y}$

Step 1 First find the LCD, using the methods of the previous section.

$12 = 2 \cdot 2 \cdot 3 = 2^2 \cdot 3$	$3y = 3 \cdot y$
$15 = 3 \cdot 5$	$4y = 2 \cdot 2 \cdot y = 2^2 \cdot y$
$\text{LCD} = 2^2 \cdot 3 \cdot 5 = 60$	$\text{LCD} = 2^2 \cdot 3 \cdot y = 12y$

Step 2 Now write each rational expression as an equivalent expression with the LCD (either 60 or $12y$) as the denominator.

$$\frac{1}{12} + \frac{7}{15} = \frac{1(5)}{12(5)} + \frac{7(4)}{15(4)} = \frac{5}{\mathbf{60}} + \frac{28}{\mathbf{60}}$$

$$\frac{2}{3y} + \frac{1}{4y} = \frac{2(4)}{3y(4)} + \frac{1(3)}{4y(3)} = \frac{8}{\mathbf{12y}} + \frac{3}{\mathbf{12y}}$$

Step 3 Add the numerators. The LCD is the denominator.

Step 4 Write in lowest terms if necessary.

$$= \frac{5 + 28}{\mathbf{60}} = \frac{33}{60}, \text{ or } \frac{11}{20}$$

$$= \frac{8 + 3}{\mathbf{12y}} = \frac{11}{12y}$$

◀ Work Problem **2** at the Side.

EXAMPLE 3 Adding Rational Expressions

Add. Write the answer in lowest terms.

$$\frac{2x}{x^2 - 1} + \frac{-1}{x + 1}$$

Step 1 Since the denominators are different, find the LCD.

$x^2 - 1 = (x + 1)(x - 1)$
$x + 1$ is prime.
} The LCD is $(x + 1)(x - 1)$.

Step 2 Write each rational expression as an equivalent expression with the LCD as the denominator.

$$\frac{2x}{x^2 - 1} + \frac{-1}{x + 1}$$ The LCD is $(x + 1)(x - 1)$.

$$= \frac{2x}{(x + 1)(x - 1)} + \frac{-1\,(x - 1)}{(x + 1)\,(x - 1)}$$ Multiply the second fraction by $\frac{x - 1}{x - 1}$.

$$= \frac{2x}{(x + 1)(x - 1)} + \frac{-x + 1}{(x + 1)(x - 1)}$$ Distributive property

Step 3 $$= \frac{2x - x + 1}{(x + 1)(x - 1)}$$ Add numerators; keep the same denominator.

Continued on Next Page

ANSWERS

2. (a) $\frac{1}{6}$ (b) $\frac{57}{10x}$ (c) $\frac{7m + 6}{21n}$

$$= \frac{x+1}{(x+1)(x-1)}$$ Combine like terms in the numerator.

Step 4 $$= \frac{1(x+1)}{(x+1)(x-1)}$$ Identity property for multiplication

Remember to write 1 in the numerator.

$$= \frac{1}{x-1}$$ Divide out the common factors.

Work Problem **3** at the Side. ▶

EXAMPLE 4 Adding Rational Expressions

Add. Write the answer in lowest terms.

$$\frac{2x}{x^2+5x+6} + \frac{x+1}{x^2+2x-3}$$

$$= \frac{2x}{(x+2)(x+3)} + \frac{x+1}{(x+3)(x-1)}$$ Factor the denominators.

$$= \frac{2x(x-1)}{(x+2)(x+3)(x-1)} + \frac{(x+1)(x+2)}{(x+2)(x+3)(x-1)}$$ The LCD is $(x+2)\cdot(x+3)(x-1)$.

$$= \frac{2x(x-1)+(x+1)(x+2)}{(x+2)(x+3)(x-1)}$$ Add numerators; keep the same denominator.

$$= \frac{2x^2-2x+x^2+3x+2}{(x+2)(x+3)(x-1)}$$ Multiply.

$$= \frac{3x^2+x+2}{(x+2)(x+3)(x-1)}$$ Combine like terms.

It is usually more convenient to leave the denominator in factored form. The numerator cannot be factored here, so the expression is in lowest terms.

Work Problem **4** at the Side. ▶

EXAMPLE 5 Adding Rational Expressions with Denominators That Are Opposites

Add. Write the answer in lowest terms.

$$\frac{y}{y-2} + \frac{8}{2-y}$$ The denominators are opposites.

$$= \frac{y}{y-2} + \frac{8(-1)}{(2-y)(-1)}$$ Multiply $\frac{8}{2-y}$ by $\frac{-1}{-1}$ to get a common denominator.

$$= \frac{y}{y-2} + \frac{-8}{-2+y}$$ Distributive property

$$= \frac{y}{y-2} + \frac{-8}{y-2}$$ Rewrite $-2+y$ as $y-2$.

$$= \frac{y-8}{y-2}$$ Add numerators; keep the same denominator.

If we had chosen to use $2-y$ as the common denominator, the final answer would be in the form $\frac{8-y}{2-y}$, which is equivalent to $\frac{y-8}{y-2}$.

Work Problem **5** at the Side. ▶

3 Add. Write each answer in lowest terms.

(a) $\frac{2p}{3p+3} + \frac{5p}{2p+2}$

(b) $\frac{4}{y^2-1} + \frac{6}{y+1}$

(c) $\frac{-2}{p+1} + \frac{4p}{p^2-1}$

4 Add. Write each answer in lowest terms.

(a) $\frac{2k}{k^2-5k+4} + \frac{3}{k^2-1}$

(b) $\frac{4m}{m^2+3m+2} + \frac{2m-1}{m^2+6m+5}$

5 Add. Write the answer in lowest terms.

$\frac{m}{2m-3n} + \frac{n}{3n-2m}$

Answers

3. (a) $\frac{19p}{6(p+1)}$ **(b)** $\frac{2(3y-1)}{(y+1)(y-1)}$

(c) $\frac{2}{p-1}$

4. (a) $\frac{(2k-3)(k+4)}{(k-4)(k-1)(k+1)}$

(b) $\frac{6m^2+23m-2}{(m+2)(m+1)(m+5)}$

5. $\frac{m-n}{2m-3n}$, or $\frac{n-m}{3n-2m}$

OBJECTIVE 3 Subtract rational expressions. To subtract rational expressions having the same denominator, use the following rule.

Subtracting Rational Expressions (Same Denominator)

If $\frac{P}{Q}$ and $\frac{R}{Q}$ $(Q \neq 0)$ are rational expressions, then

$$\frac{P}{Q} - \frac{R}{Q} = \frac{P - R}{Q}.$$

In words: To subtract rational expressions with the same denominator, subtract the numerators and keep the same denominator.

EXAMPLE 6 Subtracting Rational Expressions (Same Denominator)

Subtract. Write the answer in lowest terms.

$$\frac{2m}{m-1} - \frac{m+3}{m-1}$$

Use parentheses around the quantity being subtracted.

$$= \frac{2m - (m+3)}{m-1}$$ Subtract numerators; keep the same denominator.

Be careful with signs.

$$= \frac{2m - m - 3}{m-1}$$ Distributive property

$$= \frac{m-3}{m-1}$$ Combine like terms.

CAUTION

Sign errors often occur in problems like the one in Example 6. The numerator of the fraction being subtracted must be treated as a single quantity. ***Be sure to use parentheses after the subtraction sign.***

◀ *Work Problem 6 at the Side.*

In the remaining examples, we subtract rational expressions having different denominators.

EXAMPLE 7 Subtracting Rational Expressions (Different Denominators)

Subtract. Write the answer in lowest terms.

$$\frac{9}{x-2} - \frac{3}{x}$$ The LCD is $x(x-2)$.

$$= \frac{9x}{x(x-2)} - \frac{3(x-2)}{x(x-2)}$$ Rewrite each expression with the LCD.

$$= \frac{9x - 3(x-2)}{x(x-2)}$$ Subtract numerators; keep the same denominator.

Continued on Next Page

6 Subtract. Write each answer in lowest terms.

(a) $\frac{3}{m^2} - \frac{2}{m^2}$

(b) $\frac{x}{2x+3} - \frac{3x+4}{2x+3}$

ANSWERS

6. (a) $\frac{1}{m^2}$ **(b)** $\frac{-2(x+2)}{2x+3}$

Be careful here.

$$= \frac{9x - 3x + 6}{x(x - 2)} \quad \text{Distributive property}$$

$$= \frac{6x + 6}{x(x - 2)}, \text{ or } \frac{6(x + 1)}{x(x - 2)} \quad \text{Combine like terms; factor the numerator.}$$

Note

We factored the final numerator in Example 7 to get $\frac{6(x+1)}{x(x-2)}$. The fundamental property does not apply, however, since there are no common factors to divide out. The answer is in lowest terms.

Work Problem 7 at the Side. ▶

EXAMPLE 8 Subtracting Rational Expressions with Denominators That Are Opposites

Subtract. Write the answer in lowest terms.

$$\frac{3x}{x - 5} - \frac{2x - 25}{5 - x}$$ The denominators are opposites. We choose $x - 5$ as the common denominator.

$$= \frac{3x}{x - 5} - \frac{(2x - 25)(-1)}{(5 - x)(-1)}$$ Multiply $\frac{2x-25}{5-x}$ by $\frac{-1}{-1}$ to get a common denominator.

$$= \frac{3x}{x - 5} - \frac{-2x + 25}{x - 5}$$ $(5 - x)(-1) = -5 + x = x - 5$

Use parentheses.

$$= \frac{3x - (-2x + 25)}{x - 5}$$ Subtract numerators.

$$= \frac{3x + 2x - 25}{x - 5}$$ Distributive property

$$= \frac{5x - 25}{x - 5}$$ Combine like terms.

$$= \frac{5(x - 5)}{x - 5}$$ Factor.

$$= 5$$ Divide out the common factor.

Work Problem 8 at the Side. ▶

EXAMPLE 9 Subtracting Rational Expressions

Subtract. Write the answer in lowest terms.

$$\frac{6x}{x^2 - 2x + 1} - \frac{1}{x^2 - 1}$$

Begin by factoring the denominators.

$$x^2 - 2x + 1 = (x - 1)^2 \quad \text{and} \quad x^2 - 1 = (x - 1)(x + 1)$$

From the factored denominators, we identify the LCD, $(x - 1)^2(x + 1)$. ***We use the factor x − 1 twice*** because it appears twice in the first denominator.

Continued on Next Page

7 Subtract. Write each answer in lowest terms.

(a) $\frac{1}{k + 4} - \frac{2}{k}$

(b) $\frac{6}{a + 2} - \frac{1}{a - 3}$

8 Subtract. Write each answer in lowest terms.

(a) $\frac{5}{x - 1} - \frac{3x}{1 - x}$

(b) $\frac{2y}{y - 2} - \frac{1 + y}{2 - y}$

Answers

7. (a) $\frac{-k - 8}{k(k + 4)}$ **(b)** $\frac{5(a - 4)}{(a + 2)(a - 3)}$

8. (a) $\frac{5 + 3x}{x - 1}$, or $\frac{-5 - 3x}{1 - x}$

(b) $\frac{3y + 1}{y - 2}$, or $\frac{-3y - 1}{2 - y}$

9 Subtract. Write each answer in lowest terms.

(a) $\frac{4y}{y^2-1}-\frac{5}{y^2+2y+1}$

(b) $\frac{3r}{r^2-5r}-\frac{4}{r^2-10r+25}$

$$\frac{6x}{(x-1)^2}-\frac{1}{(x-1)(x+1)}$$ The LCD is $(x-1)^2(x+1)$.

$$=\frac{6x(x+1)}{(x-1)^2(x+1)}-\frac{1(x-1)}{(x-1)(x-1)(x+1)}$$ Fundamental property

$$=\frac{6x(x+1)-1(x-1)}{(x-1)^2(x+1)}$$ Subtract numerators.

$$=\frac{6x^2+6x-x+1}{(x-1)^2(x+1)}$$ Distributive property

$$=\frac{6x^2+5x+1}{(x-1)^2(x+1)},\quad\text{or}\quad\frac{(2x+1)(3x+1)}{(x-1)^2(x+1)}$$ Combine like terms; factor the numerator.

Verify that the final answer is in lowest terms.

◀ Work Problem 9 at the Side.

10 Subtract. Write each answer in lowest terms.

(a) $\frac{2}{p^2-5p+4}-\frac{3}{p^2-1}$

(b) $\frac{q}{2q^2+5q-3}-\frac{3q+4}{3q^2+10q+3}$

EXAMPLE 10 Subtracting Rational Expressions

Subtract. Write the answer in lowest terms.

$$\frac{q}{q^2-4q-5}-\frac{3}{2q^2-13q+15}$$

To find the LCD, factor each denominator.

$$q^2-4q-5=(q+1)(q-5)$$
$$2q^2-13q+15=(q-5)(2q-3)$$

The LCD here is $(q+1)(q-5)(2q-3)$. Write each rational expression with the LCD, using the fundamental property.

$$\frac{q}{(q+1)(q-5)}-\frac{3}{(q-5)(2q-3)}$$ The LCD is $(q+1)\cdot(q-5)(2q-3)$.

$$=\frac{q(2q-3)}{(q+1)(q-5)(2q-3)}-\frac{3(q+1)}{(q+1)(q-5)(2q-3)}$$

$$=\frac{q(2q-3)-3(q+1)}{(q+1)(q-5)(2q-3)}$$ Subtract numerators.

$$=\frac{2q^2-3q-3q-3}{(q+1)(q-5)(2q-3)}$$ Distributive property

$$=\frac{2q^2-6q-3}{(q+1)(q-5)(2q-3)}$$ Combine like terms.

Verify that the final answer is in lowest terms.

◀ Work Problem 10 at the Side.

Answers

9. (a) $\frac{4y^2-y+5}{(y+1)^2(y-1)}$
(b) $\frac{3r-19}{(r-5)^2}$

10. (a) $\frac{14-p}{(p-4)(p-1)(p+1)}$
(b) $\frac{-3q^2-4q+4}{(2q-1)(q+3)(3q+1)}$

16.4 Exercises

Match the expression in Column I with the correct sum or difference in Column II.

I		II	
1. $\frac{x}{x+6}+\frac{6}{x+6}$	**2.** $\frac{2x}{x-6}-\frac{12}{x-6}$	**A.** 2	**B.** $\frac{x-6}{x+6}$
3. $\frac{6}{x-6}-\frac{x}{x-6}$	**4.** $\frac{6}{x+6}-\frac{x}{x+6}$	**C.** -1	**D.** $\frac{6+x}{6x}$
5. $\frac{x}{x+6}-\frac{6}{x+6}$	**6.** $\frac{1}{x}+\frac{1}{6}$	**E.** 1	**F.** 0
7. $\frac{1}{6}-\frac{1}{x}$	**8.** $\frac{1}{6x}-\frac{1}{6x}$	**G.** $\frac{x-6}{6x}$	**H.** $\frac{6-x}{x+6}$

Note: When adding and subtracting rational expressions, several different equivalent forms of the answer often exist. If your answer does not look exactly like the one given in the back of the book, check to see whether you have written an equivalent form.

Add or subtract. Write each answer in lowest terms. See Examples 1 and 6.

9. $\frac{4}{m}+\frac{7}{m}$

10. $\frac{5}{p}+\frac{11}{p}$

11. $\frac{a+b}{2}-\frac{a-b}{2}$

12. $\frac{x-y}{2}-\frac{x+y}{2}$

13. $\frac{5}{y+4}-\frac{1}{y+4}$

14. $\frac{4}{y+3}-\frac{1}{y+3}$

15. $\frac{5m}{m+1}-\frac{1+4m}{m+1}$

16. $\frac{4x}{x+2}-\frac{2+3x}{x+2}$

17. $\frac{x^2}{x+5}+\frac{5x}{x+5}$

18. $\frac{t^2}{t-3}+\frac{-3t}{t-3}$

19. $\frac{y^2-3y}{y+3}+\frac{-18}{y+3}$

20. $\frac{r^2-8r}{r-5}+\frac{15}{r-5}$

21. Explain with an example how to add or subtract rational expressions with the same denominator.

22. Explain with an example how to add or subtract rational expressions with different denominators.

Add or subtract. Write each answer in lowest terms. See Examples 2, 3, 4, and 7.

23. $\frac{z}{5} + \frac{1}{3}$

24. $\frac{p}{8} + \frac{3}{5}$

25. $\frac{5}{7} - \frac{r}{2}$

26. $\frac{10}{9} - \frac{z}{3}$

27. $-\frac{3}{4} - \frac{1}{2x}$

28. $-\frac{5}{8} - \frac{3}{2a}$

29. $\frac{3}{5x} + \frac{9}{4x}$

30. $\frac{3}{2x} + \frac{4}{7x}$

31. $\frac{x+1}{6} + \frac{3x+3}{9}$

32. $\frac{2x-6}{4} + \frac{x+5}{6}$

33. $\frac{x+3}{3x} + \frac{2x+2}{4x}$

34. $\frac{x+2}{5x} + \frac{6x+3}{3x}$

35. $\frac{2}{x+3} + \frac{1}{x}$

36. $\frac{3}{x-4} + \frac{2}{x}$

37. $\frac{1}{k+5} - \frac{2}{k}$

38. $\frac{3}{m+1} - \frac{4}{m}$

39. $\frac{x}{x-2} + \frac{-8}{x^2-4}$

40. $\frac{2x}{x-1} + \frac{-4}{x^2-1}$

41. $\frac{x}{x-2} + \frac{4}{x+2}$

42. $\frac{2x}{x-1} + \frac{3}{x+1}$

43. $\frac{t}{t+2} + \frac{5-t}{t} - \frac{4}{t^2+2t}$

44. $\frac{2p}{p-3} + \frac{2+p}{p} - \frac{-6}{p^2-3p}$

45. What are the two possible LCDs that could be used for the sum

$$\frac{10}{m-2}+\frac{5}{2-m}?$$

46. If one form of the correct answer to a sum or difference of rational expressions is $\frac{4}{k-3}$, what would be an alternative form of the answer if the denominator is $3-k$?

Add or subtract. Write each answer in lowest terms. See Examples 5 and 8.

47. $\frac{4}{x-5}+\frac{6}{5-x}$

48. $\frac{10}{m-2}+\frac{5}{2-m}$

49. $\frac{-1}{1-y}+\frac{3-4y}{y-1}$

50. $\frac{-4}{p-3}-\frac{p+1}{3-p}$

51. $\frac{2}{x-y^2}+\frac{7}{y^2-x}$

52. $\frac{-8}{p-q^2}+\frac{3}{q^2-p}$

53. $\frac{x}{5x-3y}-\frac{y}{3y-5x}$

54. $\frac{t}{8t-9s}-\frac{s}{9s-8t}$

55. $\frac{3}{4p-5}+\frac{9}{5-4p}$

56. $\frac{8}{3-7y}-\frac{2}{7y-3}$

In each subtraction problem, the rational expression that follows the subtraction sign has a numerator with more than one term. ***Be careful with signs*** *and find each difference. See Examples 6–10.*

57. $\frac{2m}{m-n}-\frac{5m+n}{2m-2n}$

58. $\frac{5p}{p-q}-\frac{3p+1}{4p-4q}$

59. $\dfrac{5}{x^2 - 9} - \dfrac{x + 2}{x^2 + 4x + 3}$

60. $\dfrac{1}{a^2 - 1} - \dfrac{a - 1}{a^2 + 3a - 4}$

61. $\dfrac{2q + 1}{3q^2 + 10q - 8} - \dfrac{3q + 5}{2q^2 + 5q - 12}$

62. $\dfrac{4y - 1}{2y^2 + 5y - 3} - \dfrac{y + 3}{6y^2 + y - 2}$

Perform the indicated operations. See Examples 1–10.

63. $\dfrac{4}{r^2 - r} + \dfrac{6}{r^2 + 2r} - \dfrac{1}{r^2 + r - 2}$

64. $\dfrac{6}{k^2 + 3k} - \dfrac{1}{k^2 - k} + \dfrac{2}{k^2 + 2k - 3}$

65. $\dfrac{x + 3y}{x^2 + 2xy + y^2} + \dfrac{x - y}{x^2 + 4xy + 3y^2}$

66. $\dfrac{m}{m^2 - 1} + \dfrac{m - 1}{m^2 + 2m + 1}$

67. $\dfrac{r + y}{18r^2 + 9ry - 2y^2} + \dfrac{3r - y}{36r^2 - y^2}$

68. $\dfrac{2x - z}{2x^2 + xz - 10z^2} - \dfrac{x + z}{x^2 - 4z^2}$

69. Refer to the rectangle in the figure.

(a) Find an expression that represents its perimeter. Give the simplified form.

(b) Find an expression that represents its area. Give the simplified form.

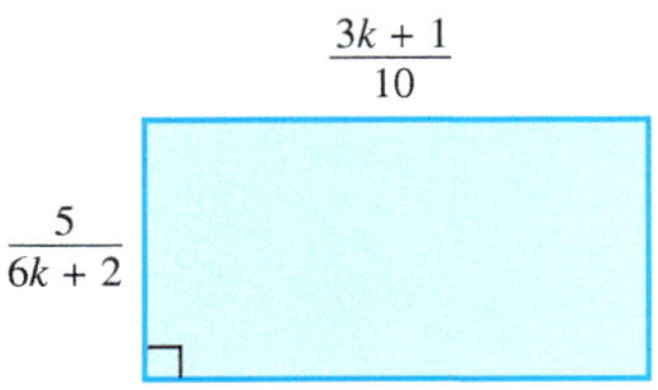

70. Refer to the triangle in the figure. Find an expression that represents its perimeter.

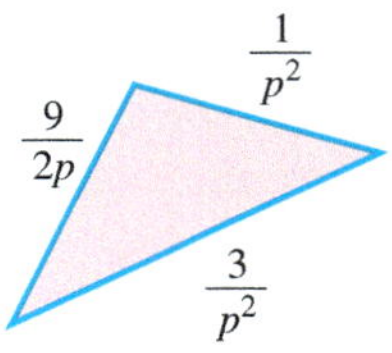

16.5 Complex Fractions

OBJECTIVES

1. Simplify a complex fraction by writing it as a division problem (Method 1).
2. Simplify a complex fraction by multiplying numerator and denominator by the least common denominator (Method 2).

The quotient of two mixed numbers in arithmetic, such as $2\frac{1}{2} \div 3\frac{1}{4}$, can be written as a fraction:

$$2\frac{1}{2} \div 3\frac{1}{4} = \frac{2\frac{1}{2}}{3\frac{1}{4}} = \frac{2 + \frac{1}{2}}{3 + \frac{1}{4}}.$$

The last expression is the quotient of expressions that involve fractions. In algebra, some rational expressions also have fractions in the numerator, or denominator, or both.

Complex Fraction

A rational expression with one or more fractions in the numerator, or denominator, or both, is called a **complex fraction.**

Examples of complex fractions include

$$\frac{2 + \frac{1}{2}}{3 + \frac{1}{4}}, \quad \frac{\frac{3x^2 - 5x}{6x^2}}{2x - \frac{1}{x}}, \quad \text{and} \quad \frac{3 + x}{5 - \frac{2}{x}}. \qquad \text{Complex fractions}$$

The parts of a complex fraction are named as follows.

$$\frac{\frac{2}{p} - \frac{1}{q}}{\frac{3}{p} + \frac{5}{q}}$$

← Numerator of complex fraction (top: $\frac{2}{p} - \frac{1}{q}$)
← Main fraction bar
← Denominator of complex fraction (bottom: $\frac{3}{p} + \frac{5}{q}$)

OBJECTIVE 1 Simplify a complex fraction by writing it as a division problem (Method 1). Since the main fraction bar represents division in a complex fraction, one method of simplifying a complex fraction involves division. This is the method introduced in **Section 4.6.**

Method 1 for Simplifying a Complex Fraction

Step 1 Write both the numerator and denominator as single fractions.

Step 2 Change the complex fraction to a division problem.

Step 3 Perform the indicated division.

Once again, the first example shows complex fractions from both arithmetic and algebra.

1 Simplify each complex fraction using Method 1.

(a) $\dfrac{\dfrac{2}{5}+\dfrac{1}{4}}{\dfrac{1}{2}+\dfrac{1}{3}}$

(b) $\dfrac{6+\dfrac{1}{x}}{5-\dfrac{2}{x}}$

(c) $\dfrac{9-\dfrac{4}{p}}{\dfrac{2}{p}+1}$

2 Simplify each complex fraction using Method 1.

(a) $\dfrac{\dfrac{rs^2}{t}}{\dfrac{r^2s}{t^2}}$

(b) $\dfrac{\dfrac{m^2n^3}{p}}{\dfrac{m^4n}{p^2}}$

Answers

1. (a) $\frac{39}{50}$ (b) $\frac{6x+1}{5x-2}$ (c) $\frac{9p-4}{2+p}$

2. (a) $\frac{st}{r}$ (b) $\frac{n^2p}{m^2}$

EXAMPLE 1 Simplifying Complex Fractions (Method 1)

Simplify each complex fraction.

(a) $\dfrac{\dfrac{2}{3}+\dfrac{5}{9}}{\dfrac{1}{4}+\dfrac{1}{12}}$ **(b)** $\dfrac{6+\dfrac{3}{x}}{\dfrac{x}{4}+\dfrac{1}{8}}$

Step 1 First, write each numerator as a single fraction.

$$\frac{2}{3}+\frac{5}{9}=\frac{2(3)}{3(3)}+\frac{5}{9}=\frac{6}{9}+\frac{5}{9}=\frac{11}{9} \qquad 6+\frac{3}{x}=\frac{6}{1}+\frac{3}{x}=\frac{6x}{x}+\frac{3}{x}=\frac{6x+3}{x}$$

Do the same thing with each denominator.

$$\frac{1}{4}+\frac{1}{12}=\frac{1(3)}{4(3)}+\frac{1}{12}=\frac{3}{12}+\frac{1}{12}=\frac{4}{12} \qquad \frac{x}{4}+\frac{1}{8}=\frac{x(2)}{4(2)}+\frac{1}{8}=\frac{2x}{8}+\frac{1}{8}=\frac{2x+1}{8}$$

Step 2 Write the equivalent complex fraction as a division problem.

$$\frac{\frac{11}{9}}{\frac{4}{12}}=\frac{11}{9}\div\frac{4}{12} \qquad \frac{\frac{6x+3}{x}}{\frac{2x+1}{8}}=\frac{6x+3}{x}\div\frac{2x+1}{8}$$

Step 3 Now use the definition of division and multiply by the reciprocal. Then write in lowest terms using the fundamental property.

$$=\frac{11}{9}\cdot\frac{12}{4} \qquad =\frac{6x+3}{x}\cdot\frac{8}{2x+1}$$

$$=\frac{11\cdot3\cdot4}{3\cdot3\cdot4} \qquad =\frac{3(2x+1)}{x}\cdot\frac{8}{2x+1}$$

$$=\frac{11}{3} \qquad =\frac{24}{x}$$

◀ Work Problem 1 at the Side.

EXAMPLE 2 Simplifying a Complex Fraction (Method 1)

Simplify the complex fraction.

$$\frac{\frac{xp}{q^3}}{\frac{p^2}{qx^2}}$$

The numerator and denominator are single fractions, so use the definition of division and then the fundamental property.

$$\frac{xp}{q^3}\div\frac{p^2}{qx^2}=\frac{xp}{q^3}\cdot\frac{qx^2}{p^2}=\frac{x^3}{q^2p}$$

◀ Work Problem 2 at the Side.

EXAMPLE 3 Simplifying a Complex Fraction (Method 1)

Simplify the complex fraction.

$$\frac{\frac{3}{x+2} - 4}{\frac{2}{x+2} + 1}$$

$$= \frac{\frac{3}{x+2} - \frac{4(x+2)}{x+2}}{\frac{2}{x+2} + \frac{1(x+2)}{x+2}}$$ Write both second terms with a denominator of $x + 2$.

$$= \frac{\frac{3 - 4(x+2)}{x+2}}{\frac{2 + 1(x+2)}{x+2}}$$ Subtract in the numerator. Add in the denominator.

$$= \frac{\frac{3 - 4x - 8}{x+2}}{\frac{2 + x + 2}{x+2}}$$ Be careful with signs. Distributive property

$$= \frac{\frac{-5 - 4x}{x+2}}{\frac{4 + x}{x+2}}$$ Combine like terms.

$$= \frac{-5 - 4x}{x+2} \cdot \frac{x+2}{4+x}$$ Multiply by the reciprocal.

$$= \frac{-5 - 4x}{4 + x}$$ Lowest terms

CAUTION

$$\frac{\frac{a}{b} + \frac{c}{d}}{\frac{e}{f} + \frac{g}{h}}$$ *does not equal* $$\left(\frac{a}{b} + \frac{c}{d}\right) \cdot \left(\frac{f}{e} + \frac{h}{g}\right).$$

Work Problem 3 at the Side. ▶

OBJECTIVE 2 Simplify a complex fraction by multiplying numerator and denominator by the least common denominator (Method 2). Since any expression can be multiplied by a form of 1 to get an equivalent expression, we can multiply both the numerator and the denominator of a complex fraction by the same nonzero expression to get an equivalent complex fraction. If we choose the expression to be the LCD of all the fractions within the complex fraction, the result will no longer be complex. This is Method 2.

3 Simplify using Method 1.

$$\frac{\frac{2}{x-1} + \frac{1}{x+1}}{\frac{3}{x-1} - \frac{4}{x+1}}$$

ANSWER

3. $\frac{3x+1}{-x+7}$

4 Simplify each complex fraction using Method 2.

(a) $\dfrac{\frac{2}{3}-\frac{1}{4}}{\frac{4}{9}+\frac{1}{2}}$

(b) $\dfrac{2-\frac{6}{a}}{3+\frac{4}{a}}$

(c) $\dfrac{\frac{p}{5-p}}{\frac{4p}{2p+1}}$

Method 2 for Simplifying a Complex Fraction

Step 1 Find the LCD of all fractions within the complex fraction.

Step 2 Multiply both the numerator and the denominator of the complex fraction by this LCD using the distributive property as necessary. Write in lowest terms.

In the next example, Method 2 is used to simplify the complex fractions from Example 1.

EXAMPLE 4 Simplifying Complex Fractions (Method 2)

Simplify each complex fraction.

(a) $\dfrac{\frac{2}{3}+\frac{5}{9}}{\frac{1}{4}+\frac{1}{12}}$ **(b)** $\dfrac{6+\frac{3}{x}}{\frac{x}{4}+\frac{1}{8}}$

Step 1 Find the LCD for all denominators in the complex fraction.

The LCD for 3, 9, 4, and 12 is 36. | The LCD for x, 4, and 8 is $8x$.

Step 2 Multiply the numerator and denominator of the complex fraction by the LCD.

(a)

$$\dfrac{\frac{2}{3}+\frac{5}{9}}{\frac{1}{4}+\frac{1}{12}} = \dfrac{36\left(\frac{2}{3}+\frac{5}{9}\right)}{36\left(\frac{1}{4}+\frac{1}{12}\right)}$$

Multiply each term by 36.

$$= \dfrac{36\left(\frac{2}{3}\right)+36\left(\frac{5}{9}\right)}{36\left(\frac{1}{4}\right)+36\left(\frac{1}{12}\right)} = \dfrac{24+20}{9+3} = \dfrac{44}{12} = \dfrac{4\cdot 11}{4\cdot 3} = \dfrac{11}{3}$$

(b)

$$\dfrac{6+\frac{3}{x}}{\frac{x}{4}+\frac{1}{8}} = \dfrac{8x\left(6+\frac{3}{x}\right)}{8x\left(\frac{x}{4}+\frac{1}{8}\right)}$$

Multiply each term by $8x$.

$$= \dfrac{8x(6)+8x\left(\frac{3}{x}\right)}{8x\left(\frac{x}{4}\right)+8x\left(\frac{1}{8}\right)} = \dfrac{48x+24}{2x^2+x} = \dfrac{24(2x+1)}{x(2x+1)} = \dfrac{24}{x}$$

◀ Work Problem **4** at the Side.

Answers

4. (a) $\frac{15}{34}$ **(b)** $\frac{2a-6}{3a+4}$ **(c)** $\frac{2p+1}{4(5-p)}$

EXAMPLE 5 Simplifying a Complex Fraction (Method 2)

Simplify the complex fraction.

$$\frac{\frac{3}{5m} - \frac{2}{m^2}}{\frac{9}{2m} + \frac{3}{4m^2}}$$ The LCD for $5m$, m^2, $2m$, and $4m^2$ is $20m^2$.

$$= \frac{20m^2\left(\frac{3}{5m} - \frac{2}{m^2}\right)}{20m^2\left(\frac{9}{2m} + \frac{3}{4m^2}\right)}$$ Multiply numerator and denominator by $20m^2$.

$$= \frac{20m^2\left(\frac{3}{5m}\right) - 20m^2\left(\frac{2}{m^2}\right)}{20m^2\left(\frac{9}{2m}\right) + 20m^2\left(\frac{3}{4m^2}\right)}$$ Distributive property

$$= \frac{12m - 40}{90m + 15}$$ Multiply and simplify.

Work Problem **5** *at the Side.* ▶

5 Simplify using Method 2.

$$\frac{\frac{2}{5x} - \frac{3}{x^2}}{\frac{7}{4x} + \frac{1}{2x^2}}$$

Some students use Method 1 for problems like Example 2, which is the quotient of two fractions, and Method 2 for problems like Examples 1, 3, 4, and 5, which have sums or differences in the numerators or denominators.

EXAMPLE 6 Simplifying Complex Fractions

Simplify each complex fraction. Use either method.

(a) $$\frac{\frac{1}{y} + \frac{2}{y+2}}{\frac{4}{y} - \frac{3}{y+2}}$$ We use Method 2, since there are sums and differences in the numerator and denominator.

$$= \frac{\left(\frac{1}{y} + \frac{2}{y+2}\right)y(y+2)}{\left(\frac{4}{y} - \frac{3}{y+2}\right)y(y+2)}$$ Multiply numerator and denominator by the LCD, $y(y+2)$.

$$= \frac{\left(\frac{1}{y}\right)y(y+2) + \left(\frac{2}{y+2}\right)y(y+2)}{\left(\frac{4}{y}\right)y(y+2) - \left(\frac{3}{y+2}\right)y(y+2)}$$ Distributive property

$$= \frac{1(y+2) + 2y}{4(y+2) - 3y}$$ Fundamental property

$$= \frac{y + 2 + 2y}{4y + 8 - 3y}$$ Distributive property

$$= \frac{3y + 2}{y + 8}$$ Combine like terms.

Continued on Next Page

ANSWER

5. $\frac{8x - 60}{35x + 10}$

6 Simplify each complex fraction. Use either method.

(a) $\dfrac{\dfrac{1}{x}+\dfrac{2}{x-1}}{\dfrac{2}{x}-\dfrac{4}{x-1}}$

(b) $\dfrac{1-\dfrac{2}{x}-\dfrac{15}{x^2}}{1+\dfrac{5}{x}+\dfrac{6}{x^2}}$

(c) $\dfrac{\dfrac{2x+3}{x-4}}{\dfrac{4x^2-9}{x^2-16}}$

(b) $\dfrac{1-\dfrac{2}{x}-\dfrac{3}{x^2}}{1-\dfrac{5}{x}+\dfrac{6}{x^2}}$ There are sums and differences in the numerator and denominator, so we use Method 2.

$$= \frac{\left(1-\frac{2}{x}-\frac{3}{x^2}\right)x^2}{\left(1-\frac{5}{x}+\frac{6}{x^2}\right)x^2}$$ Multiply numerator and denominator by the LCD, x^2.

$$= \frac{x^2-2x-3}{x^2-5x+6}$$ Distributive property

$$= \frac{(x-3)(x+1)}{(x-3)(x-2)}$$ Factor.

$$= \frac{x+1}{x-2}$$ Divide out the common factor.

You may wish to verify that in this example, like the others, *either* method can be used to simplify the complex fractions.

(c) $\dfrac{\dfrac{x+2}{x-3}}{\dfrac{x^2-4}{x^2-9}}$ This is a quotient of two rational expressions, so we use Method 1.

$$= \frac{x+2}{x-3} \div \frac{x^2-4}{x^2-9}$$ Write as a division problem.

$$= \frac{x+2}{x-3} \cdot \frac{x^2-9}{x^2-4}$$ Multiply by the reciprocal.

$$= \frac{(x+2)(x+3)(x-3)}{(x-3)(x+2)(x-2)}$$ Multiply; factor.

$$= \frac{x+3}{x-2}$$ Divide out the common factors.

◀ Work Problem **6** at the Side.

Answers

6. (a) $\dfrac{3x-1}{-2x-2}$ (b) $\dfrac{x-5}{x+2}$ (c) $\dfrac{x+4}{2x-3}$

16.5 Exercises

FOR EXTRA HELP MyMathLab Math XL PRACTICE WATCH DOWNLOAD READ REVIEW

Note: In many problems involving complex fractions, several different equivalent forms of the answer often exist. If your answer does not look exactly like the one given in the back of the book, check to see whether you have written an equivalent form.

1. Consider the complex fraction $\dfrac{\frac{1}{2}-\frac{1}{3}}{\frac{5}{6}-\frac{1}{12}}$. Answer each part, outlining Method 1 for simplifying this complex fraction.

(a) To combine the terms in the numerator, we must find the LCD of $\frac{1}{2}$ and $\frac{1}{3}$. What is this LCD? Determine the simplified form of the numerator of the complex fraction.

(b) To combine the terms in the denominator, we must find the LCD of $\frac{5}{6}$ and $\frac{1}{12}$. What is this LCD? Determine the simplified form of the denominator of the complex fraction.

(c) Now use the results from parts (a) and (b) to write the complex fraction as a division problem using the symbol ÷.

(d) Perform the operation from part (c) to obtain the final simplification.

2. Consider the same complex fraction given in Exercise 1, $\dfrac{\frac{1}{2}-\frac{1}{3}}{\frac{5}{6}-\frac{1}{12}}$. Answer each part, outlining Method 2 for simplifying this complex fraction.

(a) We must determine the LCD of all the fractions within the complex fraction. What is this LCD?

(b) Multiply every term in the complex fraction by the LCD found in part (a), but at this time do not combine the terms in the numerator and the denominator.

(c) Now combine the terms from part (b) to obtain the simplified form of the complex fraction.

Simplify each complex fraction. Use either method. See Examples 1–6.

3. $\dfrac{-\frac{4}{3}}{\frac{2}{9}}$

4. $\dfrac{-\frac{5}{6}}{\frac{5}{4}}$

5. $\dfrac{\frac{p}{q^2}}{\frac{p^2}{q}}$

6. $\dfrac{\frac{a}{x}}{\frac{a^2}{2x}}$

7. $\dfrac{\frac{x}{y^2}}{\frac{x^2}{y}}$

8. $\dfrac{\frac{p^4}{r}}{\frac{p^2}{r^2}}$

9. $\dfrac{\frac{4a^4b^3}{3a}}{\frac{2ab^4}{b^2}}$

10. $\dfrac{\frac{2r^4t^2}{3t}}{\frac{5r^2t^5}{3r}}$

11. $\dfrac{\dfrac{m+2}{3}}{\dfrac{m-4}{m}}$

12. $\dfrac{\dfrac{q-5}{q}}{\dfrac{q+5}{3}}$

13. $\dfrac{\dfrac{2}{x}-3}{\dfrac{2-3x}{2}}$

14. $\dfrac{6+\dfrac{2}{r}}{\dfrac{3r+1}{4}}$

15. $\dfrac{\dfrac{1}{x}+x}{\dfrac{x^2+1}{8}}$

16. $\dfrac{\dfrac{3}{m}-m}{\dfrac{3-m^2}{4}}$

17. $\dfrac{a-\dfrac{5}{a}}{a+\dfrac{1}{a}}$

18. $\dfrac{q+\dfrac{1}{q}}{q+\dfrac{4}{q}}$

19. $\dfrac{\dfrac{1}{2}+\dfrac{1}{p}}{\dfrac{2}{3}+\dfrac{1}{p}}$

20. $\dfrac{\dfrac{3}{4}-\dfrac{1}{r}}{\dfrac{1}{5}+\dfrac{1}{r}}$

21. $\dfrac{\dfrac{2}{p^2}-\dfrac{3}{5p}}{\dfrac{4}{p}+\dfrac{1}{4p}}$

22. $\dfrac{\dfrac{2}{m^2}-\dfrac{3}{m}}{\dfrac{2}{5m^2}+\dfrac{1}{3m}}$

23. $\dfrac{\dfrac{t}{t+2}}{\dfrac{4}{t^2-4}}$

24. $\dfrac{\dfrac{m}{m+1}}{\dfrac{3}{m^2-1}}$

25. $\dfrac{\dfrac{1}{k+1}-1}{\dfrac{1}{k+1}+1}$

26. $\dfrac{\dfrac{2}{p-1}+2}{\dfrac{3}{p-1}-2}$

27. $\dfrac{2+\dfrac{1}{x}-\dfrac{28}{x^2}}{3+\dfrac{13}{x}+\dfrac{4}{x^2}}$

28. $\dfrac{4-\dfrac{11}{x}-\dfrac{3}{x^2}}{2-\dfrac{1}{x}-\dfrac{15}{x^2}}$

29. $\dfrac{\dfrac{1}{m-1}+\dfrac{2}{m+2}}{\dfrac{2}{m+2}-\dfrac{1}{m-3}}$

30. $\dfrac{\dfrac{5}{r+3}-\dfrac{1}{r-1}}{\dfrac{2}{r+2}+\dfrac{3}{r+3}}$

31. $2-\dfrac{2}{2+\dfrac{2}{2+2}}$

32. $3-\dfrac{2}{4+\dfrac{2}{4-2}}$

16.6 Equations with Rational Expressions and Graphs

OBJECTIVES

1. Determine the domain of the variable in a rational equation.
2. Solve rational equations.
3. Recognize the graph of a rational function.

In **Section 8.1,** we defined the domain of a rational function as the set of all possible values of the variable. (We can also refer to this as "the domain of the variable.") Any value that makes the denominator 0 is excluded.

OBJECTIVE 1 Determine the domain of the variable in a rational equation. The **domain of the variable in a rational equation** is the intersection (overlap) of the domains of the rational expressions in the equation.

EXAMPLE 1 Determining Domains in Rational Equations

Find the domain of the variable in each equation.

(a) $\dfrac{2}{x} - \dfrac{3}{2} = \dfrac{7}{2x}$

The domains of the three rational terms of the equation are, in order, $\{x \mid x \neq 0\}$, $(-\infty, \infty)$, and $\{x \mid x \neq 0\}$. The intersection of these three domains is all real numbers except 0, which may be written $\{x \mid x \neq 0\}$.

(b) $\dfrac{2}{x-3} - \dfrac{3}{x+3} = \dfrac{12}{x^2-9}$

The domains of these three terms are, respectively, $\{x \mid x \neq 3\}$, $\{x \mid x \neq -3\}$, and $\{x \mid x \neq \pm 3\}$. ($\pm$ is read "positive or negative," or "plus or minus.") The domain of the variable is the intersection of the three domains, all real numbers except 3 and -3, written $\{x \mid x \neq \pm 3\}$.

Work Problem 1 at the Side. ▶

1 Find the domain of the variable in each equation.

(a) $\dfrac{3}{x} + \dfrac{1}{2} = \dfrac{5}{6x}$

(b) $\dfrac{4}{x-5} - \dfrac{2}{x+5} = \dfrac{1}{x^2-25}$

OBJECTIVE 2 Solve rational equations. The easiest way to solve most equations involving rational expressions is to multiply all terms in the equation by the least common denominator. This step will clear the equation of all denominators. ***We can do this only with equations, not expressions.***

CAUTION

When each side of an equation is multiplied by a *variable* expression, the resulting "solutions" may not satisfy the original equation. ***You must either determine and observe the domain or check all proposed solutions in the original equation. It is wise to do both.***

EXAMPLE 2 Solving a Rational Equation

Solve $\dfrac{2}{x} - \dfrac{3}{2} = \dfrac{7}{2x}$.

The domain, which excludes 0, was found in Example 1(a).

$$2x\left(\frac{2}{x} - \frac{3}{2}\right) = 2x\left(\frac{7}{2x}\right) \quad \text{Multiply by the LCD, } 2x.$$

$$2x\left(\frac{2}{x}\right) - 2x\left(\frac{3}{2}\right) = 2x\left(\frac{7}{2x}\right) \quad \text{Distributive property}$$

$$4 - 3x = 7 \quad \text{Multiply.}$$

$$-3x = 3 \quad \text{Subtract 4.}$$

$$\text{Proposed solution} \longrightarrow x = -1 \quad \text{Divide by } -3.$$

Continued on Next Page

ANSWERS

1. (a) $\{x \mid x \neq 0\}$ (b) $\{x \mid x \neq \pm 5\}$

2 Solve $-\frac{3}{20}+\frac{2}{x}=\frac{5}{4x}$.

Check Replace x with -1 in the original equation.

$$\frac{2}{x}-\frac{3}{2}=\frac{7}{2x} \quad \text{Original equation}$$
$$\frac{2}{-1}-\frac{3}{2}\stackrel{?}{=}\frac{7}{2(-1)} \quad \text{Let } x=-1.$$
$$-2-\frac{3}{2}\stackrel{?}{=}-\frac{7}{2}$$
$$-\frac{7}{2}=-\frac{7}{2} \quad \text{True}$$

The solution set is $\{-1\}$.

◀ Work Problem **2** at the Side.

EXAMPLE 3 Solving a Rational Equation with No Solution

Solve $\frac{2}{x-3}-\frac{3}{x+3}=\frac{12}{x^2-9}$.

Using the result from Example 1(b), we know that the domain excludes 3 and -3, since these values make one or more of the denominators in the equation equal 0. Multiply each side by the LCD, $(x+3)(x-3)$.

$$(x+3)(x-3)\left(\frac{2}{x-3}-\frac{3}{x+3}\right)=(x+3)(x-3)\left(\frac{12}{x^2-9}\right)$$
$$(x+3)(x-3)\left(\frac{2}{x-3}\right)-(x+3)(x-3)\left(\frac{3}{x+3}\right)$$
$$=(x+3)(x-3)\left(\frac{12}{x^2-9}\right) \quad \text{Distributive property}$$
$$2(x+3)-3(x-3)=12 \quad \text{Multiply.}$$
$$2x+6-3x+9=12 \quad \text{Distributive property}$$
$$-x+15=12 \quad \text{Combine like terms.}$$
$$-x=-3 \quad \text{Subtract 15.}$$
$$\text{Proposed solution} \longrightarrow x=3 \quad \text{Divide by } -1.$$

Since 3 is not in the domain, it cannot be a solution of the equation. Substituting 3 in the original equation shows why.

Check

$$\frac{2}{x-3}-\frac{3}{x+3}=\frac{12}{x^2-9} \quad \text{Original equation}$$
$$\frac{2}{3-3}-\frac{3}{3+3}\stackrel{?}{=}\frac{12}{3^2-9} \quad \text{Let } x=3.$$
$$\frac{2}{0}-\frac{3}{6}\stackrel{?}{=}\frac{12}{0}$$

Since division by 0 is undefined, the given equation has no solution, and the solution set is $\emptyset$.

◀ Work Problem **3** at the Side.

3 Solve each equation.

(a) $\frac{3}{x+1}=\frac{1}{x-1}-\frac{2}{x^2-1}$

(b) $\frac{1}{x-3}+\frac{1}{x+3}=\frac{6}{x^2-9}$

ANSWERS

2. $\{5\}$

3. **(a)** $\emptyset$ **(b)** $\emptyset$

EXAMPLE 4 **Solving a Rational Equation**

Solve $\frac{3}{p^2+p-2} - \frac{1}{p^2-1} = \frac{7}{2(p^2+3p+2)}$.

Factor each denominator to find the LCD, $2(p-1)(p+2)(p+1)$. The domain excludes 1, -2, and -1. Multiply each side by the LCD.

$$2(p-1)(p+2)(p+1)\left(\frac{3}{(p+2)(p-1)} - \frac{1}{(p+1)(p-1)}\right)$$
$$= 2(p-1)(p+2)(p+1)\left(\frac{7}{2(p+2)(p+1)}\right)$$

$$2 \cdot 3(p+1) - 2(p+2) = 7(p-1) \quad \text{Distributive property}$$
$$6p + 6 - 2p - 4 = 7p - 7 \quad \text{Distributive property}$$
$$4p + 2 = 7p - 7 \quad \text{Combine like terms.}$$
$$9 = 3p \quad \text{Subtract } 4p\text{; add 7.}$$
$$\text{Proposed solution} \longrightarrow 3 = p \quad \text{Divide by 3.}$$

Note that 3 is in the domain. Substitute 3 for p in the original equation to check that the solution set is $\{3\}$.

Work Problem **4** *at the Side.* ▶

4 Solve

$$\frac{4}{x^2+x-6} - \frac{1}{x^2-4} = \frac{2}{x^2+5x+6}.$$

EXAMPLE 5 **Solving a Rational Equation That Leads to a Quadratic Equation**

Solve $\frac{2}{3x+1} = \frac{1}{x} - \frac{6x}{3x+1}$.

Since the denominator $3x+1$ cannot equal 0, $-\frac{1}{3}$ is excluded from the domain, as is 0. Multiply each side by the LCD, $x(3x+1)$.

$$x(3x+1)\left(\frac{2}{3x+1}\right) = x(3x+1)\left(\frac{1}{x} - \frac{6x}{3x+1}\right)$$
$$x(3x+1)\left(\frac{2}{3x+1}\right) = x(3x+1)\left(\frac{1}{x}\right) - x(3x+1)\left(\frac{6x}{3x+1}\right)$$

Distributive property

$$2x = 3x + 1 - 6x^2$$

Write this quadratic equation in standard form with 0 on the right side.

$$6x^2 - 3x + 2x - 1 = 0$$
$$6x^2 - x - 1 = 0 \quad \text{Standard form}$$
$$(3x+1)(2x-1) = 0 \quad \text{Factor.}$$
$$3x + 1 = 0 \quad \text{or} \quad 2x - 1 = 0 \quad \text{Zero-factor property}$$
$$x = -\frac{1}{3} \quad \text{or} \quad x = \frac{1}{2} \quad \text{Proposed solutions}$$

Because $-\frac{1}{3}$ is not in the domain, it is not a solution. Check that the solution set is $\{\frac{1}{2}\}$.

Work Problem **5** *at the Side.* ▶

5 Solve

$$\frac{1}{x+4} + \frac{x}{x-4} = \frac{-8}{x^2-16}.$$

ANSWERS

4. $\{-9\}$

5. $\{-1\}$

6 Graph each rational function, and give the equations of the vertical and horizontal asymptotes.

(a) $f(x) = -\frac{1}{x}$

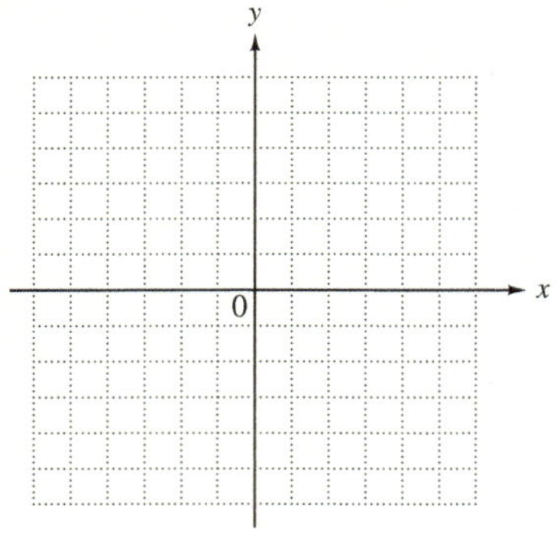

(b) $f(x) = \frac{2}{x+3}$

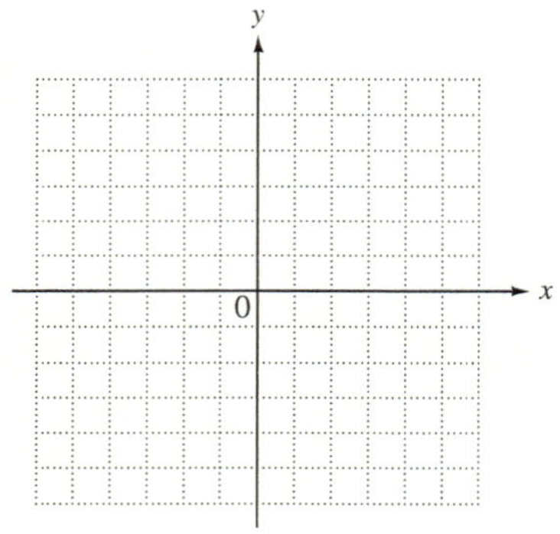

ANSWERS

6. **(a)** vertical asymptote: $x = 0$; horizontal asymptote: $y = 0$

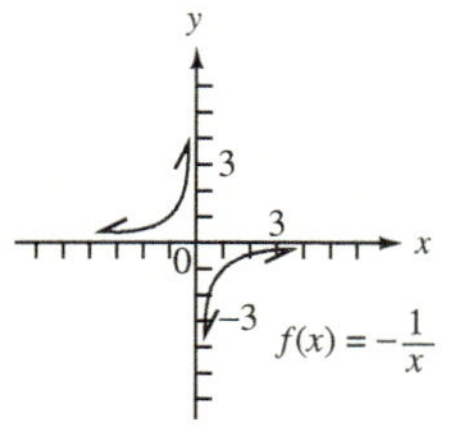

(b) vertical asymptote: $x = -3$; horizontal asymptote: $y = 0$

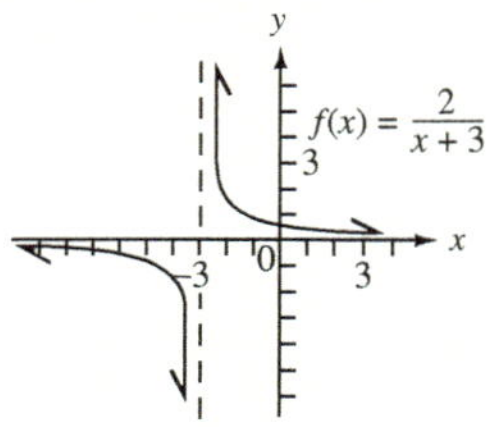

OBJECTIVE 3 Recognize the graph of a rational function. As mentioned in **Section 8.1,** a function defined by a quotient of polynomials is a **rational function.** Because one or more values of x may be excluded from the domain of most rational functions, their graphs are often **discontinuous.** That is, there will be one or more breaks in the graph. For example, we use point plotting and observing the domain to graph the simple rational function defined by

$$f(x) = \frac{1}{x}.$$

The domain of this function includes all real numbers except 0. Thus, there will be no point on the graph with $x = 0$. The vertical line with equation $x = 0$ is called a **vertical asymptote** of the graph. The horizontal line with equation $y = 0$ is called a **horizontal asymptote.** We show some typical ordered pairs in the table for both negative and positive x-values.

x	-3	-2	-1	$-\frac{1}{2}$	$-\frac{1}{4}$	$-\frac{1}{10}$	$\frac{1}{10}$	$\frac{1}{4}$	$\frac{1}{2}$	1	2	3
y	$-\frac{1}{3}$	$-\frac{1}{2}$	-1	-2	-4	-10	10	4	2	1	$\frac{1}{2}$	$\frac{1}{3}$

Notice that the closer positive values of x are to 0, the larger y is. Similarly, the closer negative values of x are to 0, the smaller (more negative) y is. Using this observation, excluding 0 from the domain, and plotting the points in the table, we obtain the graph in Figure 2.

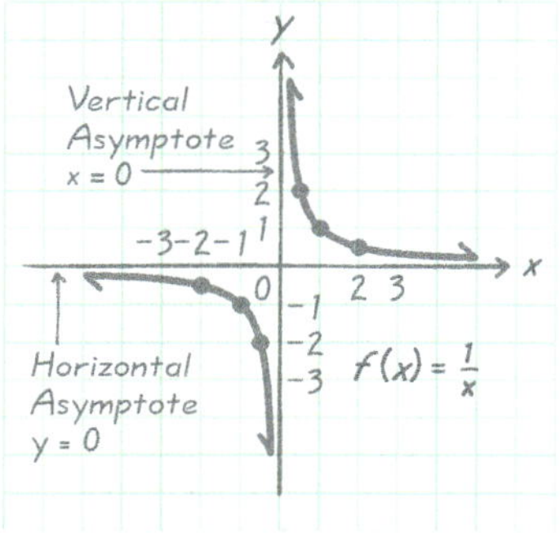

Figure 2

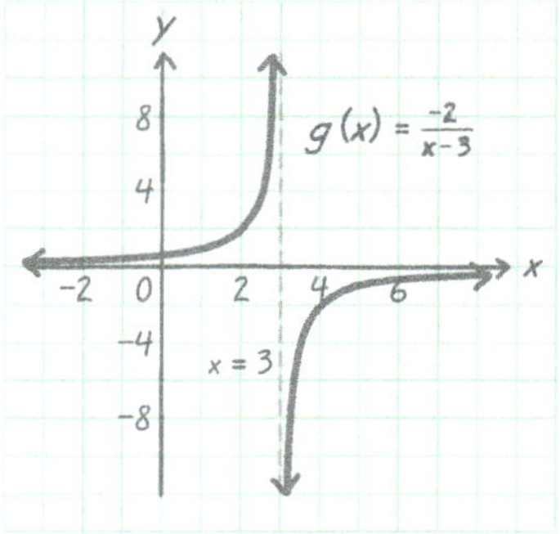

Figure 3

The graph of

$$g(x) = \frac{-2}{x-3}$$

is shown in Figure 3. Some ordered pairs are shown in the table.

x	-2	-1	0	1	2	2.5	2.75	3.25	3.5	4	5	6
y	$\frac{2}{5}$	$\frac{1}{2}$	$\frac{2}{3}$	1	2	4	8	-8	-4	-2	-1	$-\frac{2}{3}$

There is no point on the graph for $x = 3$ because 3 is excluded from the domain. The dashed line $x = 3$ represents the vertical asymptote and is not part of the graph. As suggested by the points from the table, the graph gets closer to the vertical asymptote as the x-values get closer to 3. Again, $y = 0$ is a horizontal asymptote.

◀ *Work Problem* **6** *at the Side.*

16.6 Exercises

FOR EXTRA HELP MyMathLab Math XL PRACTICE WATCH DOWNLOAD READ REVIEW

As explained in this section, any values that would cause a denominator to equal 0 *must be excluded from the domain and consequently as solutions of an equation that has variable expressions in the denominators.* ***(a)*** *Without actually solving the equation, list all possible numbers that would have to be rejected if they appeared as potential solutions.* ***(b)*** *Then give the domain using set notation. See Example 1.*

1. $\frac{1}{x+1} - \frac{1}{x-2} = 0$

2. $\frac{3}{x+4} - \frac{2}{x-9} = 0$

3. $\frac{5}{3x+5} - \frac{1}{x} = \frac{1}{2x+3}$

4. $\frac{6}{4x+7} - \frac{3}{x} = \frac{5}{6x-13}$

5. $\frac{1}{3x} + \frac{1}{2x} = \frac{x}{3}$

6. $\frac{5}{6x} - \frac{8}{2x} = \frac{x}{4}$

7. $\frac{3x+1}{x-4} = \frac{6x+5}{2x-7}$

8. $\frac{4x-1}{2x+3} = \frac{12x-25}{6x-2}$

9. $\frac{2}{x^2-x} + \frac{1}{x+3} = \frac{4}{x-2}$

10. Suppose that in solving the following equation, all of your algebraic steps are correct. Is it possible that your proposed solution would have to be rejected? Explain.

$$\frac{x+7}{4} - \frac{x+3}{3} = \frac{x}{12}$$

Solve each equation. See Examples 2–5.

11. $\frac{-5}{2x} + \frac{3}{4x} = \frac{-7}{4}$

12. $\frac{6}{5x} - \frac{2}{3x} = \frac{-8}{45}$

13. $x - \frac{24}{x} = -2$

14. $p + \frac{15}{p} = -8$

15. $\frac{x - 4}{x + 6} = \frac{2x + 3}{2x - 1}$

16. $\frac{5x - 8}{x + 2} = \frac{5x - 1}{x + 3}$

17. $\frac{3x + 1}{x - 4} = \frac{6x + 5}{2x - 7}$

18. $\frac{4x - 1}{2x + 3} = \frac{12x - 25}{6x - 2}$

19. $\frac{1}{y - 1} + \frac{5}{12} = \frac{-2}{3y - 3}$

20. $\frac{4}{m + 2} - \frac{11}{9} = \frac{1}{3m + 6}$

21. $\frac{-2}{3t - 6} - \frac{1}{36} = \frac{-3}{4t - 8}$

22. $\frac{3}{4m + 2} = \frac{17}{2} - \frac{7}{2m + 1}$

23. $\frac{3}{k + 2} - \frac{2}{k^2 - 4} = \frac{1}{k - 2}$

24. $\frac{3}{x - 2} + \frac{21}{x^2 - 4} = \frac{14}{x + 2}$

25. $\frac{1}{y + 2} + \frac{3}{y + 7} = \frac{5}{y^2 + 9y + 14}$

26. $\frac{1}{t + 3} + \frac{4}{t + 5} = \frac{2}{t^2 + 8t + 15}$

27. $\frac{9}{x} + \frac{4}{6x - 3} = \frac{2}{6x - 3}$

28. $\frac{5}{n} + \frac{4}{6 - 3n} = \frac{2n}{6 - 3n}$

29. $\frac{6}{w + 3} + \frac{-7}{w - 5} = \frac{-48}{w^2 - 2w - 15}$

30. $\frac{2}{r - 5} + \frac{3}{2r + 1} = \frac{22}{2r^2 - 9r - 5}$

31. $\frac{x}{x - 3} + \frac{4}{x + 3} = \frac{18}{x^2 - 9}$

32. $\frac{2x}{x - 3} + \frac{4}{x + 3} = \frac{-24}{x^2 - 9}$

33. $\frac{6}{x - 4} + \frac{5}{x} = \frac{-20}{x^2 - 4x}$

34. $\dfrac{7}{x-4}+\dfrac{3}{x}=\dfrac{-12}{x^2-4x}$

35. $\dfrac{2}{4x+7}+\dfrac{x}{3}=\dfrac{6}{12x+21}$

36. $\dfrac{5x+14}{x^2-9}=\dfrac{-2x^2-5x+2}{x^2-9}+\dfrac{2x+4}{x-3}$

37. $\dfrac{4x-7}{4x^2-9}=\dfrac{-2x^2+5x-4}{4x^2-9}+\dfrac{x+1}{2x+3}$

38. What is wrong with the following problem? "Solve $\dfrac{2x+1}{3x-4}+\dfrac{1}{2x+3}$."

Graph each rational function. Give the equations of the vertical and horizontal asymptotes. See Objective 3 and Figures 2 and 3.

39. $f(x)=\dfrac{2}{x}$

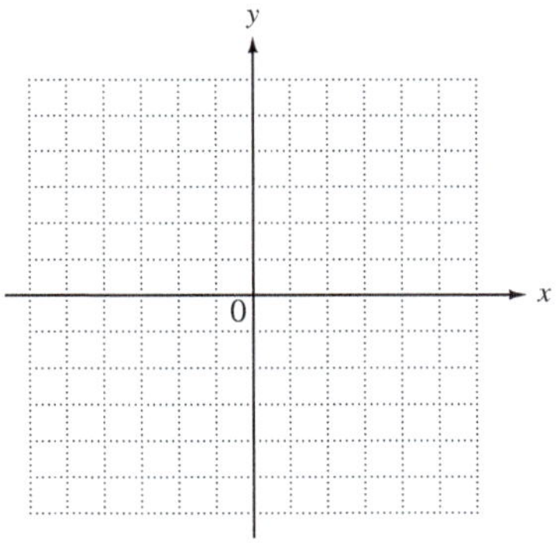

40. $f(x)=\dfrac{3}{x}$

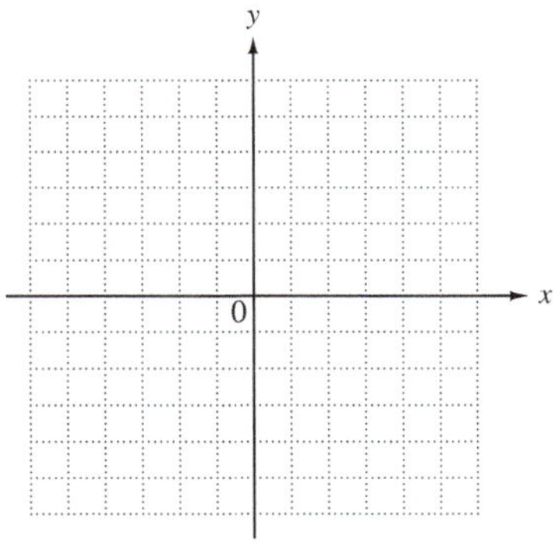

41. $f(x)=\dfrac{1}{x-2}$

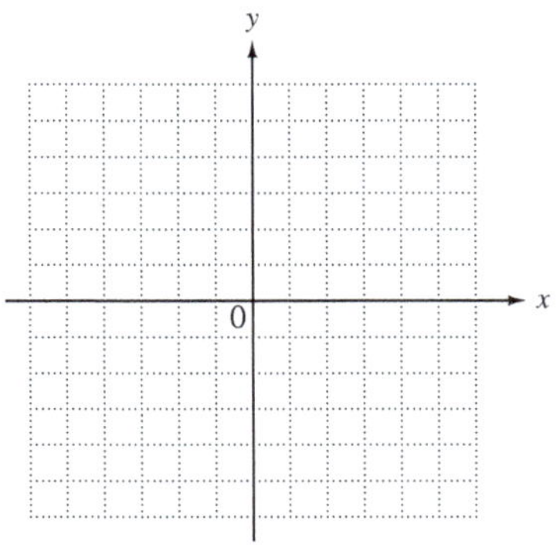

42. $f(x)=\dfrac{1}{x+2}$

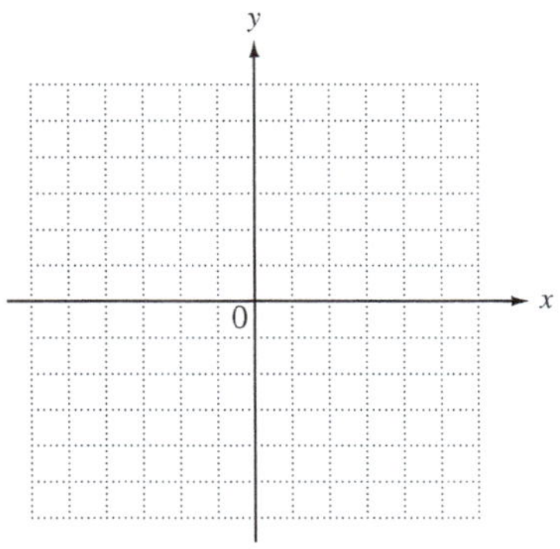

Solve each problem.

43. The average number of vehicles waiting in line to enter a parking area is modeled by the rational function defined by

$$w(x) = \frac{x^2}{2(1 - x)},$$

where x is a quantity between 0 and 1 known as the **traffic intensity.** (*Source*: Mannering, F. and W. Kilareski, *Principles of Highway Engineering and Traffic Control,* John Wiley and Sons, 1990.) To the nearest tenth, find the average number of vehicles waiting for each traffic intensity.

(a) 0.1

(b) 0.8

(c) 0.9

(d) What happens to waiting time as traffic intensity increases?

44. The force required to keep a 2000-lb car going 30 mph from skidding on a curve, where r is the radius of the curve in feet, is given by

$$F(r) = \frac{225{,}000}{r}.$$

(a) What radius must a curve have if a force of 450 lb is needed to keep the car from skidding?

(b) As the radius of the curve is lengthened, how is the force affected?

Relating Concepts (Exercises 45–48) For Individual or Group Work

An equation of the form

$$\frac{A}{x + B} + \frac{x}{x - B} = \frac{C}{x^2 - B^2}$$

will have one rejected solution if the relationship $C = -2AB$ holds true. (This can be proved using methods not covered in intermediate algebra.) For example, if $A = 1$ and $B = 2$, then $C = -2AB = -2(1)(2) = -4$, and the equation becomes

$$\frac{1}{x + 2} + \frac{x}{x - 2} = \frac{-4}{x^2 - 4}.$$

This equation has solution set $\{-1\}$; the potential solution -2 must be rejected. To further understand this idea, ***work Exercises 45–48 in order.***

45. Show that the second equation does indeed have solution set $\{-1\}$ and -2 must be rejected.

46. Let $A = 2$ and let $B = 1$. What is the corresponding value of C? Solve the equation determined by A, B, and C. What is the solution set? What value must be rejected?

47. Let $A = 4$ and let $B = -3$. What is the corresponding value of C? Solve the equation determined by A, B, and C. What is the solution set? What value must be rejected?

48. Choose two numbers of your own, letting one be A and the other be B. Repeat the process described in Exercises 46 and 47.

Summary Exercises on Rational Expressions and Equations

A common student error is to confuse an equation, *such as* $\frac{x}{2} + \frac{x}{3} = -5$, *with an expression, such as* $\frac{x}{2} + \frac{x}{3}$. *Look for the equals sign to distinguish between them. Equations are solved for a numerical answer, while problems involving operations result in simplified expressions.*

Solving an Equation

Solve: $\frac{x}{2} + \frac{x}{3} = -5$.

Multiply each side by the LCD, 6.

$$6\left(\frac{x}{2} + \frac{x}{3}\right) = 6(-5)$$

$$6\left(\frac{x}{2}\right) + 6\left(\frac{x}{3}\right) = 6(-5)$$

$$3x + 2x = -30$$

$$5x = -30$$

$$x = -6$$

Check that the solution set is $\{-6\}$.

Simplifying an Expression Involving an Operation

Add: $\frac{x}{2} + \frac{x}{3}$.

Write both fractions with the LCD, 6.

$$\frac{x}{2} + \frac{x}{3}$$

$$= \frac{x \cdot 3}{2 \cdot 3} + \frac{x \cdot 2}{3 \cdot 2}$$

$$= \frac{3x}{6} + \frac{2x}{6}$$

$$= \frac{3x + 2x}{6}$$

$$= \frac{5x}{6}$$

Identify each exercise as an expression *or an* equation. *Then simplify the expression by performing the indicated operation, or solve the given equation, as appropriate.*

1. $\frac{x}{2} - \frac{x}{4} = 5$

2. $\frac{4x - 20}{x^2 - 25} \cdot \frac{(x + 5)^2}{10}$

3. $\frac{6}{7x} - \frac{4}{x}$

4. $\dfrac{\frac{1}{x} + \frac{1}{y}}{\frac{1}{x} - \frac{1}{y}}$

5. $\frac{5}{7t} = \frac{52}{7} - \frac{3}{t}$

6. $\frac{x - 5}{3} + \frac{1}{3} = \frac{x - 2}{5}$

7. $\frac{7}{6x} + \frac{5}{8x}$

8. $\frac{4}{x} - \frac{8}{x + 1} = 0$

9. $\dfrac{\frac{6}{x + 1} - \frac{1}{x}}{\frac{2}{x} - \frac{4}{x + 1}}$

10. $\frac{8}{r + 2} - \frac{7}{4r + 8}$

11. $\frac{x}{x + y} + \frac{2y}{x - y}$

12. $\frac{3p^2 - 6p}{p + 5} \div \frac{p^2 - 4}{8p + 40}$

13. $\frac{x-2}{9} \cdot \frac{5}{8-4x}$

14. $\frac{a-4}{3} + \frac{11}{6} = \frac{a+1}{2}$

15. $\frac{b^2+b-6}{b^2+2b-8} \cdot \frac{b^2+8b+16}{3b+12}$

16. $\frac{10z^2-5z}{3z^3-6z^2} \div \frac{2z^2+5z-3}{z^2+z-6}$

17. $\frac{5}{x^2-2x} - \frac{3}{x^2-4}$

18. $\frac{6}{t+1} + \frac{4}{5t+5} = \frac{34}{15}$

19. $\dfrac{\frac{5}{x} - \frac{3}{y}}{\frac{9x^2-25y^2}{x^2y}}$

20. $\frac{-2}{a^2+2a-3} - \frac{5}{3-3a} = \frac{4}{3a+9}$

21. $\frac{4y^2-13y+3}{2y^2-9y+9} \div \frac{4y^2+11y-3}{6y^2-5y-6}$

22. $\frac{8}{3k+9} - \frac{8}{15} = \frac{2}{5k+15}$

23. $\frac{3r}{r-2} = 1 + \frac{6}{r-2}$

24. $\frac{6z^2-5z-6}{6z^2+5z-6} \cdot \frac{12z^2-17z+6}{12z^2-z-6}$

25. $\frac{-1}{3-x} - \frac{2}{x-3}$

26. $\dfrac{\frac{t}{4} - \frac{1}{t}}{1 + \frac{t+4}{t}}$

27. $\frac{2}{y+1} - \frac{3}{y^2-y-2} = \frac{3}{y-2}$

28. $\frac{7}{2x^2-8x} + \frac{3}{x^2-16}$

29. $\frac{3}{y-3} - \frac{3}{y^2-5y+6} = \frac{2}{y-2}$

30. $\dfrac{2k + \frac{5}{k-1}}{3k - \frac{2}{k}}$

16.7 Applications of Rational Expressions

OBJECTIVES

1. Find the value of an unknown variable in a formula.
2. Solve a formula for a specified variable.
3. Solve applications using proportions.
4. Solve applications about distance, rate, and time.
5. Solve applications about work rates.

OBJECTIVE 1 Find the value of an unknown variable in a formula. Formulas may contain rational expressions, such as $t = \frac{d}{r}$ and $\frac{1}{f} = \frac{1}{p} + \frac{1}{q}$.

EXAMPLE 1 Finding the Value of a Variable in a Formula

In physics, the focal length, f, of a lens is given by the formula

$$\frac{1}{f} = \frac{1}{p} + \frac{1}{q},$$

where p is the distance from the object to the lens and q is the distance from the lens to the image. See Figure 4. Find q if $p = 20$ cm and $f = 10$ cm.

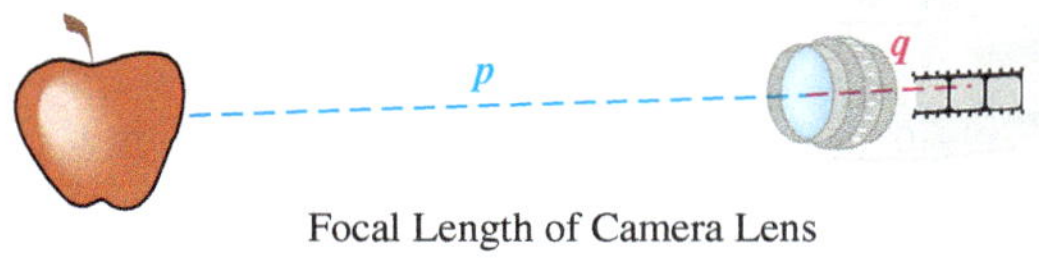

Focal Length of Camera Lens

Figure 4

Replace f with 10 and p with 20.

$$\frac{1}{f} = \frac{1}{p} + \frac{1}{q} \qquad \text{Solve this equation for } q.$$

$$\frac{1}{10} = \frac{1}{20} + \frac{1}{q} \qquad \text{Let } f = 10,\ p = 20.$$

$$20q \cdot \frac{1}{10} = 20q\left(\frac{1}{20} + \frac{1}{q}\right) \qquad \text{Multiply by the LCD, } 20q.$$

$$20q \cdot \frac{1}{10} = 20q\left(\frac{1}{20}\right) + 20q\left(\frac{1}{q}\right) \qquad \text{Distributive property.}$$

$$2q = q + 20 \qquad \text{Multiply.}$$

$$q = 20 \qquad \text{Subtract } q.$$

The distance from the lens to the image is 20 cm.

Work Problem 1 at the Side. ▶

1 Use the formula given in Example 1 to answer each part.

(a) Find p if $f = 15$ and $q = 25$.

(b) Find f if $p = 6$ and $q = 9$.

(c) Find q if $f = 12$ and $p = 16$.

OBJECTIVE 2 Solve a formula for a specified variable. The goal in solving for a specified variable is to isolate it on one side of the equals sign.

EXAMPLE 2 Solving a Formula for a Specified Variable

Solve $\frac{1}{f} = \frac{1}{p} + \frac{1}{q}$ for p.

$$\frac{1}{f} = \frac{1}{p} + \frac{1}{q}$$

$$fpq \cdot \frac{1}{f} = fpq\left(\frac{1}{p} + \frac{1}{q}\right) \qquad \text{Multiply by the LCD, } fpq.$$

$$pq = fq + fp \qquad \text{Distributive property}$$

Continued on Next Page

ANSWERS

1. (a) $\frac{75}{2}$ (b) $\frac{18}{5}$ (c) 48

2 Solve $\frac{3}{p} + \frac{3}{q} = \frac{5}{r}$ for q.

Transform the equation so that the terms with p (the specified variable) are on the same side. One way to do this is to subtract fp from each side.

$$pq = fq + fp$$

$$pq - fp = fq \quad \text{Subtract } fp.$$

This is a key step.

$$p(q - f) = fq \quad \text{Factor out } p.$$

$$p = \frac{fq}{q - f} \quad \text{Divide by } q - f.$$

◀ Work Problem **2** at the Side.

EXAMPLE 3 Solving a Formula for a Specified Variable

Solve $I = \frac{nE}{R + nr}$ for n.

$$I = \frac{nE}{R + nr}$$

$$(R + nr)I = (R + nr)\frac{nE}{R + nr} \quad \text{Multiply by } R + nr.$$

$$RI + nrI = nE$$

$$RI = nE - nrI \quad \text{Subtract } nrI.$$

$$RI = n(E - rI) \quad \text{Factor out } n.$$

$$\frac{RI}{E - rI} = n \quad \text{Divide by } E - rI.$$

3 Solve $A = \frac{Rr}{R + r}$ for R.

CAUTION

Refer to the steps in Examples 2 and 3 that factor out the desired variable. ***The variable for which you are solving must be a factor on only one side of the equation,*** so that each side can be divided by the remaining factor in the last step.

◀ Work Problem **3** at the Side.

We can now solve problems that translate into equations with rational expressions. To do so, we continue to use the six-step problem-solving method from **Section 2.3.**

OBJECTIVE 3 Solve applications using proportions. A **ratio** is a comparison of two quantities. The ratio of a to b may be written in any of the following ways:

$$a \text{ to } b, \quad a : b, \quad \text{or} \quad \frac{a}{b}. \qquad \text{Ratio of } a \text{ to } b$$

Ratios are usually written as quotients in algebra. A **proportion** is a statement that two ratios are equal, such as

$$\frac{a}{b} = \frac{c}{d}. \qquad \text{Proportion}$$

Proportions are a useful and important type of rational equation.

ANSWERS

2. $q = \frac{3rp}{5p - 3r}$, or $q = \frac{-3rp}{3r - 5p}$

3. $R = \frac{-Ar}{A - r}$, or $R = \frac{Ar}{r - A}$

EXAMPLE 4 **Solving a Proportion**

In 2005, about 15 of every 100 Americans had no health insurance coverage. The population at that time was about 296 million. How many million Americans had no health insurance? (*Source*: U.S. Census Bureau.)

Step 1 **Read** the problem.

Step 2 **Assign a variable.** Let $x =$ the number (in millions) who had no health insurance.

Step 3 **Write an equation.** To get an equation, set up a proportion. The ratio 15 to 100 should equal the ratio x to 296.

$$\frac{15}{100} = \frac{x}{296} \quad \text{Write a proportion.}$$

Step 4 **Solve.**

$$29{,}600\left(\frac{15}{100}\right) = 29{,}600\left(\frac{x}{296}\right) \quad \text{Multiply by a common denominator.}$$

$$4440 = 100x \quad \text{Simplify.}$$

$$x = 44.4 \quad \text{Divide by 100.}$$

Step 5 **State the answer.** There were about 44.4 million Americans with no health insurance in 2005.

Step 6 **Check** that the ratio of 44.4 million to 296 million equals $\frac{15}{100}$.

Work Problem **4** at the Side. ▶

4 Solve the problem.

In 2006, approximately 11.7% (that is, 11.7 of every 100) of the 73,740,000 children under 18 yr of age in the United States had no health insurance. How many such children were uninsured? (*Source*: U.S. Census Bureau.)

EXAMPLE 5 **Solving a Proportion Involving Rates**

Marissa's car uses 10 gal of gas to travel 210 mi. She has 5 gal of gas in the car, and she still needs to drive 640 mi. If we assume the car continues to use gas at the same rate, how many more gallons will she need?

Step 1 **Read** the problem.

Step 2 **Assign a variable.** Let $x =$ the additional number of gallons of gas.

Step 3 **Write an equation.** To get an equation, set up a proportion.

$$\frac{\text{gallons} \rightarrow 10}{\text{miles} \rightarrow 210} = \frac{5 + x \leftarrow \text{gallons}}{640 \leftarrow \text{miles}}$$

Step 4 **Solve.** We could multiply by the LCD $10 \cdot 21 \cdot 64$. Instead we use an alternative method that involves *cross products:* For $\frac{a}{b} = \frac{c}{d}$ to be true, then the cross products ad and bc must be equal. Thus,

$$10 \cdot 640 = 210(5 + x) \quad \text{If } \tfrac{a}{b} = \tfrac{c}{d}, \text{ then } ad = bc.$$

$$6400 = 1050 + 210x \quad \text{Multiply; distributive property}$$

$$5350 = 210x \quad \text{Subtract 1050.}$$

$$25.5 \approx x. \quad \text{Divide by 210.}$$

Step 5 **State the answer.** Marissa will need about 25.5 more gallons of gas.

Step 6 **Check.** The 25.5 gal plus the 5 gal equals 30.5 gal.

$$\frac{30.5}{640} \approx \mathbf{0.048} \quad \text{and} \quad \frac{10}{210} \approx \mathbf{0.048}$$

Since the ratios are equal, the answer is correct.

Work Problem **5** at the Side. ▶

5 Solve the problem.

Lauren's car uses 15 gal of gasoline to drive 495 mi. She has 6 gal of gasoline in the car, and she wants to know how much more gasoline she will need to drive 600 mi. If we assume that the car continues to use gasoline at the same rate, how many more gallons will she need? (Round your answer to the nearest tenth.)

ANSWERS

4. 8,627,580
5. 12.2 more gallons

OBJECTIVE 4 Solve applications about distance, rate, and time. The next examples use the distance formula $d = rt$ that was first introduced in **Section 2.2.** A familiar example of a rate is speed, which is the ratio of distance to time, or $r = \frac{d}{t}$.

EXAMPLE 6 Solving a Problem about Distance, Rate, and Time

A paddle wheeler goes 10 mi against the current in a river in the same time that it goes 15 mi with the current. If the speed of the current is 3 mph, find the speed of the boat in still water.

Step 1 **Read** the problem. We must find the speed of the boat in still water.

Step 2 **Assign a variable.**

Let $x =$ the speed of the boat in still water.

When the boat is traveling *against* the current, the current slows the boat down, and the speed of the boat is the difference between its speed in still water and the speed of the current. So, the speed against the current is $(x - 3)$ mph.

When the boat is traveling *with* the current, the current speeds the boat up, and the speed of the boat is the sum of its speed in still water and the speed of the current, that is, $(x + 3)$ mph.

Thus, $x - 3 =$ the speed of the boat *against* the current,

and $x + 3 =$ the speed of the boat *with* the current.

Because the time is the same going against the current as with the current, find time in terms of distance and rate (speed) for each situation. Start with the distance formula,

$$d = rt,$$

and divide each side by r to get $t = \frac{d}{r}$. Against the current, the distance is 10 mi and the rate is $(x - 3)$ mph, giving

$$t = \frac{d}{r} = \frac{10}{x-3}. \quad \text{Time against the current}$$

With the current, the distance is 15 mi and the rate is $(x + 3)$ mph, so

$$t = \frac{d}{r} = \frac{15}{x+3}. \quad \text{Time with the current}$$

This information is summarized in the following table.

	Distance	Rate	Time
Against Current	10	$x - 3$	$\frac{10}{x-3}$
With Current	15	$x + 3$	$\frac{15}{x+3}$

Times are equal.

Step 3 **Write an equation.** Because the times are equal,

$$\frac{10}{x-3} = \frac{15}{x+3}.$$

Continued on Next Page

Step 4 **Solve.** $\dfrac{10}{x-3} = \dfrac{15}{x+3}$

$$(x+3)(x-3)\left(\frac{10}{x-3}\right) = (x+3)(x-3)\left(\frac{15}{x+3}\right) \quad \text{Multiply by the LCD.}$$

$$10(x+3) = 15(x-3) \quad \text{Multiply.}$$

$$10x + 30 = 15x - 45 \quad \text{Distributive property}$$

$$30 = 5x - 45 \quad \text{Subtract } 10x.$$

$$75 = 5x \quad \text{Add 45.}$$

$$15 = x \quad \text{Divide by 5.}$$

Step 5 **State the answer.** The speed of the boat in still water is 15 mph.

Step 6 **Check** the answer: $\dfrac{10}{15-3} = \dfrac{15}{15+3}$ is true.

Work Problem **6** at the Side. ▶

6 Solve the problem.
A plane travels 100 mi against the wind in the same time that it takes to travel 120 mi with the wind. The wind speed is 20 mph.

(a) Complete this table.

	d	r	t
Against Wind	100	$x - 20$	
With Wind	120	$x + 20$	

(b) Find the speed of the plane in still air.

EXAMPLE 7 Solving a Problem about Distance, Rate, and Time

At O'Hare International Airport in Chicago, Cheryl and Bill are walking to the gate (at the same speed) to catch their flight to Denver. Bill steps onto the moving sidewalk and continues to walk while Cheryl uses the stationary sidewalk. If the sidewalk moves at 1 m per sec and Bill saves 50 sec covering the 300-m distance, what is their walking speed?

Step 1 **Read** the problem. We must find their walking speed.

Step 2 **Assign a variable.** Let x represent their walking speed in meters per second. Thus Cheryl travels at x meters per second and Bill travels at $(x + 1)$ meters per second. Express their times in terms of the known distances and the variable rates. As in Example 6, start with $d = rt$ and divide each side by r to get $t = \frac{d}{r}$. For Cheryl, the distance is 300 m and the rate is x, so Cheryl's time is

$$t = \frac{d}{r} = \frac{300}{x}. \quad \text{Cheryl's time}$$

Bill travels 300 m at a rate of $x + 1$, so his time is

$$t = \frac{d}{r} = \frac{300}{x+1}. \quad \text{Bill's time}$$

This information is summarized in the following table.

	Distance	Rate	Time
Cheryl	300	x	$\dfrac{300}{x}$
Bill	300	$x + 1$	$\dfrac{300}{x+1}$

Step 3 **Write an equation** using the times from the table.

Bill's time is Cheryl's time less 50 seconds.

$$\frac{300}{x+1} = \frac{300}{x} - 50$$

Continued on Next Page

Answers

6. **(a)** $\dfrac{100}{x-20}$; $\dfrac{120}{x+20}$ **(b)** 220 mph

7 Solve the problem.

Kathy Manley drove 300 mi north from San Antonio, mostly on the freeway. She usually averaged 55 mph, but an accident slowed her speed through Dallas to 15 mph. If her trip took 6 hr, how many miles did she drive at reduced speed?

	d	r	t
Normal Speed	$300 - x$	55	
Reduced Speed	x	15	

Step 4 **Solve.**

$$\frac{300}{x+1} = \frac{300}{x} - 50$$

$$x(x+1)\left(\frac{300}{x+1}\right) = x(x+1)\left(\frac{300}{x} - 50\right) \quad \text{Multiply by the LCD, } x(x+1).$$

$$x(x+1)\left(\frac{300}{x+1}\right) = x(x+1)\left(\frac{300}{x}\right) - x(x+1)(50) \quad \text{Distributive property}$$

$$300x = 300(x+1) - 50x(x+1) \quad \text{Multiply.}$$

$$300x = 300x + 300 - 50x^2 - 50x \quad \text{Distributive property}$$

$$50x^2 + 50x - 300 = 0 \quad \text{Standard form}$$

$$x^2 + x - 6 = 0 \quad \text{Divide by 50.}$$

$$(x+3)(x-2) = 0 \quad \text{Factor.}$$

$$x + 3 = 0 \quad \text{or} \quad x - 2 = 0 \quad \text{Zero-factor property}$$

$$x = -3 \quad \text{or} \quad x = 2 \quad \text{Solve each equation.}$$

Discard the negative answer, since speed cannot be negative.

Step 5 **State the answer.** Their walking speed is 2 m per sec.

Step 6 **Check** the answer in the words of the original problem.

◀ *Work Problem* 7 *at the Side.*

OBJECTIVE 5 Solve applications about work rates. Problems about work are closely related to distance problems.

Problem-Solving Hint

People work at different rates. If the letters r, t, and A represent the rate at which the work is done, the time required, and the amount of work accomplished, respectively, then $\boldsymbol{A = rt}$. Notice the similarity to the distance formula, $d = rt$.

Amount of work can be measured in terms of jobs accomplished. Thus, if 1 job is completed, $A = 1$, and the formula gives the rate as

$$1 = rt$$

$$r = \frac{1}{t}.$$

To solve a work problem, we begin by using the following fact to express all rates of work.

Rate of Work

If a job can be accomplished in t units of time, then the rate of work is

$$\frac{1}{t} \textbf{ job per unit of time.}$$

See if you can identify the six problem-solving steps in the next example.

Answer

7. $11\frac{1}{4}$ mi

EXAMPLE 8 **Solving a Problem about Work**

Letitia and Kareem are working on a neighborhood cleanup. Kareem can clean up all the trash in the area in 7 hr, while Letitia can do the same job in 5 hr. How long will it take them if they work together?

Let $x =$ the number of hours it will take the two people working together. Just as we made a table for the distance formula, $d = rt$, make a table here for $A = rt$, with $A = 1$. Since $A = 1$, the rate for each person will be $\frac{1}{t}$, where t is the time it takes the person to complete the job alone. For example, since Kareem can clean up all the trash in 7 hr, his rate is $\frac{1}{7}$ of the job per hour. Similarly, Letitia's rate is $\frac{1}{5}$ of the job per hour.

	Rate	Time Working Together	Fractional Part of the Job Done
Kareem	$\frac{1}{7}$	x	$\frac{1}{7}x$
Letitia	$\frac{1}{5}$	x	$\frac{1}{5}x$

Since together they complete 1 job, the sum of the fractional parts accomplished by them should equal 1.

$$\underbrace{\frac{1}{7}x}_{\text{Part done by Kareem}} + \underbrace{\frac{1}{5}x}_{\text{part done by Letitia}} \;\underset{\text{is}}{=}\; \underbrace{1}_{\text{1 whole job.}}$$

$$35\left(\frac{1}{7}x + \frac{1}{5}x\right) = 35 \cdot 1 \qquad \text{Multiply by the LCD, 35.}$$

$$5x + 7x = 35 \qquad \text{Distributive property}$$

$$12x = 35 \qquad \text{Combine like terms.}$$

$$x = \frac{35}{12} \qquad \text{Divide by 12.}$$

Working together, Kareem and Letitia can do the entire job in $\frac{35}{12}$ hr, or 2 hr, 55 min. Check this result in the original problem.

Work Problem **8** at the Side. ▶

8 Solve each problem.

(a) Stan needs 45 min to do the dishes, while Deb can do them in 30 min. How long will it take them if they work together?

	Rate	Time Working Together	Fractional Part of the Job Done
Stan	$\frac{1}{45}$	x	
Deb	$\frac{1}{30}$	x	

(b) Suppose it takes Stan 35 min to do the dishes, and together they can do them in 15 min. How long will it take Deb to do them alone?

There is another way to approach problems about work. For instance, in Example 8, x represents the number of hours it will take the two people working together to complete the entire job. In one hour, $\frac{1}{x}$ of the entire job will be completed. Kareem completes $\frac{1}{7}$ of the job in one hour, and Letitia completes $\frac{1}{5}$ of the job, so the sum of their rates should equal $\frac{1}{x}$. Thus,

$$\frac{1}{7} + \frac{1}{5} = \frac{1}{x}.$$

Multiplying each side of this equation by $35x$ gives $5x + 7x = 35$. This is the same equation we got in Example 8 in the third line from the bottom. Thus, the solution of the equation is the same using either approach.

Answers

8. **(a)** 18 min **(b)** $26\frac{1}{4}$ min

Math in the Media

A "BIG LEAGUE" APPLICATION OF RATIONAL EXPRESSIONS AND EQUATIONS

In the 1994 movie *Little Big League*, young Billy Heywood (Luke Edwards) inherits the Minnesota Twins baseball team and becomes manager. He leads the team to the Division Championship and then to the playoffs. But before the final playoff game, the biggest game of the year, he can't keep his mind on his job because a homework problem is giving him trouble:

> *If Joe can paint a house in 3 hours and Sam can paint the same house in 5 hours, how long does it take for them to do it together?*

With the help of one of his players, he is able to solve the problem.

1. Use the method described in Example 8 of **Section 8.5** to solve this problem.

2. Before the player was able to solve the problem correctly, Billy got "help" from some of the other players. The incorrect answers they gave him were

   (a) 15 hr **(b)** 8 hr **(c)** 4 hr.

   Explain the faulty reasoning behind each of these incorrect answers.

3. The player who gave Billy the correct answer solved the problem as follows:

   Using the simple formula a times b over a plus b, we get our answer of one and seven-eighths.

   Show that if it takes one person a hours to complete one job and another b hours to complete the same job, then the expression stated by the player,

   $$\frac{a \cdot b}{a + b}$$

   actually does give the number of hours it would take them to do the job together. (*Hint:* Refer to Example 8, and use a and b rather than 7 and 5. Then solve the resulting formula for x.)

16.7 Exercises

FOR EXTRA HELP

MyMathLab

Math XL PRACTICE WATCH DOWNLOAD READ REVIEW

In Exercises 1–4, a familiar formula is given. Give the letter of the choice that is an equivalent form of the given formula.

1. $p = br$ (percent)

A. $b = \dfrac{p}{r}$ **B.** $r = \dfrac{b}{p}$

C. $b = \dfrac{r}{p}$ **D.** $p = \dfrac{r}{b}$

2. $V = LWH$ (geometry)

A. $H = \dfrac{LW}{V}$ **B.** $L = \dfrac{V}{WH}$

C. $L = \dfrac{WH}{V}$ **D.** $W = \dfrac{H}{VL}$

3. $m = \dfrac{F}{a}$ (physics)

A. $a = mF$ **B.** $F = \dfrac{m}{a}$

C. $F = \dfrac{a}{m}$ **D.** $F = ma$

4. $I = \dfrac{E}{R}$ (electricity)

A. $R = \dfrac{I}{E}$ **B.** $R = IE$

C. $E = \dfrac{I}{R}$ **D.** $E = RI$

Solve each problem. See Example 1.

5. A gas law in chemistry says that

$$\frac{PV}{T} = \frac{pv}{t}.$$

Suppose that $T = 300$, $t = 350$, $V = 9$, $P = 50$, and $v = 8$. Find p.

6. In work with electric circuits, the formula

$$\frac{1}{a} = \frac{1}{b} + \frac{1}{c}$$

occurs. Find b if $a = 8$ and $c = 12$.

7. A formula from anthropology says that

$$c = \frac{100b}{L}.$$

Find L if $c = 80$ and $b = 5$.

8. The gravitational force between two masses is given by

$$F = \frac{GMm}{d^2}.$$

Find M to the nearest thousandth if $F = 10$, $G = 6.67 \times 10^{-11}$, $m = 1$, and $d = 3 \times 10^{-6}$.

Solve each formula for the specified variable. See Examples 2 and 3.

9. $F = \dfrac{GMm}{d^2}$ for G (physics)

10. $F = \dfrac{GMm}{d^2}$ for M (physics)

11. $\frac{1}{a} = \frac{1}{b} + \frac{1}{c}$ for a (electricity)

12. $\frac{1}{a} = \frac{1}{b} + \frac{1}{c}$ for b (electricity)

13. $\frac{PV}{T} = \frac{pv}{t}$ for v (chemistry)

14. $\frac{PV}{T} = \frac{pv}{t}$ for T (chemistry)

15. $I = \frac{nE}{R + nr}$ for r (engineering)

16. $a = \frac{V - v}{t}$ for V (physics)

17. $A = \frac{1}{2}h(b + B)$ for b (mathematics)

18. $S = \frac{n}{2}(a + \ell)d$ for n (mathematics)

19. $\frac{E}{e} = \frac{R + r}{r}$ for r (engineering)

20. $y = \frac{x + z}{a - x}$ for x

21. To solve the equation $m = \frac{ab}{a - b}$ for a, what is the first step?

22. Suppose you are asked to solve the equation

$$rp - rq = p + q$$

for r. What is the first step?

Solve each problem mentally. Use proportions in Exercises 23 and 24.

23. In a mathematics class, 3 of every 4 students are girls. If there are 28 students in the class, how many are girls? How many are boys?

24. In a certain southern state, sales tax on a purchase of \$1.50 is \$0.12. What is the sales tax on a purchase of \$9.00?

25. If Marin can mow her yard in 2 hr, what is her rate (in job per hour)?

26. A van traveling from Atlanta to Detroit averages 50 mph and takes 14 hr to make the trip. What is the driving distance from Atlanta to Detroit?

Use a proportion to solve each problem. Give answers to the nearest tenth if an approximation is needed. See Examples 4 and 5.

27. On a map of the United States, the distance between Seattle and Durango is 4.125 in. The two cities are actually 1238 miles apart. On this same map, what would be the distance between Chicago and El Paso, two cities that are actually 1606 mi apart? (*Source:* Universal Map Atlas.)

28. On a map of the United States, the distance between Reno and Phoenix is 2.5 in. The two cities are actually 768 miles apart. On this same map, what would be the distance between St. Louis and Jacksonville, two cities that are actually 919 mi apart? (*Source:* Universal Map Atlas.)

29. On a world globe, the distance between New York and Cairo, two cities that are actually 5619 mi apart, is 8.5 in. On this same globe, how far apart are Madrid and Rio de Janeiro, two cities that are actually 5045 mi apart? (*Source:* Author's globe, *World Almanac and Book of Facts.*)

30. On a world globe, the distance between San Francisco and Melbourne, two cities that are actually 7856 mi apart, is 11.875 in. On this same globe, how far apart are Mexico City and Singapore, two cities that are actually 10,327 mi apart? (*Source:* Author's globe, *World Almanac and Book of Facts.*)

Solve each problem. See Examples 4 and 5.

31. On May 23, 2008, the Boston Red Sox were in first place in the East Division of the American League, having won 31 of their first 50 regular season games. If the team continued to win the same fraction of its games, how many games would the Red Sox win for the complete 162-game season? Round your answer to the nearest whole number. (*Source:* www.mlb.com)

32. During 2004–2005, the ratio of teachers to students in public elementary and secondary schools was approximately 1 to 16. If a public school had 846 students, how many teachers would be at the school if this ratio was valid for that school? Round your answer to the nearest whole number. (*Source*: U.S. National Center for Education Statistics.)

33. Biologists tagged 500 fish in a lake on January 1. On February 1 they returned and collected a random sample of 400 fish, 8 of which had been previously tagged. Approximately how many fish does the lake have based on this experiment?

34. Suppose that in the experiment of Exercise 33, 10 of the previously tagged fish were collected on February 1. What would be the estimate of the fish population?

35. Bruce Johnston's Shelby Cobra uses 5 gal of gasoline to drive 156 mi. He has 3 gal of gasoline in the car, and he wants to know how much more gasoline he will need to drive 300 mi. If we assume that the car continues to use gasoline at the same rate, how many more gallons will he need?

36. Mike Love's T-bird uses 6 gal of gasoline to drive 141 miles. He has 4 gal of gasoline in the car, and he wants to know how much more gasoline he will need to drive 275 mi. If we assume that the car continues to use gasoline at the same rate, how many more gallons will he need?

Nurses use proportions to determine the amount of a drug to administer when the dose of the drug is measured in milligrams but the drug is packaged in a diluted form in milliliters. (*Source:* Hoyles, Celia, Richard Noss, and Stefano Pozzi, "Proportional Reasoning in Nursing Practice," *Journal for Research in Mathematics Education,* January 2001.) *For example, to find the number of milliliters of fluid needed to administer* 300 mg *of a drug that comes packaged as* 120 mg *in* 2 mL *of fluid, a nurse sets up the proportion*

$$\frac{120 \text{ mg}}{2 \text{ mL}} = \frac{300 \text{ mg}}{x \text{ mL}},$$

where x represents the amount to administer in milliliters. Use this method to find the correct dose for each prescription.

37. 120 mg of Amakacine packaged as 100 mg in 2-mL vials

38. 1.5 mg of morphine packaged as 20 mg ampules diluted in 10 mL of fluid

In geometry, it is shown that two triangles with corresponding angle measures equal, called **similar triangles,** *have corresponding sides proportional. For example, in the figure, angle A = angle D, angle B = angle E, and angle C = angle F, so the triangles are similar. Then the following ratios of corresponding sides are equal.*

$$\frac{4}{6} = \frac{6}{9} = \frac{2x + 1}{2x + 5}$$

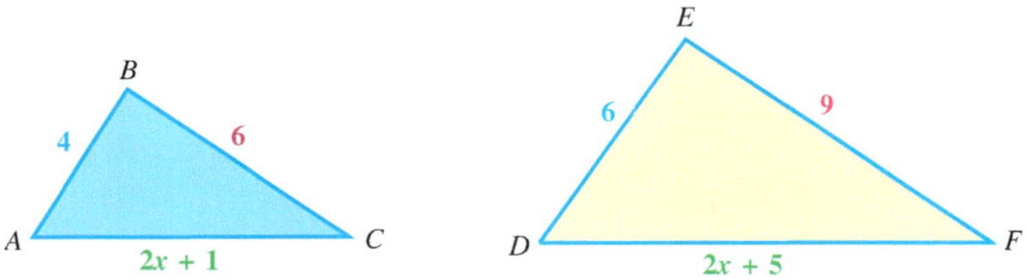

39. Solve for x using the given proportion to find the lengths of the third sides of the triangles.

40. Suppose the following triangles are similar. Find y and the lengths of the two longest sides of each triangle.

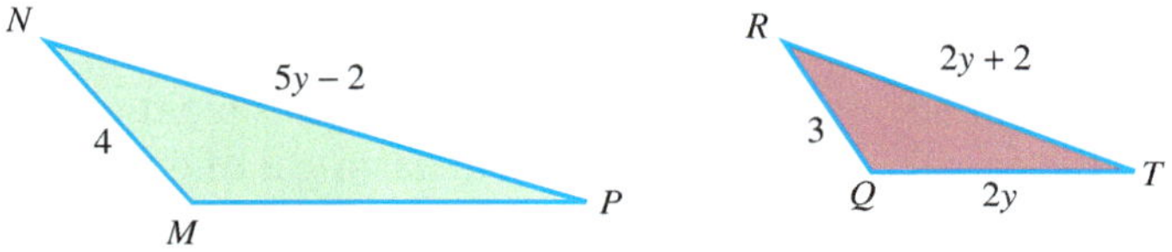

Solve each problem. See Examples 6 and 7.

41. Kellen's boat goes 12 mph. Find the rate of the current of the river if she can go 6 mi upstream in the same amount of time she can go 10 mi downstream.

	Distance	Rate	Time
Downstream	10	$12 + x$	
Upstream	6	$12 - x$	

42. Kasey can travel 8 mi upstream in the same time it takes her to go 12 mi downstream. Her boat goes 15 mph in still water. What is the rate of the current?

	Distance	Rate	Time
Downstream			
Upstream			

43. On his drive from Montpelier, Vermont, to Columbia, South Carolina, Dylan Davis averaged 51 mph. If he had been able to average 60 mph, he would have reached his destination 3 hr earlier. What is the driving distance between Montpelier and Columbia?

44. Leah drove from her apartment to her parents' house for the weekend. Driving to their house on Saturday morning, she was able to average 60 mph because traffic was light. However, returning on Sunday night, she was able to average only 45 mph on the same route, because traffic was heavy. The drive on Sunday took her 1.5 hr longer than the drive on Saturday. What is the distance between Leah's apartment and her parents' house?

45. A private plane traveled from San Francisco to a secret rendezvous. It averaged 200 mph. On the return trip, the average speed was 300 mph. If the total traveling time was 4 hr, how far from San Francisco was the secret rendezvous?

46. Johnny averages 30 mph when he drives on the old highway to his favorite fishing hole, and he averages 50 mph when most of his route is on the interstate. If both routes are the same length, and he saves 2 hr by traveling on the interstate, how far away is the fishing hole?

47. On the first part of a trip to Carmel traveling on the freeway, Marge averaged 60 mph. On the rest of the trip, which was 10 mi longer than the first part, she averaged 50 mph. Find the total distance to Carmel if the second part of the trip took 30 min more than the first part.

48. While on vacation, Jim and Annie decided to drive all day. During the first part of their trip on the highway, they averaged 60 mph. When they got to Houston, traffic caused them to average only 30 mph. The distance they drove in Houston was 100 mi less than their distance on the highway. What was their total driving distance if they spent 50 min more on the highway than they did in Houston?

Solve each problem. See Example 8.

49. Butch and Peggy want to pick up the mess that their grandson, Grant, has made in his playroom. Butch could do it in 15 min working alone. Peggy, working alone, could clean it in 12 min. How long will it take them if they work together?

	Rate	Time Working Together	Fractional Part of the Job Done
Butch	$\frac{1}{15}$	x	
Peggy	$\frac{1}{12}$	x	

50. Lou can groom Jay Beckenstein's dogs in 8 hr, but it takes his business partner, Janet, only 5 hr to groom the same dogs. How long will it take them to groom Jay's dogs if they work together?

	Rate	Time Working Together	Fractional Part of the Job Done
Lou	$\frac{1}{8}$	x	
Janet	$\frac{1}{5}$	x	

51. Jerry and Kuba are laying a hardwood floor. Working alone, Jerry can do the job in 20 hr. If the two of them work together, they can complete the job in 12 hr. How long would it take Kuba to lay the floor working alone?

52. Mrs. Disher is a high school mathematics teacher. She can grade a set of chapter tests in 5 hr working alone. If her student teacher Mr. Howes helps her, it will take 3 hr to grade the tests. How long would it take Mr. Howes to grade the tests if he worked alone?

53. If a vat of acid can be filled by an inlet pipe in 10 hr and emptied by an outlet pipe in 20 hr, how long will it take to fill the vat if both pipes are open?

54. A winery has a vat to hold Chardonnay. An inlet pipe can fill the vat in 9 hr, while an outlet pipe can empty it in 12 hr. How long will it take to fill the vat if both the outlet and the inlet pipes are open?

55. Suppose that Hortense and Mort can clean their entire house in 7 hr, while their toddler, Mimi, just by being around, can completely mess it up in only 2 hr. If Hortense and Mort clean the house while Mimi is at her grandma's, and then start cleaning up after Mimi the minute she gets home, how long does it take from the time Mimi gets home until the whole place is a shambles?

56. An inlet pipe can fill an artificial lily pond in 60 min, while an outlet pipe can empty it in 80 min. Through an error, both pipes are left open. How long will it take for the pond to fill?

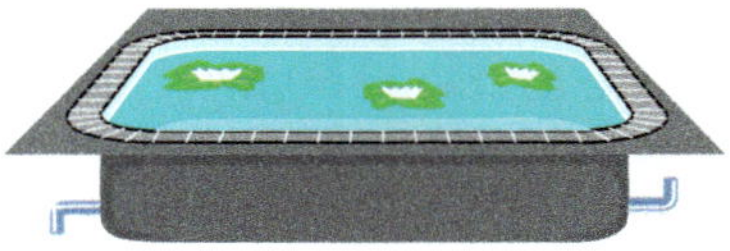

17

Roots & Radicals

17.1 Evaluating Roots

OBJECTIVES

1. **Find square roots.**
2. **Decide whether a given root is rational, irrational, or not a real number.**
3. **Find decimal approximations for irrational square roots.**
4. **Use the Pythagorean formula.**
5. **Find cube, fourth, and other roots.**

In **Section 1.8,** we discussed the idea of the *square* of a number. Recall that squaring a number means multiplying the number by itself.

$$\text{If } a = 8, \quad \text{then} \quad a^2 = 8 \cdot 8 = 64.$$
$$\text{If } a = -4, \quad \text{then} \quad a^2 = (-4)(-4) = 16.$$
$$\text{If } a = -\frac{1}{2}, \quad \text{then} \quad a^2 = \left(-\frac{1}{2}\right)\left(-\frac{1}{2}\right) = \frac{1}{4}.$$

In this chapter, we consider the opposite process.

$$\text{If } a^2 = 64, \quad \text{then} \quad a = ?$$
$$\text{If } a^2 = 16, \quad \text{then} \quad a = ?$$
$$\text{If } a^2 = \frac{1}{4}, \quad \text{then} \quad a = ?$$

OBJECTIVE 1 Find square roots. To find a in the three preceding statements, we must find a number that when multiplied by itself results in the given number. The number a is called a **square root** of the number a^2.

EXAMPLE 1 Finding All Square Roots of a Number

Find all square roots of 49.

To find a square root of 49, think of a number that when multiplied by itself gives 49. One square root is 7 because $7 \cdot 7 = 49$. Another square root of 49 is -7 because $(-7)(-7) = 49$. The number 49 has two square roots, 7 and -7; one is positive, and one is negative.

◀ Work Problem 1 at the Side.

1 Find all square roots.

(a) 100

(b) 25

(c) 36

(d) $\frac{25}{36}$

The **positive** or **principal square root** of a number is written with the symbol $\sqrt{}$. For example, the positive square root of 121 is 11, written

$$\sqrt{121} = 11.$$

The symbol $-\sqrt{}$ is used for the **negative square root** of a number. For example, the negative square root of 121 is -11, written

$$-\sqrt{121} = -11.$$

The symbol $\sqrt{}$, called a **radical sign,** always represents the positive square root (except that $\sqrt{0} = 0$). The number inside the radical sign is called the **radicand,** and the entire expression, radical sign and radicand, is called a **radical.**

Radical sign ↘ ↙ Radicand

$$\underbrace{\sqrt{a}}_{\text{Radical}}$$

An algebraic expression containing a radical is called a **radical expression.**

Radicals have a long mathematical history. The radical sign $\sqrt{}$ has been used since sixteenth-century Germany and was probably derived from the letter *R*. The radical symbol in the margin comes from the Latin word for root, *radix*. It was first used by Leonardo da Pisa (Fibonnaci) in 1220.

Early radical symbol

ANSWERS

1. **(a)** 10, -10 **(b)** 5, -5 **(c)** 6, -6 **(d)** $\frac{5}{6}, -\frac{5}{6}$

We summarize our discussion of square roots as follows.

Square Roots of a

If a is a positive real number, then

$\sqrt{a}$ is the positive or principal square root of a,

and $-\sqrt{a}$ is the negative square root of a.

For nonnegative a,

$$\sqrt{a} \cdot \sqrt{a} = \left(\sqrt{a}\right)^2 = a \quad \text{and} \quad -\sqrt{a} \cdot \left(-\sqrt{a}\right) = \left(-\sqrt{a}\right)^2 = a.$$

Also, $\sqrt{0} = 0$.

Calculator Tip Recall from **Section 5.8** that most calculators have a square root key, usually labeled $\sqrt{x}$, that allows us to find the square root of a number. On some models, the square root key must be used in conjunction with the key marked INV or 2nd.

EXAMPLE 2 Finding Square Roots

Find each square root.

(a) $\sqrt{144}$

The radical $\sqrt{144}$ represents the positive or principal square root of 144. Think of a positive number whose square is 144.

$$12^2 = 144, \quad \text{so} \quad \sqrt{144} = 12.$$

(b) $-\sqrt{1024}$

This symbol represents the negative square root of 1024. A calculator with a square root key can be used to find $\sqrt{1024} = 32$. Then, $-\sqrt{1024} = -32$.

(c) $\sqrt{\frac{4}{9}} = \frac{2}{3}$ **(d)** $-\sqrt{\frac{16}{49}} = -\frac{4}{7}$ **(e)** $\sqrt{0.81} = 0.9$

Work Problem **2** at the Side. ▶

As noted above, when the square root of a positive real number is squared, the result is that positive real number. (Also, $\left(\sqrt{0}\right)^2 = 0$.)

EXAMPLE 3 Squaring Radical Expressions

Find the *square* of each radical expression.

(a) $\sqrt{13}$

$\left(\sqrt{13}\right)^2 = 13$ Definition of square root

(b) $-\sqrt{29}$

$\left(-\sqrt{29}\right)^2 = 29$ The square of a *negative* number is positive.

(c) $\sqrt{p^2+1}$

$\left(\sqrt{p^2+1}\right)^2 = p^2 + 1$

Work Problem **3** at the Side. ▶

2 Find each square root.

(a) $\sqrt{16}$

(b) $-\sqrt{169}$

(c) $-\sqrt{225}$

(d) $\sqrt{729}$

(e) $\sqrt{\frac{36}{25}}$

(f) $\sqrt{0.49}$

3 Find the *square* of each radical expression.

(a) $\sqrt{41}$

(b) $-\sqrt{39}$

(c) $\sqrt{2x^2+3}$

Answers

2. (a) 4 **(b)** -13 **(c)** -15 **(d)** 27 **(e)** $\frac{6}{5}$ **(f)** 0.7

3. (a) 41 **(b)** 39 **(c)** $2x^2 + 3$

OBJECTIVE 2 Decide whether a given root is rational, irrational, or not a real number. All numbers with square roots that are rational are called **perfect squares.**

Perfect squares		Rational square roots
25		$\sqrt{25} = 5$
144	are perfect squares since	$\sqrt{144} = 12$
$\frac{4}{9}$		$\sqrt{\frac{4}{9}} = \frac{2}{3}$

A number that is not a perfect square has a square root that is not a rational number. For example, $\sqrt{5}$ is not a rational number because it cannot be written as the ratio of two integers. Its decimal equivalent (or approximation) neither terminates nor repeats. However, $\sqrt{5}$ is a real number and corresponds to a point on the number line. As mentioned in **Section 10.1,** a real number that is not rational is called an **irrational number.** The number $\sqrt{5}$ is irrational. Many square roots of integers are irrational.

> If a is a *positive* real number that is *not* a perfect square, then $\sqrt{a}$ is irrational.

Not every number has a real number square root. For example, there is no real number that can be squared to obtain -36. (The square of a real number can never be negative.) Because of this, $\sqrt{-36}$ ***is not a real number.***

> If a is a *negative* real number, then $\sqrt{a}$ is *not* a real number.

CAUTION
Be careful not to confuse $\sqrt{-36}$ and $-\sqrt{36}$. $\sqrt{-36}$ is not a real number since there is no real number that can be squared to obtain -36. However, $-\sqrt{36}$ is the negative square root of 36, which is -6.

EXAMPLE 4 Identifying Types of Square Roots

Tell whether each square root is *rational, irrational,* or *not a real number.*

(a) $\sqrt{17}$
Because 17 is not a perfect square, $\sqrt{17}$ is irrational.

(b) $\sqrt{64}$
The number 64 is a perfect square, 8^2, so $\sqrt{64} = 8$ is a rational number.

(c) $\sqrt{-25}$
There is no real number whose square is -25. Therefore, $\sqrt{-25}$ is not a real number.

◄ *Work Problem 4 at the Side.*

4 Tell whether each square root is *rational, irrational,* or *not a real number.*

(a) $\sqrt{9}$

(b) $\sqrt{7}$

(c) $\sqrt{\frac{4}{9}}$

(d) $\sqrt{72}$

(e) $\sqrt{-43}$

ANSWERS

4. **(a)** rational **(b)** irrational **(c)** rational **(d)** irrational **(e)** not a real number

Note

Not all irrational numbers are square roots of integers. For example, π (approximately 3.14159) is an irrational number that is not a square root of any integer.

OBJECTIVE 3 Find decimal approximations for irrational square roots. Even if a number is irrational, a decimal that approximates the number can be found using a calculator, as we did in **Section 5.8.** For example, if we use a calculator to find $\sqrt{10}$, the display might show 3.16227766, which is only an *approximation* of $\sqrt{10}$, not an exact rational value.

EXAMPLE 5 Approximating Irrational Square Roots

Find a decimal approximation for each square root. Round answers to the nearest thousandth.

(a) $\sqrt{11}$

Using the square root key of a calculator gives $3.31662479 \approx 3.317$, where $\approx$ means "is approximately equal to."

(b) $\sqrt{39} \approx 6.245$ Use a calculator. **(c)** $-\sqrt{740} \approx -27.203$

Work Problem 5 at the Side. ▶

5 Find a decimal approximation for each square root. Round answers to the nearest thousandth.

(a) $\sqrt{28}$

(b) $\sqrt{63}$

(c) $-\sqrt{190}$

(d) $\sqrt{1000}$

OBJECTIVE 4 Use the Pythagorean formula. Many applications of square roots use the Pythagorean formula. Recall from **Section 14.7** that by this formula if c is the length of the hypotenuse of a right triangle, and a and b are the lengths of the two legs, as shown in Figure 1, then

$$a^2 + b^2 = c^2$$

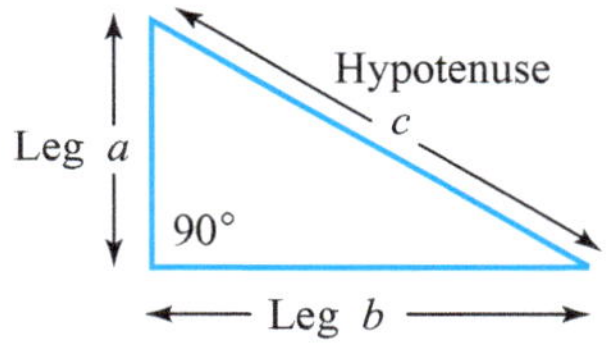

Figure 1

EXAMPLE 6 Using the Pythagorean Formula

Find the length of the unknown side of each right triangle with sides a, b, and c, where c is the hypotenuse.

(a) $a = 3, b = 4$

Use the Pythagorean formula to find c^2 first.

$$c^2 = a^2 + b^2$$

$$c^2 = 3^2 + 4^2 \quad \text{Let } a = 3 \text{ and } b = 4.$$

$$c^2 = 9 + 16 \quad \text{Square.}$$

$$c^2 = 25 \quad \text{Add.}$$

Since the length of a side of a triangle must be a positive number, find the positive square root of 25 to get c.

$$c = \sqrt{25} = 5$$

Continued on Next Page

ANSWERS

5. **(a)** 5.292 **(b)** 7.937 **(c)** −13.784 **(d)** 31.623

6 Find the length of the unknown side in each right triangle. Give any decimal approximations to the nearest thousandth.

(a) $a = 7, b = 24$

(b) $c = 15, b = 13$

(c)

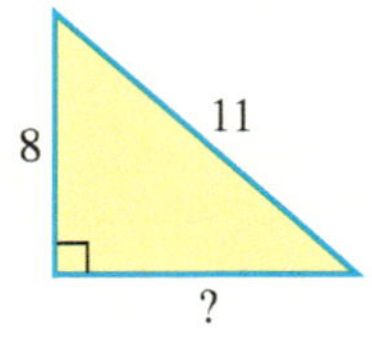

(b) $c = 9, b = 5$

Substitute the given values in the Pythagorean formula. Then solve for a^2.

$$c^2 = a^2 + b^2$$

$$9^2 = a^2 + 5^2 \quad \text{Let } c = 9 \text{ and } b = 5.$$

$$81 = a^2 + 25 \quad \text{Square.}$$

$$56 = a^2 \quad \text{Subtract 25.}$$

Use a calculator to find the positive square root of 56 to get a.

$$a = \sqrt{56} \approx 7.483$$

CAUTION

Be careful not to make the common mistake of thinking that $\sqrt{a^2 + b^2}$ equals $a + b$. As Example 6(a) shows, $\sqrt{9 + 16} = \sqrt{25} = 5$. However, $\sqrt{9} + \sqrt{16} = 3 + 4 = 7$. Since $5 \neq 7$, in general,

$$\sqrt{a^2 + b^2} \neq a + b.$$

◀ *Work Problem* **6** *at the Side.*

The Pythagorean formula can be used to solve applied problems that involve right triangles. Use the same six problem-solving steps that we have been using throughout the text.

EXAMPLE 7 Using the Pythagorean Formula to Solve an Application

A ladder 10 ft long leans against a wall. The foot of the ladder is 6 ft from the base of the wall. How high up the wall does the top of the ladder rest?

Step 1 **Read** the problem again.

Step 2 **Assign a variable.** As shown in Figure 2, a right triangle is formed with the ladder as the hypotenuse. Let a represent the height of the top of the ladder when measured straight down to the ground.

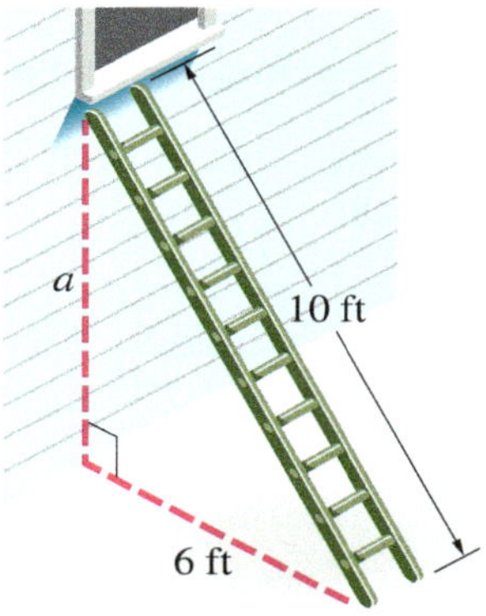

Figure 2

Continued on Next Page

ANSWERS

6. **(a)** 25 **(b)** $\sqrt{56} \approx 7.483$
(c) $\sqrt{57} \approx 7.550$

Step 3 **Write an equation** using the Pythagorean formula.

Substitute carefully.

$$c^2 = a^2 + b^2$$

$$10^2 = a^2 + 6^2 \quad \text{Let } c = 10 \text{ and } b = 6.$$

Step 4 **Solve.**

$$100 = a^2 + 36 \quad \text{Square.}$$

$$64 = a^2 \quad \text{Subtract 36.}$$

$$\sqrt{64} = a$$

$$a = 8 \quad \sqrt{64} = 8$$

Choose the positive square root of 64 since a represents a length.

Step 5 **State the answer.** The top of the ladder rests 8 ft up the wall.

Step 6 **Check.** From Figure 2, we see that we must have

$$8^2 + 6^2 \stackrel{?}{=} 10^2$$

$$64 + 36 = 100. \quad \text{True}$$

The check confirms that the top of the ladder rests 8 ft up the wall.

Work Problem **7** at the Side. ▶

7 A rectangle has dimensions 5 ft by 12 ft. Find the length of its diagonal.

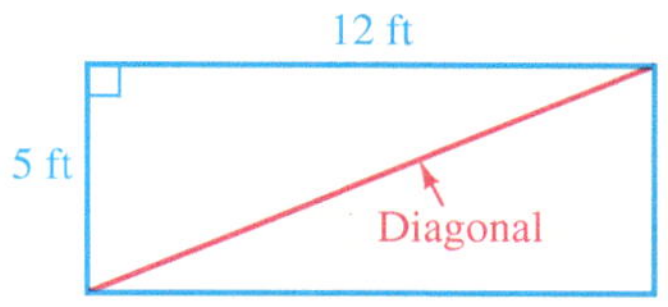

(Note that the diagonal divides the rectangle into two right triangles with itself as the hypotenuse.)

OBJECTIVE 5 Find cube, fourth, and other roots. Finding the square root of a number is the inverse (reverse) of squaring a number. In a similar way, there are inverses to finding the cube of a number, or finding the fourth or higher power of a number. These inverses are the **cube root,** written $\sqrt[3]{a}$, and the **fourth root,** written $\sqrt[4]{a}$. Similar symbols are used for higher roots. In general, we have the following.

> **$\sqrt[n]{a}$**
>
> The nth root of a is written: $\sqrt[n]{a}$

In $\sqrt[n]{a}$, the number n is the **index,** or **order,** of the radical.

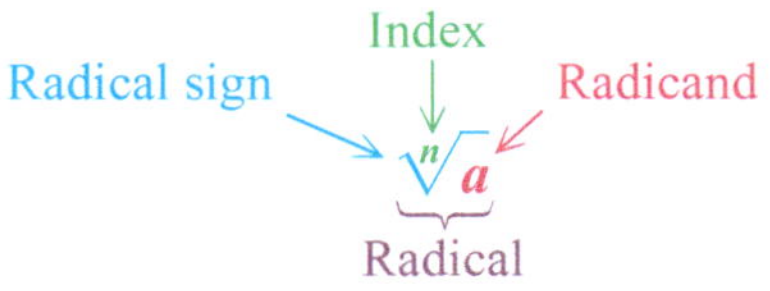

We could write $\sqrt[2]{a}$ instead of $\sqrt{a}$, but the simpler symbol $\sqrt{a}$ is customary since the square root is the most commonly used root.

> **Calculator Tip** A calculator that has a key marked $\sqrt[x]{y}$, x^y, or y^x (again perhaps in conjunction with the INV or 2nd key) can be used to find other roots.

When working with cube roots or fourth roots, it is helpful to memorize the first few *perfect cubes* ($1^3 = 1$, $2^3 = 8$, $3^3 = 27$, and so on) and the first few *perfect fourth powers* ($1^4 = 1$, $2^4 = 16$, $3^4 = 81$, and so on).

Work Problem **8** at the Side. ▶

8 Complete the following list of perfect cubes and perfect fourth powers.

Perfect Cubes	Perfect Fourth Powers
$1^3 = 1$	$1^4 = 1$
$2^3 = 8$	$2^4 = 16$
$3^3 = 27$	$3^4 = 81$
$4^3 =$ ____	$4^4 =$ ____
$5^3 =$ ____	$5^4 =$ ____
$6^3 =$ ____	$6^4 =$ ____
$7^3 =$ ____	$7^4 =$ ____
$8^3 =$ ____	$8^4 =$ ____
$9^3 =$ ____	$9^4 =$ ____
$10^3 =$ ____	$10^4 =$ ____

ANSWERS

7. 13 ft

8. Perfect cubes: 64; 125; 216; 343; 512; 729; 1000
Perfect fourth powers: 256; 625; 1296; 2401; 4096; 6561; 10,000

9 Find each cube root.

(a) $\sqrt[3]{27}$

(b) $\sqrt[3]{64}$

(c) $\sqrt[3]{-125}$

10 Find each root.

(a) $\sqrt[4]{81}$

(b) $\sqrt[4]{-81}$

(c) $-\sqrt[4]{81}$

(d) $\sqrt[5]{243}$

(e) $\sqrt[5]{-243}$

ANSWERS

9. (a) 3 **(b)** 4 **(c)** -5

10. (a) 3 **(b)** not a real number **(c)** -3 **(d)** 3 **(e)** -3

EXAMPLE 8 Finding Cube Roots

Find each cube root.

(a) $\sqrt[3]{8}$

Look for a number that can be cubed to give 8. Because $2^3 = 8, \sqrt[3]{8} = 2$.

(b) $\sqrt[3]{-8} = -2$ because $(-2)^3 = -8$.

(c) $\sqrt[3]{216} = 6$ because $6^3 = 216$.

Notice in Example 8(b) that we can find the cube root of a negative number. (Contrast this with the square root of a negative number, which is not real.) In fact, the cube root of a positive number is positive, and the cube root of a negative number is negative. ***There is only one real number cube root for each real number.***

◀ Work Problem 9 at the Side.

When a radical has an ***even index*** (square root, fourth root, and so on), ***the radicand must be nonnegative*** to yield a real number root. Also,

$$\sqrt{a}, \sqrt[4]{a}, \sqrt[6]{a}, \text{ and so on are positive (principal) roots;}$$

$$-\sqrt{a}, -\sqrt[4]{a}, -\sqrt[6]{a}, \text{ and so on are negative roots.}$$

EXAMPLE 9 Finding Other Roots

Find each root.

(a) $\sqrt[4]{16} = 2$ because 2 is positive and $2^4 = 16$.

(b) $-\sqrt[4]{16}$

From part (a), $\sqrt[4]{16} = 2$, so the negative root is $-\sqrt[4]{16} = -2$.

(c) $\sqrt[4]{-16}$

For a real number fourth root, the radicand must be nonnegative. There is no real number that equals $\sqrt[4]{-16}$.

(d) $-\sqrt[5]{32}$

First find $\sqrt[5]{32}$. Because 2 is the number whose fifth power is 32, $\sqrt[5]{32} = 2$. Since $\sqrt[5]{32} = 2$, it follows that

$$-\sqrt[5]{32} = -2.$$

(e) $\sqrt[5]{-32}$

Because $(-2)^5 = -32, \sqrt[5]{-32} = -2$.

◀ Work Problem 10 at the Side.

17.1 Exercises

FOR EXTRA HELP MyMathLab MathXL PRACTICE WATCH DOWNLOAD READ REVIEW

Decide whether each statement is true *or* false. *If* false, *tell why.*

1. Every positive number has two real square roots.
2. A negative number has negative square roots.
3. Every nonnegative number has two real square roots.
4. The positive square root of a positive number is its principal square root.
5. The cube root of every real number has the same sign as the number itself.
6. Every positive number has three real cube roots.

Find all square roots of each number. See Example 1.

7. 9
8. 16
9. 64
10. 100
11. 169
12. 225
13. $\frac{25}{196}$
14. $\frac{81}{400}$
15. 900
16. 1600

Find each square root. See Examples 2 and 4(c).

17. $\sqrt{1}$
18. $\sqrt{4}$
19. $\sqrt{49}$
20. $\sqrt{81}$
21. $-\sqrt{256}$
22. $-\sqrt{196}$
23. $-\sqrt{\frac{144}{121}}$
24. $-\sqrt{\frac{49}{36}}$
25. $\sqrt{0.64}$
26. $\sqrt{0.16}$
27. $\sqrt{-121}$
28. $\sqrt{-64}$
29. $-\sqrt{-49}$
30. $-\sqrt{-100}$

Find the square of each radical expression. See Example 3.

31. $\sqrt{100}$
32. $\sqrt{36}$
33. $-\sqrt{19}$
34. $-\sqrt{99}$
35. $\sqrt{\frac{2}{3}}$
36. $\sqrt{\frac{5}{7}}$
37. $\sqrt{3x^2 + 4}$
38. $\sqrt{9y^2 + 3}$

What must be true about the value of a for each statement in Exercises 39–42 to be true?

39. $\sqrt{a}$ represents a positive number.

40. $-\sqrt{a}$ represents a negative number.

41. $\sqrt{a}$ is not a real number.

42. $-\sqrt{a}$ is not a real number.

Write rational, irrational, *or* not a real number *for each number. If a number is rational, give its exact value. If a number is irrational, give a decimal approximation to the nearest thousandth. Use a calculator as necessary. See Examples 4 and 5.*

43. $\sqrt{25}$

44. $\sqrt{169}$

45. $\sqrt{29}$

46. $\sqrt{33}$

47. $-\sqrt{64}$

48. $-\sqrt{81}$

49. $-\sqrt{300}$

50. $-\sqrt{500}$

51. $\sqrt{-29}$

52. $\sqrt{-47}$

53. $\sqrt{1200}$

54. $\sqrt{1500}$

Work Exercises 55 and 56 without using a calculator.

55. Choose the best estimate for the length and width (in meters) of this rectangle.

A. 11 by 6 **B.** 11 by 7 **C.** 10 by 7 **D.** 10 by 6

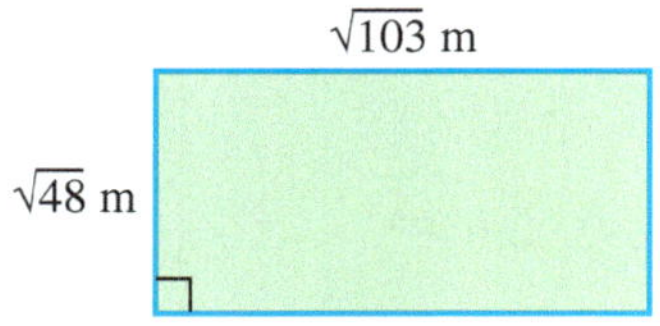

56. Choose the best estimate for the base and height (in feet) of this triangle.

A. $b = 8, h = 5$ **B.** $b = 8, h = 4$
C. $b = 9, h = 5$ **D.** $b = 9, h = 4$

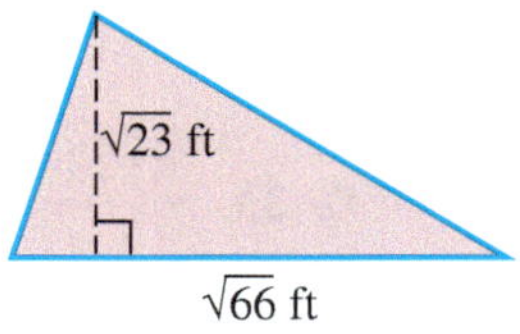

Find the length of the unknown side of each right triangle with sides a, b, and c, where c is the hypotenuse. See Figure 1 and Example 6. Give any decimal approximations to the nearest thousandth.

57. $a = 8, b = 15$

58. $a = 24, b = 10$

59. $a = 6, c = 10$

60. $b = 12, c = 13$

61. $a = 11, b = 4$

62. $a = 13, b = 9$

Solve each problem. See Example 7.

63. The diagonal of a rectangle measures 25 cm. The width of the rectangle is 7 cm. Find the length of the rectangle.

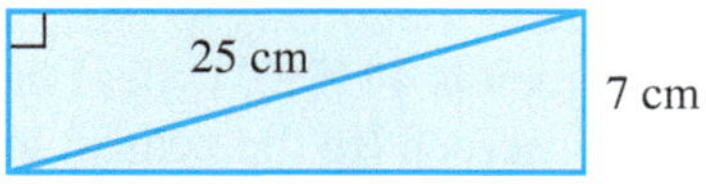

64. The length of a rectangle is 40 m, and the width is 9 m. Find the measure of the diagonal of the rectangle.

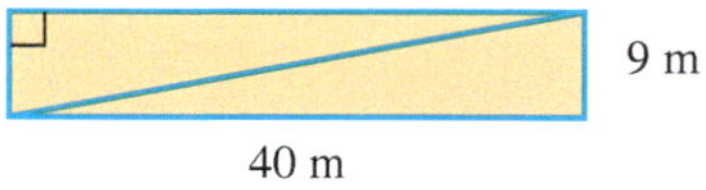

65. Tyler is flying a kite on 100 ft of string. How high is it above his hand (vertically) if the horizontal distance between Tyler and the kite is 60 ft?

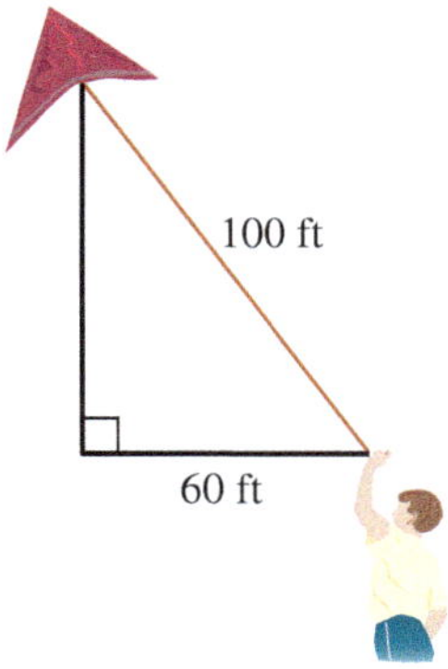

66. A guy wire is attached to the mast of a short-wave transmitting antenna. It is attached 96 ft above ground level. If the wire is staked to the ground 72 ft from the base of the mast, how long is the wire?

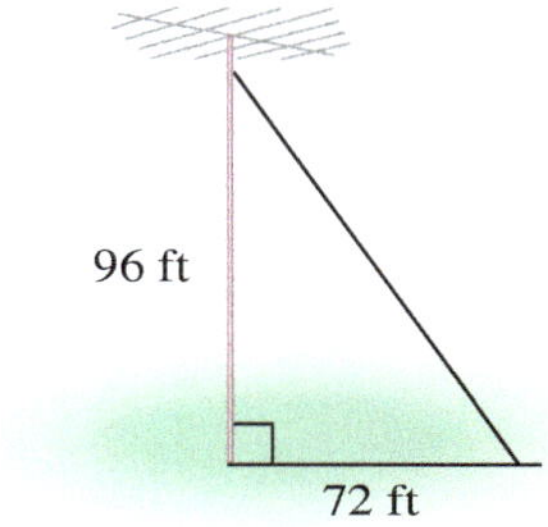

67. A surveyor measured the distances shown in the figure. Find the distance across the lake between points *R* and *S*.

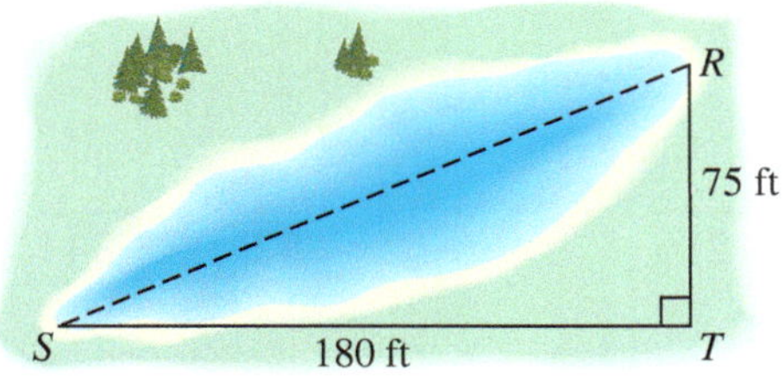

68. A boat is being pulled toward a dock with a rope attached at water level. When the boat is 24 ft from the dock, 30 ft of rope is extended. What is the height of the dock above the water?

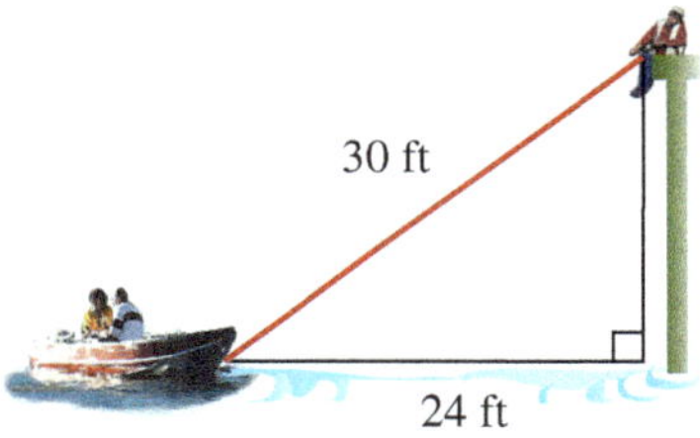

69. A surveyor wants to find the height of a building. At a point 110.0 ft from the base of the building he sights to the top of the building and finds the distance to be 193.0 ft. How high is the building (to the nearest tenth)?

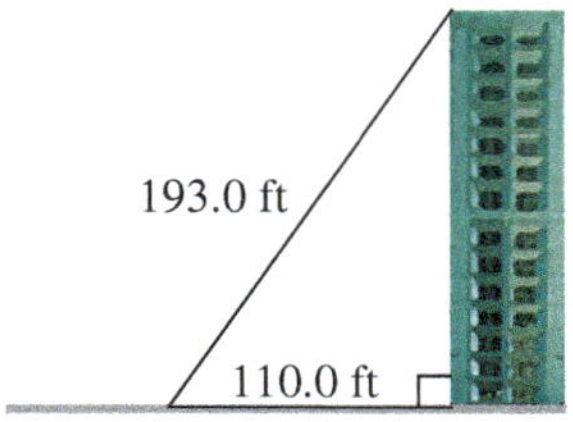

70. Two towns are separated by a dense forest. To go from Town B to Town A, it is necessary to travel due west for 19.0 mi, then turn due north and travel for 14.0 mi. How far apart are the towns (to the nearest tenth)?

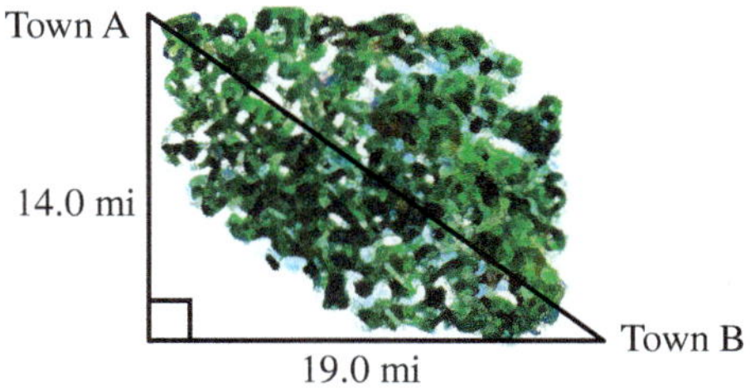

71. Following Hurricane Katrina, thousands of pine trees in southeastern Louisiana formed right triangles as shown in the photo. Suppose that, for a small such tree, the vertical distance from the base of the broken tree to the point of the break is 4.5 ft. The length of the broken part is 12.0 ft. How far along the ground (to the nearest tenth) is it from the base of the tree to the point where the broken part touches the ground?

72. One of the authors of this text purchased a new rear-projection Toshiba 51H84 television. A television set is "sized" according to the diagonal measurement of the viewing screen. The author purchased a 51-in. TV, so the TV measures 51 in. from one corner of the viewing screen diagonally to the other corner. The viewing screen is 44.5 in. wide. Find the height of the viewing screen (to the nearest tenth).

73. What is the value of x (to the nearest thousandth) in the figure?

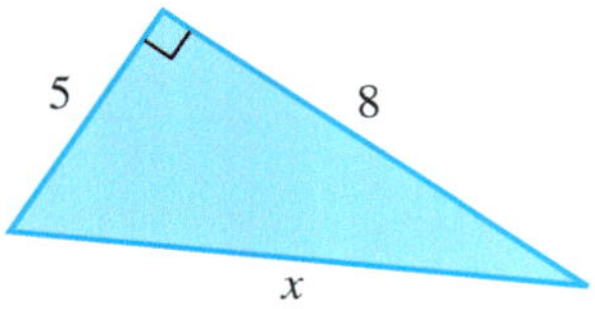

74. What is the value of y (to the nearest thousandth) in the figure?

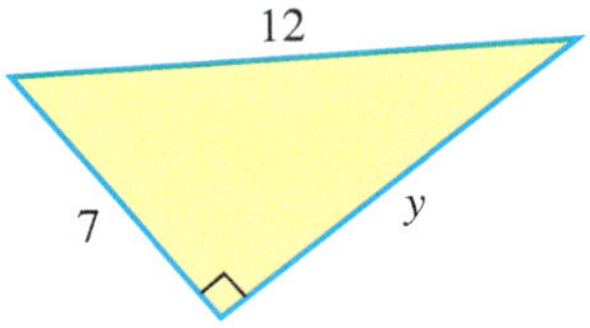

Find each root. See Examples 8 and 9.

75. $\sqrt[3]{1}$

76. $\sqrt[3]{729}$

77. $\sqrt[3]{125}$

78. $\sqrt[3]{1000}$

79. $\sqrt[3]{-27}$

80. $\sqrt[3]{-64}$

81. $\sqrt[3]{-216}$

82. $\sqrt[3]{-343}$

83. $-\sqrt[3]{-8}$

84. $-\sqrt[3]{-216}$

85. $\sqrt[4]{256}$

86. $\sqrt[4]{625}$

87. $\sqrt[4]{1296}$

88. $\sqrt[4]{10,000}$

89. $\sqrt[4]{-1}$

90. $\sqrt[4]{-625}$

91. $-\sqrt[4]{625}$

92. $-\sqrt[4]{256}$

93. $\sqrt[5]{-1024}$

94. $\sqrt[5]{-100,000}$

17.2 Radical Expressions and Graphs

OBJECTIVES

1. **Find roots of numbers.**
2. **Find principal roots.**
3. **Graph functions defined by radical expressions.**
4. **Find *n*th roots of *n*th powers.**
5. **Use a calculator to find roots.**

OBJECTIVE 1 Find roots of numbers. Recall from **Section 1.3** that $6^2 = 36$; that is, 6 *squared* is 36. The opposite (or inverse) of *squaring* a number is taking its *square root*. Thus,

It is customary to write $\sqrt{}$, rather than $\sqrt[2]{}$.

$$\sqrt{36} = 6, \quad \text{because} \quad 6^2 = 36.$$

We now extend our discussion of roots to *cube roots* $\sqrt[3]{}$, *fourth roots* $\sqrt[4]{}$, and higher roots. In general, $\sqrt[n]{a}$ is a number whose nth power equals a. That is,

$$\sqrt[n]{a} = b \quad \textbf{means} \quad b^n = a.$$

The number a is the **radicand,** n is the **index,** or **order,** and the expression $\sqrt[n]{a}$ is a **radical.**

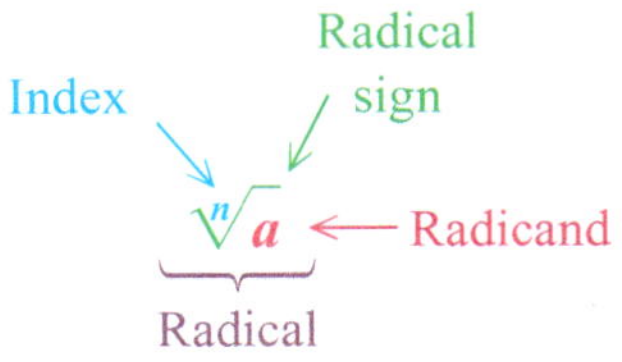

Example 1 Simplifying Higher Roots

Simplify.

(a) $\sqrt[3]{64} = 4$, because $4^3 = 64$.

(b) $\sqrt[3]{125} = 5$, because $5^3 = 125$.

(c) $\sqrt[4]{16} = 2$, because $2^4 = 16$.

(d) $\sqrt[5]{32} = 2$, because $2^5 = 32$.

(e) $\sqrt[3]{\dfrac{8}{27}} = \dfrac{2}{3}$, because $\left(\dfrac{2}{3}\right)^3 = \dfrac{8}{27}$.

(f) $\sqrt[4]{0.0016} = 0.2$, because $(0.2)^4 = 0.0016$.

◀ Work Problem 1 at the Side.

1 Simplify.

(a) $\sqrt[3]{27}$

(b) $\sqrt[3]{1000}$

(c) $\sqrt[4]{256}$

(d) $\sqrt[5]{243}$

(e) $\sqrt[4]{\dfrac{16}{81}}$

(f) $\sqrt[3]{0.064}$

OBJECTIVE 2 Find principal roots. If n is even, positive numbers have two nth roots. For example, both 4 and -4 are square roots of 16, and 2 and -2 are fourth roots of 16. The notation $\sqrt[n]{a}$ represents the positive root, called the **principal root,** and $-\sqrt[n]{a}$ represents the negative root.

nth Root

Case 1 If n is ***even*** and a is ***positive or 0,*** then

$\sqrt[n]{a}$ represents the **principal nth root** of a, and

$-\sqrt[n]{a}$ represents the **negative nth root** of a.

Case 2 If n is ***even*** and a is ***negative,*** then

$\sqrt[n]{a}$ is not a real number.

Case 3 If n is ***odd,*** then

there is exactly one nth root of a, written $\sqrt[n]{a}$.

If n is even, then the two nth roots of a are often written together as $\pm\sqrt[n]{a}$, with $\pm$ read "positive or negative," or "plus or minus."

ANSWERS

1. (a) 3 (b) 10 (c) 4 (d) 3 (e) $\frac{2}{3}$ (f) 0.4

Example 2 **Finding Roots**

Find each root.

(a) $\sqrt{100} = 10$

While 100 has two square roots, $\sqrt{100}$ represents the principal square root, which is 10.

(b) $-\sqrt{100} = -10$

Here, we want the negative square root, -10.

(c) $\sqrt[4]{81} = 3$ Principal 4th root

(d) $-\sqrt[4]{81} = -3$ Negative 4th root

Parts (a)–(d) illustrate Case 1 in the preceding box.

(e) $\sqrt[4]{-81}$

The index is *even* and the radicand is *negative,* so $\sqrt[4]{-81}$ is not a real number. This is Case 2 in the preceding box.

(f) $\sqrt[3]{8} = 2$, because $2^3 = 8$.

(g) $\sqrt[3]{-8} = -2$, because $(-2)^3 = -8$.

Parts (f) and (g) illustrate Case 3 in the box. The index is *odd,* so each radical represents exactly one *n*th root (regardless of whether the radicand is positive, negative, or 0).

Work Problem **2** *at the Side.* ▶

2 Find each root.

(a) $\sqrt{36}$

(b) $-\sqrt{36}$

(c) $\sqrt[4]{16}$

(d) $-\sqrt[4]{16}$

(e) $\sqrt[4]{-16}$

(f) $\sqrt[5]{1024}$

(g) $\sqrt[5]{-1024}$

OBJECTIVE 3 **Graph functions defined by radical expressions.** A **radical expression** is an algebraic expression that contains radicals.

$$3 - \sqrt{x}, \quad \sqrt[3]{x}, \quad \text{and} \quad \sqrt{2x - 1} \qquad \text{Radical expressions}$$

In earlier chapters we graphed functions defined by polynomial and rational expressions. Now we examine the graphs of functions defined by the radical expressions $f(x) = \sqrt{x}$ and $f(x) = \sqrt[3]{x}$.

Figure 1 shows the graph of the **square root function** defined by $f(x) = \sqrt{x}$, together with a table of selected points. Only nonnegative values can be used for x, so the domain is $[0, \infty)$. Because $\sqrt{x}$ is the principal square root of x, it always has a nonnegative value, so the range is also $[0, \infty)$.

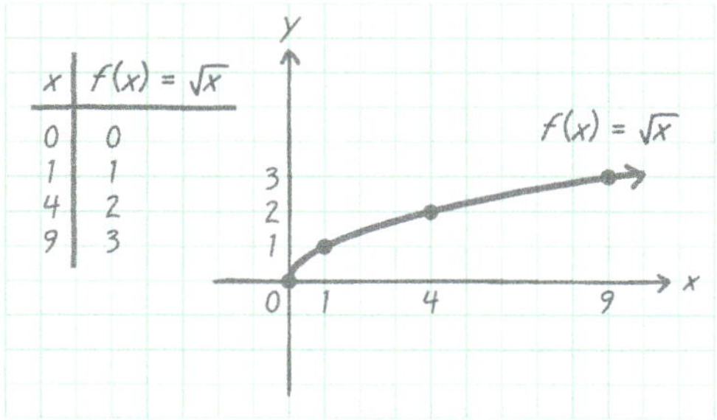

Figure 1

Figure 2 on the next page shows the graph of the **cube root function** defined by $f(x) = \sqrt[3]{x}$, together with a table of selected points. Since any real number (positive, negative, or 0) can be used for x in the cube root function, $\sqrt[3]{x}$ can be positive, negative, or 0. Thus, both the domain and the range of the cube root function are $(-\infty, \infty)$.

ANSWERS

2. **(a)** 6 **(b)** -6 **(c)** 2 **(d)** -2 **(e)** not a real number **(f)** 4 **(g)** -4

3 Graph each function by creating a table of values. Give the domain and range.

(a) $f(x) = \sqrt{x} + 2$

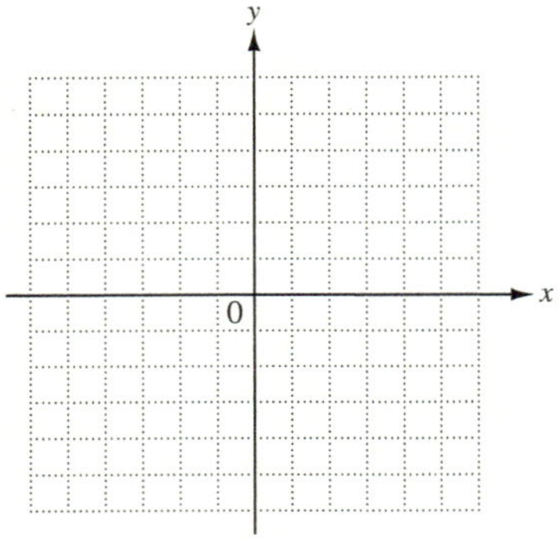

(b) $f(x) = \sqrt[3]{x - 1}$

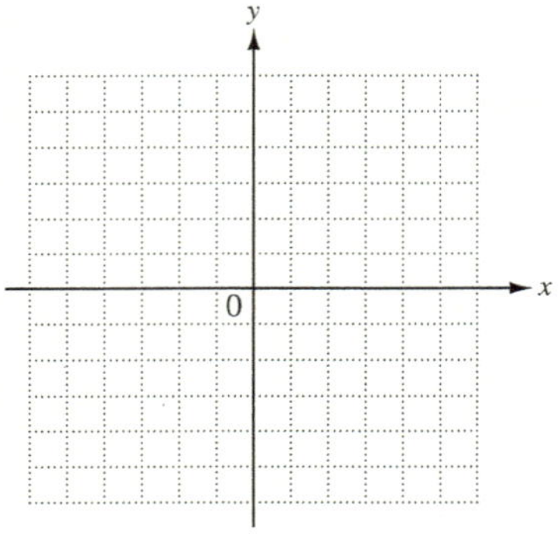

ANSWERS

3. (a) domain: $[0, \infty)$; range: $[2, \infty)$

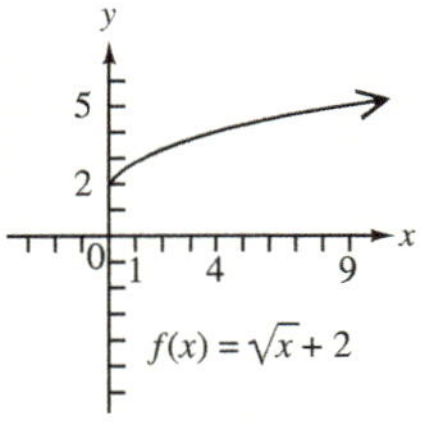

(b) domain: $(-\infty, \infty)$; range: $(-\infty, \infty)$

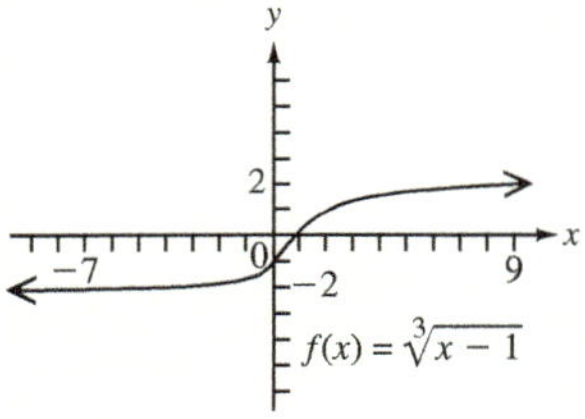

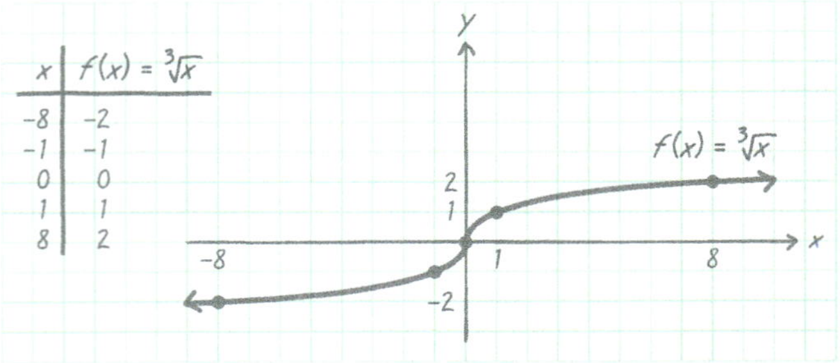

Figure 2

Example 3 **Graphing Functions Defined with Radicals**

Graph each function by creating a table of values. Give the domain and the range.

(a) $f(x) = \sqrt{x - 3}$

A table of values is shown. The x-values were chosen so that the function values are all integers. For the radicand to be nonnegative, we must have $x - 3 \geq 0$, or $x \geq 3$. Therefore, the domain is $[3, \infty)$. Again, function values are positive or 0, so the range is $[0, \infty)$. See the graph in Figure 3.

x	$f(x) = \sqrt{x - 3}$
3	$\sqrt{3 - 3} = 0$
4	$\sqrt{4 - 3} = 1$
7	$\sqrt{7 - 3} = 2$

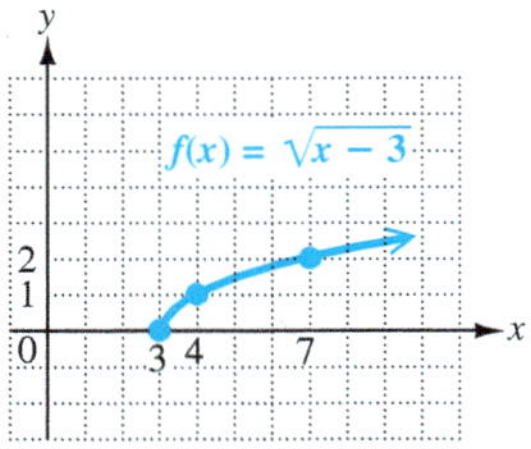

Figure 3

(b) $f(x) = \sqrt[3]{x} + 2$

See the table and Figure 4. Both the domain and the range are $(-\infty, \infty)$.

x	$f(x) = \sqrt[3]{x} + 2$
−8	$\sqrt[3]{-8} + 2 = 0$
−1	$\sqrt[3]{-1} + 2 = 1$
0	$\sqrt[3]{0} + 2 = 2$
1	$\sqrt[3]{1} + 2 = 3$
8	$\sqrt[3]{8} + 2 = 4$

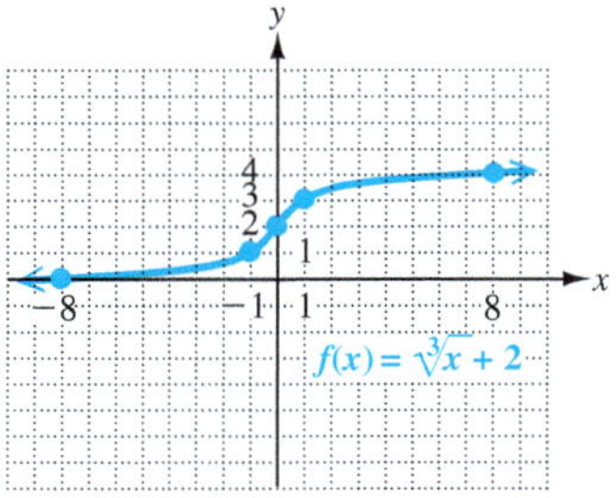

Figure 4

◀ Work Problem **3** at the Side.

OBJECTIVE 4 **Find *n*th roots of *n*th powers.** What does $\sqrt{a^2}$ equal? Your first answer might be a, but this is not necessarily true. For example, consider the following:

If $a = 6$, then $\sqrt{a^2} = \sqrt{6^2} = \sqrt{36} = 6$.

If $a = -6$, then $\sqrt{a^2} = \sqrt{(-6)^2} = \sqrt{36} = 6$. ← Instead of −6, we get 6, the *absolute value* of −6.

Since the symbol $\sqrt{a^2}$ represents the *nonnegative* square root, we write $\sqrt{a^2}$ with absolute value bars, as $|a|$, because a may be a negative number.

$\sqrt{a^2}$

For any real number a, $\sqrt{a^2} = |a|$.

In words, the principal square root of a^2 is the absolute value of a.

Example 4 Simplifying Square Roots Using Absolute Value

Find each square root that is a real number.

(a) $\sqrt{7^2} = |7| = 7$ (b) $\sqrt{(-7)^2} = |-7| = 7$

(c) $\sqrt{k^2} = |k|$ (d) $\sqrt{(-k)^2} = |-k| = |k|$

Work Problem 4 at the Side. ▶

We can generalize this idea to any nth root.

$\sqrt[n]{a^n}$

If n is an *even* positive integer, then $\sqrt[n]{a^n} = |a|$.

If n is an *odd* positive integer, then $\sqrt[n]{a^n} = a$.

In words, use absolute value when n is even; absolute value is not necessary when n is odd.

Example 5 Simplifying Higher Roots Using Absolute Value

Simplify each root.

(a) $\sqrt[6]{(-3)^6} = |-3| = 3$ n is even; use absolute value.

(b) $\sqrt[5]{(-4)^5} = -4$ n is odd.

(c) $-\sqrt[4]{(-9)^4} = -|-9| = -9$ n is even; use absolute value.

(d) $-\sqrt{m^4} = -|m^2| = -m^2$ For all m, $|m^2| = m^2$.

No absolute value bars are needed here because m^2 is nonnegative for any real number value of m.

(e) $\sqrt[3]{a^{12}} = a^4$, because $a^{12} = (a^4)^3$.

(f) $\sqrt[4]{x^{12}} = |x^3|$

We use absolute value bars to guarantee that the result is not negative (because x^3 can be either positive or negative, depending on x). If desired, $|x^3|$ can be written as $x^2 \cdot |x|$.

Work Problem 5 at the Side. ▶

OBJECTIVE 5 Use a calculator to find roots. Radical expressions often represent irrational numbers. To find approximations of such radicals, we usually use a calculator. For example,

$$\sqrt{15} \approx 3.872983346, \quad \sqrt[3]{10} \approx 2.15443469, \quad \text{and} \quad \sqrt[4]{2} \approx 1.189207115,$$

where the symbol $\approx$ means "is approximately equal to." In this book, we often give approximations rounded to three decimal places. Thus,

$$\sqrt{15} \approx 3.873, \quad \sqrt[3]{10} \approx 2.154, \quad \text{and} \quad \sqrt[4]{2} \approx 1.189.$$

4 Find each square root that is a real number.

(a) $\sqrt{15^2}$

(b) $\sqrt{(-12)^2}$

(c) $\sqrt{r^2}$

(d) $\sqrt{(-r)^2}$

5 Simplify.

(a) $\sqrt[4]{(-5)^4}$

(b) $\sqrt[5]{(-7)^5}$

(c) $-\sqrt[6]{(-3)^6}$

(d) $-\sqrt[4]{m^8}$

(e) $\sqrt[3]{x^{24}}$

(f) $\sqrt[6]{y^{18}}$

Answers

4. (a) 15 (b) 12 (c) $|r|$ (d) $|r|$

5. (a) 5 (b) -7 (c) -3 (d) $-m^2$ (e) x^8 (f) $|y^3|$

6 Use a calculator to approximate each radical to three decimal places.

(a) $\sqrt{17}$

(b) $-\sqrt{362}$

(c) $\sqrt[3]{9482}$

(d) $\sqrt[4]{6825}$

Calculator Tip The methods for finding approximations differ among makes and models, and you should always consult your owner's manual for keystroke instructions. Be aware that graphing calculators often differ from scientific calculators in the order in which keystrokes are made.

Figure 5 shows how the preceding approximations are displayed on a TI-83/84 Plus graphing calculator. In Figure 5(a), eight or nine decimal places are shown, while in Figure 5(b), the number of decimal places is fixed at three.

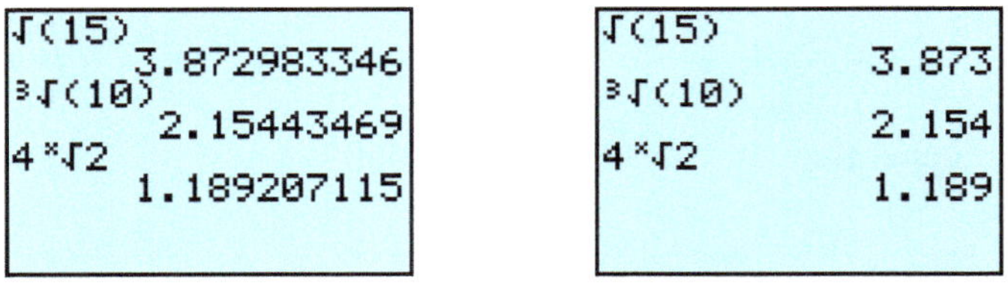

(a) (b)

Figure 5

There is a simple way to check that a calculator approximation is "in the ballpark." Because 16 is a little larger than 15, $\sqrt{16} = 4$ should be a little larger than $\sqrt{15}$. Thus, 3.873 is a reasonable approximation for $\sqrt{15}$.

Example 6 Finding Approximations for Roots

Use a calculator to verify that each approximation is correct.

(a) $\sqrt{39} \approx 6.245$ **(b)** $-\sqrt{72} \approx -8.485$

(c) $\sqrt[3]{93} \approx 4.531$ **(d)** $\sqrt[4]{39} \approx 2.499$

*Work Problem **6** at the Side.*

7 Use the formula in Example 7 to approximate f to the nearest thousand if

$L = 6 \times 10^{-5}$
and $C = 4 \times 10^{-9}$.

Example 7 Using Roots to Calculate Resonant Frequency

In electronics, the resonant frequency f of a circuit may be found by the formula

$$f = \frac{1}{2\pi\sqrt{LC}},$$

where f is in cycles per second, L is in henrys, and C is in farads.* Find the resonant frequency f if $L = 5 \times 10^{-4}$ henry and $C = 3 \times 10^{-10}$ farad. Give your answer to the nearest thousand.

Find the value of f when $L = 5 \times 10^{-4}$ and $C = 3 \times 10^{-10}$.

$$f = \frac{1}{2\pi\sqrt{LC}} \quad \text{Given formula}$$

$$f = \frac{1}{2\pi\sqrt{(5 \times 10^{-4})(3 \times 10^{-10})}} \quad \text{Substitute for } L \text{ and } C.$$

$$f \approx 411{,}000 \quad \text{Use a calculator.}$$

The resonant frequency f is approximately 411,000 cycles per sec.

*Work Problem **7** at the Side.*

Answers

6. **(a)** 4.123 **(b)** −19.026 **(c)** 21.166 **(d)** 9.089
7. 325,000 cycles per sec

*Henrys and farads are units of measure in electronics.

17.2 Exercises

FOR EXTRA HELP MyMathLab | Math XL PRACTICE | WATCH | DOWNLOAD | READ | REVIEW

Match each expression from Column I with the equivalent choice from Column II. Answers may be used once, more than once, or not at all. See Examples 1 and 2.

I

1. $-\sqrt{16}$ **2.** $\sqrt{-16}$

3. $\sqrt[3]{-27}$ **4.** $\sqrt[5]{-32}$

5. $\sqrt[4]{16}$ **6.** $-\sqrt[3]{64}$

II

A. 3 **B.** -2

C. 2 **D.** -3

E. -4 **F.** Not a real number

Choose the closest approximation of each square root.

7. $\sqrt{123.5}$

A. 9 **B.** 10 **C.** 11 **D.** 12

8. $\sqrt{67.8}$

A. 7 **B.** 8 **C.** 9 **D.** 10

Refer to the figure to answer the questions in Exercises 9–10.

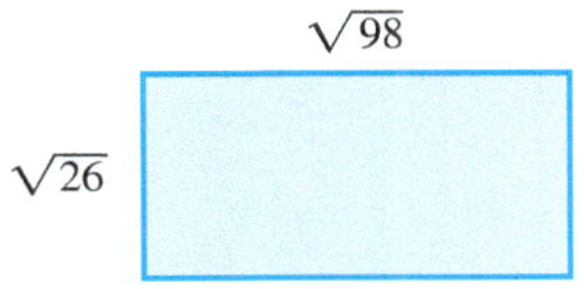

9. Which one of the following is the best estimate of its area?

A. 2500 **B.** 250 **C.** 50 **D.** 100

10. Which one of the following is the best estimate of its perimeter?

A. 15 **B.** 250 **C.** 100 **D.** 30

11. Consider the expression $-\sqrt{-a}$. Decide whether it is positive, negative, 0, or not a real number if

(a) $a > 0$, **(b)** $a < 0$, **(c)** $a = 0$.

12. If n is odd, under what conditions is $\sqrt[n]{a}$

(a) positive, **(b)** negative, **(c)** 0?

Find each root that is a real number. Use a calculator as necessary. See Examples 1 and 2.

13. $-\sqrt{81}$ **14.** $-\sqrt{121}$ **15.** $\sqrt[3]{216}$ **16.** $\sqrt[3]{343}$

17. $\sqrt[3]{-64}$ **18.** $\sqrt[3]{-125}$ **19.** $-\sqrt[3]{512}$ **20.** $-\sqrt[3]{1000}$

21. $\sqrt[4]{1296}$ **22.** $\sqrt[4]{625}$ **23.** $-\sqrt[4]{16}$ **24.** $-\sqrt[4]{256}$

25. $\sqrt[4]{-625}$ **26.** $\sqrt[4]{-256}$ **27.** $\sqrt[6]{729}$ **28.** $\sqrt[6]{64}$

29. $\sqrt[6]{-64}$ **30.** $\sqrt[6]{-1}$ **31.** $\sqrt{\dfrac{64}{81}}$ **32.** $\sqrt{\dfrac{100}{9}}$

33. $\sqrt{0.49}$ **34.** $\sqrt{0.81}$ **35.** $\sqrt[3]{\dfrac{64}{27}}$ **36.** $\sqrt[4]{\dfrac{81}{16}}$

37. $-\sqrt[6]{\dfrac{1}{64}}$ **38.** $-\sqrt[5]{\dfrac{1}{32}}$ **39.** $\sqrt[3]{0.001}$ **40.** $\sqrt[3]{0.125}$

Graph each function and give its domain and range. See Example 3.

41. $f(x) = \sqrt{x+3}$

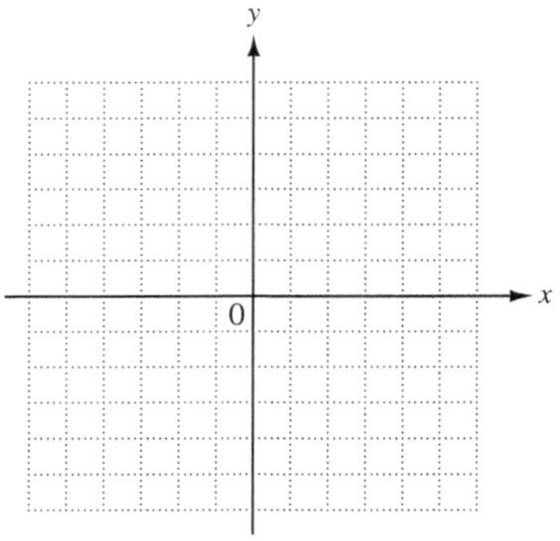

42. $f(x) = \sqrt{x-5}$

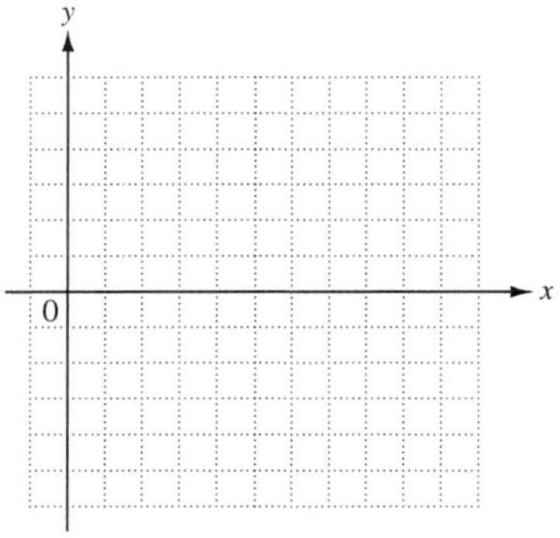

43. $f(x) = \sqrt{x} - 2$

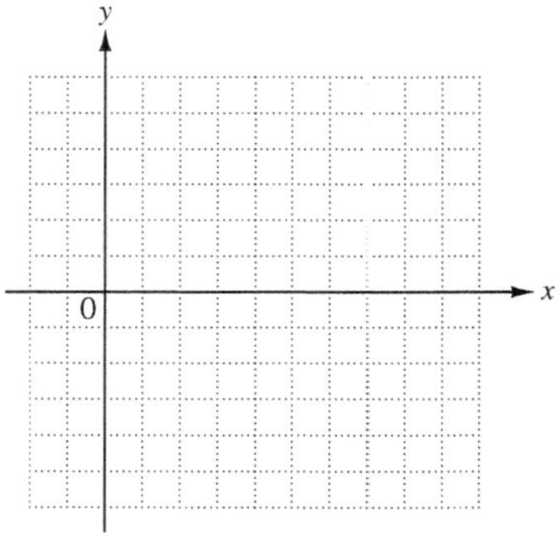

44. $f(x) = \sqrt{x} + 4$

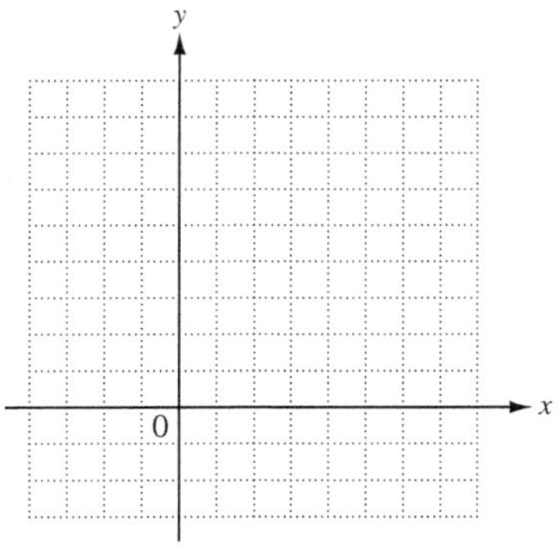

45. $f(x) = \sqrt[3]{x} - 3$

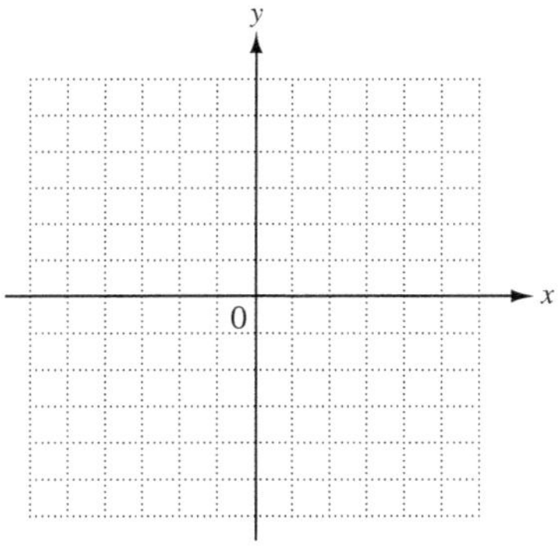

46. $f(x) = \sqrt[3]{x} + 1$

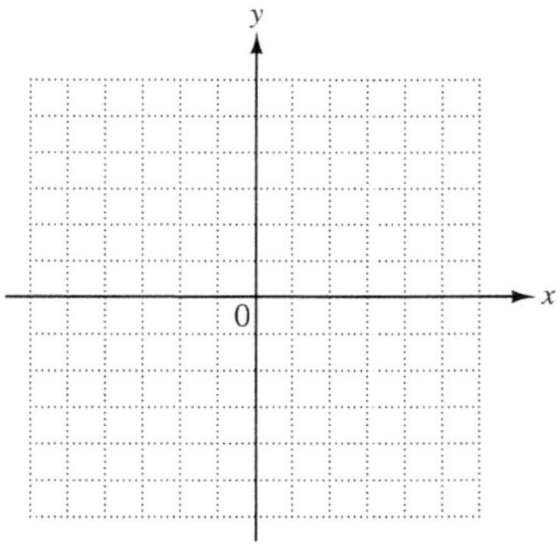

Simplify each root. See Examples 4 and 5.

47. $\sqrt{12^2}$
48. $\sqrt{19^2}$
49. $\sqrt{(-10)^2}$
50. $\sqrt{(-13)^2}$
51. $\sqrt[6]{(-2)^6}$
52. $\sqrt[6]{(-4)^6}$
53. $\sqrt[5]{(-9)^5}$
54. $\sqrt[5]{(-8)^5}$
55. $-\sqrt[6]{(-5)^6}$
56. $-\sqrt[6]{(-7)^6}$
57. $\sqrt{x^2}$
58. $-\sqrt{x^2}$
59. $\sqrt{(-z)^2}$
60. $\sqrt{(-q)^2}$
61. $\sqrt[3]{x^3}$
62. $-\sqrt[3]{x^3}$
63. $\sqrt[3]{x^{15}}$
64. $\sqrt[3]{m^9}$
65. $\sqrt[6]{x^{30}}$
66. $\sqrt[4]{k^{20}}$

Use a calculator to find a decimal approximation for each radical. Round answers to three decimal places. See Example 6.

67. $\sqrt{9483}$
68. $\sqrt{6825}$
69. $\sqrt{284.361}$
70. $\sqrt{846.104}$
71. $-\sqrt{82}$
72. $-\sqrt{91}$
73. $\sqrt[3]{423}$
74. $\sqrt[3]{555}$
75. $\sqrt[4]{100}$
76. $\sqrt[4]{250}$
77. $\sqrt[5]{23.8}$
78. $\sqrt[5]{98.4}$

Solve each problem. See Example 7.

79. Use the formula in Example 7 to calculate the resonant frequency of a circuit to the nearest thousand if $L = 7.237 \times 10^{-5}$ henry and $C = 2.5 \times 10^{-10}$ farad.

80. The threshold weight T for a person is the weight above which the risk of death increases greatly. The threshold weight in pounds for men aged 40–49 is related to height in inches by the formula

$$h = 12.3\sqrt[3]{T}.$$

What height corresponds to a threshold weight of 216 lb for a 43-yr-old man? Round your answer to the nearest inch, and then to the nearest tenth of a foot.

81. According to an article in *The World Scanner Report,* the distance D, in miles, to the horizon from an observer's point of view over water or "flat" earth is given by

$$D = \sqrt{2H},$$

where H is the height of the point of view, in feet. If a person whose eyes are 6 ft above ground level is standing at the top of a hill 44 ft above "flat" earth, approximately how far to the horizon will she be able to see?

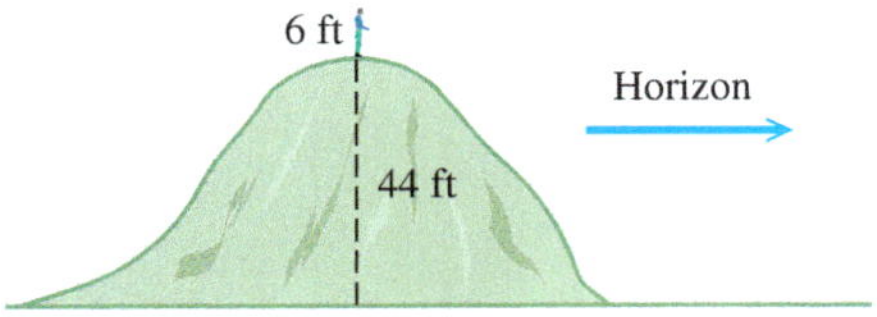

82. The time for one complete swing of a simple pendulum is

$$t = 2\pi\sqrt{\frac{L}{g}},$$

where t is time in seconds, L is the length of the pendulum in feet, and g, the force due to gravity, is about 32 ft per sec^2. Find the time of a complete swing of a 2-ft pendulum to the nearest tenth of a second.

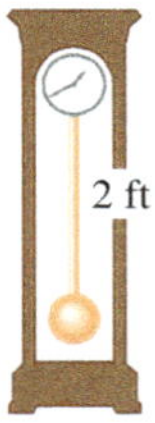

83. **Heron's formula** gives a method of finding the area of a triangle if the lengths of its sides are known. Suppose that a, b, and c are the lengths of the sides. Let s denote one-half of the perimeter of the triangle (called the **semiperimeter**); that is,

$$s = \frac{1}{2}(a + b + c).$$

Then the area of the triangle is

$$A = \sqrt{s(s - a)(s - b)(s - c)}.$$

Find the area of the Bermuda Triangle, if the "sides" of this triangle measure approximately 850 mi, 925 mi, and 1300 mi. Give your answer to the nearest thousand square miles.

84. The Vietnam Veterans' Memorial in Washington, D.C., is in the shape of an unenclosed isosceles triangle with equal sides of length 246.75 ft. If the triangle were enclosed, the third side would have length 438.14 ft. Use Heron's formula from the previous exercise to find the area of this enclosure to the nearest hundred square feet. (*Source:* Information pamphlet obtained at the Vietnam Veterans' Memorial.)

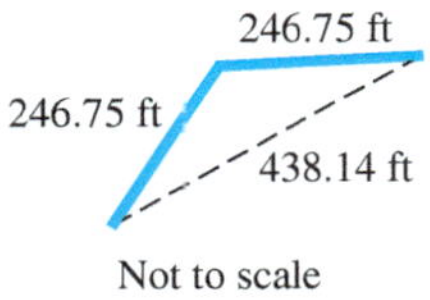

The coefficient of self-induction L (in henrys), the energy P stored in an electronic circuit (in joules), and the current I (in amps) are related by the formula

$$I = \sqrt{\frac{2P}{L}}.$$

Round your answers in Exercises 85 and 86 to the nearest thousandth.

85. Find I if $P = 120$ and $L = 80$.

86. Find I if $P = 100$ and $L = 40$.

17.3 Rational Exponents

OBJECTIVE 1 Use exponential notation for *n*th roots. We now look at exponents that are rational numbers of the form $\frac{1}{n}$, or $1/n$, where n is a natural number.

Consider the product $(3^{1/2})^2 = 3^{1/2} \cdot 3^{1/2}$. Using the rules of exponents from **Section 6.1**, extended to rational exponents, we can simplify this product as follows:

$$(3^{1/2})^2 = 3^{1/2} \cdot 3^{1/2}$$
$$= 3^{1/2+1/2} \quad \text{Product rule: } a^m \cdot a^n = a^{m+n}$$
$$= 3^1 \quad \text{Add exponents.}$$
$$= 3.$$

Also, by definition,

$$\left(\sqrt{3}\right)^2 = \sqrt{3} \cdot \sqrt{3} = 3.$$

Since both $(3^{1/2})^2$ and $\left(\sqrt{3}\right)^2$ are equal to 3, it seems reasonable to define

$$3^{1/2} = \sqrt{3}.$$

This suggests the following generalization.

$a^{1/n}$

If $\sqrt[n]{a}$ is a real number, then $\quad \boldsymbol{a^{1/n} = \sqrt[n]{a}}.$

Example 1 Evaluating Exponentials of the Form $a^{1/n}$

Evaluate each exponential.

(The denominator is the index.) (The denominator is the index. $\sqrt{\ }$ means $\sqrt[2]{\ }$.)

(a) $64^{1/3} = \sqrt[3]{64} = 4$ **(b)** $100^{1/2} = \sqrt{100} = 10$

(c) $-256^{1/4} = -\sqrt[4]{256} = -4$

(d) $(-256)^{1/4} = \sqrt[4]{-256}$ is not a real number, because the radicand, -256, is negative and the index is even.

(e) $(-32)^{1/5} = \sqrt[5]{-32} = -2$ **(f)** $\left(\frac{1}{8}\right)^{1/3} = \sqrt[3]{\frac{1}{8}} = \frac{1}{2}$

CAUTION

Notice the difference between parts (c) and (d) in Example 1. The radical in part (c) is the ***negative fourth root of a positive number,*** while the radical in part (d) is the ***principal fourth root of a negative number, which is not a real number.***

Work Problem 1 at the Side. ▶

OBJECTIVES

1. Use exponential notation for *n*th roots.
2. Define and use expressions of the form $a^{m/n}$.
3. Convert between radicals and rational exponents.
4. Use the rules for exponents with rational exponents.

1 Evaluate each exponential.

(a) $8^{1/3}$

(b) $9^{1/2}$

(c) $-81^{1/4}$

(d) $(-81)^{1/4}$

(e) $(-64)^{1/3}$

(f) $\left(\frac{1}{32}\right)^{1/5}$

ANSWERS

1. (a) 2 **(b)** 3 **(c)** -3 **(d)** not a real number **(e)** -4 **(f)** $\frac{1}{2}$

OBJECTIVE 2 Define and use expressions of the form $a^{m/n}$.

We know that $8^{1/3} = \sqrt[3]{8}$. How should we define a number like $8^{2/3}$? For past rules of exponents to be valid,

$$8^{2/3} = 8^{(1/3)2} = (8^{1/3})^2.$$

Since $8^{1/3} = \sqrt[3]{8}$,

$$8^{2/3} = \left(\sqrt[3]{8}\right)^2 = 2^2 = 4.$$

Generalizing from this example, we define $a^{m/n}$ as follows.

> $a^{m/n}$
>
> If m and n are positive integers with m/n in lowest terms, then
>
> $$a^{m/n} = (a^{1/n})^m,$$
>
> provided that $a^{1/n}$ is a real number. If $a^{1/n}$ is not a real number, then $a^{m/n}$ is not a real number.

Example 2 Evaluating Exponentials of the Form $a^{m/n}$

Evaluate each exponential.

(Think: $36^{1/2} = \sqrt{36} = 6$)

(a) $36^{3/2} = (36^{1/2})^3 = 6^3 = 216$

(Think: $125^{1/3} = \sqrt[3]{125} = 5$)

(b) $125^{2/3} = (125^{1/3})^2 = 5^2 = 25$

(Be careful. The base is 4.)

(c) $-4^{5/2} = -(4^{5/2}) = -(4^{1/2})^5 = -(2)^5 = -32$

Because the base here is 4, the negative sign is *not* affected by the exponent.

(d) $(-27)^{2/3} = [(-27)^{1/3}]^2 = (-3)^2 = 9$

Notice how the $-$ sign is used in parts (c) and (d). In part (c), we first evaluate the exponential and then find its negative. In part (d), the $-$ sign is part of the base, -27.

(e) $(-100)^{3/2} = [(-100)^{1/2}]^3$, which is not a real number, since $(-100)^{1/2}$, or $\sqrt{-100}$, is not a real number.

◀ Work Problem 2 at the Side.

Recall from **Section 6.1** that for any natural number n,

$$a^{-n} = \frac{1}{a^n} \quad (a \neq 0).$$

When a rational exponent is negative, we apply this interpretation of negative exponents.

> $a^{-m/n}$
>
> If $a^{m/n}$ is a real number, then
>
> $$a^{-m/n} = \frac{1}{a^{m/n}} \quad (a \neq 0).$$

2 Evaluate each exponential.

(a) $25^{3/2}$

(b) $27^{2/3}$

(c) $-16^{3/2}$

(d) $(-64)^{2/3}$

(e) $(-36)^{3/2}$

ANSWERS

2. **(a)** 125 **(b)** 9 **(c)** -64 **(d)** 16 **(e)** not a real number

Example 3 **Evaluating Exponentials with Negative Rational Exponents**

Evaluate each exponential.

(a) $16^{-3/4} = \frac{1}{16^{3/4}} = \frac{1}{(16^{1/4})^3} = \frac{1}{(\sqrt[4]{16})^3} = \frac{1}{2^3} = \frac{1}{8}$

The denominator of 3/4 is the index and the numerator is the exponent.

(b) $25^{-3/2} = \frac{1}{25^{3/2}} = \frac{1}{(25^{1/2})^3} = \frac{1}{(\sqrt{25})^3} = \frac{1}{5^3} = \frac{1}{125}$

(c) $\left(\frac{8}{27}\right)^{-2/3} = \frac{1}{\left(\frac{8}{27}\right)^{2/3}} = \frac{1}{\left(\sqrt[3]{\frac{8}{27}}\right)^2} = \frac{1}{\left(\frac{2}{3}\right)^2} = \frac{1}{\frac{4}{9}} = \frac{9}{4}$

$\frac{1}{\frac{4}{9}} = 1 \div \frac{4}{9} = 1 \cdot \frac{9}{4}$

We could also use the rule $\left(\frac{b}{a}\right)^{-m} = \left(\frac{a}{b}\right)^m$ here, as follows:

$$\left(\frac{8}{27}\right)^{-2/3} = \left(\frac{27}{8}\right)^{2/3} = \left(\sqrt[3]{\frac{27}{8}}\right)^2 = \left(\frac{3}{2}\right)^2 = \frac{9}{4}.$$

Take the reciprocal only of the base, *not* the exponent.

Work Problem **3** at the Side. ▶

CAUTION

Be careful to distinguish between exponential expressions such as

$16^{-1/4}$, which equals $\frac{1}{2}$; $-16^{1/4}$, which equals -2; and

$-16^{-1/4}$, which equals $-\frac{1}{2}$.

A negative exponent does not necessarily lead to a negative result. Negative exponents lead to reciprocals, which may be positive.

We obtain an alternative definition of $a^{m/n}$ by applying the power rule a little differently than earlier. If all indicated roots are real numbers, then

$$a^{m/n} = a^{m(1/n)} = (a^m)^{1/n}, \quad \text{so} \quad a^{m/n} = (a^m)^{1/n}.$$

$a^{m/n}$

If all indicated roots are real numbers, then

$$a^{m/n} = (a^{1/n})^m = (a^m)^{1/n}.$$

We can now evaluate an expression such as $27^{2/3}$ in two ways:

$27^{2/3} = (27^{1/3})^2 = 3^2 = 9$

or $27^{2/3} = (27^2)^{1/3} = 729^{1/3} = 9$.

The result is the same.

In most cases, it is easier to use $(a^{1/n})^m$.

3 Evaluate each exponential.

(a) $36^{-3/2}$

(b) $32^{-4/5}$

(c) $\left(\frac{4}{9}\right)^{-5/2}$

ANSWERS

3. (a) $\frac{1}{216}$ (b) $\frac{1}{16}$ (c) $\frac{243}{32}$

4 Write each exponential as a radical. Assume that all variables represent positive real numbers. Use the definition that takes the root first.

(a) $19^{1/2}$

(b) $5^{2/3}$

(c) $4k^{3/5}$

(d) $5x^{3/5} - (2x)^{3/5}$

(e) $x^{-5/7}$

(f) $(m^3 + n^3)^{1/3}$

5 Write each radical as an exponential and simplify. Assume that all variables represent positive real numbers.

(a) $\sqrt{37}$

(b) $\sqrt[4]{9^8}$

(c) $\sqrt[4]{t^4}$

Answers

4. (a) $\sqrt{19}$ **(b)** $(\sqrt[3]{5})^2$ **(c)** $4(\sqrt[5]{k})^3$
(d) $5(\sqrt[5]{x})^3 - (\sqrt[5]{2x})^3$
(e) $\frac{1}{(\sqrt[7]{x})^5}$ **(f)** $\sqrt[3]{m^3 + n^3}$

5. (a) $37^{1/2}$ **(b)** 9^2, or 81 **(c)** t

This rule can also be expressed with radicals as follows.

Radical Form of $a^{m/n}$

If all indicated roots are real numbers, then

$$a^{m/n} = \sqrt[n]{a^m} = \left(\sqrt[n]{a}\right)^m.$$

In words, raise a to the mth power and then take the nth root, or take the nth root of a and then raise to the mth power.

For example,

$$8^{2/3} = \sqrt[3]{8^2} = \sqrt[3]{64} = 4, \quad \text{and} \quad 8^{2/3} = \left(\sqrt[3]{8}\right)^2 = 2^2 = 4,$$

so

$$8^{2/3} = \sqrt[3]{8^2} = \left(\sqrt[3]{8}\right)^2.$$

OBJECTIVE 3 Convert between radicals and rational exponents. Using the definition of rational exponents, we can simplify many problems involving radicals by converting the radicals to numbers with rational exponents. After simplifying, we convert the answer back to radical form.

Example 4 Converting between Rational Exponents and Radicals

In (a)–(f), write each exponential as a radical. Assume that all variables represent positive real numbers. Use the definition that takes the root first.

(a) $13^{1/2} = \sqrt{13}$ **(b)** $6^{3/4} = \left(\sqrt[4]{6}\right)^3$ **(c)** $9m^{5/8} = 9\left(\sqrt[8]{m}\right)^5$

(d) $6x^{2/3} - (4x)^{3/5} = 6\left(\sqrt[3]{x}\right)^2 - \left(\sqrt[5]{4x}\right)^3$

(e) $r^{-2/3} = \frac{1}{r^{2/3}} = \frac{1}{\left(\sqrt[3]{r}\right)^2}$

(f) $(a^2 + b^2)^{1/2} = \sqrt{a^2 + b^2}$ ← $\sqrt{a^2 + b^2} \neq a + b$

In (g)–(i), write each radical as an exponential. Simplify. Assume that all variables represent positive real numbers.

(g) $\sqrt{10} = 10^{1/2}$

(h) $\sqrt[4]{3^8} = 3^{8/4} = 3^2 = 9$

(i) $\sqrt[6]{z^6} = z^{6/6} = z^1 = z$, since z is positive.

◀ Work Problem **4** at the Side.

Note

In Example 4(i), it was not necessary to use absolute value bars, since the directions specifically stated that the variable represents a positive real number. Because the absolute value of the positive real number z is z itself, the answer is simply z.

◀ Work Problem **5** at the Side.

OBJECTIVE 4 Use the rules for exponents with rational exponents. The definition of rational exponents allows us to apply the rules for exponents first introduced in **Section 6.1.**

Rules for Rational Exponents

Let r and s be rational numbers. For all real numbers a and b for which the indicated expressions exist:

$$a^r \cdot a^s = a^{r+s} \qquad a^{-r} = \frac{1}{a^r} \qquad \frac{a^r}{a^s} = a^{r-s} \qquad \left(\frac{a}{b}\right)^{-r} = \frac{b^r}{a^r}$$

$$(a^r)^s = a^{rs} \qquad (ab)^r = a^r b^r \qquad \left(\frac{a}{b}\right)^r = \frac{a^r}{b^r} \qquad a^{-r} = \left(\frac{1}{a}\right)^r.$$

Example 5 Applying Rules for Rational Exponents

Write with only positive exponents. Assume that all variables represent positive real numbers.

(a) $2^{1/2} \cdot 2^{1/4}$

$= 2^{1/2+1/4}$ Product rule

$= 2^{3/4}$ Add exponents.

(b) $\dfrac{5^{2/3}}{5^{7/3}}$

$= 5^{2/3-7/3}$ Quotient rule

$= 5^{-5/3}$ Subtract exponents.

$= \dfrac{1}{5^{5/3}}$ $a^{-r} = \frac{1}{a^r}$

(c) $\dfrac{(x^{1/2}y^{2/3})^4}{y}$

$= \dfrac{(x^{1/2})^4(y^{2/3})^4}{y}$ Power rule

$= \dfrac{x^2y^{8/3}}{y^1}$ Power rule

$= x^2y^{8/3-1}$ Quotient rule

$= x^2y^{5/3}$ $\frac{8}{3} - 1 = \frac{8}{3} - \frac{3}{3} = \frac{5}{3}$

(d) $\left(\dfrac{x^4y^{-6}}{x^{-2}y^{1/3}}\right)^{-2/3}$

$= \dfrac{(x^4)^{-2/3}(y^{-6})^{-2/3}}{(x^{-2})^{-2/3}(y^{1/3})^{-2/3}}$ Power rule

$= \dfrac{x^{-8/3}y^4}{x^{4/3}y^{-2/9}}$ Power rule

$= x^{-8/3-4/3}y^{4-(-2/9)}$ Quotient rule

$= x^{-4}y^{38/9}$ Use parentheses to avoid errors. $4 - \left(-\frac{2}{9}\right) = \frac{36}{9} + \frac{2}{9} = \frac{38}{9}$

$= \dfrac{y^{38/9}}{x^4}$ Definition of negative exponent

Continued on Next Page

The same result is obtained if we simplify within the parentheses first.

$$\left(\frac{x^4y^{-6}}{x^{-2}y^{1/3}}\right)^{-2/3}$$

$= (x^{4-(-2)}y^{-6-1/3})^{-2/3}$ Quotient rule

$= (x^6y^{-19/3})^{-2/3}$ $-6 - \frac{1}{3} = -\frac{18}{3} - \frac{1}{3} = -\frac{19}{3}$

$= (x^6)^{-2/3}(y^{-19/3})^{-2/3}$ Power rule

$= x^{-4}y^{38/9}$ Power rule

$= \frac{y^{38/9}}{x^4}$ Definition of negative exponent

(e) $m^{3/4}(m^{5/4} - m^{1/4})$

$= m^{3/4}(m^{5/4}) - m^{3/4}(m^{1/4})$ Distributive property

$= m^{3/4+5/4} - m^{3/4+1/4}$ Product rule

$= m^{8/4} - m^{4/4}$

$= m^2 - m$

Do not make the common mistake of multiplying exponents in the first step.

◀ Work Problem 6 at the Side.

CAUTION

Use the rules of exponents in problems like those in Example 5. Do not convert the expressions to radical form.

Example 6 Applying Rules for Rational Exponents

Write all radicals as exponentials, and then apply the rules for rational exponents. Give answers in exponential form. Assume that all variables represent positive real numbers.

(a) $\sqrt[3]{x^2} \cdot \sqrt[4]{x}$

$= x^{2/3} \cdot x^{1/4}$ Convert to rational exponents.

$= x^{2/3+1/4}$ Product rule

$= x^{8/12+3/12}$ Write exponents with a common denominator.

$= x^{11/12}$

(b) $\frac{\sqrt{x^3}}{\sqrt[3]{x^2}}$

$= \frac{x^{3/2}}{x^{2/3}}$ Convert to rational exponents.

$= x^{3/2-2/3}$ Quotient rule

$= x^{5/6}$ $\frac{3}{2} - \frac{2}{3} = \frac{9}{6} - \frac{4}{6} = \frac{5}{6}$

(c) $\sqrt{\sqrt[4]{z}}$

$= \sqrt{z^{1/4}}$ Convert the inside radical to rational exponents.

$= (z^{1/4})^{1/2}$ Convert to rational exponents.

$= z^{1/8}$ Power rule

◀ Work Problem 7 at the Side.

6 Write with only positive exponents. Assume that all variables represent positive real numbers.

(a) $11^{3/4} \cdot 11^{5/4}$

(b) $\frac{7^{3/4}}{7^{7/4}}$

(c) $\frac{9^{2/3}(x^{1/3})^4}{9^{-1/3}}$

(d) $\left(\frac{a^3b^{-4}}{a^{-2}b^{1/5}}\right)^{-1/2}$

(e) $a^{2/3}(a^{7/3} + a^{1/3})$

7 Write all radicals as exponentials, and then apply the rules for rational exponents. Give answers in exponential form. Assume that all variables represent positive real numbers.

(a) $\sqrt[5]{m^3} \cdot \sqrt{m}$

(b) $\frac{\sqrt[3]{p^5}}{\sqrt{p^3}}$

(c) $\sqrt[4]{\sqrt[3]{x}}$

ANSWERS

6. (a) 11^2, or 121 (b) $\frac{1}{7}$ (c) $9x^{4/3}$ (d) $\frac{b^{21/10}}{a^{5/2}}$ (e) $a^3 + a$

7. (a) $m^{11/10}$ (b) $p^{1/6}$ (c) $x^{1/12}$

17.3 Exercises

FOR EXTRA HELP MyMathLab Math XL PRACTICE WATCH DOWNLOAD READ REVIEW

Match each expression from Column I with the equivalent choice from Column II.

I		II	
1. $2^{1/2}$	**2.** $(-27)^{1/3}$	**A.** -4	**B.** 8
3. $-16^{1/2}$	**4.** $(-16)^{1/2}$	**C.** $\sqrt{2}$	**D.** $-\sqrt{6}$
5. $(-32)^{1/5}$	**6.** $(-32)^{2/5}$	**E.** -3	**F.** $\sqrt{6}$
7. $4^{3/2}$	**8.** $6^{2/4}$	**G.** 4	**H.** -2
9. $-6^{2/4}$	**10.** $36^{0.5}$	**I.** 6	**J.** Not a real number

Evaluate each exponential. See Examples 1–3.

11. $169^{1/2}$ **12.** $121^{1/2}$ **13.** $729^{1/3}$ **14.** $512^{1/3}$ **15.** $16^{1/4}$

16. $625^{1/4}$ **17.** $\left(\frac{64}{81}\right)^{1/2}$ **18.** $\left(\frac{8}{27}\right)^{1/3}$ **19.** $(-27)^{1/3}$ **20.** $(-32)^{1/5}$

21. $(-144)^{1/2}$ **22.** $(-36)^{1/2}$ **23.** $100^{3/2}$ **24.** $64^{3/2}$

25. $81^{3/4}$ **26.** $216^{2/3}$ **27.** $-16^{5/2}$ **28.** $-32^{3/5}$

29. $(-8)^{4/3}$ **30.** $(-243)^{2/5}$ **31.** $32^{-3/5}$ **32.** $27^{-4/3}$

33. $64^{-3/2}$ **34.** $81^{-3/2}$ **35.** $\left(\frac{125}{27}\right)^{-2/3}$ **36.** $\left(\frac{64}{125}\right)^{-2/3}$

Write with radicals. Assume that all variables represent positive real numbers. Use the definition that takes the root first. See Example 4.

37. $12^{1/2}$

38. $3^{1/2}$

39. $8^{3/4}$

40. $7^{2/3}$

41. $(9q)^{5/8} - (2x)^{2/3}$

42. $(3p)^{3/4} + (4x)^{1/3}$

43. $(2m)^{-3/2}$

44. $(5y)^{-3/5}$

45. $(2y + x)^{2/3}$

46. $(r + 2z)^{3/2}$

47. $(3m^4 + 2k^2)^{-2/3}$

48. $(5x^2 + 3z^3)^{-5/6}$

49. Show that, in general, $\sqrt{a^2 + b^2} \neq a + b$ by replacing a with 3 and b with 4.

50. Suppose someone claims that $\sqrt[n]{a^n + b^n}$ must equal $a + b$, since when $a = 1$ and $b = 0$, a true statement results:

$$\sqrt[n]{a^n + b^n} = \sqrt[n]{1^n + 0^n} = \sqrt[n]{1^n} = 1 = 1 + 0 = a + b.$$

Explain why this is faulty reasoning.

Simplify by first converting to rational exponents. Assume that all variables represent positive real numbers. See Example 4.

51. $\sqrt{2^{12}}$

52. $\sqrt{5^{10}}$

53. $\sqrt[3]{4^9}$

54. $\sqrt[4]{6^8}$

55. $\sqrt{x^{20}}$

56. $\sqrt{r^{50}}$

57. $\sqrt[3]{x} \cdot \sqrt{x}$

58. $\sqrt[4]{y} \cdot \sqrt[5]{y^2}$

59. $\dfrac{\sqrt[3]{t^4}}{\sqrt[5]{t^4}}$

60. $\dfrac{\sqrt[4]{w^3}}{\sqrt[6]{w}}$

Simplify each expression. Write all answers with positive exponents. Assume that all variables represent positive real numbers. See Example 5.

61. $3^{1/2} \cdot 3^{3/2}$

62. $6^{4/3} \cdot 6^{2/3}$

63. $\dfrac{64^{5/3}}{64^{4/3}}$

64. $\dfrac{125^{7/3}}{125^{5/3}}$

65. $y^{7/3} \cdot y^{-4/3}$

66. $r^{-8/9} \cdot r^{17/9}$

67. $x^{2/3} \cdot x^{-1/4}$

68. $x^{2/5} \cdot x^{-1/3}$

69. $\dfrac{k^{1/3}}{k^{2/3} \cdot k^{-1}}$

70. $\dfrac{z^{3/4}}{z^{5/4} \cdot z^{-2}}$

71. $\dfrac{(x^{1/4}y^{2/5})^{20}}{x^2}$

72. $\dfrac{(r^{1/5}s^{2/3})^{15}}{r^2}$

73. $\dfrac{(x^{2/3})^2}{(x^2)^{7/3}}$

74. $\dfrac{(p^3)^{1/4}}{(p^{5/4})^2}$

75. $\dfrac{m^{3/4}n^{-1/4}}{(m^2n)^{1/2}}$

76. $\dfrac{(a^2b^5)^{-1/4}}{(a^{-3}b^2)^{1/6}}$

77. $\dfrac{p^{1/5}p^{7/10}p^{1/2}}{(p^3)^{-1/5}}$

78. $\dfrac{z^{1/3}z^{-2/3}z^{1/6}}{(z^{-1/6})^3}$

79. $\left(\dfrac{b^{-3/2}}{c^{-5/3}}\right)^2(b^{-1/4}c^{-1/3})^{-1}$

80. $\left(\dfrac{m^{-2/3}}{a^{-3/4}}\right)^4(m^{-3/8}a^{1/4})^{-2}$

81. $\left(\dfrac{p^{-1/4}q^{-3/2}}{3^{-1}p^{-2}q^{-2/3}}\right)^{-2}$

82. $\left(\dfrac{2^{-2}w^{-3/4}x^{-5/8}}{w^{3/4}x^{-1/2}}\right)^{-3}$

83. $p^{2/3}(p^{1/3} + 2p^{4/3})$

84. $z^{5/8}(3z^{5/8} + 5z^{11/8})$

85. $k^{1/4}(k^{3/2} - k^{1/2})$

86. $r^{3/5}(r^{1/2} + r^{3/4})$

87. $6a^{7/4}(a^{-7/4} + 3a^{-3/4})$

88. $4m^{5/3}(m^{-2/3} - 4m^{-5/3})$

89. $5m^{-2/3}(m^{2/3} + m^{-7/3})$

Write radicals as exponentials, and then apply the rules for rational exponents. Give answers in exponential form. Assume that all radicands represent positive real numbers. See Example 6.

90. $\sqrt[5]{x^3} \cdot \sqrt[4]{x}$

91. $\sqrt[6]{y^5} \cdot \sqrt[3]{y^2}$

92. $\dfrac{\sqrt{x^5}}{\sqrt{x^8}}$

93. $\dfrac{\sqrt[3]{k^5}}{\sqrt[3]{k^7}}$

94. $\sqrt{y} \cdot \sqrt[3]{yz}$

95. $\sqrt[3]{xz} \cdot \sqrt{z}$

96. $\sqrt[4]{\sqrt[3]{m}}$

97. $\sqrt[3]{\sqrt{k}}$

98. $\sqrt{\sqrt[3]{\sqrt[4]{x}}}$

99. $\sqrt[3]{\sqrt[5]{\sqrt{y}}}$

100. $\sqrt{y^{5/4}}$

101. $\sqrt[3]{x^{5/9}}$

Solve each problem.

102. Meteorologists can determine the duration of a storm by using the function defined by

$$T(D) = 0.07D^{3/2},$$

where D is the diameter of the storm in miles and T is the time in hours. Find the duration of a storm with a diameter of 16 mi. Round your answer to the nearest tenth of an hour.

103. The threshold weight T, in pounds, for a person is the weight above which the risk of death increases greatly. The threshold weight in pounds for men aged 40–49 is related to height in inches by the function defined by

$$h(T) = (1860.867T)^{1/3}.$$

What height corresponds to a threshold weight of 200 lb for a 46-yr-old man? Round your answer to the nearest inch, and then to the nearest tenth of a foot.

*The **windchill factor** is a measure of the cooling effect that the wind has on a person's skin. It calculates the equivalent cooling temperature if there were no wind. The National Weather Service uses the formula*

$$\text{Windchill temperature} = 35.74 + 0.6215T - 35.75V^{4/25} + 0.4275TV^{4/25},$$

where T is the temperature in °F and V is the wind speed in miles per hour, to calculate windchill. The chart gives the windchill factor for various wind speeds and temperatures at which frostbite is a risk, and how quickly it may occur.

Temperature (°F)

Wind speed (mph) / Calm	40	30	20	10	0	−10	−20	−30	−40
5	36	25	13	1	−11	−22	−34	−46	−57
10	34	21	9	−4	−16	−28	−41	−53	−66
15	32	19	6	−7	−19	−32	−45	−58	−71
20	30	17	4	−9	−22	−35	−48	−61	−74
25	29	16	3	−11	−24	−37	−51	−64	−78
30	28	15	1	−12	−26	−39	−53	−67	−80
35	28	14	0	−14	−27	−41	−55	−69	−82
40	27	13	−1	−15	−29	−43	−57	−71	−84

Frostbites times: 30 minutes 10 minutes 5 minutes

Source: National Oceanic and Atmospheric Administration, National Weather Service.

Use the formula to determine the windchill to the nearest tenth of a degree, given the following conditions. Compare your answers with the appropriate entries in the table.

104. 30°F, 15-mph wind

105. 10°F, 30-mph wind

17.4 Simplifying Radical Expressions

OBJECTIVES

1. Use the product rule for radicals.
2. Use the quotient rule for radicals.
3. Simplify radicals.
4. Simplify products and quotients of radicals with different indexes.
5. Use the Pythagorean formula.
6. Use the distance formula.

OBJECTIVE 1 Use the product rule for radicals. Is the product of two *n*th-root radicals equal to the *n*th root of the product of the radicands? For example, are $\sqrt{36 \cdot 4}$ and $\sqrt{36} \cdot \sqrt{4}$ equal?

$$\sqrt{36 \cdot 4} = \sqrt{144} = 12$$
$$\sqrt{36} \cdot \sqrt{4} = 6 \cdot 2 = 12$$

The result is the same.

This is an example of the **product rule for radicals.**

Product Rule for Radicals

If $\sqrt[n]{a}$ and $\sqrt[n]{b}$ are real numbers and n is a natural number, then

$$\sqrt[n]{a} \cdot \sqrt[n]{b} = \sqrt[n]{ab}.$$

In words, the product of two *n*th roots is the *n*th root of the product.

We justify the product rule using the rules for rational exponents. Since $\sqrt[n]{a} = a^{1/n}$ and $\sqrt[n]{b} = b^{1/n}$,

$$\sqrt[n]{a} \cdot \sqrt[n]{b} = a^{1/n} \cdot b^{1/n} = (ab)^{1/n} = \sqrt[n]{ab}.$$

CAUTION

Use the product rule only when the radicals have the same index.

Example 1 Using the Product Rule

Multiply. Assume that all variables represent positive real numbers.

(a) $\sqrt{5} \cdot \sqrt{7} = \sqrt{5 \cdot 7} = \sqrt{35}$

(b) $\sqrt{2} \cdot \sqrt{19} = \sqrt{2 \cdot 19} = \sqrt{38}$

(c) $\sqrt{11} \cdot \sqrt{p} = \sqrt{11p}$

(d) $\sqrt{7} \cdot \sqrt{11xyz} = \sqrt{77xyz}$

Work Problem 1 at the Side. ▶

Example 2 Using the Product Rule

Multiply. Assume that all variables represent positive real numbers.

(a) $\sqrt[3]{3} \cdot \sqrt[3]{12} = \sqrt[3]{3 \cdot 12} = \sqrt[3]{36}$

(b) $\sqrt[4]{8y} \cdot \sqrt[4]{3r^2} = \sqrt[4]{24yr^2}$

(c) $\sqrt[6]{10m^4} \cdot \sqrt[6]{5m} = \sqrt[6]{50m^5}$

(d) $\sqrt[4]{2} \cdot \sqrt[5]{2}$ cannot be simplified using the product rule for radicals, because the indexes (4 and 5) are different.

Work Problem 2 at the Side. ▶

1 Multiply. Assume that all variables represent positive real numbers.

(a) $\sqrt{5} \cdot \sqrt{13}$

(b) $\sqrt{10y} \cdot \sqrt{3k}$

2 Multiply. Assume that all variables represent positive real numbers.

(a) $\sqrt[3]{2} \cdot \sqrt[3]{7}$

(b) $\sqrt[6]{8r^2} \cdot \sqrt[6]{2r^3}$

(c) $\sqrt[5]{9y^2x} \cdot \sqrt[5]{8xy^2}$

(d) $\sqrt{7} \cdot \sqrt[3]{5}$

ANSWERS

1. **(a)** $\sqrt{65}$ **(b)** $\sqrt{30yk}$

2. **(a)** $\sqrt[3]{14}$ **(b)** $\sqrt[6]{16r^5}$ **(c)** $\sqrt[5]{72y^4x^2}$
(d) cannot be simplified using the product rule

OBJECTIVE 2 Use the quotient rule for radicals. The **quotient rule for radicals** is similar to the product rule.

Quotient Rule for Radicals

If $\sqrt[n]{a}$ and $\sqrt[n]{b}$ are real numbers, $b \neq 0$, and n is a natural number, then

$$\sqrt[n]{\frac{a}{b}} = \frac{\sqrt[n]{a}}{\sqrt[n]{b}}.$$

In words, the nth root of a quotient is the quotient of the nth roots.

Example 3 Using the Quotient Rule

Simplify. Assume that all variables represent positive real numbers.

(a) $\sqrt{\frac{16}{25}} = \frac{\sqrt{16}}{\sqrt{25}} = \frac{4}{5}$

(b) $\sqrt{\frac{7}{36}} = \frac{\sqrt{7}}{\sqrt{36}} = \frac{\sqrt{7}}{6}$

(c) $\sqrt[3]{-\frac{8}{125}} = \sqrt[3]{\frac{-8}{125}} = \frac{\sqrt[3]{-8}}{\sqrt[3]{125}} = \frac{-2}{5} = -\frac{2}{5}$ $\quad \frac{-a}{b} = -\frac{a}{b}$

(d) $\sqrt[3]{\frac{7}{216}} = \frac{\sqrt[3]{7}}{\sqrt[3]{216}} = \frac{\sqrt[3]{7}}{6}$

(e) $\sqrt[5]{\frac{x}{32}} = \frac{\sqrt[5]{x}}{\sqrt[5]{32}} = \frac{\sqrt[5]{x}}{2}$

(f) $-\sqrt[3]{\frac{m^6}{125}} = -\frac{\sqrt[3]{m^6}}{\sqrt[3]{125}} = -\frac{m^2}{5}$

Think: $\sqrt[3]{m^6} = m^{6/3} = m^2$

◀ Work Problem 3 at the Side.

OBJECTIVE 3 Simplify radicals. We use the product and quotient rules to simplify radicals. A radical is **simplified** if the following four conditions are met.

Conditions for a Simplified Radical

1. The radicand has no factor raised to a power greater than or equal to the index.
2. The radicand has no fractions.
3. No denominator has a radical.
4. Exponents in the radicand and the index of the radical have greatest common factor 1.

3 Simplify. Assume that all variables represent positive real numbers.

(a) $\sqrt{\frac{100}{81}}$

(b) $\sqrt{\frac{11}{25}}$

(c) $\sqrt[3]{-\frac{125}{216}}$

(d) $\sqrt{\frac{y^8}{16}}$

(e) $-\sqrt[3]{\frac{x^2}{r^{12}}}$

ANSWERS

3. (a) $\frac{10}{9}$ (b) $\frac{\sqrt{11}}{5}$ (c) $-\frac{5}{6}$
(d) $\frac{y^4}{4}$ (e) $-\frac{\sqrt[3]{x^2}}{r^4}$

Example 4 **Simplifying Roots of Numbers**

Simplify.

(a) $\sqrt{24}$

Check to see whether 24 is divisible by a perfect square (the square of a natural number) such as 4, 9, 16, ... The largest perfect square that divides into 24 is 4.

$$\sqrt{24}$$
$$= \sqrt{4 \cdot 6} \quad \text{Factor; 4 is a perfect square.}$$
$$= \sqrt{4} \cdot \sqrt{6} \quad \text{Product rule}$$
$$= 2\sqrt{6} \quad \sqrt{4} = 2$$

(b) $\sqrt{108}$

As shown on the left, the number 108 is divisible by the perfect square 36. If this perfect square is not immediately clear, try factoring 108 into its prime factors, as shown on the right.

$$\sqrt{108}$$
$$= \sqrt{36 \cdot 3}$$
$$= \sqrt{36} \cdot \sqrt{3}$$
$$= 6\sqrt{3}$$

$$\sqrt{108}$$
$$= \sqrt{2^2 \cdot 3^3}$$
$$= \sqrt{2^2 \cdot 3^2 \cdot 3}$$
$$= \sqrt{2^2} \cdot \sqrt{3^2} \cdot \sqrt{3} \quad \text{Product rule}$$
$$= 2 \cdot 3 \cdot \sqrt{3} \quad \sqrt{2^2} = 2,\ \sqrt{3^2} = 3$$
$$= 6\sqrt{3} \quad \text{Multiply.}$$

(c) $\sqrt{10}$ No perfect square (other than 1) divides into 10, so $\sqrt{10}$ cannot be simplified further.

(d) $\sqrt[3]{16}$

The largest perfect *cube* that divides into 16 is 8, so factor 16 as $8 \cdot 2$.

$$\sqrt[3]{16}$$

Remember to write the index.

$$= \sqrt[3]{8 \cdot 2} \quad \text{8 is a perfect cube.}$$
$$= \sqrt[3]{8} \cdot \sqrt[3]{2} \quad \text{Product rule}$$
$$= 2\sqrt[3]{2} \quad \sqrt[3]{8} = 2$$

(e)
$$-\sqrt[4]{162}$$
$$= -\sqrt[4]{81 \cdot 2} \quad \text{81 is a perfect 4th power.}$$
$$= -\sqrt[4]{81} \cdot \sqrt[4]{2} \quad \text{Product rule}$$
$$= -3\sqrt[4]{2} \quad \sqrt[4]{81} = 3$$

Remember the negative sign in each line.

Work Problem **4** *at the Side.* ▶

CAUTION

Be careful with which factors belong outside the radical sign and which belong inside. Note in Example 4(b) how $2 \cdot 3$ is written outside because $\sqrt{2^2} = 2$ and $\sqrt{3^2} = 3$. The remaining 3 is left inside the radical.

4 Simplify.

(a) $\sqrt{32}$

(b) $\sqrt{45}$

(c) $\sqrt{300}$

(d) $\sqrt{35}$

(e) $-\sqrt[3]{54}$

(f) $\sqrt[4]{243}$

Answers

4. **(a)** $4\sqrt{2}$ **(b)** $3\sqrt{5}$ **(c)** $10\sqrt{3}$
(d) cannot be simplified further
(e) $-3\sqrt[3]{2}$ **(f)** $3\sqrt[4]{3}$

5 Simplify. Assume that all variables represent positive real numbers.

(a) $\sqrt{25p^7}$

(b) $\sqrt{72y^3x}$

(c) $\sqrt[3]{-27y^7x^5z^6}$

(d) $-\sqrt[4]{32a^5b^7}$

Example 5 Simplifying Radicals Involving Variables

Simplify. Assume that all variables represent positive real numbers.

(a) $\sqrt{16m^3}$

$= \sqrt{16m^2 \cdot m}$ Factor.

$= \sqrt{16m^2} \cdot \sqrt{m}$ Product rule

$= 4m\sqrt{m}$

Absolute value bars are not needed around the m in color because of the assumption that all the variables represent *positive* real numbers.

(b) $\sqrt{200k^7q^8}$

$= \sqrt{10^2 \cdot 2 \cdot (k^3)^2 \cdot k \cdot (q^4)^2}$ Factor.

$= 10k^3q^4\sqrt{2k}$ Remove perfect square factors.

(c) $\sqrt[3]{-8x^4y^5}$

$= \sqrt[3]{(-8x^3y^3)(xy^2)}$ Choose $-8x^3y^3$ as the perfect cube that divides into $-8x^4y^5$.

$= \sqrt[3]{-8x^3y^3} \cdot \sqrt[3]{xy^2}$ Product rule

$= -2xy\sqrt[3]{xy^2}$

(d) $-\sqrt[4]{32y^9}$

$= -\sqrt[4]{(16y^8)(2y)}$ $16y^8$ is the largest 4th power that divides into $32y^9$.

$= -\sqrt[4]{16y^8} \cdot \sqrt[4]{2y}$ Product rule

$= -2y^2\sqrt[4]{2y}$

◀ Work Problem 5 at the Side.

Note

From Example 5 we see that if a variable is raised to a power with an exponent divisible by 2, it is a perfect square. If it is raised to a power with an exponent divisible by 3, it is a perfect cube. ***In general, if it is raised to a power with an exponent divisible by n, it is a perfect nth power.***

The conditions for a simplified radical given earlier state that an exponent in the radicand and the index of the radical should have greatest common factor 1. The next example applies this condition.

Example 6 Simplifying Radicals by Using Lesser Indexes

Simplify. Assume that all variables represent positive real numbers.

(a) $\sqrt[9]{5^6}$

We can write this radical using rational exponents and then write the exponent in lowest terms. We then express the answer as a radical.

$$\sqrt[9]{5^6} = 5^{6/9} = 5^{2/3} = \sqrt[3]{5^2}, \quad \text{or} \quad \sqrt[3]{25}$$

(b) $\sqrt[4]{p^2} = p^{2/4} = p^{1/2} = \sqrt{p}$ (Recall the assumption that $p > 0$.)

Answers

5. **(a)** $5p^3\sqrt{p}$ **(b)** $6y\sqrt{2yx}$ **(c)** $-3y^2xz^2\sqrt[3]{yx^2}$ **(d)** $-2ab\sqrt[4]{2ab^3}$

These examples suggest the following rule.

> If m is an integer, n and k are natural numbers, and all indicated roots exist, then
>
> $$\sqrt[kn]{a^{km}} = \sqrt[n]{a^m}.$$

Work Problem **6** at the Side. ▶

OBJECTIVE 4 Simplify products and quotients of radicals with different indexes. Since the product and quotient rules for radicals apply only when they have the same index, we multiply and divide radicals with different indexes by using rational exponents.

Example 7 Multiplying Radicals with Different Indexes

Simplify $\sqrt{7} \cdot \sqrt[3]{2}$.

Because the different indexes, 2 and 3, have a least common index of 6, we use rational exponents to write each radical as a sixth root.

$$\sqrt{7} = 7^{1/2} = 7^{3/6} = \sqrt[6]{7^3} = \sqrt[6]{343}$$

$$\sqrt[3]{2} = 2^{1/3} = 2^{2/6} = \sqrt[6]{2^2} = \sqrt[6]{4}$$

Therefore,

$$\sqrt{7} \cdot \sqrt[3]{2}$$

$= \sqrt[6]{343} \cdot \sqrt[6]{4}$ Substitute; $\sqrt{7} = \sqrt[6]{343}$, $\sqrt[3]{2} = \sqrt[6]{4}$

$= \sqrt[6]{1372}.$ Product rule

Work Problem **7** at the Side. ▶

OBJECTIVE 5 Use the Pythagorean formula. The **Pythagorean formula** relates the lengths of the three sides of a right triangle.

Pythagorean Formula

If a and b are the lengths of the shorter sides of a right triangle and c is the length of the longest side, then

$$a^2 + b^2 = c^2.$$

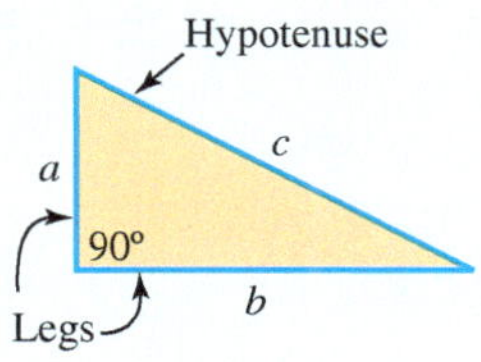

The two shorter sides are the **legs** of the triangle, and the longest side is the **hypotenuse.** The hypotenuse is the side opposite the right angle. Thus,

$$\textbf{leg}^2 + \textbf{leg}^2 = \textbf{hypotenuse}^2.$$

In **Section 10.1,** we will see that an equation such as $x^2 = 7$ has two solutions: $\sqrt{7}$ (the principal, or positive, square root of 7) and $-\sqrt{7}$. Similarly, $c^2 = 52$ has two solutions, $\pm\sqrt{52}$, or $\pm 2\sqrt{13}$. In applications we often choose only the positive square root.

6 Simplify. Assume that all variables represent positive real numbers.

(a) $\sqrt[12]{2^3}$

(b) $\sqrt[6]{t^2}$

7 Simplify $\sqrt{5} \cdot \sqrt[3]{4}$.

ANSWERS

6. (a) $\sqrt[4]{2}$ **(b)** $\sqrt[3]{t}$

7. $\sqrt[6]{2000}$

8 Find the length of the unknown side in each triangle.

(a)

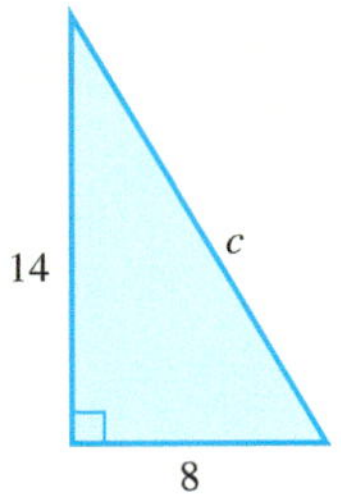

(b)

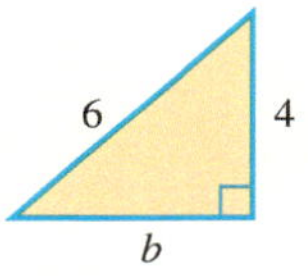

(*Hint:* Write the Pythagorean formula as $b^2 = c^2 - a^2$ here.)

Example 8 Using the Pythagorean Formula

Use the Pythagorean formula to find the length of the hypotenuse in the triangle in Figure 6.

To find the length of the hypotenuse c, let $a = 4$ and $b = 6$. Then, use the formula.

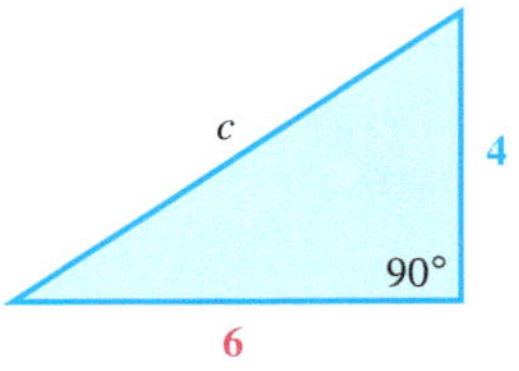

Figure 6

$$a^2 + b^2 = c^2$$

Substitute carefully. $4^2 + 6^2 = c^2$ Let $a = 4$ and $b = 6$.

$16 + 36 = c^2$ Apply the exponents.

$c^2 = 52$ Add; interchange sides.

$c = \sqrt{52}$ Choose the principal root.

$c = \sqrt{4 \cdot 13}$ Factor.

$c = \sqrt{4} \cdot \sqrt{13}$ Product rule

$c = 2\sqrt{13}$ Simplify.

The length of the hypotenuse is $2\sqrt{13}$.

Work Problem **8** *at the Side.*

CAUTION

When substituting in the Pythagorean formula $a^2 + b^2 = c^2$, be sure that the lengths of the legs are substituted for a and b and the length of the hypotenuse is substituted for c.

OBJECTIVE 6 Use the distance formula. The *distance formula* allows us to find the distance between two points in the coordinate plane, or the length of the line segment joining those two points.

Figure 7 shows the points $(3, -4)$ and $(-5, 3)$. The vertical line through $(-5, 3)$ and the horizontal line through $(3, -4)$ intersect at the point $(-5, -4)$. Thus, the point $(-5, -4)$ becomes the vertex of the right angle in a right triangle. By the Pythagorean formula, the sum of the squares of the lengths of the two legs a and b of the right triangle in Figure 7 is equal to the square of the length of the hypotenuse, d:

$$\mathbf{a^2 + b^2 = d^2, \quad \text{or} \quad d^2 = a^2 + b^2.}$$

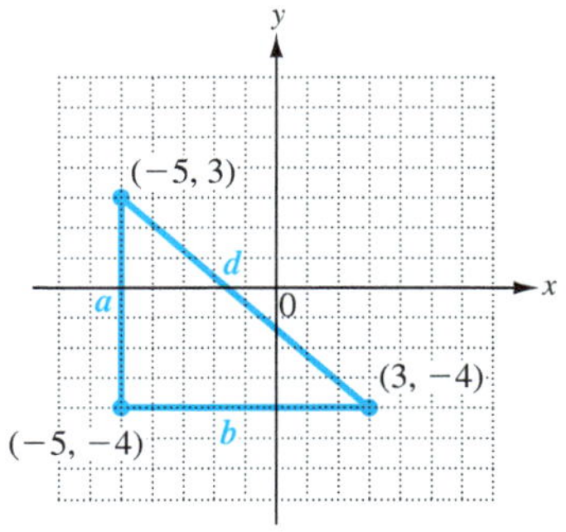

Figure 7

The length a is the difference between the y-coordinates of the endpoints. Since the x-coordinate of both of these points in Figure 7 is -5, the side is vertical, and we can find a by finding the difference between the y-coordinates. We subtract -4 from 3 to get a positive value for a.

$$a = 3 - (-4) = 7$$

ANSWERS

8. (a) $2\sqrt{65}$ **(b)** $2\sqrt{5}$

Similarly, we find b by subtracting -5 from 3.

$$b = 3 - (-5) = 8$$

Substituting these values into the formula, we have

$$d^2 = a^2 + b^2$$

$$d^2 = 7^2 + 8^2 \qquad \text{Let } a = 7 \text{ and } b = 8.$$

$$d^2 = 49 + 64 \qquad \text{Apply the exponents.}$$

$$d^2 = 113 \qquad \text{Add.}$$

$$d = \sqrt{113}. \qquad \text{Choose the principal root.}$$

We choose the principal root since distance cannot be negative. Therefore, the distance between $(-5, 3)$ and $(3, -4)$ is $\sqrt{113}$.

Note

It is customary to leave the distance in radical form. Do not use a calculator to get an approximation, unless you are specifically directed to do so.

This result can be generalized. Figure 8 shows the two points (x_1, y_1) and (x_2, y_2). The distance a between (x_1, y_1) and (x_2, y_1) is

$$a = x_2 - x_1,$$

and the distance b between (x_2, y_2) and (x_2, y_1) is

$$b = y_2 - y_1.$$

From the Pythagorean formula,

$$d^2 = a^2 + b^2$$

$$d^2 = (x_2 - x_1)^2 + (y_2 - y_1)^2.$$

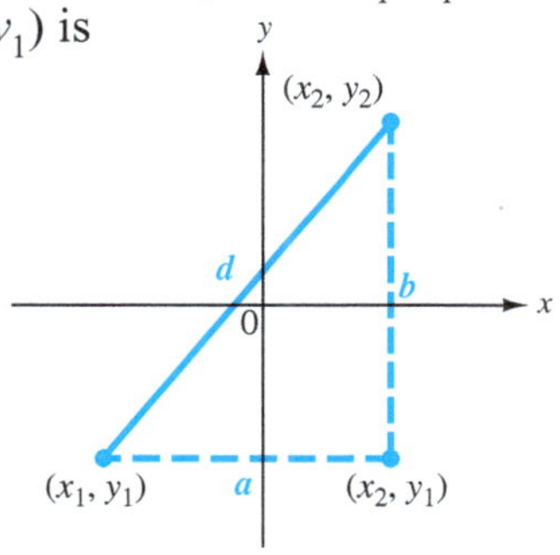

Figure 8

Choosing the principal square root gives the **distance formula.**

Distance Formula

The distance between the points (x_1, y_1) and (x_2, y_2) is

$$d = \sqrt{(x_2 - x_1)^2 + (y_2 - y_1)^2}.$$

Example 9 Using the Distance Formula

Find the distance between $(-3, 5)$ and $(6, 4)$.

When using the distance formula to find the distance between two points, designating the points as (x_1, y_1) and (x_2, y_2) is arbitrary. We choose $(x_1, y_1) = (-3, 5)$ and $(x_2, y_2) = (6, 4)$.

$$d = \sqrt{(x_2 - x_1)^2 + (y_2 - y_1)^2} \qquad \text{Distance formula}$$

$$d = \sqrt{(6 - (-3))^2 + (4 - 5)^2} \qquad x_2 = 6, y_2 = 4, x_1 = -3, y_1 = 5$$

Substitute carefully.

$$d = \sqrt{9^2 + (-1)^2}$$

$$d = \sqrt{82} \qquad \text{Leave in radical form.}$$

Work Problem 9 at the Side. ▶

9 Find the distance between each pair of points.

(a) $(2, -1)$ and $(5, 3)$

(b) $(-3, 2)$ and $(0, -4)$

ANSWERS

9. (a) 5 **(b)** $\sqrt{45}$, or $3\sqrt{5}$

Math in the Media

SO, DID THE SCARECROW REALLY GET A BRAIN?

Probably the most famous mathematical statement in the history of motion pictures is heard in the 1939 classic *The Wizard of Oz*. Ray Bolger's character, the Scarecrow, wants a brain. When the Wizard grants him his "Th.D." (Doctor of Thinkology), the Scarecrow replies with a statement that has made mathematics teachers shudder for 70 years.

Scarecrow: *The sum of the square roots of any two sides of an isosceles triangle is equal to the square root of the remaining side.*

His statement is quite impressive and sounds like the formula for the *Pythagorean Theorem* (page 531). Let's see why it is incorrect.

1. To what kind of triangle does the Scarecrow refer in his statement? To what kind of triangle does the Pythagorean Theorem actually refer?

2. In the Scarecrow's statement, he refers to square roots. In applying the formula for the Pythagorean Theorem, do you find square roots of the sides? If not, what do you find?

3. An isosceles triangle has two sides of equal length. Draw an isosceles triangle with two sides of length 9 units and remaining side of length 4 units. Now show that this triangle does not satisfy the Scarecrow's statement.

   (This is called a *counterexample* and is sufficient to show that his statement is false in general.)

4. Use wording similar to that of the Scarecrow, but state the Pythagorean Theorem correctly.

17.4 Exercises

FOR EXTRA HELP MyMathLab Math XL PRACTICE WATCH DOWNLOAD READ REVIEW

Decide whether each statement is true *or* false *by using the product rule explained in this section. Then support your answer by finding a calculator approximation for each expression.*

1. $2\sqrt{12} = \sqrt{48}$

2. $\sqrt{72} = 2\sqrt{18}$

3. $3\sqrt{8} = 2\sqrt{18}$

4. $5\sqrt{72} = 6\sqrt{50}$

5. Explain why $\sqrt[3]{x} \cdot \sqrt[3]{x}$ is not equal to x. What is it equal to?

6. Explain why $\sqrt[4]{x} \cdot \sqrt[4]{x}$ is not equal to x, but *is* equal to $\sqrt{x}$, for $x \geq 0$.

7. Which one of the following is *not* equal to $\sqrt{\frac{1}{2}}$? (Do not use calculator approximations.)

A. $\sqrt{0.5}$ **B.** $\sqrt{\frac{2}{4}}$ **C.** $\sqrt{\frac{3}{6}}$ **D.** $\frac{\sqrt{4}}{\sqrt{16}}$

8. Use the π key on your calculator to get a value for π. Now find an approximation for $\sqrt[4]{\frac{2143}{22}}$. Does the result mean that π is actually equal to $\sqrt[4]{\frac{2143}{22}}$? Why or why not?

Multiply using the product rule. Assume all variables represent positive real numbers. See Examples 1 and 2.

9. $\sqrt{5} \cdot \sqrt{6}$

10. $\sqrt{10} \cdot \sqrt{3}$

11. $\sqrt{14} \cdot \sqrt{x}$

12. $\sqrt{23} \cdot \sqrt{t}$

13. $\sqrt{14} \cdot \sqrt{3pqr}$

14. $\sqrt{7} \cdot \sqrt{5xt}$

15. $\sqrt[3]{7x} \cdot \sqrt[3]{2y}$

16. $\sqrt[3]{9x} \cdot \sqrt[3]{4y}$

17. $\sqrt[4]{11} \cdot \sqrt[4]{3}$

18. $\sqrt[4]{6} \cdot \sqrt[4]{9}$

19. $\sqrt[4]{2x} \cdot \sqrt[4]{3y^2}$

20. $\sqrt[4]{3y^2} \cdot \sqrt[4]{6yz}$

21. $\sqrt[3]{7} \cdot \sqrt[4]{3}$

22. $\sqrt[5]{8} \cdot \sqrt[6]{12}$

Simplify. Assume that all variables represent positive real numbers. See Example 3.

23. $\sqrt{\frac{64}{121}}$

24. $\sqrt{\frac{16}{49}}$

25. $\sqrt{\frac{3}{25}}$

26. $\sqrt{\frac{13}{49}}$

27. $\sqrt{\frac{x}{25}}$

28. $\sqrt{\frac{k}{100}}$

29. $\sqrt{\frac{p^6}{81}}$

30. $\sqrt{\frac{w^{10}}{36}}$

31. $\sqrt[3]{-\frac{27}{64}}$

32. $\sqrt[3]{-\frac{216}{125}}$

33. $\sqrt[3]{\frac{r^2}{8}}$

34. $\sqrt[3]{\frac{t}{125}}$

35. $-\sqrt[4]{\frac{81}{x^4}}$

36. $-\sqrt[4]{\frac{625}{y^4}}$

37. $\sqrt[5]{\frac{1}{x^{15}}}$

38. $\sqrt[5]{\frac{32}{y^{20}}}$

Express each radical in simplified form. See Example 4.

39. $\sqrt{12}$

40. $\sqrt{18}$

41. $\sqrt{288}$

42. $\sqrt{72}$

43. $-\sqrt{32}$

44. $-\sqrt{48}$

45. $-\sqrt{28}$

46. $-\sqrt{24}$

47. $\sqrt{30}$

48. $\sqrt{46}$

49. $\sqrt[3]{128}$

50. $\sqrt[3]{24}$

51. $\sqrt[3]{-16}$ 52. $\sqrt[3]{-250}$ 53. $\sqrt[3]{40}$ 54. $\sqrt[3]{375}$

55. $-\sqrt[4]{512}$ 56. $-\sqrt[4]{1250}$ 57. $\sqrt[5]{64}$ 58. $\sqrt[5]{128}$

59. A student claimed that $\sqrt[3]{14}$ is not in simplified form, since $14 = 8 + 6$, and 8 is a perfect cube. Was his reasoning correct? Why or why not?

60. Explain in your own words why $\sqrt[3]{k^4}$ is not a simplified radical.

Express each radical in simplified form. Assume that all variables represent positive real numbers. See Example 5.

61. $\sqrt{72k^2}$ 62. $\sqrt{18m^2}$ 63. $\sqrt{144x^3y^9}$ 64. $\sqrt{169s^5t^{10}}$

65. $\sqrt{121x^6}$ 66. $\sqrt{256z^{12}}$ 67. $-\sqrt[3]{27t^{12}}$ 68. $-\sqrt[3]{64y^{18}}$

69. $-\sqrt{100m^8z^4}$ 70. $-\sqrt{25t^6s^{20}}$ 71. $-\sqrt[3]{-125a^6b^9c^{12}}$ 72. $-\sqrt[3]{-216y^{15}x^6z^3}$

73. $\sqrt[4]{\frac{1}{16}r^8t^{20}}$ 74. $\sqrt[4]{\frac{81}{256}t^{12}u^8}$ 75. $\sqrt{50x^3}$ 76. $\sqrt{300z^3}$

77. $-\sqrt{500r^{11}}$ 78. $-\sqrt{200p^{13}}$ 79. $\sqrt{13x^7y^8}$ 80. $\sqrt{23k^9p^{14}}$

81. $\sqrt[3]{8z^6w^9}$

82. $\sqrt[3]{64a^{15}b^{12}}$

83. $\sqrt[3]{-16z^5t^7}$

84. $\sqrt[3]{-81m^4n^{10}}$

85. $\sqrt[4]{81x^{12}y^{16}}$

86. $\sqrt[4]{81t^8u^{28}}$

87. $-\sqrt[4]{162r^{15}s^{10}}$

88. $-\sqrt[4]{32k^5m^{10}}$

89. $\sqrt{\dfrac{y^{11}}{36}}$

90. $\sqrt{\dfrac{v^{13}}{49}}$

91. $\sqrt[3]{\dfrac{x^{16}}{27}}$

92. $\sqrt[3]{\dfrac{y^{17}}{125}}$

Simplify. Assume that x ≥ 0. See Example 6.

93. $\sqrt[4]{48^2}$

94. $\sqrt[4]{50^2}$

95. $\sqrt[4]{25}$

96. $\sqrt[6]{8}$

97. $\sqrt[10]{x^{25}}$

98. $\sqrt[12]{x^{44}}$

Simplify by first writing the radicals as radicals with the same index. Then multiply. Assume that x ≥ 0. See Example 7.

99. $\sqrt[3]{4} \cdot \sqrt{3}$

100. $\sqrt[3]{5} \cdot \sqrt{6}$

101. $\sqrt[4]{3} \cdot \sqrt[3]{4}$

102. $\sqrt[3]{2} \cdot \sqrt[5]{3}$

103. $\sqrt{x} \cdot \sqrt[3]{x}$

104. $\sqrt[3]{x} \cdot \sqrt[4]{x}$

Find the unknown length in each right triangle. Simplify the answer if necessary. See Example 8.

105.

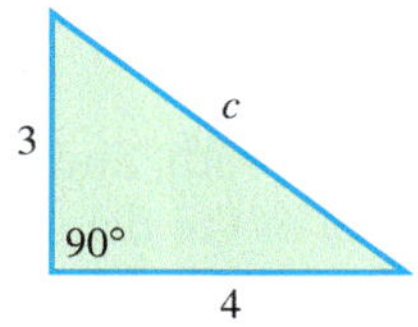

106.

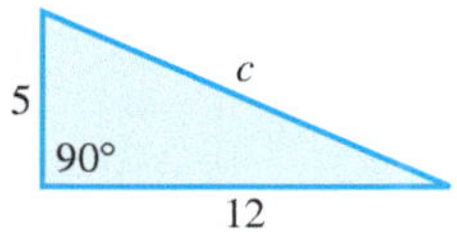

107.

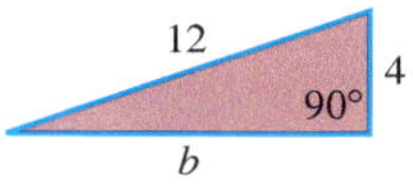

108.

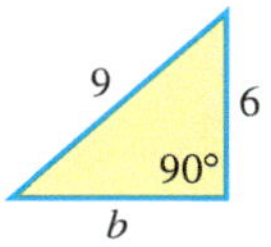

109.

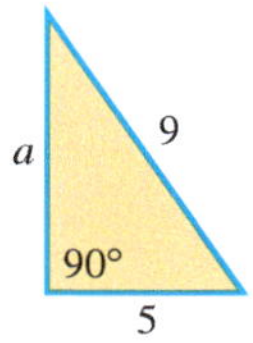

110.

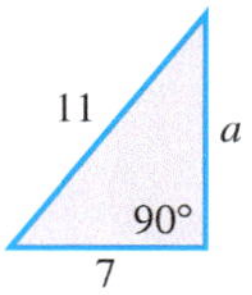

Find the distance between each pair of points. See Example 9.

111. $(6, 13)$ and $(1, 1)$

112. $(8, 13)$ and $(2, 5)$

113. $(-6, 5)$ and $(3, -4)$

114. $(-1, 5)$ and $(-7, 7)$

115. $(-8, 2)$ and $(-4, 1)$

116. $(-1, 2)$ and $(5, 3)$

117. $(4.7, 2.3)$ and $(1.7, -1.7)$

118. $(-2.9, 18.2)$ and $(2.1, 6.2)$

119. $(\sqrt{2}, \sqrt{6})$ and $(-2\sqrt{2}, 4\sqrt{6})$

120. $(\sqrt{7}, 9\sqrt{3})$ and $(-\sqrt{7}, 4\sqrt{3})$

121. $(x + y, y)$ and $(x - y, x)$

122. $(c, c - d)$ and $(d, c + d)$

Solve each problem.

123. A Sanyo color television, model AVM-2755, has a rectangular screen with a 21.7-in. width. Its height is 16 in. What is the diagonal of the screen to the nearest tenth of an inch? (*Source:* Actual measurements of the author's television.)

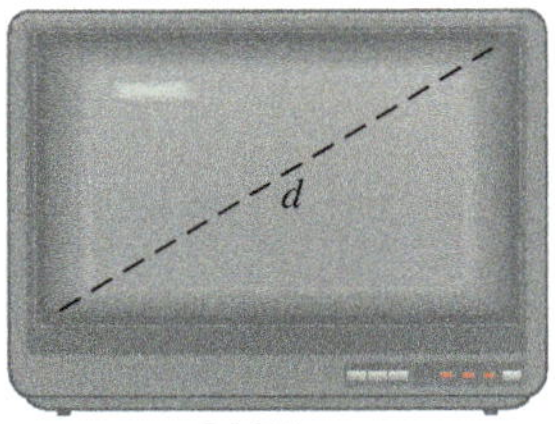

124. The length of the diagonal of a box is given by

$$D = \sqrt{L^2 + W^2 + H^2},$$

where L, W, and H are the length, width, and height of the box. Find the length of the diagonal, D, of a box that is 4 ft long, 3 ft high, and 2 ft wide. Give the exact value, and then round to the nearest tenth of a foot.

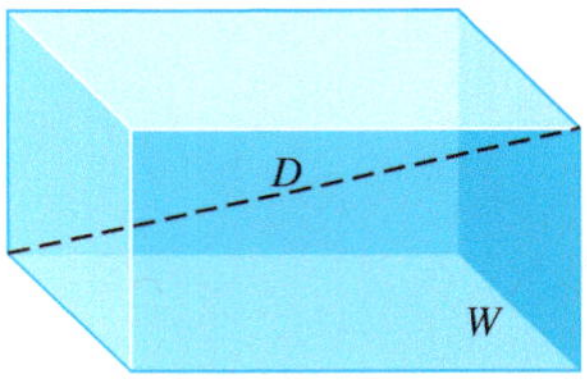

125. A formula from electronics dealing with impedance of parallel resonant circuits is

$$I = \frac{E}{\sqrt{R^2 + \omega^2 L^2}},$$

where the variables are in appropriate units. Find I if $E = 282$, $R = 100$, $L = 264$, and $\omega = 120\pi$. Give your answer to the nearest thousandth.

126. In the study of sound, one version of the law of tensions is

$$f_1 = f_2\sqrt{\frac{F_1}{F_2}}.$$

If $F_1 = 300$, $F_2 = 60$, and $f_2 = 260$, find f_1 to the nearest unit.

127. The following letter appeared in the column "Ask Tom Why," written by Tom Skilling of the *Chicago Tribune*.

Dear Tom,

I cannot remember the formula to calculate the distance to the horizon. I have a stunning view from my 14th floor condo, 150 feet above the ground. How far can I see?

Ted Fleischaker; Indianapolis, Ind.

Skilling's answer was as follows.

To find the distance to the horizon in miles, take the square root of the height of your view in feet and multiply that result by 1.224. Your answer will be the number of miles to the horizon. (*Source: Chicago Tribune*, August 17, 2002.)

Assuming Ted's eyes are 6 ft above the ground, the total height from the ground is $150 + 6 = 156$ ft. To the nearest tenth of a mile, how far can he see to the horizon?

17.5 Adding and Subtracting Radical Expressions

OBJECTIVE 1 Simplify radical expressions involving addition and subtraction. Expressions such as $4\sqrt{2} + 3\sqrt{2}$ and $2\sqrt{3} - 5\sqrt{3}$ can be simplified by using the distributive property.

$$4\sqrt{2} + 3\sqrt{2} = (4 + 3)\sqrt{2} = 7\sqrt{2}$$
This is similar to simplifying $4x + 3x$ as $7x$.

$$2\sqrt{3} - 5\sqrt{3} = (2 - 5)\sqrt{3} = -3\sqrt{3}$$
This is similar to simplifying $2x - 5x$ as $-3x$.

CAUTION
Only radical expressions with the same index and the same radicand may be combined. Expressions such as $5\sqrt{3} + 2\sqrt{2}$ or $3\sqrt{3} + 2\sqrt[3]{3}$ cannot be simplified by combining terms.

Example 1 Adding and Subtracting Radicals

Add or subtract to simplify each radical expression.

(a) $3\sqrt{24} + \sqrt{54}$

Simplify each radical; then use the distributive property to combine terms.

$$3\sqrt{24} + \sqrt{54}$$
$$= 3\sqrt{4} \cdot \sqrt{6} + \sqrt{9} \cdot \sqrt{6} \quad \text{Product rule}$$
$$= 3 \cdot 2\sqrt{6} + 3\sqrt{6}$$
$$= 6\sqrt{6} + 3\sqrt{6}$$
$$= 9\sqrt{6} \quad \text{Combine like terms.}$$

(b) $2\sqrt{20x} - \sqrt{45x}, \quad x \geq 0$

$$= 2\sqrt{4} \cdot \sqrt{5x} - \sqrt{9} \cdot \sqrt{5x} \quad \text{Product rule}$$
$$= 2 \cdot 2\sqrt{5x} - 3\sqrt{5x}$$
$$= 4\sqrt{5x} - 3\sqrt{5x}$$
$$= \sqrt{5x} \quad \text{Combine like terms.}$$

(c) $2\sqrt{3} - 4\sqrt{5}$

Here the radicals differ and are already simplified, so $2\sqrt{3} - 4\sqrt{5}$ cannot be simplified further.

Work Problem 1 at the Side. ▶

CAUTION
Do not confuse the product rule with combining like terms. ***The root of a sum does not equal the sum of the roots.*** For example,

$$\sqrt{9 + 16} \neq \sqrt{9} + \sqrt{16},$$

since $\sqrt{9 + 16} = \sqrt{25} = 5$, but $\sqrt{9} + \sqrt{16} = 3 + 4 = 7$.

OBJECTIVE

1 Simplify radical expressions involving addition and subtraction.

1 Add or subtract to simplify each radical expression.

(a) $3\sqrt{5} + 7\sqrt{5}$

(b) $2\sqrt{11} - \sqrt{11} + 3\sqrt{44}$

(c) $5\sqrt{12y} + 6\sqrt{75y}, \ y \geq 0$

(d) $3\sqrt{8} - 6\sqrt{50} + 2\sqrt{200}$

(e) $9\sqrt{5} - 4\sqrt{10}$

ANSWERS

1. (a) $10\sqrt{5}$ **(b)** $7\sqrt{11}$
(c) $40\sqrt{3y}$ **(d)** $-4\sqrt{2}$
(e) cannot be simplified further

2 Simplify. Assume that all variables represent positive real numbers.

(a) $7\sqrt[3]{81} + 3\sqrt[3]{24}$

(b) $-2\sqrt[4]{32} - 7\sqrt[4]{162}$

(c) $\sqrt[3]{p^4q^7} - \sqrt[3]{64pq}$

3 Simplify. Assume that all variables represent positive real numbers.

(a) $2\sqrt{\frac{8}{9}} - 2\frac{\sqrt{27}}{\sqrt{108}}$

(b) $\sqrt{\frac{80}{y^4}} + \sqrt{\frac{81}{y^{10}}}$

Answers

2. (a) $27\sqrt[3]{3}$ (b) $-25\sqrt[4]{2}$
(c) $(pq^2 - 4)\sqrt[3]{pq}$

3. (a) $\frac{4\sqrt{2} - 3}{3}$ (b) $\frac{4y^3\sqrt{5} + 9}{y^5}$

Example 2 Adding and Subtracting Radicals

Simplify. Assume that all variables represent positive real numbers.

(a) $2\sqrt[3]{16} - 5\sqrt[3]{54}$ — Remember to write the index with each radical.

$= 2\sqrt[3]{8 \cdot 2} - 5\sqrt[3]{27 \cdot 2}$ Factor.

$= 2\sqrt[3]{8} \cdot \sqrt[3]{2} - 5\sqrt[3]{27} \cdot \sqrt[3]{2}$ Product rule

$= 2 \cdot 2 \cdot \sqrt[3]{2} - 5 \cdot 3 \cdot \sqrt[3]{2}$

$= 4\sqrt[3]{2} - 15\sqrt[3]{2}$

$= (4 - 15)\sqrt[3]{2}$ Distributive property

$= -11\sqrt[3]{2}$ Combine like terms.

(b) $2\sqrt[3]{x^2y} + \sqrt[3]{8x^5y^4}$

$= 2\sqrt[3]{x^2y} + \sqrt[3]{(8x^3y^3)x^2y}$ Factor.

$= 2\sqrt[3]{x^2y} + \sqrt[3]{8x^3y^3} \cdot \sqrt[3]{x^2y}$ Product rule

$= 2\sqrt[3]{x^2y} + 2xy\sqrt[3]{x^2y}$

$= (2 + 2xy)\sqrt[3]{x^2y}$ Distributive property — This result cannot be simplified further.

◄ Work Problem 2 at the Side.

Example 3 Adding and Subtracting Radicals with Fractions

Simplify. Assume that all variables represent positive real numbers.

(a) $2\sqrt{\frac{75}{16}} + 4\frac{\sqrt{8}}{\sqrt{32}}$

$= 2\frac{\sqrt{25 \cdot 3}}{\sqrt{16}} + 4\frac{\sqrt{4 \cdot 2}}{\sqrt{16 \cdot 2}}$ Quotient rule; factor.

$= 2\left(\frac{5\sqrt{3}}{4}\right) + 4\left(\frac{2\sqrt{2}}{4\sqrt{2}}\right)$ Product rule; take square roots.

$= \frac{5\sqrt{3}}{2} + 2$ Multiply; $\frac{\sqrt{2}}{\sqrt{2}} = 1$.

$= \frac{5\sqrt{3}}{2} + \frac{4}{2}$ Write with a common denominator.

$= \frac{5\sqrt{3} + 4}{2}$ Add fractions.

(b) $10\sqrt[3]{\frac{5}{x^6}} - 3\sqrt[3]{\frac{4}{x^9}}$

$= 10\frac{\sqrt[3]{5}}{\sqrt[3]{x^6}} - 3\frac{\sqrt[3]{4}}{\sqrt[3]{x^9}}$ Quotient rule

$= \frac{10\sqrt[3]{5}}{x^2} - \frac{3\sqrt[3]{4}}{x^3}$ Simplify denominators.

$= \frac{10\sqrt[3]{5} \cdot x}{x^2 \cdot x} - \frac{3\sqrt[3]{4}}{x^3}$ Write with a common denominator.

$= \frac{10x\sqrt[3]{5} - 3\sqrt[3]{4}}{x^3}$ Subtract fractions.

◄ Work Problem 3 at the Side.

17.5 Exercises

FOR EXTRA HELP MyMathLab — MathXL PRACTICE · WATCH · DOWNLOAD · READ · REVIEW

1. Which one of the following sums could be simplified without first simplifying the individual radical expressions?
   A. $\sqrt{50} + \sqrt{32}$ **B.** $3\sqrt{6} + 9\sqrt{6}$ **C.** $\sqrt[3]{32} - \sqrt[3]{108}$ **D.** $\sqrt[5]{6} - \sqrt[5]{192}$

2. Let $a = 1$ and $b = 64$.
   (a) Evaluate $\sqrt{a} + \sqrt{b}$. Then find $\sqrt{a+b}$. Are they equal?
   (b) Evaluate $\sqrt[3]{a} + \sqrt[3]{b}$. Then find $\sqrt[3]{a+b}$. Are they equal?
   (c) Complete the following: In general, $\sqrt[n]{a} + \sqrt[n]{b} \neq$ ______________, based on the observations in parts (a) and (b) of this exercise.

3. Even though the indexes of the terms are not equal, the sum $\sqrt{64} + \sqrt[3]{125} + \sqrt[4]{16}$ can be simplified quite easily. What is this sum? Why can these terms be combined so easily?

4. Explain why $28 - 4\sqrt{2}$ ***is not equal to*** $24\sqrt{2}$. (This is a common error among algebra students.)

Simplify. Assume that all variables represent positive real numbers. See Examples 1 and 2.

5. $\sqrt{36} - \sqrt{100}$
6. $\sqrt{25} - \sqrt{81}$
7. $-2\sqrt{48} + 3\sqrt{75}$
8. $4\sqrt{32} - 2\sqrt{8}$
9. $\sqrt[3]{16} + 4\sqrt[3]{54}$
10. $3\sqrt[3]{24} - 2\sqrt[3]{192}$
11. $\sqrt[4]{32} + 3\sqrt[4]{2}$
12. $\sqrt[4]{405} - 2\sqrt[4]{5}$
13. $6\sqrt{18} - \sqrt{32} + 2\sqrt{50}$
14. $5\sqrt{8} + 3\sqrt{72} - 3\sqrt{50}$
15. $5\sqrt{6} + 2\sqrt{10}$
16. $3\sqrt{11} - 5\sqrt{13}$
17. $2\sqrt{5} + 3\sqrt{20} + 4\sqrt{45}$
18. $5\sqrt{54} - 2\sqrt{24} - 2\sqrt{96}$
19. $8\sqrt{2x} - \sqrt{8x} + \sqrt{72x}$
20. $4\sqrt{18k} - \sqrt{72k} + \sqrt{50k}$
21. $3\sqrt{72m^2} - 5\sqrt{32m^2}$
22. $9\sqrt{27p^2} - 14\sqrt{108p^2}$
23. $-\sqrt[3]{54} + 2\sqrt[3]{16}$
24. $15\sqrt[3]{81} - 4\sqrt[3]{24}$
25. $2\sqrt[3]{27x} - 2\sqrt[3]{8x}$
26. $6\sqrt[3]{128m} + 3\sqrt[3]{16m}$
27. $\sqrt[3]{x^2y} - \sqrt[3]{8x^2y}$
28. $3\sqrt[3]{x^2y^2} - 2\sqrt[3]{64x^2y^2}$
29. $3x\sqrt[3]{xy^2} - 2\sqrt[3]{8x^4y^2}$
30. $6q^2\sqrt[3]{5q} - 2q\sqrt[3]{40q^4}$

31. $5\sqrt[4]{32} + 3\sqrt[4]{162}$

32. $2\sqrt[4]{512} + 4\sqrt[4]{32}$

33. $3\sqrt[4]{x^5y} - 2x\sqrt[4]{xy}$

34. $2\sqrt[4]{m^9p^6} - 3m^2p\sqrt[4]{mp^2}$

35. $2\sqrt[4]{32a^3} + 5\sqrt[4]{2a^3}$

36. $-\sqrt[4]{16r} + 5\sqrt[4]{r}$

Simplify. Assume that all variables represent positive real numbers. See Example 3.

37. $\dfrac{2\sqrt{5}}{3} + \dfrac{\sqrt{5}}{6}$

38. $\dfrac{4\sqrt{3}}{3} + \dfrac{2\sqrt{3}}{9}$

39. $\sqrt{\dfrac{8}{9}} + \sqrt{\dfrac{18}{36}}$

40. $\sqrt{\dfrac{12}{16}} + \sqrt{\dfrac{48}{64}}$

41. $\dfrac{\sqrt{32}}{3} + \dfrac{2\sqrt{2}}{3} - \dfrac{\sqrt{2}}{\sqrt{9}}$

42. $\dfrac{\sqrt{27}}{2} - \dfrac{3\sqrt{3}}{2} + \dfrac{\sqrt{3}}{\sqrt{4}}$

43. $3\sqrt{\dfrac{50}{9}} + 8\dfrac{\sqrt{2}}{\sqrt{8}}$

44. $9\sqrt{\dfrac{48}{25}} - 2\dfrac{\sqrt{2}}{\sqrt{98}}$

45. $\sqrt{\dfrac{25}{x^8}} - \sqrt{\dfrac{9}{x^6}}$

46. $\sqrt{\dfrac{100}{y^4}} + \sqrt{\dfrac{81}{y^{10}}}$

47. $3\sqrt[3]{\dfrac{m^5}{27}} - 2m\sqrt[3]{\dfrac{m^2}{64}}$

48. $2a\sqrt[4]{\dfrac{a}{16}} - 5a\sqrt[4]{\dfrac{a}{81}}$

49. $3\sqrt[3]{\dfrac{2}{x^6}} - 4\sqrt[3]{\dfrac{5}{x^9}}$

50. $-4\sqrt[3]{\dfrac{4}{t^9}} + 3\sqrt[3]{\dfrac{9}{t^{12}}}$

Solve each problem. Give answers as simplified radical expressions.

51. Find the perimeter of the triangle.

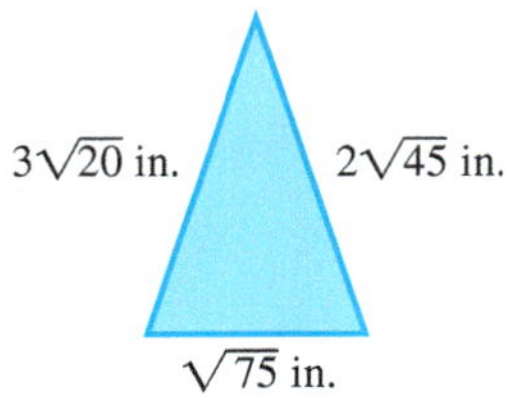

52. Find the perimeter of the rectangle.

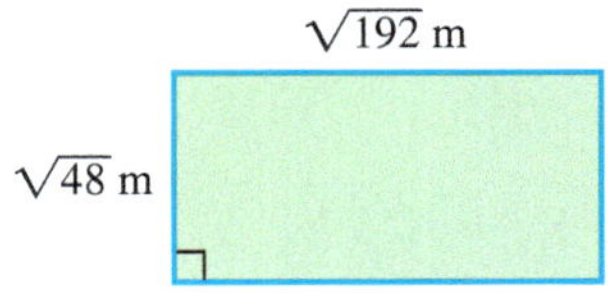

53. What is the perimeter of the computer graphic?

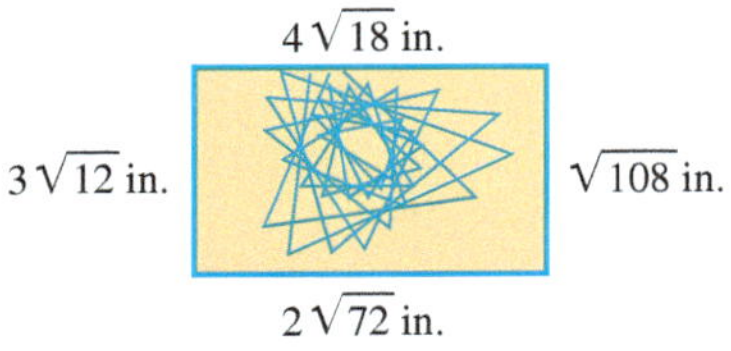

54. Find the area of the trapezoid.

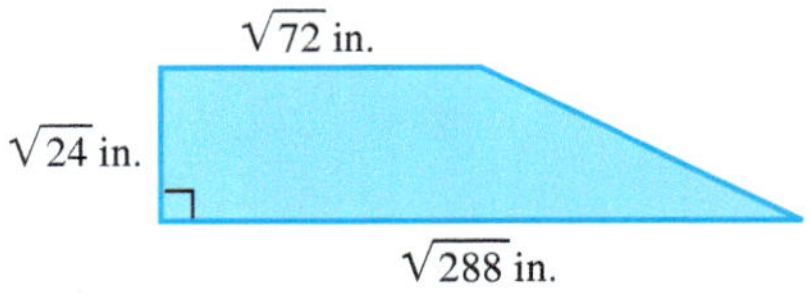

17.6 Multiplying and Dividing Radical Expressions

OBJECTIVES

1. Multiply radical expressions.
2. Rationalize denominators with one radical term.
3. Rationalize denominators with binomials involving radicals.
4. Write radical quotients in lowest terms.

OBJECTIVE 1 Multiply radical expressions. We multiply binomial expressions involving radicals by using the FOIL method from **Section 6.4.** For example, we find the product of $\sqrt{5} + 3$ and $\sqrt{6} + 1$ as follows:

$$\left(\sqrt{5}+3\right)\left(\sqrt{6}+1\right)$$

$$= \underbrace{\sqrt{5}\cdot\sqrt{6}}_{\text{First}} + \underbrace{\sqrt{5}\cdot 1}_{\text{Outer}} + \underbrace{3\cdot\sqrt{6}}_{\text{Inner}} + \underbrace{3\cdot 1}_{\text{Last}}$$

$$= \sqrt{30} + \sqrt{5} + 3\sqrt{6} + 3.$$

This result cannot be simplified further.

Example 1 Multiplying Binomials Involving Radical Expressions

Multiply, using the FOIL method.

(a) $\left(7-\sqrt{3}\right)\left(\sqrt{5}+\sqrt{2}\right)$

$$= \overset{\text{F}}{7\sqrt{5}} + \overset{\text{O}}{7\sqrt{2}} - \overset{\text{I}}{\sqrt{3}\cdot\sqrt{5}} - \overset{\text{L}}{\sqrt{3}\cdot\sqrt{2}}$$

$$= 7\sqrt{5} + 7\sqrt{2} - \sqrt{15} - \sqrt{6}$$

(b) $\left(\sqrt{10}+\sqrt{3}\right)\left(\sqrt{10}-\sqrt{3}\right)$

$$= \sqrt{10}\cdot\sqrt{10} - \sqrt{10}\cdot\sqrt{3} + \sqrt{3}\cdot\sqrt{10} - \sqrt{3}\cdot\sqrt{3}$$

$$= 10 - 3$$

$$= 7$$

The product $\left(\sqrt{10}+\sqrt{3}\right)\left(\sqrt{10}-\sqrt{3}\right) = \left(\sqrt{10}\right)^2 - \left(\sqrt{3}\right)^2$ is the difference of squares:

$$(x+y)(x-y) = x^2 - y^2. \quad \text{Here, } x = \sqrt{10} \text{ and } y = \sqrt{3}.$$

(c) $\left(\sqrt{7}-3\right)^2$

$$= \left(\sqrt{7}-3\right)\left(\sqrt{7}-3\right)$$

$$= \sqrt{7}\cdot\sqrt{7} - 3\sqrt{7} - 3\sqrt{7} + 3\cdot 3$$

$$= 7 - 6\sqrt{7} + 9$$

$$= 16 - 6\sqrt{7}$$

Be careful! These terms cannot be combined.

(d) $\left(5-\sqrt[3]{3}\right)\left(5+\sqrt[3]{3}\right)$

$$= 5\cdot 5 + 5\sqrt[3]{3} - 5\sqrt[3]{3} - \sqrt[3]{3}\cdot\sqrt[3]{3}$$

$$= 25 - \sqrt[3]{3^2}$$

$$= 25 - \sqrt[3]{9}$$

Remember to write the index 3 in *each* radical.

(e) $\left(\sqrt{k}+\sqrt{y}\right)\left(\sqrt{k}-\sqrt{y}\right)$

$$= \left(\sqrt{k}\right)^2 - \left(\sqrt{y}\right)^2 \quad \text{Difference of squares}$$

$$= k - y, \quad k \geq 0 \text{ and } y \geq 0$$

1 Multiply, using the FOIL method.

(a) $\left(2+\sqrt{3}\right)\left(1+\sqrt{5}\right)$

(b) $\left(4+\sqrt{3}\right)\left(4-\sqrt{3}\right)$

(c) $\left(\sqrt{13}-2\right)^2$

(d) $\left(4+\sqrt[3]{7}\right)\left(4-\sqrt[3]{7}\right)$

(e) $\left(\sqrt{p}+\sqrt{s}\right)\left(\sqrt{p}-\sqrt{s}\right)$, $p \geq 0$ and $s \geq 0$

Note

In Example 1(c) we could have used the formula for the square of a binomial,

$$(x-y)^2 = x^2 - 2xy + y^2,$$

to obtain the same result:

$$\begin{aligned}&\left(\sqrt{7}-3\right)^2\\ &= \left(\sqrt{7}\right)^2 - 2\left(\sqrt{7}\right)(3) + 3^2\\ &= 7 - 6\sqrt{7} + 9\\ &= 16 - 6\sqrt{7}.\end{aligned}$$

◀ *Work Problem* **1** *at the Side.*

OBJECTIVE 2 Rationalize denominators with one radical term. As defined earlier, a simplified radical expression will have no radical in the denominator. The origin of this agreement no doubt occurred before the days of high-speed calculation, when computation was a tedious process performed by hand.

For example, consider the radical expression $\frac{1}{\sqrt{2}}$. To find a decimal approximation by hand, it would be necessary to divide 1 by a decimal approximation for $\sqrt{2}$, such as 1.414. It would be much easier if the divisor were a whole number. This can be accomplished by multiplying $\frac{1}{\sqrt{2}}$ by 1 in the form $\frac{\sqrt{2}}{\sqrt{2}}$. ***Multiplying by 1 in any form does not change the value of the original expression.***

$$\frac{1}{\sqrt{2}} \cdot \frac{\sqrt{2}}{\sqrt{2}} = \frac{\sqrt{2}}{2} \qquad \text{Multiply by 1; } \frac{\sqrt{2}}{\sqrt{2}} = 1.$$

Now the computation would require dividing 1.414 by 2 to obtain 0.707, a much easier task.

With current technology, either form of this fraction can be approximated with the same number of keystrokes. See Figure 9, which shows how a calculator gives the same approximation for both forms of the expression.

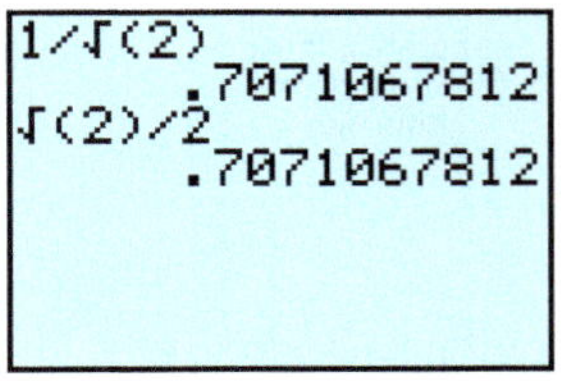

Figure 9

Rationalizing a Denominator

A common way of "standardizing" the form of a radical expression is to have the denominator contain no radicals. The process of removing radicals from a denominator so that the denominator contains only rational numbers is called **rationalizing the denominator.**

ANSWERS

1. **(a)** $2+2\sqrt{5}+\sqrt{3}+\sqrt{15}$
(b) 13 **(c)** $17-4\sqrt{13}$
(d) $16-\sqrt[3]{49}$ **(e)** $p-s$

Example 2 Rationalizing Denominators with Square Roots

Rationalize each denominator.

(a) $\dfrac{3}{\sqrt{7}}$

Multiply by $\frac{\sqrt{7}}{\sqrt{7}}$. This is an application of the multiplicative identity property.

$$\frac{3}{\sqrt{7}} = \frac{3 \cdot \sqrt{7}}{\sqrt{7} \cdot \sqrt{7}} = \frac{3\sqrt{7}}{7}$$

In the denominator, $\sqrt{7} \cdot \sqrt{7} = \sqrt{7 \cdot 7} = \sqrt{49} = 7$. The final denominator is now a rational number.

(b) $\dfrac{5\sqrt{2}}{\sqrt{5}} = \dfrac{5\sqrt{2} \cdot \sqrt{5}}{\sqrt{5} \cdot \sqrt{5}} = \dfrac{5\sqrt{10}}{5} = \sqrt{10}$

(c) $\dfrac{-6}{\sqrt{12}}$

Less work is involved if the radical in the denominator is simplified first.

$$\frac{-6}{\sqrt{12}} = \frac{-6}{\sqrt{4 \cdot 3}} = \frac{-6}{2\sqrt{3}} = \frac{-3}{\sqrt{3}}$$

Now rationalize the denominator.

$$\frac{-3}{\sqrt{3}} = \frac{-3 \cdot \sqrt{3}}{\sqrt{3} \cdot \sqrt{3}} = \frac{-3\sqrt{3}}{3} = -\sqrt{3}$$

Work Problem 2 at the Side. ▶

2 Rationalize each denominator.

(a) $\dfrac{8}{\sqrt{3}}$

(b) $\dfrac{5\sqrt{6}}{\sqrt{5}}$

(c) $\dfrac{3}{\sqrt{48}}$

(d) $\dfrac{-16}{\sqrt{32}}$

Example 3 Rationalizing Denominators in Roots of Fractions

Simplify each radical. Assume that all variables represent positive real numbers.

(a) $\sqrt{\dfrac{18}{125}}$

$= \dfrac{\sqrt{18}}{\sqrt{125}}$ Quotient rule

$= \dfrac{\sqrt{9 \cdot 2}}{\sqrt{25 \cdot 5}}$ Factor.

$= \dfrac{3\sqrt{2}}{5\sqrt{5}}$ Product rule

$= \dfrac{3\sqrt{2} \cdot \sqrt{5}}{5\sqrt{5} \cdot \sqrt{5}}$ Multiply by $\frac{\sqrt{5}}{\sqrt{5}}$.

$= \dfrac{3\sqrt{10}}{5 \cdot 5}$ Product rule

$= \dfrac{3\sqrt{10}}{25}$ Multiply.

Continued on Next Page

ANSWERS

2. (a) $\dfrac{8\sqrt{3}}{3}$ (b) $\sqrt{30}$ (c) $\dfrac{\sqrt{3}}{4}$ (d) $-2\sqrt{2}$

3 Simplify each radical. Assume that all variables represent positive real numbers.

(a) $\sqrt{\frac{8}{45}}$

(b) $\sqrt{\frac{72}{y}}$

(c) $\sqrt{\frac{200k^6}{y^7}}$

4 Simplify.

(a) $\sqrt[3]{\frac{15}{32}}$

(b) $\sqrt[3]{\frac{m^{12}}{n}}, \quad n \neq 0$

(c) $\sqrt[4]{\frac{6y}{w^2}}, \quad y \geq 0, w \neq 0$

(b) $\sqrt{\frac{50m^4}{p^5}}$

$= \frac{\sqrt{50m^4}}{\sqrt{p^5}}$ Quotient rule

$= \frac{\sqrt{25m^4 \cdot 2}}{\sqrt{p^4 \cdot p}}$ Factor.

$= \frac{5m^2\sqrt{2}}{p^2\sqrt{p}}$ Product rule

$= \frac{5m^2\sqrt{2} \cdot \sqrt{p}}{p^2\sqrt{p} \cdot \sqrt{p}}$ Multiply by $\frac{\sqrt{p}}{\sqrt{p}}$.

$= \frac{5m^2\sqrt{2p}}{p^2 \cdot p}$ Product rule

$= \frac{5m^2\sqrt{2p}}{p^3}$ Multiply.

◀ Work Problem **3** at the Side.

Example 4 Rationalizing Denominators with Higher Roots

Simplify.

(a) $\sqrt[3]{\frac{27}{16}}$

Use the quotient rule and simplify the numerator and denominator.

$$\sqrt[3]{\frac{27}{16}} = \frac{\sqrt[3]{27}}{\sqrt[3]{16}} = \frac{3}{\sqrt[3]{8} \cdot \sqrt[3]{2}} = \frac{3}{2\sqrt[3]{2}}$$

To get a rational denominator, multiply the numerator and denominator by a number that will result in a perfect cube in the radicand in the denominator. Since $2 \cdot 4 = 8$, a perfect cube, apply the multiplicative identity property and multiply the numerator and denominator by $\sqrt[3]{4}$.

$$\sqrt[3]{\frac{27}{16}} = \frac{3}{2\sqrt[3]{2}} = \frac{3 \cdot \sqrt[3]{4}}{2\sqrt[3]{2} \cdot \sqrt[3]{4}} = \frac{3\sqrt[3]{4}}{2\sqrt[3]{8}} = \frac{3\sqrt[3]{4}}{2 \cdot 2} = \frac{3\sqrt[3]{4}}{4}$$

(b) $\sqrt[4]{\frac{5x}{z}} = \frac{\sqrt[4]{5x} \cdot \sqrt[4]{z^3}}{\sqrt[4]{z} \cdot \sqrt[4]{z^3}} = \frac{\sqrt[4]{5xz^3}}{\sqrt[4]{z^4}} = \frac{\sqrt[4]{5xz^3}}{z}, \quad x \geq 0, z > 0$

CAUTION

In Example 4(a), a typical error is to multiply the numerator and denominator by $\sqrt[3]{2}$, forgetting that $\sqrt[3]{2} \cdot \sqrt[3]{2} = \sqrt[3]{2^2}$, which does ***not*** equal 2. We need ***three*** factors of 2 to get 2^3 under the radical.

$$\sqrt[3]{2} \cdot \sqrt[3]{2} \cdot \sqrt[3]{2} = \sqrt[3]{2^3}, \quad \text{which does equal} \quad 2.$$

◀ Work Problem **4** at the Side.

Answers

3. (a) $\frac{2\sqrt{10}}{15}$ (b) $\frac{6\sqrt{2y}}{y}$ (c) $\frac{10k^3\sqrt{2y}}{y^4}$

4. (a) $\frac{\sqrt[3]{30}}{4}$ (b) $\frac{m^4\sqrt[3]{n^2}}{n}$ (c) $\frac{\sqrt[4]{6yw^2}}{w}$

OBJECTIVE 3 Rationalize denominators with binomials involving radicals. Recall the special product

$$(x+y)(x-y) = x^2 - y^2.$$

To rationalize a denominator that contains a binomial expression (one that contains exactly two terms) involving radicals, such as

$$\frac{3}{1+\sqrt{2}},$$

we must use *conjugates*. The conjugate of $1+\sqrt{2}$ is $1-\sqrt{2}$. In general, $x+y$ and $x-y$ are **conjugates.**

> **Rationalizing a Binomial Denominator**
> If a radical expression has a sum or difference with square root radicals in the denominator, rationalize the denominator by multiplying both the numerator and denominator by the conjugate of the denominator.

For $\frac{3}{1+\sqrt{2}}$, we rationalize the denominator by multiplying both the numerator and denominator by $1-\sqrt{2}$, the conjugate of the denominator.

$$\frac{3}{1+\sqrt{2}}$$

$$= \frac{3(1-\sqrt{2})}{(1+\sqrt{2})(1-\sqrt{2})}$$

$$= \frac{3(1-\sqrt{2})}{-1}$$

The denominator is now a rational number.

$$(1+\sqrt{2})(1-\sqrt{2}) = 1^2 - (\sqrt{2})^2 = 1 - 2 = -1$$

$$= \frac{3}{-1}(1-\sqrt{2})$$

$$= -3(1-\sqrt{2}), \quad \text{or} \quad -3+3\sqrt{2}$$

Example 5 Rationalizing Binomial Denominators

Rationalize each denominator.

(a) $\dfrac{5}{4-\sqrt{3}}$

$$= \frac{5(4+\sqrt{3})}{(4-\sqrt{3})(4+\sqrt{3})}$$ Multiply the numerator and denominator by $4+\sqrt{3}$.

$$= \frac{5(4+\sqrt{3})}{16-3}$$ Multiply in the denominator.

$$= \frac{5(4+\sqrt{3})}{13}$$ Subtract.

Notice that the numerator is left in factored form. This makes it easier to determine whether the expression is written in lowest terms.

Continued on Next Page

5 Rationalize each denominator.

(a) $\dfrac{-4}{\sqrt{5}+2}$

(b) $\dfrac{15}{\sqrt{7}+\sqrt{2}}$

(c) $\dfrac{\sqrt{3}+\sqrt{5}}{\sqrt{2}-\sqrt{7}}$

(d) $\dfrac{2}{\sqrt{k}+\sqrt{z}}$, $k \neq z, k > 0, z > 0$

6 Write each quotient in lowest terms.

(a) $\dfrac{24-36\sqrt{7}}{16}$

(b) $\dfrac{2x+\sqrt{32x^2}}{6x}$, $x > 0$

Answers

5. (a) $-4(\sqrt{5}-2)$ (b) $3(\sqrt{7}-\sqrt{2})$
(c) $\dfrac{-(\sqrt{6}+\sqrt{21}+\sqrt{10}+\sqrt{35})}{5}$
(d) $\dfrac{2(\sqrt{k}-\sqrt{z})}{k-z}$

6. (a) $\dfrac{6-9\sqrt{7}}{4}$ (b) $\dfrac{1+2\sqrt{2}}{3}$

(b) $\dfrac{\sqrt{2}-\sqrt{3}}{\sqrt{5}+\sqrt{3}}$

$$= \frac{(\sqrt{2}-\sqrt{3})(\sqrt{5}-\sqrt{3})}{(\sqrt{5}+\sqrt{3})(\sqrt{5}-\sqrt{3})}$$ Multiply the numerator and denominator by $\sqrt{5}-\sqrt{3}$.

$$= \frac{\sqrt{10}-\sqrt{6}-\sqrt{15}+3}{5-3}$$ Multiply.

$$= \frac{\sqrt{10}-\sqrt{6}-\sqrt{15}+3}{2}$$ Subtract in the denominator.

(c) $\dfrac{3}{\sqrt{5m}-\sqrt{p}}$, $5m \neq p, m > 0, p > 0$

$$= \frac{3(\sqrt{5m}+\sqrt{p})}{(\sqrt{5m}-\sqrt{p})(\sqrt{5m}+\sqrt{p})}$$

$$= \frac{3(\sqrt{5m}+\sqrt{p})}{5m-p}$$

◀ Work Problem 5 at the Side.

OBJECTIVE 4 Write radical quotients in lowest terms.

Example 6 Writing Radical Quotients in Lowest Terms

Write each quotient in lowest terms.

(a) $\dfrac{6+2\sqrt{5}}{4}$

$$= \frac{2(3+\sqrt{5})}{2 \cdot 2}$$ Factor the numerator and denominator. (This is a key step.)

$$= \frac{3+\sqrt{5}}{2}$$ Divide out the common factor.

Here is an alternative method for writing this expression in lowest terms.

$$\frac{6+2\sqrt{5}}{4} = \frac{6}{4} + \frac{2\sqrt{5}}{4} = \frac{3}{2} + \frac{\sqrt{5}}{2} = \frac{3+\sqrt{5}}{2}$$

(b) $\dfrac{5y-\sqrt{8y^2}}{6y}$, $y > 0$

$$= \frac{5y-2y\sqrt{2}}{6y}$$ $\sqrt{8y^2} = \sqrt{4y^2 \cdot 2} = 2y\sqrt{2}$

$$= \frac{y(5-2\sqrt{2})}{6y}$$ Factor the numerator.

$$= \frac{5-2\sqrt{2}}{6}$$ Divide out the common factor.

◀ Work Problem 6 at the Side.

17.6 Exercises

FOR EXTRA HELP

MyMathLab — Math XL PRACTICE — WATCH — DOWNLOAD — READ — REVIEW

Match each part of a rule for a special product in Column I with the part it equals in Column II.

I	II
1. $(x+\sqrt{y})(x-\sqrt{y})$	**A.** $x-y$
2. $(\sqrt{x}+y)(\sqrt{x}-y)$	**B.** $x+2y\sqrt{x}+y^2$
3. $(\sqrt{x}+\sqrt{y})(\sqrt{x}-\sqrt{y})$	**C.** $x-y^2$
4. $(\sqrt{x}+\sqrt{y})^2$	**D.** $x-2\sqrt{xy}+y$
5. $(\sqrt{x}-\sqrt{y})^2$	**E.** x^2-y
6. $(\sqrt{x}+y)^2$	**F.** $x+2\sqrt{xy}+y$

Multiply, and then simplify each product. Assume that all variables represent positive real numbers. See Example 1.

7. $\sqrt{3}(\sqrt{12}-4)$

8. $\sqrt{5}(\sqrt{125}-6)$

9. $\sqrt{2}(\sqrt{18}-\sqrt{3})$

10. $\sqrt{5}(\sqrt{15}+\sqrt{5})$

11. $(\sqrt{6}+2)(\sqrt{6}-2)$

12. $(\sqrt{7}+8)(\sqrt{7}-8)$

13. $(\sqrt{12}-\sqrt{3})(\sqrt{12}+\sqrt{3})$

14. $(\sqrt{18}+\sqrt{8})(\sqrt{18}-\sqrt{8})$

15. $(\sqrt{3}+2)(\sqrt{6}-5)$

16. $(\sqrt{7}+1)(\sqrt{2}-4)$

17. $(\sqrt{3x}+2)(\sqrt{3x}-2)$

18. $(\sqrt{6y}-4)(\sqrt{6y}+4)$

19. $(2\sqrt{x}+\sqrt{y})(2\sqrt{x}-\sqrt{y})$

20. $(\sqrt{p}+5\sqrt{s})(\sqrt{p}-5\sqrt{s})$

21. $(4\sqrt{x}+3)^2$

22. $(5\sqrt{p}-6)^2$

23. $(9-\sqrt[3]{2})(9+\sqrt[3]{2})$

24. $(7+\sqrt[3]{6})(7-\sqrt[3]{6})$

25. The correct answer to Exercise 7 is $6 - 4\sqrt{3}$. Explain why this is not equal to $2\sqrt{3}$.

26. When we rationalize the denominator in the radical expression $\frac{1}{\sqrt{2}}$, we multiply both the numerator and denominator by $\sqrt{2}$. What property of real numbers covered in **Section 1.4** justifies this procedure?

Rationalize the denominator in each expression. Assume that all variables represent positive real numbers. See Example 2.

27. $\frac{7}{\sqrt{7}}$ **28.** $\frac{11}{\sqrt{11}}$ **29.** $\frac{15}{\sqrt{3}}$ **30.** $\frac{12}{\sqrt{6}}$ **31.** $\frac{\sqrt{3}}{\sqrt{2}}$

32. $\frac{\sqrt{7}}{\sqrt{6}}$ **33.** $\frac{9\sqrt{3}}{\sqrt{5}}$ **34.** $\frac{3\sqrt{2}}{\sqrt{11}}$ **35.** $\frac{-6}{\sqrt{18}}$ **36.** $\frac{-5}{\sqrt{24}}$

37. $\frac{-8\sqrt{3}}{\sqrt{k}}$ **38.** $\frac{-4\sqrt{13}}{\sqrt{m}}$ **39.** $\frac{6\sqrt{3y}}{\sqrt{y^3}}$ **40.** $\frac{-8\sqrt{5y}}{\sqrt{y^5}}$

41. Explain why $\frac{1}{\sqrt[3]{2}}$ would not be written with the denominator rationalized if you begin by multiplying both the numerator and denominator by $\sqrt[3]{2}$. By what should you multiply them both to achieve the desired result?

42. Look again at the expression in Exercise 39. Start by multiplying both the numerator and the denominator by $\sqrt{y}$, to obtain the final answer. Then start over, multiplying both the numerator and denominator by $\sqrt{y^3}$, to obtain the same answer. Which method do you prefer? Why?

Simplify. Assume that all variables represent positive real numbers. See Examples 3 and 4.

43. $\sqrt{\frac{7}{2}}$ **44.** $\sqrt{\frac{10}{3}}$ **45.** $-\sqrt{\frac{7}{50}}$ **46.** $-\sqrt{\frac{13}{75}}$ **47.** $\sqrt{\frac{24}{x}}$

48. $\sqrt{\frac{52}{y}}$ **49.** $-\sqrt{\frac{98r^3}{s}}$ **50.** $-\sqrt{\frac{150m^5}{n}}$ **51.** $\sqrt{\frac{288x^7}{y^9}}$ **52.** $\sqrt{\frac{242t^9}{u^{11}}}$

53. $\sqrt[3]{\frac{2}{3}}$ **54.** $\sqrt[3]{\frac{4}{5}}$ **55.** $\sqrt[3]{\frac{4}{9}}$ **56.** $\sqrt[3]{\frac{5}{16}}$ **57.** $-\sqrt[3]{\frac{2p}{r^2}}$

58. $-\sqrt[3]{\frac{6x}{y^2}}$ **59.** $\sqrt[4]{\frac{16}{x}}$ **60.** $\sqrt[4]{\frac{81}{y}}$ **61.** $\sqrt[4]{\frac{2y}{z}}$ **62.** $\sqrt[4]{\frac{7t}{s^2}}$

Rationalize the denominator in each expression. Assume that all variables represent positive real numbers and that no denominators are 0. See Example 5.

63. $\frac{2}{4+\sqrt{3}}$ **64.** $\frac{6}{5+\sqrt{2}}$ **65.** $\frac{6}{\sqrt{5}+\sqrt{3}}$ **66.** $\frac{12}{\sqrt{6}+\sqrt{3}}$

67. $\frac{-4}{\sqrt{3}-\sqrt{7}}$ **68.** $\frac{-3}{\sqrt{2}+\sqrt{5}}$ **69.** $\frac{1-\sqrt{2}}{\sqrt{7}+\sqrt{6}}$ **70.** $\frac{-1-\sqrt{3}}{\sqrt{6}+\sqrt{5}}$

71. $\frac{\sqrt{2}-\sqrt{3}}{\sqrt{6}-\sqrt{5}}$ **72.** $\frac{\sqrt{5}+\sqrt{6}}{\sqrt{3}-\sqrt{2}}$ **73.** $\frac{4}{\sqrt{x}-2\sqrt{y}}$

74. $\frac{5}{3\sqrt{r}+\sqrt{s}}$ **75.** $\frac{\sqrt{x}-\sqrt{y}}{\sqrt{2x}+\sqrt{3y}}$ **76.** $\frac{\sqrt{a}+\sqrt{b}}{\sqrt{5a}-\sqrt{2b}}$

Write each quotient in lowest terms. Assume that all variables represent positive real numbers. See Example 6.

77. $\dfrac{25 + 10\sqrt{6}}{20}$

78. $\dfrac{12 - 6\sqrt{2}}{24}$

79. $\dfrac{16 + 4\sqrt{8}}{12}$

80. $\dfrac{12 + 9\sqrt{72}}{18}$

81. $\dfrac{6x + \sqrt{24x^3}}{3x}$

82. $\dfrac{11y + \sqrt{242y^5}}{22y}$

Relating Concepts (Exercises 83–86) For Individual or Group Work

Sometimes it is desirable to ***rationalize the numerator*** *in an expression. The procedure is similar to rationalizing the denominator. For example, to rationalize the numerator in the following expression, we multiply both the numerator and denominator by the conjugate of the numerator,* $6 + \sqrt{2}$.

$$\frac{6 - \sqrt{2}}{3}$$

$$= \frac{\left(6 - \sqrt{2}\right)\left(6 + \sqrt{2}\right)}{3\left(6 + \sqrt{2}\right)}$$

$$= \frac{36 - 2}{3\left(6 + \sqrt{2}\right)}$$

$$= \frac{34}{3\left(6 + \sqrt{2}\right)}$$

In the final expression, the numerator is rationalized. ***Work Exercises 83–86 in order.***

83. Rationalize the numerator of $\dfrac{8\sqrt{5} - 1}{6}$.

84. Rationalize the numerator of $\dfrac{3\sqrt{a} + \sqrt{b}}{\sqrt{b} - \sqrt{a}}$. Assume a and b are positive and $a \neq b$.

85. Rationalize the denominator of the expression in Exercise 84.

86. Describe the difference in the procedures used in Exercises 84 and 85.

Summary Exercises on Operations with Radicals and Rational Exponents

Recall that a simplified radical satisfies the following conditions.

Conditions for a Simplified Radical

1. The radicand has no factor raised to a power greater than or equal to the index.
2. The radicand has no fractions.
3. No denominator has a radical.
4. Exponents in the radicand and the index of the radical have greatest common factor 1.

Perform all indicated operations, and express each answer in simplest form with positive exponents. Assume that all variables represent positive real numbers.

1. $6\sqrt{10} - 12\sqrt{10}$

2. $\sqrt{7}(\sqrt{7} - \sqrt{2})$

3. $(1 - \sqrt{3})(2 + \sqrt{6})$

4. $\sqrt{50} - \sqrt{98} + \sqrt{72}$

5. $(3\sqrt{5} + 2\sqrt{7})^2$

6. $\dfrac{-3}{\sqrt{6}}$

7. $\dfrac{8}{\sqrt{7} + \sqrt{5}}$

8. $\sqrt[3]{16x^2} - \sqrt[3]{54x^2} + \sqrt[3]{128x^2}$

9. $\dfrac{1 - \sqrt{2}}{1 + \sqrt{2}}$

10. $(1 - \sqrt[3]{3})(1 + \sqrt[3]{3} + \sqrt[3]{9})$

11. $(\sqrt{5} + 7)(\sqrt{5} - 7)$

12. $\dfrac{1}{\sqrt{x} - \sqrt{5}}, \quad x \neq 5$

13. $\sqrt[3]{8a^3b^5c^9}$

14. $\dfrac{15}{\sqrt[3]{9}}$

15. $\dfrac{3}{\sqrt{5} + 2}$

16. $\sqrt{\dfrac{3}{5x}}$

17. $\dfrac{16\sqrt{3}}{5\sqrt{12}}$

18. $\dfrac{2\sqrt{25}}{8\sqrt{50}}$

19. $\dfrac{-10}{\sqrt[3]{10}}$

20. $\dfrac{\sqrt{6}+\sqrt{5}}{\sqrt{6}-\sqrt{5}}$

21. $\sqrt{12x}-\sqrt{75x}$

22. $(5-3\sqrt{3})^2$

23. $(\sqrt{74}-\sqrt{73})(\sqrt{74}+\sqrt{73})$

24. $\sqrt[3]{\dfrac{13}{81}}$

25. $-t^2\sqrt[4]{t}+3\sqrt[4]{t^9}-t\sqrt[4]{t^5}$

26. $\dfrac{\sqrt{3}+\sqrt{7}}{\sqrt{6}-\sqrt{5}}$

27. $\dfrac{6}{\sqrt[4]{3}}$

28. $\dfrac{1}{1-\sqrt[3]{3}}$

29. $\sqrt[3]{\dfrac{x^2y}{x^{-3}y^4}}$

30. $\sqrt{12}-\sqrt{108}-\sqrt[3]{27}$

31. $\dfrac{x^{-2/3}y^{4/5}}{x^{-5/3}y^{-2/5}}$

32. $\left(\dfrac{x^{3/4}y^{2/3}}{x^{1/3}y^{5/8}}\right)^{24}$

33. $(125x^3)^{-2/3}$

34. $(3x^{-2/3}y^{1/2})(-2x^{5/8}y^{-1/3})$

35. $\dfrac{4^{1/2}+3^{1/2}}{4^{1/2}-3^{1/2}}$

36. $(\sqrt{6}-\sqrt{5})^2(\sqrt{6}+\sqrt{5})^2$

17.7 Solving Equations with Radicals

OBJECTIVES

1. Solve radical equations using the power rule.
2. Solve radical equations that require additional steps.
3. Solve radical equations with indexes greater than 2.

An equation that includes one or more radical expressions with a variable is called a **radical equation.** Some examples of radical equations are

$$\sqrt{x-4} = 8, \quad \sqrt{5x+12} = 3\sqrt{2x-1}, \quad \text{and} \quad \sqrt[3]{6+x} = 27.$$

Radical equations

OBJECTIVE 1 Solve radical equations using the power rule. The equation $x = 1$ has only one solution. Its solution set is $\{1\}$. If we square both sides of this equation, we get $x^2 = 1$. This new equation has two solutions: -1 and 1. Notice that the solution of the original equation is also a solution of the squared equation. However, the squared equation has another solution, -1, that is *not* a solution of the original equation. When solving equations with radicals, we use this idea of raising both sides to a power. This is an application of the **power rule.**

Power Rule for Solving Equations with Radicals

If both sides of an equation are raised to the same power, all solutions of the original equation are also solutions of the new equation.

Read the power rule carefully; it does not say that all solutions of the new equation are solutions of the original equation. They may or may not be. Solutions that do not satisfy the original equation are called **extraneous solutions;** they must be discarded.

CAUTION

When the power rule is used to solve an equation, ***every solution of the new equation* must *be checked in the original equation.***

Example 1 Using the Power Rule

Solve $\sqrt{3x+4} = 8$.

Use the power rule and square both sides to get

$$\left(\sqrt{3x+4}\right)^2 = 8^2$$

$$3x + 4 = 64$$

$$3x = 60 \quad \text{Subtract 4.}$$

$$x = 20. \quad \text{Divide by 3.}$$

To check, substitute the proposed solution in the *original* equation.

Check

$$\sqrt{3x+4} = 8 \quad \text{Original equation}$$

$$\sqrt{3 \cdot 20 + 4} \stackrel{?}{=} 8 \quad \text{Let } x = 20.$$

$$\sqrt{64} \stackrel{?}{=} 8$$

$$8 = 8 \quad \text{True}$$

Since 20 satisfies the *original* equation, the solution set is $\{20\}$.

Work Problem 1 at the Side. ▶

1 Solve each equation.

(a) $\sqrt{r} = 3$

(b) $\sqrt{5x+1} = 4$

ANSWERS

1. (a) $\{9\}$ **(b)** $\{3\}$

The solution of the equation in Example 1 can be generalized.

2 Solve each equation.

(a) $\sqrt{5x+3}+2=0$

(b) $\sqrt{x-9}-3=0$

Solving an Equation with Radicals

Step 1 **Isolate the radical.** Make sure that one radical term is alone on one side of the equation.

Step 2 **Apply the power rule.** Raise both sides of the equation to a power that is the same as the index of the radical.

Step 3 **Solve.** Solve the resulting equation; if it still contains a radical, repeat Steps 1 and 2.

Step 4 **Check** all proposed solutions in the original equation.

CAUTION

Remember to check (Step 4) or you may get an incorrect solution set.

Example 2 **Using the Power Rule**

Solve $\sqrt{5x-1}+3=0$.

Step 1 To isolate the radical on one side, subtract 3 from each side.

$$\sqrt{5x-1}=-3$$

Step 2 Now square both sides.

$$\left(\sqrt{5x-1}\right)^2=(-3)^2$$

Step 3

$$5x-1=9$$

$$5x=10 \quad \text{Add 1.}$$

$$x=2 \quad \text{Divide by 5.}$$

Step 4 Check the proposed solution, 2, by substituting it in the original equation.

Check

$$\sqrt{5x-1}+3=0 \quad \text{Original equation}$$

$$\sqrt{5\cdot 2-1}+3\stackrel{?}{=}0 \quad \text{Let } x=2.$$

$$3+3=0 \quad \text{False}$$

This false result shows that 2 is *not* a solution of the original equation; it is extraneous. The solution set is $\emptyset$.

Note

We could have determined after Step 1 that the equation in Example 2 has no solution because the expression on the left cannot equal a negative number.

◀ *Work Problem* **2** *at the Side.*

OBJECTIVE 2 **Solve radical equations that require additional steps.** The next examples involve finding the square of a binomial. Recall that

$$(x+y)^2=x^2+2xy+y^2.$$

Answers

2. **(a)** $\emptyset$ **(b)** $\{18\}$

Example 3 Using the Power Rule; Squaring a Binomial

Solve $\sqrt{4 - x} = x + 2$.

Step 1 The radical is alone on the left side of the equation.

Step 2 Square both sides. On the right, $(x + 2)^2 = x^2 + 2(x)(2) + 2^2$.

$$\left(\sqrt{4 - x}\right)^2 = (x + 2)^2$$ Remember the middle term.

$$4 - x = x^2 + 4x + 4$$

Pay careful attention here. Twice the product of 2 and x

Step 3 The new equation is quadratic, so get 0 on one side.

$0 = x^2 + 5x$ Subtract 4; add x.

$0 = x(x + 5)$ Factor.

Set *each* factor equal to 0. $x = 0$ or $x + 5 = 0$ Zero-factor property

$x = -5$ Solve.

Step 4 Check each proposed solution in the original equation.

Check If $x = 0$, then

$\sqrt{4 - x} = x + 2$

$\sqrt{4 - 0} \stackrel{?}{=} 0 + 2$

$\sqrt{4} \stackrel{?}{=} 2$

$2 = 2.$ True

If $x = -5$, then

$\sqrt{4 - x} = x + 2$

$\sqrt{4 - (-5)} \stackrel{?}{=} -5 + 2$

$\sqrt{9} \stackrel{?}{=} -3$

$3 = -3.$ False

The solution set is $\{0\}$. The other proposed solution, -5, is extraneous.

Work Problem 3 at the Side. ▶

3 Solve.

(a) $\sqrt{3x - 5} = x - 1$

(b) $x + 1 = \sqrt{-2x - 2}$

Example 4 Using the Power Rule; Squaring a Binomial

Solve $\sqrt{x^2 - 4x + 9} = x - 1$.

Square both sides. On the right, $(x - 1)^2 = x^2 - 2(x)(1) + 1^2$.

$$\left(\sqrt{x^2 - 4x + 9}\right)^2 = (x - 1)^2$$ Remember the middle term.

$$x^2 - 4x + 9 = x^2 - 2x + 1$$

Twice the product of x and -1

$-2x = -8$ Subtract x^2 and 9; add $2x$.

$x = 4$ Divide by -2.

Check $\sqrt{x^2 - 4x + 9} = x - 1$ Original equation

$\sqrt{4^2 - 4 \cdot 4 + 9} \stackrel{?}{=} 4 - 1$ Let $x = 4$.

$3 = 3$ True

The solution set of the original equation is $\{4\}$.

Work Problem 4 at the Side. ▶

4 Solve.

$\sqrt{4x^2 + 2x - 3} = 2x + 7$

CAUTION

When a radical equation requires squaring a binomial, as in Examples 3 and 4, ***remember to include the middle term.***

Answers

3. (a) $\{2, 3\}$ **(b)** $\{-1\}$

4. $\{-2\}$

Example 5 **Using the Power Rule; Squaring Twice**

Solve $\sqrt{5x + 6} + \sqrt{3x + 4} = 2$.

Start by isolating one radical on one side of the equation. Do this by subtracting $\sqrt{3x + 4}$ from each side.

$$\sqrt{5x + 6} = 2 - \sqrt{3x + 4} \qquad \text{Subtract } \sqrt{3x+4}.$$

$$\left(\sqrt{5x + 6}\right)^2 = \left(2 - \sqrt{3x + 4}\right)^2 \qquad \text{Square both sides.}$$

$$5x + 6 = 4 - 4\sqrt{3x + 4} + (3x + 4)$$

Be careful here.

Remember the middle term.

Twice the product of 2 and $-\sqrt{3x + 4}$

This equation still contains a radical, so isolate the radical term on the right and square both sides again.

$$5x + 6 = 8 + 3x - 4\sqrt{3x + 4} \qquad \text{Combine like terms.}$$

$$2x - 2 = -4\sqrt{3x + 4} \qquad \text{Subtract 8 and } 3x.$$

$$x - 1 = -2\sqrt{3x + 4} \qquad \text{Divide by 2 to make the numbers smaller.}$$

Divide each term by 2.

$$(x - 1)^2 = \left(-2\sqrt{3x + 4}\right)^2 \qquad \text{Square both sides again.}$$

$$x^2 - 2x + 1 = (-2)^2\left(\sqrt{3x + 4}\right)^2 \qquad \text{On the right, } (ab)^2 = a^2b^2.$$

$$x^2 - 2x + 1 = 4(3x + 4) \qquad \text{Apply the exponents.}$$

$$x^2 - 2x + 1 = 12x + 16 \qquad \text{Distributive property}$$

$$x^2 - 14x - 15 = 0 \qquad \text{Standard form}$$

$$(x + 1)(x - 15) = 0 \qquad \text{Factor.}$$

$$x + 1 = 0 \quad \text{or} \quad x - 15 = 0 \qquad \text{Zero-factor property}$$

$$x = -1 \quad \text{or} \quad x = 15 \qquad \text{Solve each equation.}$$

Check each of these proposed solutions in the original equation. Only -1 satisfies the equation, so the solution set, $\{-1\}$, has only one element.

◀ Work Problem **5** at the Side.

5 **(a)** Verify that 15 is an extraneous solution of the equation in Example 5 and must be discarded.

(b) Solve.

$$\sqrt{2x + 3} + \sqrt{x + 1} = 1$$

OBJECTIVE 3 **Solve radical equations with indexes greater than 2.** The power rule also works for powers greater than 2.

Example 6 **Using the Power Rule for a Power Greater than 2**

Solve $\sqrt[3]{x + 5} = \sqrt[3]{2x - 6}$.

Raise both sides to the third power.

$$\left(\sqrt[3]{x + 5}\right)^3 = \left(\sqrt[3]{2x - 6}\right)^3$$

$$x + 5 = 2x - 6$$

$$11 = x \qquad \text{Subtract } x\text{; add 6.}$$

Check $$\sqrt[3]{x + 5} = \sqrt[3]{2x - 6} \qquad \text{Original equation}$$

$$\sqrt[3]{11 + 5} \stackrel{?}{=} \sqrt[3]{2 \cdot 11 - 6} \qquad \text{Let } x = 11.$$

$$\sqrt[3]{16} = \sqrt[3]{16} \qquad \text{True}$$

The solution set is $\{11\}$.

◀ Work Problem **6** at the Side.

6 Solve each equation.

(a) $\sqrt[3]{2x + 7} = \sqrt[3]{3x - 2}$

(b) $\sqrt[4]{2x + 5} + 1 = 0$

Answers

5. **(a)** The final step in the check leads to $16 = 2$, which is false.
(b) $\{-1\}$

6. **(a)** $\{9\}$ **(b)** $\emptyset$

17.7 Exercises

Check each equation to see if the given value for x is a solution.

1. $\sqrt{3x + 18} = x$
(a) 6 **(b)** -3

2. $\sqrt{3x - 3} = x - 1$
(a) 1 **(b)** 4

3. $\sqrt{x + 2} = \sqrt{9x - 2} - 2\sqrt{x - 1}$
(a) 2 **(b)** 7

4. $\sqrt{8x - 3} = 2x$
(a) $\frac{3}{2}$ **(b)** $\frac{1}{2}$

5. Is 9 a solution of the equation $\sqrt{x} = -3$? If not, what is the solution of this equation? Explain.

6. Before even attempting to solve $\sqrt{3x + 18} = x$, how can you be sure that the equation cannot have a negative solution?

Solve each equation. See Examples 1–4.

7. $\sqrt{x - 2} = 3$

8. $\sqrt{x + 1} = 7$

9. $\sqrt{6x - 1} = 1$

10. $\sqrt{7x - 3} = 5$

11. $\sqrt{4x + 3} + 1 = 0$

12. $\sqrt{5x - 3} + 2 = 0$

13. $\sqrt{3k + 1} - 4 = 0$

14. $\sqrt{5z + 1} - 11 = 0$

15. $4 - \sqrt{x - 2} = 0$

16. $9 - \sqrt{4k + 1} = 0$

17. $\sqrt{9a - 4} = \sqrt{8a + 1}$

18. $\sqrt{4p - 2} = \sqrt{3p + 5}$

19. $2\sqrt{x} = \sqrt{3x + 4}$

20. $2\sqrt{m} = \sqrt{5m - 16}$

21. $3\sqrt{z - 1} = 2\sqrt{2z + 2}$

22. $5\sqrt{4x + 1} = 3\sqrt{10x + 25}$

23. $k = \sqrt{k^2 + 4k - 20}$

24. $p = \sqrt{p^2 - 3p + 18}$

25. $x = \sqrt{x^2 + 3x + 9}$

26. $z = \sqrt{z^2 - 4z - 8}$

27. $\sqrt{9 - x} = x + 3$

28. $\sqrt{5 - x} = x + 1$

29. $\sqrt{k^2 + 2k + 9} = k + 3$

30. $\sqrt{x^2 - 3x + 3} = x - 1$

31. $\sqrt{r^2 + 9r + 3} = -r$

32. $\sqrt{p^2 - 15p + 15} = p - 5$

33. $\sqrt{z^2 + 12z - 4} + 4 - z = 0$

34. $\sqrt{m^2 + 3m + 12} - m - 2 = 0$

35. A student wrote the following as his first step in solving $\sqrt{3x + 4} = 8 - x$.

$$3x + 4 = 64 + x^2$$

WHAT WENT WRONG? Solve the given equation correctly.

36. A student wrote the following as her first step in solving $\sqrt{5x + 6} = \sqrt{x + 3} + 3$.

$$5x + 6 = x + 3 + 9$$

WHAT WENT WRONG? Solve the given equation correctly.

Solve each equation. See Examples 5 and 6.

37. $\sqrt[3]{2x+5} = \sqrt[3]{6x+1}$

38. $\sqrt[3]{p-1} = 2$

39. $\sqrt[3]{a^2+5a+1} = \sqrt[3]{a^2+4a}$

40. $\sqrt[3]{r^2+2r+8} = \sqrt[3]{r^2}$

41. $\sqrt[3]{2m-1} = \sqrt[3]{m+13}$

42. $\sqrt[3]{2k-11} - \sqrt[3]{5k+1} = 0$

43. $\sqrt[4]{a+8} = \sqrt[4]{2a}$

44. $\sqrt[4]{z+11} = \sqrt[4]{2z+6}$

45. $\sqrt[3]{x-8} + 2 = 0$

46. $\sqrt[3]{r+1} + 1 = 0$

47. $\sqrt[4]{2k-5} + 4 = 0$

48. $\sqrt[4]{8z-3} + 2 = 0$

49. $\sqrt{k+2} - \sqrt{k-3} = 1$

50. $\sqrt{r+6} - \sqrt{r-2} = 2$

51. $\sqrt{2r+11} - \sqrt{5r+1} = -1$

52. $\sqrt{3x-2} - \sqrt{x+3} = 1$

53. $\sqrt{3p+4} - \sqrt{2p-4} = 2$

54. $\sqrt{4x+5} - \sqrt{2x+2} = 1$

55. $\sqrt{3-3p} - 3 = \sqrt{3p+2}$

56. $\sqrt{4x+7} - 4 = \sqrt{4x-1}$

57. $\sqrt{2\sqrt{x+11}} = \sqrt{4x+2}$

58. $\sqrt{1+\sqrt{24-10x}} = \sqrt{3x+5}$

For each equation, rewrite the expressions with rational exponents as radical expressions, and then solve using the procedures explained in this section.

59. $(2x - 9)^{1/2} = 2 + (x - 8)^{1/2}$

60. $(3w + 7)^{1/2} = 1 + (w + 2)^{1/2}$

61. $(2w - 1)^{2/3} - w^{1/3} = 0$

62. $(x^2 - 2x)^{1/3} - x^{1/3} = 0$

Solve each formula from electricity and radio for the indicated variable. (Source: Cooke, Nelson M., and Joseph B. Orleans, Mathematics Essential to Electricity and Radio, *McGraw-Hill, 1943.)*

63. $V = \sqrt{\frac{2K}{m}}$ for K

64. $V = \sqrt{\frac{2K}{m}}$ for m

65. $Z = \sqrt{\frac{L}{C}}$ for C

66. $Z = \sqrt{\frac{L}{C}}$ for L

67. $f = \frac{1}{2\pi\sqrt{LC}}$ for L

68. $r = \sqrt{\frac{Mm}{F}}$ for F

A number of useful formulas involve radicals or radical expressions. Many occur in the mathematics needed for working with objects in space. The formula

$$N = \frac{1}{2\pi}\sqrt{\frac{a}{r}}$$

is used to find the rotational rate N of a space station. Here a is the acceleration and r represents the radius of the space station in meters. To find the value of r that will make N simulate the effect of gravity on Earth, the equation must be solved for r, using the required value of N. (Source: Kastner, Bernice, Space Mathematics, *NASA, 1972.)*

69. Solve the equation for r.

70. If $a = 9.8$ m per sec^2, find the value of r (to the nearest tenth) using each value of N.

(a) $N = 0.063$ rotation per sec

(b) $N = 0.04$ rotation per sec

17.8 Complex Numbers

As we saw in **Section 1.1,** the set of real numbers includes many other number sets (the rational numbers, integers, and natural numbers, for example). In this section, a new set of numbers is introduced that includes the set of real numbers, as well as numbers that are even roots of negative numbers, like $\sqrt{-2}$.

OBJECTIVES

1. Simplify numbers of the form $\sqrt{-b}$, where $b > 0$.
2. Recognize subsets of the complex numbers.
3. Add and subtract complex numbers.
4. Multiply complex numbers.
5. Divide complex numbers.
6. Find powers of i.

OBJECTIVE 1 Simplify numbers of the form $\sqrt{-b}$, where $b > 0$.

The equation $x^2 + 1 = 0$ has no real number solution since any solution must be a number whose square is -1. In the set of real numbers, all squares are nonnegative numbers because the product of two positive numbers or two negative numbers is positive and $0^2 = 0$. To provide a solution for the equation $x^2 + 1 = 0$, we introduce a new number i.

Imaginary Unit i

The **imaginary unit i** is defined as

$$i = \sqrt{-1}, \quad \text{where} \quad i^2 = -1.$$

In words, i is the principal square root of -1.

This definition of i makes it possible to define any square root of a negative number as follows.

$\sqrt{-b}$

For any positive number b, $\quad \sqrt{-b} = i\sqrt{b}.$

Example 1 Simplifying Square Roots of Negative Numbers

Write each number as a product of a real number and i.

(a) $\sqrt{-100} = i\sqrt{100} = 10i$

(b) $-\sqrt{-36} = -i\sqrt{36} = -6i$

(c) $\sqrt{-2} = i\sqrt{2}$

(d) $\sqrt{-8} = i\sqrt{8} = i\sqrt{4 \cdot 2} = 2i\sqrt{2}$

CAUTION

It is easy to mistake $\sqrt{2}i$ for $\sqrt{2i}$, with the i under the radical. For this reason, we usually write $\sqrt{2}i$ as $i\sqrt{2}$, as in the definition of $\sqrt{-b}$.

Work Problem 1 at the Side. ▶

1 Write each number as a product of a real number and i.

(a) $\sqrt{-16}$

(b) $-\sqrt{-81}$

(c) $\sqrt{-7}$

(d) $\sqrt{-32}$

When finding a product such as $\sqrt{-4} \cdot \sqrt{-9}$, we cannot use the product rule for radicals because it applies only to nonnegative radicands. ***For this reason, we change $\sqrt{-b}$ to the form $i\sqrt{b}$ before performing any multiplications or divisions.***

ANSWERS

1. (a) $4i$ (b) $-9i$ (c) $i\sqrt{7}$ (d) $4i\sqrt{2}$

2 Multiply.

(a) $\sqrt{-7} \cdot \sqrt{-5}$

(b) $\sqrt{-5} \cdot \sqrt{-10}$

(c) $\sqrt{-15} \cdot \sqrt{2}$

3 Divide.

(a) $\dfrac{\sqrt{-32}}{\sqrt{-2}}$

(b) $\dfrac{\sqrt{-27}}{\sqrt{-3}}$

(c) $\dfrac{\sqrt{-40}}{\sqrt{10}}$

Answers

2. (a) $-\sqrt{35}$ (b) $-5\sqrt{2}$ (c) $i\sqrt{30}$
3. (a) 4 (b) 3 (c) $2i$

For example,

$$\begin{aligned}
&\sqrt{-4} \cdot \sqrt{-9} \\
&= i\sqrt{4} \cdot i\sqrt{9} && \sqrt{-b} = i\sqrt{b} \\
&= i \cdot 2 \cdot i \cdot 3 \\
&= 6i^2 \\
&= 6(-1) && \text{Substitute: } i^2 = -1. \\
&= -6.
\end{aligned}$$

First write all square roots in terms of i.

CAUTION

Using the product rule for radicals *before* using the definition of $\sqrt{-b}$ gives a *wrong* answer. The preceding example shows that

$$\sqrt{-4} \cdot \sqrt{-9} = -6, \quad \text{Correct}$$

but

$$\sqrt{-4(-9)} = \sqrt{36} = 6, \quad \text{Incorrect}$$

so

$$\sqrt{-4} \cdot \sqrt{-9} \neq \sqrt{-4(-9)}.$$

Example 2 Multiplying Square Roots of Negative Numbers

Multiply.

(a)

$$\begin{aligned}
&\sqrt{-3} \cdot \sqrt{-7} \\
&= i\sqrt{3} \cdot i\sqrt{7} && \sqrt{-b} = i\sqrt{b} \\
&= i^2\sqrt{3 \cdot 7} && \text{Product rule} \\
&= (-1)\sqrt{21} && \text{Substitute: } i^2 = -1. \\
&= -\sqrt{21}
\end{aligned}$$

First write all square roots in terms of i.

(b) $\sqrt{-2} \cdot \sqrt{-8}$

$$\begin{aligned}
&= i\sqrt{2} \cdot i\sqrt{8} \\
&= i^2\sqrt{2 \cdot 8} \\
&= (-1)\sqrt{16} \\
&= (-1)4, \quad \text{or} \quad -4
\end{aligned}$$

(c) $\sqrt{-5} \cdot \sqrt{6}$

$$\begin{aligned}
&= i\sqrt{5} \cdot \sqrt{6} \\
&= i\sqrt{30}
\end{aligned}$$

◀ Work Problem **2** at the Side.

Example 3 Dividing Square Roots of Negative Numbers

Divide.

(a)

$$\begin{aligned}
&\frac{\sqrt{-75}}{\sqrt{-3}} \\
&= \frac{i\sqrt{75}}{i\sqrt{3}} \\
&= \sqrt{\frac{75}{3}} && \text{Quotient rule} \\
&= \sqrt{25} && \text{Divide.} \\
&= 5
\end{aligned}$$

First write all square roots in terms of i.

(b)

$$\begin{aligned}
&\frac{\sqrt{-32}}{\sqrt{8}} \\
&= \frac{i\sqrt{32}}{\sqrt{8}} \\
&= i\sqrt{\frac{32}{8}} \\
&= i\sqrt{4} \\
&= 2i
\end{aligned}$$

◀ Work Problem **3** at the Side.

OBJECTIVE 2 Recognize subsets of the complex numbers. With the imaginary unit i and the real numbers, a new set of numbers can be formed that includes the real numbers as a subset. The *complex numbers* are defined as follows.

Complex Number

If a and b are real numbers, then any number of the form $\boldsymbol{a + bi}$ is called a **complex number.** In the complex number $a + bi$, the number a is called the **real part** and b is called the **imaginary part.***

For a complex number $a + bi$, if $b = 0$, then $a + bi = a$, which is a real number. ***Thus, the set of real numbers is a subset of the set of complex numbers.*** If $a = 0$ and $b \neq 0$, the complex number is said to be a **pure imaginary number.** For example, $3i$ is a pure imaginary number. A number such as $7 + 2i$ is a **nonreal complex number.** These numbers are very useful in applications, particularly in work with electricity.

The relationships among the sets of numbers are shown in Figure 10.

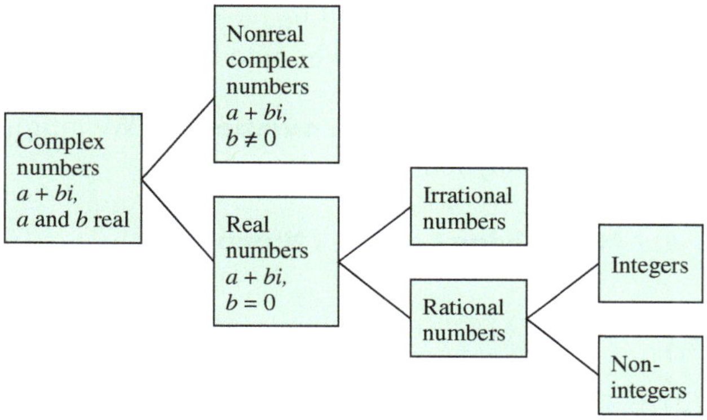

Figure 10

OBJECTIVE 3 Add and subtract complex numbers. The commutative, associative, and distributive properties for real numbers are also valid for complex numbers. ***Thus, to add complex numbers, we add their real parts and add their imaginary parts.***

Example 4 Adding Complex Numbers

Add.

(a) $(2 + 3i) + (6 + 4i)$

$= (2 + 6) + (3 + 4)i$ Commutative, associative, and distributive properties

$= 8 + 7i$ Add real parts; add imaginary parts.

(b) $5 + (9 - 3i)$

$= (5 + 9) - 3i$ Associative property

$= 14 - 3i$ Add real parts.

Work Problem 4 at the Side. ▶

4 Add.

(a) $(4 + 6i) + (-3 + 5i)$

(b) $(-1 + 8i) + (9 - 3i)$

*Some texts define bi as the imaginary part of the complex number $a + bi$.

ANSWERS

4. (a) $1 + 11i$ **(b)** $8 + 5i$

5 Subtract.

(a) $(7 + 3i) - (4 + 2i)$

(b) $(-6 - i) - (-5 - 4i)$

(c) $8 - (3 - 2i)$

6 Multiply.

(a) $6i(4 + 3i)$

(b) $(6 - 4i)(2 + 4i)$

(c) $(3 - 2i)(3 + 2i)$

ANSWERS

5. **(a)** $3 + i$ **(b)** $-1 + 3i$ **(c)** $5 + 2i$

6. **(a)** $-18 + 24i$ **(b)** $28 + 16i$ **(c)** 13

To subtract complex numbers, we subtract their real parts and subtract their imaginary parts.

Example 5 **Subtracting Complex Numbers**

Subtract.

(a) $(6 + 5i) - (3 + 2i)$

$= (6 - 3) + (5 - 2)i$ Properties of real numbers

$= 3 + 3i$ Subtract real parts; subtract imaginary parts.

(b) $(7 - 3i) - (8 - 6i)$

$= (7 - 8) + [-3 - (-6)]i$

$= -1 + 3i$

(c) $(-9 + 4i) - (-9 + 8i)$

$= (-9 + 9) + (4 - 8)i$

$= 0 - 4i$

$= -4i$

◀ Work Problem **5** at the Side.

In Example 5(c), the answer was written as $0 - 4i$ and then as just $-4i$. A complex number written in the form $a + bi$, like $0 - 4i$, is in **standard form.** In this section, most answers will be given in standard form, but if a or b is 0, we consider answers such as a or bi to be in standard form.

OBJECTIVE 4 **Multiply complex numbers.** We multiply complex numbers as we multiply polynomials.

Example 6 **Multiplying Complex Numbers**

Multiply.

(a) $4i(2 + 3i)$

$= 4i(2) + 4i(3i)$ Distributive property

$= 8i + 12i^2$ Multiply.

$= 8i + 12(-1)$ Substitute: $i^2 = -1$.

$= -12 + 8i$ Standard form

(b) $(3 + 5i)(4 - 2i)$

$$= \underbrace{3(4)}_{\text{First}} + \underbrace{3(-2i)}_{\text{Outer}} + \underbrace{5i(4)}_{\text{Inner}} + \underbrace{5i(-2i)}_{\text{Last}}$$

Use the FOIL method for multiplying binomials. **(Section 6.4)**

$= 12 - 6i + 20i - 10i^2$ Multiply.

$= 12 + 14i - 10(-1)$ Combine imaginary terms; $i^2 = -1$.

$= 12 + 14i + 10$ Multiply.

$= 22 + 14i$ Combine real terms.

(c) $(2 + 3i)(1 - 5i)$

$= 2(1) + 2(-5i) + 3i(1) + 3i(-5i)$ FOIL

$= 2 - 10i + 3i - 15i^2$

$= 2 - 7i - 15(-1)$ Use parentheses around -1 to avoid errors.

$= 2 - 7i + 15$

$= 17 - 7i$

◀ Work Problem **6** at the Side.

The two complex numbers $a + bi$ and $a - bi$ are called *complex conjugates,* or simply *conjugates,* of each other. ***The product of a complex number and its conjugate is always a real number,*** as shown here.

$$(a + bi)(a - bi) = a^2 - abi + abi - b^2i^2$$
$$= a^2 - b^2(-1)$$
$$(a + bi)(a - bi) = a^2 + b^2$$

The product eliminates i.

For example, $(3 + 7i)(3 - 7i) = 3^2 + 7^2 = 9 + 49 = 58.$

OBJECTIVE 5 Divide complex numbers. The quotient of two complex numbers should be a complex number. To write the quotient as a complex number, we need to eliminate i in the denominator. We use conjugates and a process like that for rationalizing a denominator to do this.

Example 7 Dividing Complex Numbers

Find each quotient.

(a) $\dfrac{8 + 9i}{5 + 2i}$

Multiply both the numerator and denominator by the conjugate of the denominator. The conjugate of $5 + 2i$ is $5 - 2i$.

$$\frac{8 + 9i}{5 + 2i}$$

$$= \frac{(8 + 9i)(5 - 2i)}{(5 + 2i)(5 - 2i)} \qquad \frac{5 - 2i}{5 - 2i} = 1$$

$$= \frac{40 - 16i + 45i - 18i^2}{5^2 + 2^2} \qquad \text{In the denominator, } (a + bi)(a - bi) = a^2 + b^2.$$

$$= \frac{58 + 29i}{29} \qquad -18i^2 = -18(-1) = 18; \text{ Combine like terms.}$$

$$= \frac{29(2 + i)}{29} \qquad \text{Factor the numerator.}$$

Factor first; then divide out the common factor.

$$= 2 + i \qquad \text{Lowest terms}$$

(b) $\dfrac{1 + i}{i}$

$$= \frac{(1 + i)(-i)}{i(-i)} \qquad \text{Multiply numerator and denominator by } -i, \text{ the conjugate of } i.$$

$$= \frac{-i - i^2}{-i^2} \qquad \text{Distributive property; multiply.}$$

$$= \frac{-i - (-1)}{-(-1)} \qquad \text{Substitute: } i^2 = -1.$$

Use parentheses to avoid errors.

$$= \frac{-i + 1}{1}$$

$$= 1 - i$$

Work Problem 7 at the Side. ▶

7 Find each quotient.

(a) $\dfrac{2 + i}{3 - i}$

(b) $\dfrac{8 - 4i}{1 - i}$

(c) $\dfrac{5}{3 - 2i}$

(d) $\dfrac{5 - i}{i}$

Answers

7. (a) $\frac{1}{2} + \frac{1}{2}i$ (b) $6 + 2i$
(c) $\frac{15}{13} + \frac{10}{13}i$ (d) $-1 - 5i$

8 Find each power of i.

(a) i^{21}

(b) i^{36}

(c) i^{50}

(d) i^{-9}

Calculator Tip In Examples 4–7, we showed how complex numbers can be added, subtracted, multiplied, and divided algebraically. Many current models of graphing calculators can perform these operations. Figure 11 shows how the computations in parts of Examples 4–7 are displayed on a TI-83/84 Plus calculator. Be sure to use parentheses as shown.

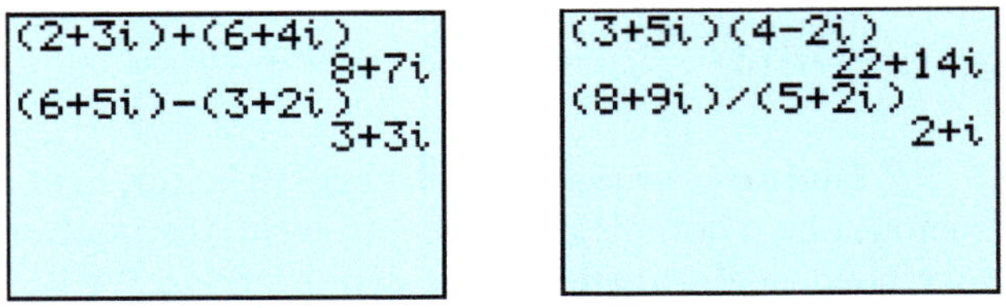

Figure 11

OBJECTIVE 6 Find powers of *i*. Because i^2 is defined to be -1, we can find higher powers of i as shown in the following examples.

$$i^3 = i \cdot i^2 = i(-1) = -i \qquad i^6 = i^2 \cdot i^4 = (-1) \cdot 1 = -1$$

$$i^4 = i^2 \cdot i^2 = (-1)(-1) = 1 \qquad i^7 = i^3 \cdot i^4 = (-i) \cdot 1 = -i$$

$$i^5 = i \cdot i^4 = i \cdot 1 = i \qquad i^8 = i^4 \cdot i^4 = 1 \cdot 1 = 1$$

As these examples suggest, the powers of i rotate through the four numbers i, -1, $-i$, and 1. Larger powers of i can be simplified by using the fact that $i^4 = 1$. For example,

$$i^{75} = (i^4)^{18} \cdot i^3 = 1^{18} \cdot i^3 = 1 \cdot i^3 = i^3 = -i.$$

Example 8 Simplifying Powers of *i*

Find each power of i.

(a) $i^{12} = (i^4)^3 = 1^3 = 1$

(b) $i^{39} = i^{36} \cdot i^3 = (i^4)^9 \cdot i^3 = 1^9 \cdot (-i) = -i$

(c) $i^{-2} = \dfrac{1}{i^2} = \dfrac{1}{-1} = -1$

(d) $i^{-1} = \dfrac{1}{i} = \dfrac{1(-i)}{i(-i)} = \dfrac{-i}{-i^2} = \dfrac{-i}{-(-1)} = \dfrac{-i}{1} = -i$

◀ Work Problem **8** at the Side.

Answers

8. **(a)** i **(b)** 1 **(c)** -1 **(d)** $-i$

17.8 Exercises

FOR EXTRA HELP MyMathLab Math XL PRACTICE WATCH DOWNLOAD READ REVIEW

Decide whether each expression is equal to 1, −1, i, or −i.

1. $\sqrt{-1}$

2. $-i^2$

3. $\frac{1}{i}$

4. $(-i)^2$

5. Every real number is a complex number. Explain why this is so.

6. Not every complex number is a real number. Give an example of this, and explain why this statement is true.

Write each number as a product of a real number and i. Simplify all radical expressions. See Example 1.

7. $\sqrt{-169}$

8. $\sqrt{-225}$

9. $-\sqrt{-144}$

10. $-\sqrt{-196}$

11. $\sqrt{-5}$

12. $\sqrt{-21}$

13. $\sqrt{-48}$

14. $\sqrt{-96}$

Multiply or divide as indicated. See Examples 2 and 3.

15. $\sqrt{-15} \cdot \sqrt{-15}$

16. $\sqrt{-19} \cdot \sqrt{-19}$

17. $\sqrt{-3} \cdot \sqrt{-19}$

18. $\sqrt{-7} \cdot \sqrt{-15}$

19. $\sqrt{-4} \cdot \sqrt{-25}$

20. $\sqrt{-9} \cdot \sqrt{-81}$

21. $\sqrt{-3} \cdot \sqrt{11}$

22. $\sqrt{-5} \cdot \sqrt{13}$

23. $\frac{\sqrt{-300}}{\sqrt{-100}}$

24. $\frac{\sqrt{-40}}{\sqrt{-10}}$

25. $\frac{\sqrt{-75}}{\sqrt{3}}$

26. $\frac{\sqrt{-160}}{\sqrt{10}}$

Add or subtract as indicated. Write your answers in standard form. See Examples 4 and 5.

27. $(3 + 2i) + (-4 + 5i)$

28. $(7 + 15i) + (-11 + 14i)$

29. $(5 - i) + (-5 + i)$

30. $(-2 + 6i) + (2 - 6i)$

31. $(4 + i) - (-3 - 2i)$

32. $(9 + i) - (3 + 2i)$

33. $(-3 - 4i) - (-1 - 4i)$

34. $(-2 - 3i) - (-5 - 3i)$

35. $(-4 + 11i) + (-2 - 4i) + (7 + 6i)$

36. $(-1 + i) + (2 + 5i) + (3 + 2i)$

37. $[(7 + 3i) - (4 - 2i)] + (3 + i)$

38. $[(7 + 2i) + (-4 - i)] - (2 + 5i)$

39. Fill in the blank with the correct response: Because $(4 + 2i) - (3 + i) = 1 + i$, using the definition of subtraction we can check this to find that

$$(1 + i) + (3 + i) = \_\_\_\_\_\_.$$

40. Fill in the blank with the correct response: Because $\frac{-5}{2 - i} = -2 - i$, using the definition of division we can check this to find that

$$(-2 - i)(2 - i) = \_\_\_\_\_\_.$$

Multiply. See Example 6.

41. $(3i)(27i)$

42. $(5i)(125i)$

43. $(-8i)(-2i)$

44. $(-32i)(-2i)$

45. $5i(-6 + 2i)$

46. $3i(4 + 9i)$

47. $(4 + 3i)(1 - 2i)$

48. $(7 - 2i)(3 + i)$

49. $(4 + 5i)^2$

50. $(3 + 2i)^2$

51. $(12 + 3i)(12 - 3i)$

52. $(6 + 7i)(6 - 7i)$

53. **(a)** What is the conjugate of $a + bi$?

(b) If we multiply $a + bi$ by its conjugate, we get, ______ + ______, which is always a real number.

54. Explain the procedure you would use to find the quotient

$$\frac{-1 + 5i}{3 + 2i}.$$

Write each quotient in the form $a + bi$*. See Example 7.*

55. $\frac{2}{1 - i}$

56. $\frac{29}{5 + 2i}$

57. $\frac{-7 + 4i}{3 + 2i}$

58. $\frac{-38 - 8i}{7 + 3i}$

59. $\frac{8i}{2 + 2i}$

60. $\frac{-8i}{1 + i}$

61. $\frac{2 - 3i}{2 + 3i}$

62. $\frac{-1 + 5i}{3 + 2i}$

Relating Concepts (Exercises 63–68) For Individual or Group Work

Consider these expressions:

Binomials	**Complex Numbers**
$x + 2, \quad 3x - 1$	$1 + 2i, \quad 3 - i.$

When we add, subtract, or multiply complex numbers in standard form, the rules are the same as those for the corresponding operations on binomials. That is, we add or subtract like terms, and we use FOIL to multiply. Division, however, is comparable to division by the sum or difference of radicals, where we multiply by the conjugate of the denominator to get a rational denominator. To express the quotient of two complex numbers in standard form, we also multiply by the conjugate of the denominator. ***Work Exercises 63–68 in order,*** *to better understand these ideas.*

63. **(a)** Add the two binomials.

(b) Add the two complex numbers.

64. **(a)** Subtract the second binomial from the first.

(b) Subtract the second complex number from the first.

65. **(a)** Multiply the two binomials.

(b) Multiply the two complex numbers.

66. **(a)** Rationalize the denominator: $\dfrac{\sqrt{3} - 1}{1 + \sqrt{2}}$.

(b) Write in standard form: $\dfrac{3 - i}{1 + 2i}$.

67. Explain why the answers for parts (a) and (b) in Exercise 65 do not correspond as the answers in Exercises 63 and 64 do.

68. Explain why the answers for parts (a) and (b) in Exercise 66 do not correspond as the answers in Exercises 63 and 64 do.

69. Recall that if $a \neq 0$, then $\frac{1}{a}$ is called the reciprocal of a. Use this definition to express the reciprocal of $5 - 4i$ in the form $a + bi$.

70. Recall that if $a \neq 0$, then a^{-1} is defined to be $\frac{1}{a}$. Use this definition to express $(4 - 3i)^{-1}$ in the form $a + bi$.

Find each power of i. See Example 8.

71. i^{18}

72. i^{26}

73. i^{89}

74. i^{45}

75. i^{96}

76. i^{48}

77. i^{-5}

78. i^{-17}

79. A student simplified i^{-18} as follows:

$$i^{-18} = i^{-18} \cdot i^{20} = i^{-18+20} = i^2 = -1.$$

Explain the mathematical justification for this correct work.

80. Explain why

$$(46 + 25i)(3 - 6i) \quad \text{and} \quad (46 + 25i)(3 - 6i)i^{12}$$

must be equal. (Do not actually perform the computation.)

Ohm's law *for the current I in a circuit with voltage E, resistance R, capacitance reactance* X_c*, and inductive reactance* X_L *is*

$$I = \frac{E}{R + (X_L - X_c)i}.$$

Use this law to work Exercises 81 and 82.

81. Find I if $E = 2 + 3i$, $R = 5$, $X_L = 4$, and $X_c = 3$.

82. Find E if $I = 1 - i$, $R = 2$, $X_L = 3$, and $X_c = 1$.

83. Show that $1 + 5i$ is a solution of

$$x^2 - 2x + 26 = 0.$$

84. Show that $3 + 2i$ is a solution of

$$x^2 - 6x + 13 = 0.$$

18

Quadratics

18.1 Solving Quadratic Equations by the Square Root Property

OBJECTIVES

1. Solve equations of the form $x^2 = k$, where $k > 0$.
2. Solve equations of the form $(ax + b)^2 = k$, where $k > 0$.
3. Use formulas involving squared variables.

A **quadratic equation** is an equation that can be written in the form

$$ax^2 + bx + c = 0 \quad \text{Standard form}$$

for real numbers a, b, and c, with $a \neq 0$. As we saw in **Section 14.6**, we can solve a quadratic equation such as $x^2 + 4x + 3 = 0$ by first factoring and then applying the zero-factor property.

$$x^2 + 4x + 3 = 0$$

$$(x + 3)(x + 1) = 0 \quad \text{Factor.}$$

$$x + 3 = 0 \quad \text{or} \quad x + 1 = 0 \quad \text{Zero-factor property}$$

$$x = -3 \quad \text{or} \quad x = -1 \quad \text{Solve each equation.}$$

The solution set is $\{-3, -1\}$.

◀ Work Problem 1 at the Side.

1 Solve each equation by factoring.

(a) $x^2 - 2x - 15 = 0$

(b) $2x^2 - 3x + 1 = 0$

OBJECTIVE 1 Solve equations of the form $x^2 = k$, where $k > 0$. We can solve an equation such as $x^2 = 9$ by factoring as follows.

$$x^2 = 9$$

$$x^2 - 9 = 0 \quad \text{Subtract 9.}$$

$$(x + 3)(x - 3) = 0 \quad \text{Factor.}$$

$$x + 3 = 0 \quad \text{or} \quad x - 3 = 0 \quad \text{Zero-factor property}$$

$$x = -3 \quad \text{or} \quad x = 3 \quad \text{Solve each equation.}$$

We might also solve $x^2 = 9$ by noticing that x must be a number whose square is 9. Thus, $x = \sqrt{9} = 3$ or $x = -\sqrt{9} = -3$. This approach is generalized as the **square root property.**

Square Root Property

If k is a positive number and if $x^2 = k$, then

$$x = \sqrt{k} \quad \text{or} \quad x = -\sqrt{k}$$

EXAMPLE 1 Solving Quadratic Equations of the Form $x^2 = k$

Solve each equation. Write radicals in simplified form.

(a) $x^2 = 16$

By the square root property, if $x^2 = 16$, then

$$x = \sqrt{16} = 4 \quad \text{or} \quad x = -\sqrt{16} = -4.$$

An abbreviation for $x = 4$ or $x = -4$ is $x = \pm 4$ (read "positive or negative 4"). Check each solution by substituting it for x in the original equation. The solution set is $\{-4, 4\}$, or $\{\pm 4\}$.

(b) $z^2 = 5$

The solutions are $z = \sqrt{5}$ or $z = -\sqrt{5}$, so the solution set is $\{-\sqrt{5}, \sqrt{5}\}$, or $\{\pm\sqrt{5}\}$.

Continued on Next Page

ANSWERS

1. (a) $\{-3, 5\}$
   (b) $\left\{\frac{1}{2}, 1\right\}$

(c) $5m^2 - 32 = 8$

$5m^2 = 40$ Add 32.

$m^2 = 8$ Divide by 5.

Don't stop here. Simplify the radicals.

$m = \sqrt{8}$ or $m = -\sqrt{8}$ Square root property

$m = 2\sqrt{2}$ or $m = -2\sqrt{2}$ $\sqrt{8} = \sqrt{4} \cdot \sqrt{2} = 2\sqrt{2}$

The solution set is $\{-2\sqrt{2}, 2\sqrt{2}\}$, or $\{\pm 2\sqrt{2}\}$.

(d) $x^2 = -4$

Because -4 is a negative number and because the square of a real number cannot be negative, ***there is no real number solution*** of this equation. (In this book, we are concerned with finding only *real number* solutions. To use the square root property to find both real number solutions, k must be positive.) The solution set is $\emptyset$.

Work Problem 2 at the Side. ▶

2 Solve each equation. Write radicals in simplified form.

(a) $x^2 = 49$

(b) $x^2 = 11$

(c) $2x^2 + 8 = 32$

(d) $x^2 = -9$

OBJECTIVE 2 Solve equations of the form $(ax + b)^2 = k$, where $k > 0$. In each equation in Example 1, the exponent 2 had a single variable as its base. We can extend the square root property to solve equations in which the base is a binomial, as shown in the next example.

EXAMPLE 2 Solving Quadratic Equations of the Form $(x + b)^2 = k$

Solve each equation.

(a) Use $x - 3$ as the base. $(x - 3)^2 = 16$

$x - 3 = \sqrt{16}$ or $x - 3 = -\sqrt{16}$ Square root property

$x - 3 = 4$ or $x - 3 = -4$ $\sqrt{16} = 4$

$x = 7$ or $x = -1$ Add 3.

Check Substitute each solution in the original equation.

$(x - 3)^2 = 16$		$(x - 3)^2 = 16$	
$(7 - 3)^2 \stackrel{?}{=} 16$	Let $x = 7$.	$(-1 - 3)^2 \stackrel{?}{=} 16$	Let $x = -1$.
$4^2 \stackrel{?}{=} 16$		$(-4)^2 \stackrel{?}{=} 16$	
$16 = 16$	True	$16 = 16$	True

The solutions are 7 and -1, and the solution set is $\{-1, 7\}$.

(b) $(x + 1)^2 = 6$

$x + 1 = \sqrt{6}$ or $x + 1 = -\sqrt{6}$ Square root property

$x = -1 + \sqrt{6}$ or $x = -1 - \sqrt{6}$ Add -1.

Check $\left(-1 + \sqrt{6} + 1\right)^2 = \left(\sqrt{6}\right)^2 = 6;$

$\left(-1 - \sqrt{6} + 1\right)^2 = \left(-\sqrt{6}\right)^2 = 6.$

The solution set is $\{-1 + \sqrt{6}, -1 - \sqrt{6}\}$, or $\{-1 \pm \sqrt{6}\}$.

Work Problem 3 at the Side. ▶

3 Solve each equation.

(a) $(x + 2)^2 = 36$

(b) $(x - 4)^2 = 3$

ANSWERS

2. (a) $\{-7, 7\}$ (b) $\{-\sqrt{11}, \sqrt{11}\}$ (c) $\{-2\sqrt{3}, 2\sqrt{3}\}$ (d) $\emptyset$

3. (a) $\{-8, 4\}$ (b) $\{4 + \sqrt{3}, 4 - \sqrt{3}\}$

4 Solve $(2x - 5)^2 = 18$.

5 Solve each equation.

(a) $(5x + 1)^2 = 7$

(b) $(7x - 1)^2 = -1$

6 Use the formula in Example 5 to approximate the length of a bass weighing 2.80 lb and having a girth of 11 in.

Answers

4. $\left\{\frac{5 + 3\sqrt{2}}{2}, \frac{5 - 3\sqrt{2}}{2}\right\}$

5. (a) $\left\{\frac{-1 + \sqrt{7}}{5}, \frac{-1 - \sqrt{7}}{5}\right\}$

(b) $\emptyset$

6. approximately 17.48 in.

EXAMPLE 3 Solving a Quadratic Equation of the Form $(ax + b)^2 = k$

Solve $(3r - 2)^2 = 27$.

$3r - 2 = \sqrt{27}$ or $3r - 2 = -\sqrt{27}$ — Square root property

$3r - 2 = 3\sqrt{3}$ or $3r - 2 = -3\sqrt{3}$ — $\sqrt{27} = \sqrt{9} \cdot \sqrt{3} = 3\sqrt{3}$

$3r = 2 + 3\sqrt{3}$ or $3r = 2 - 3\sqrt{3}$ — Add 2.

$r = \frac{2 + 3\sqrt{3}}{3}$ or $r = \frac{2 - 3\sqrt{3}}{3}$ — Divide by 3.

The solution set is $\left\{\frac{2 + 3\sqrt{3}}{3}, \frac{2 - 3\sqrt{3}}{3}\right\}$.

◀ Work Problem **4** at the Side.

CAUTION
The solutions in Example 3 are fractions that cannot be simplified. Note that 3 is *not* a common factor in the numerator.

EXAMPLE 4 Recognizing When There Is No Real Solution

Solve $(x + 3)^2 = -9$.

Because the square root of -9 is not a real number, there is no real number solution for this equation. The solution set is $\emptyset$.

◀ Work Problem **5** at the Side.

OBJECTIVE 3 Use formulas involving squared variables.

EXAMPLE 5 Finding the Length of a Bass

We can approximate the weight of a bass, in pounds, given its length L and its girth (distance around) g, where both are measured in inches, using this formula.

$$w = \frac{L^2 g}{1200}$$

Approximate the length of a bass weighing 2.20 lb and having a girth of 10 in. (*Source: Sacramento Bee.*)

$w = \frac{L^2 g}{1200}$ — Given formula

$2.20 = \frac{L^2 \cdot 10}{1200}$ — $w = 2.20$, $g = 10$

$2640 = 10L^2$ — Multiply by 1200.

$L^2 = 264$ — Divide by 10; interchange the sides.

$L = \sqrt{264}$ or $L = -\sqrt{264}$ — Square root property

A calculator shows that $\sqrt{264} \approx 16.25$, so the length of the bass is a little more than 16 in. (We discard the negative solution $-\sqrt{264} \approx -16.25$, since L represents length.)

◀ Work Problem **6** at the Side.

18.1 Exercises

FOR EXTRA HELP MyMathLab Math XL PRACTICE WATCH DOWNLOAD READ REVIEW

Decide whether each statement is true *or* false. *If* false, *tell why.*

1. If k is a prime number, then $x^2 = k$ has two irrational solutions.

2. If k is a positive perfect square, then $x^2 = k$ has two rational solutions.

3. If k is a positive integer, then $x^2 = k$ must have two rational solutions.

4. If $-10 < k < 0$, then $x^2 = k$ has no real solution.

5. If $-10 < k < 10$, then $x^2 = k$ has no real solution.

6. If k is an integer greater than 24 and less than 26, then $x^2 = k$ has two solutions, -5 and 5.

Solve each equation by using the square root property. Write all radicals in simplest form. See Example 1.

7. $x^2 = 81$

8. $x^2 = 121$

9. $k^2 = 14$

10. $m^2 = 22$

11. $t^2 = 48$

12. $x^2 = 54$

13. $x^2 = \frac{25}{4}$

14. $m^2 = \frac{36}{121}$

15. $x^2 = -100$

16. $x^2 = -64$

17. $z^2 = 2.25$

18. $w^2 = 56.25$

19. $r^2 - 3 = 0$

20. $x^2 - 13 = 0$

21. $7x^2 = 4$

22. $3x^2 = 10$

23. $4x^2 - 72 = 0$

24. $5z^2 - 200 = 0$

25. $3x^2 - 8 = 64$

26. $2x^2 + 7 = 61$

27. $5x^2 + 4 = 8$

28. $4x^2 - 3 = 7$

Solve each equation by using the square root property. Express all radicals in simplest form. See Examples 2–4.

29. $(x - 3)^2 = 25$

30. $(x - 7)^2 = 16$

31. $(x + 5)^2 = -13$

32. $(x + 2)^2 = -17$

33. $(x - 8)^2 = 27$

34. $(x - 5)^2 = 40$

35. $(3x + 2)^2 = 49$

36. $(5x + 3)^2 = 36$

37. $(4x - 3)^2 = 9$

38. $(7x - 5)^2 = 25$

39. $(5 - 2x)^2 = 30$

40. $(3 - 2x)^2 = 70$

41. $(3x + 1)^2 = 18$

42. $(5x + 6)^2 = 75$

43. $\left(\frac{1}{2}x + 5\right)^2 = 12$

44. $\left(\frac{1}{3}x + 4\right)^2 = 27$

45. $(4x - 1)^2 - 48 = 0$

46. $(2x - 5)^2 - 180 = 0$

47. Johnny solved the equation in Exercise 39 and wrote his answer as $\left\{\frac{5 + \sqrt{30}}{2}, \frac{5 - \sqrt{30}}{2}\right\}$. Terry solved the same equation and wrote her answer as $\left\{\frac{-5 + \sqrt{30}}{-2}, \frac{-5 - \sqrt{30}}{-2}\right\}$. The teacher gave them both full credit. Explain why both students were correct, although their answers seem to differ.

48. In the solutions found in Example 3 of this section, why is it not valid to simplify by dividing out the 3s in the numerators and denominators?

Solve each problem. See Example 5.

49. One expert at marksmanship can hold a silver dollar at forehead level, drop it, draw his gun, and shoot the coin as it passes waist level. The distance traveled by a falling object is given by

$$d = 16t^2,$$

where d is the distance (in feet) the object falls in t seconds. If the coin falls about 4 ft, use the formula to estimate the time that elapses between the dropping of the coin and the shot.

50. The illumination produced by a light source depends on the distance from the source. For a particular light source, this relationship can be expressed as

$$d^2 = \frac{4050}{I},$$

where d is the distance from the source (in feet) and I is the amount of illumination in foot-candles. How far from the source is the illumination equal to 50 foot-candles?

51. The area A of a circle with radius r is given by the formula

$$A = \pi r^2.$$

If a circle has area 81π in.^{2}, what is its radius?

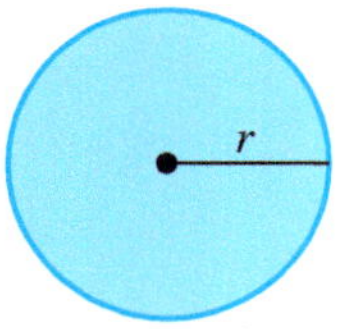

$A = \pi r^2$

52. The surface area S of a sphere with radius r is given by the formula

$$S = 4\pi r^2.$$

If a sphere has surface area 36π ft^2, what is its radius?

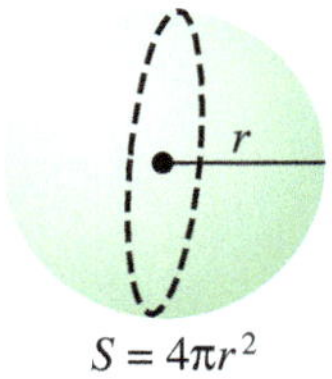

$S = 4\pi r^2$

The amount A that P dollars invested at an annual rate of interest r will grow to in 2 yr is $A = P(1 + r)^2$.

53. At what interest rate will \$100 grow to \$110.25 in 2 yr?

54. At what interest rate will \$500 grow to \$572.45 in 2 yr?

55. At what interest rate will \$200 grow to \$208.08 in 2 yr?

18.2 Solving Quadratic Equations by Completing the Square

OBJECTIVE 1 Solve quadratic equations by completing the square when the coefficient of the second-degree term is 1. The methods we have studied so far are not enough to solve this equation.

$$x^2 + 6x + 7 = 0$$

If we could write the equation in the form $(x + 3)^2$ equals a constant, we could solve it with the square root property discussed in **Section 18.1.** To do that, we need to have a perfect square trinomial on one side of the equation.

Recall from **Section 14.5** that a perfect square trinomial has the form

$$x^2 + 2kx + k^2 \quad \text{or} \quad x^2 - 2kx + k^2,$$

where k represents a positive number.

EXAMPLE 1 Creating Perfect Square Trinomials

Complete each trinomial so that it is a perfect square. Then factor the trinomial.

(a) $x^2 + 8x +$ ______

The perfect square trinomial will have the form $x^2 + 2kx + k^2$. Thus, the middle term, $8x$, must equal $2kx$.

$8x = 2kx$ ⟵ Solve this equation for k.

$4 = k$ Divide each side by $2x$.

Therefore, $k = 4$ and $k^2 = 4^2 = 16$. The required perfect square trinomial is

$$x^2 + 8x + 16, \quad \text{which factors as} \quad (x + 4)^2.$$

(b) $x^2 - 18x +$ ______

Here the perfect square trinomial will have the form $x^2 - 2kx + k^2$. The middle term, $-18x$, must equal $-2kx$.

$-18x = -2kx$ ⟵ Solve this equation for k.

$9 = k$ Divide each side by $-2x$.

Thus, $k = 9$ and $k^2 = 9^2 = 81$. The required perfect square trinomial is

$$x^2 - 18x + 81, \quad \text{which factors as} \quad (x - 9)^2.$$

Work Problem 1 at the Side. ▶

EXAMPLE 2 Rewriting an Equation to Use the Square Root Property

Solve $x^2 + 6x + 7 = 0$.

$x^2 + 6x = -7$ Subtract 7 from each side.

To solve this equation with the square root property, the quantity on the left side, $x^2 + 6x$, must be written as a perfect square trinomial in the form $x^2 + 2kx + k^2$.

$$x^2 + 6x + \underline{\quad\quad}$$

Here, $2kx = 6x$, so $k = 3$ and $k^2 = 9$. The required perfect square trinomial is

$$x^2 + 6x + 9, \quad \text{which factors as} \quad (x + 3)^2.$$

Continued on Next Page

OBJECTIVES

1. Solve quadratic equations by completing the square when the coefficient of the second-degree term is 1.
2. Solve quadratic equations by completing the square when the coefficient of the second-degree term is not 1.
3. Simplify the terms of an equation before solving.
4. Solve applied problems that require quadratic equations.

1 Complete each trinomial so that it is a perfect square. Then factor the trinomial.

(a) $x^2 + 12x +$ ______

(b) $x^2 - 14x +$ ______

(c) $x^2 - 2x +$ ______

Answers

1. (a) 36; $(x + 6)^2$
(b) 49; $(x - 7)^2$
(c) 1; $(x - 1)^2$

2 Solve $x^2 - 4x - 1 = 0$.

Therefore, if we add 9 to *each* side of $x^2 + 6x = -7$, the equation will have a perfect square trinomial on the left side, as needed.

This is a key step.

$$x^2 + 6x + 9 = -7 + 9 \quad \text{Add 9.}$$

$$(x + 3)^2 = 2 \quad \text{Factor; add.}$$

Now use the square root property to complete the solution.

$$x + 3 = \sqrt{2} \quad \text{or} \quad x + 3 = -\sqrt{2}$$

$$x = -3 + \sqrt{2} \quad \text{or} \quad x = -3 - \sqrt{2}$$

Check by substituting $-3 + \sqrt{2}$ and $-3 - \sqrt{2}$ for x in the original equation. The solution set is $\{-3 + \sqrt{2}, -3 - \sqrt{2}\}$.

◀ Work Problem **2** at the Side.

The process of changing the form of the equation in Example 2 from

$$x^2 + 6x + 7 = 0 \quad \text{to} \quad (x + 3)^2 = 2$$

is called **completing the square.** Completing the square changes only the form of the equation. To see this, multiply out the left side of $(x + 3)^2 = 2$. Then write the equation in standard form to get $x^2 + 6x + 7 = 0$.

Look again at the original equation in Example 2,

$$x^2 + 6x + 7 = 0.$$

If we take half of 6, the coefficient of x, and square it, we get 9.

$$\frac{1}{2} \cdot 6 = 3 \quad \text{and} \quad 3^2 = 9$$

Coefficient of x — Quantity added to each side

To complete the square in Example 2, we added 9 to each side.

3 Solve each equation by completing the square.

(a) $x^2 + 4x = 1$

(b) $z^2 + 6z - 3 = 0$

EXAMPLE 3 Completing the Square to Solve a Quadratic Equation

Complete the square to solve $x^2 - 8x = 5$.

To complete the square on $x^2 - 8x$, take half of -8, the coefficient of x, and square it.

$$\frac{1}{2}(-8) = -4 \quad \text{and} \quad (-4)^2 = 16$$

Coefficient of x

Add the result, 16, to each side of the equation.

$$x^2 - 8x = 5 \quad \text{Given equation}$$

$$x^2 - 8x + 16 = 5 + 16 \quad \text{Add 16.}$$

$$(x - 4)^2 = 21 \quad \text{Factor; add.}$$

$$x - 4 = \sqrt{21} \quad \text{or} \quad x - 4 = -\sqrt{21} \quad \text{Square root property}$$

$$x = 4 + \sqrt{21} \quad \text{or} \quad x = 4 - \sqrt{21} \quad \text{Add 4.}$$

A check indicates that the solution set is $\{4 + \sqrt{21}, 4 - \sqrt{21}\}$.

◀ Work Problem **3** at the Side.

ANSWERS

2. $\{2 + \sqrt{5}, 2 - \sqrt{5}\}$

3. **(a)** $\{-2 + \sqrt{5}, -2 - \sqrt{5}\}$

(b) $\{-3 + 2\sqrt{3}, -3 - 2\sqrt{3}\}$

OBJECTIVE 2 Solve quadratic equations by completing the square when the coefficient of the second-degree term is not 1. If a quadratic equation has the form

$$ax^2 + bx + c = 0, \quad \text{where } a \neq 1,$$

then to obtain 1 as the coefficient of x^2, we first divide each side of the equation by a.

EXAMPLE 4 Solving a Quadratic Equation by Completing the Square

Solve $4x^2 + 16x = 9$.

Before completing the square, the coefficient of x^2 must be 1, not 4. We get 1 as the coefficient of x^2 here by dividing each side by 4.

$$4x^2 + 16x = 9$$

$$x^2 + 4x = \frac{9}{4} \qquad \text{Divide by 4.}$$

The coefficient of x^2 must be 1.

Next, we begin to complete the square by taking half the coefficient of x, and squaring it:

$$\frac{1}{2}(4) = 2 \quad \text{and} \quad 2^2 = 4.$$

We add the result, 4, to each side of the equation.

$$x^2 + 4x + 4 = \frac{9}{4} + 4 \qquad \text{Add 4.}$$

$$(x+2)^2 = \frac{25}{4} \qquad \text{Factor; } \tfrac{9}{4} + 4 = \tfrac{9}{4} + \tfrac{16}{4} = \tfrac{25}{4}.$$

$$x + 2 = \sqrt{\frac{25}{4}} \quad \text{or} \quad x + 2 = -\sqrt{\frac{25}{4}} \qquad \text{Square root property}$$

$$x + 2 = \frac{5}{2} \quad \text{or} \quad x + 2 = -\frac{5}{2} \qquad \text{Take square roots.}$$

$$x = -2 + \frac{5}{2} \quad \text{or} \quad x = -2 - \frac{5}{2} \qquad \text{Subtract 2.}$$

$$x = \frac{1}{2} \quad \text{or} \quad x = -\frac{9}{2} \qquad -2 = -\tfrac{4}{2}$$

Check

$4x^2 + 16x = 9$	$4x^2 + 16x = 9$
$4\left(\frac{1}{2}\right)^2 + 16\left(\frac{1}{2}\right) \stackrel{?}{=} 9$	$4\left(-\frac{9}{2}\right)^2 + 16\left(-\frac{9}{2}\right) \stackrel{?}{=} 9$
$4\left(\frac{1}{4}\right) + 8 \stackrel{?}{=} 9$	$4\left(\frac{81}{4}\right) - 72 \stackrel{?}{=} 9$
$1 + 8 \stackrel{?}{=} 9$	$81 - 72 \stackrel{?}{=} 9$
$9 = 9$ True	$9 = 9$ True

The two solutions, $\frac{1}{2}$ and $-\frac{9}{2}$, check, so the solution set is $\left\{-\frac{9}{2}, \frac{1}{2}\right\}$.

Work Problem 4 at the Side. ▶

4 Solve each equation by completing the square.

(a) $9x^2 + 18x = -5$

(b) $4t^2 - 24t + 11 = 0$

ANSWERS

4. (a) $\left\{-\frac{1}{3}, -\frac{5}{3}\right\}$ (b) $\left\{\frac{11}{2}, \frac{1}{2}\right\}$

5 Solve each equation by completing the square.

(a) $3x^2 + 5x - 2 = 0$

(b) $2x^2 - 4x - 1 = 0$

The steps used to solve a quadratic equation $ax^2 + bx + c = 0$ by completing the square are summarized here.

Solving a Quadratic Equation by Completing the Square

Step 1 **Be sure the second-degree term has coefficient 1.** If the coefficient of the second-degree term is 1, proceed to Step 2. If it is not 1, but some other nonzero number a, divide each side of the equation by a.

Step 2 **Write in correct form.** Make sure that all terms with variables are on one side of the equals sign and that all constant terms are on the other side.

Step 3 **Complete the square.** Take half the coefficient of the first-degree term, and square it. Add the square to each side of the equation. Factor the variable side, and simplify on the other side.

Step 4 **Solve** the equation by using the square root property.

EXAMPLE 5 Solving a Quadratic Equation by Completing the Square

Solve $2x^2 - 7x - 9 = 0$.

Step 1 Get 1 as the coefficient of the x^2-term.

$$x^2 - \frac{7}{2}x - \frac{9}{2} = 0 \qquad \text{Divide by 2.}$$

Step 2 Add $\frac{9}{2}$ to each side to get the variable terms on the left and the constant on the right.

$$x^2 - \frac{7}{2}x = \frac{9}{2} \qquad \text{Add } \tfrac{9}{2}.$$

Step 3 To complete the square, take half the coefficient of x and square it: $\left[\frac{1}{2}\left(-\frac{7}{2}\right)\right]^2 = \left(-\frac{7}{4}\right)^2 = \frac{49}{16}$.

$$x^2 - \frac{7}{2}x + \frac{49}{16} = \frac{9}{2} + \frac{49}{16}$$

Be sure to add $\frac{49}{16}$ to *each* side.

$$\left(x - \frac{7}{4}\right)^2 = \frac{121}{16} \qquad \text{Factor; } \tfrac{9}{2} + \tfrac{49}{16} = \tfrac{72}{16} + \tfrac{49}{16} = \tfrac{121}{16}.$$

Step 4 Solve by using the square root property.

$$x - \frac{7}{4} = \sqrt{\frac{121}{16}} \quad \text{or} \quad x - \frac{7}{4} = -\sqrt{\frac{121}{16}} \qquad \text{Square root property}$$

$$x = \frac{7}{4} + \frac{11}{4} \quad \text{or} \quad x = \frac{7}{4} - \frac{11}{4} \qquad \text{Add } \tfrac{7}{4};\ \sqrt{\tfrac{121}{16}} = \tfrac{11}{4}.$$

$$x = \frac{18}{4} = \frac{9}{2} \quad \text{or} \quad x = -\frac{4}{4} = -1 \qquad \text{Simplify.}$$

A check confirms that the solution set is $\left\{-1, \frac{9}{2}\right\}$.

◀ *Work Problem* **5** *at the Side.*

Answers

5. (a) $\left\{-2, \frac{1}{3}\right\}$ **(b)** $\left\{\frac{2+\sqrt{6}}{2}, \frac{2-\sqrt{6}}{2}\right\}$

EXAMPLE 6 **Solving a Quadratic Equation by Completing the Square**

Solve $4p^2 + 8p + 5 = 0$.

$$4p^2 + 8p + 5 = 0$$

The coefficient of the second-degree term must be 1.

$$p^2 + 2p + \frac{5}{4} = 0 \qquad \text{Divide by 4.}$$

$$p^2 + 2p = -\frac{5}{4} \qquad \text{Subtract } \tfrac{5}{4}.$$

The coefficient of p is 2. Take half of 2; square the result: $\left[\frac{1}{2}(2)\right]^2 = 1^2 = 1$. Add this result to each side. Then write the left side as a perfect square.

$$p^2 + 2p + 1 = -\frac{5}{4} + 1 \qquad \text{Add 1.}$$

$$(p + 1)^2 = -\frac{1}{4} \qquad \text{Factor; add.}$$

We cannot use the square root property to solve this equation, because the square root of $-\frac{1}{4}$ is not a real number. This equation has no real number solution.* The solution set is $\emptyset$.

Work Problem **6** *at the Side.* ▶

6 Solve $5x^2 + 3x + 1 = 0$ by completing the square.

7 Solve each equation.

(a) $r(r - 3) = -1$

(b) $(x + 2)(x + 1) = 5$

OBJECTIVE 3 **Simplify the terms of an equation before solving.**

EXAMPLE 7 **Simplifying before Completing the Square**

Solve $(x + 3)(x - 1) = 2$.

$$(x + 3)(x - 1) = 2 \qquad \text{Given equation}$$

$$x^2 + 2x - 3 = 2 \qquad \text{Multiply using the FOIL method.}$$

$$x^2 + 2x = 5 \qquad \text{Add 3.}$$

$$x^2 + 2x + 1 = 5 + 1 \qquad \text{Complete the square—add } \left[\tfrac{1}{2}(2)\right]^2 = 1.$$

$$(x + 1)^2 = 6 \qquad \text{Factor on the left; add on the right.}$$

$$x + 1 = \sqrt{6} \quad \text{or} \quad x + 1 = -\sqrt{6} \qquad \text{Square root property}$$

$$x = -1 + \sqrt{6} \quad \text{or} \quad x = -1 - \sqrt{6} \qquad \text{Add } -1.$$

The solution set is $\{-1 + \sqrt{6}, -1 - \sqrt{6}\}$.

Work Problem **7** *at the Side.* ▶

Note

The solutions $-1 \pm \sqrt{6}$ given in Example 7 are *exact*. In applications, decimal solutions are more appropriate. Using the square root key of a calculator, $\sqrt{6} \approx 2.449$. Approximating the two solutions gives

$$x \approx 1.449 \quad \text{and} \quad x \approx -3.449.$$

*The equation in Example 6 has no solution over the *real number system*. In the **complex number system,** however, this equation does have solutions. The complex numbers include numbers whose squares are negative. These numbers are discussed in intermediate and college algebra courses.

ANSWERS

6. $\emptyset$

7. (a) $\left\{\frac{3 + \sqrt{5}}{2}, \frac{3 - \sqrt{5}}{2}\right\}$

(b) $\left\{\frac{-3 + \sqrt{21}}{2}, \frac{-3 - \sqrt{21}}{2}\right\}$

OBJECTIVE 4 Solve applied problems that require quadratic equations. There are many practical applications of quadratic equations. The next example illustrates an application from physics.

8 Suppose a ball is projected upward from ground level with an initial velocity of 128 ft per sec. Its height at time t (in seconds) is given by

$$s = -16t^2 + 128t,$$

where s is in feet. At what times will it be 48 ft above the ground? Give answers to the nearest tenth.

EXAMPLE 8 Solving a Velocity Problem

If a ball is projected into the air from ground level with an initial velocity of 64 ft per sec, its altitude (height) s in feet in t seconds is given by the formula

$$s = -16t^2 + 64t.$$

How long will it take the ball to be 48 ft above the ground?

Since s represents the height, we substitute **48** for s in the formula and then solve this equation for time t by completing the square.

$48 = -16t^2 + 64t$	Let $s = 48$.
$-3 = t^2 - 4t$	Divide by -16.
$t^2 - 4t = -3$	Interchange the sides.
$t^2 - 4t + 4 = -3 + 4$	Add $\left[\frac{1}{2}(-4)\right]^2 = 4$.
$(t-2)^2 = 1$	Factor; add.
$t - 2 = 1$ or $t - 2 = -1$	Square root property
$t = 3$ or $t = 1$	Add 2.

The ball reaches a height of 48 ft twice, once on the way up and again on the way down. It takes 1 sec to reach 48 ft on the way up, and then after 3 sec, the ball reaches 48 ft again on the way down.

◀ Work Problem 8 at the Side.

ANSWER

8. 0.4 sec and 7.6 sec

18.2 Exercises

FOR EXTRA HELP

MyMathLab

MathXL PRACTICE WATCH DOWNLOAD READ REVIEW

Complete each trinomial so that it is a perfect square. Then factor the trinomial. See Example 1.

1. $x^2 + 10x +$ ______

2. $x^2 + 16x +$ ______

3. $x^2 + 2x +$ ______

4. $m^2 - 2m +$ ______

5. $p^2 - 5p +$ ______

6. $x^2 + 3x +$ ______

7. Which step is an appropriate way to begin solving the quadratic equation

$$2x^2 - 4x = 9$$

by completing the square?

A. Add 4 to each side of the equation.

B. Factor the left side as $2x(x - 2)$.

C. Factor the left side as $x(2x - 4)$.

D. Divide each side by 2.

8. In Example 3 of **Section 14.6,** we solved the quadratic equation

$$4p^2 - 26p + 40 = 0$$

by factoring. If we were to solve by completing the square, would we get the same solutions, $\frac{5}{2}$ and 4?

Solve each equation by completing the square. See Examples 2 and 3.

9. $x^2 - 4x = -3$

10. $x^2 - 2x = 8$

11. $x^2 + 5x + 6 = 0$

12. $x^2 + 6x + 5 = 0$

13. $x^2 + 2x - 5 = 0$

14. $x^2 + 4x + 1 = 0$

15. $x^2 - 8x = -4$

16. $m^2 - 4m = 14$

17. $t^2 + 6t + 9 = 0$

18. $k^2 - 8k + 16 = 0$

19. $x^2 + x - 1 = 0$

20. $x^2 + x - 3 = 0$

Solve each equation by completing the square. See Examples 4–7.

21. $4x^2 + 4x - 3 = 0$

22. $9x^2 + 3x - 2 = 0$

23. $2x^2 - 4x = 5$

24. $2x^2 - 6x = 3$

25. $2p^2 - 2p + 3 = 0$

26. $3q^2 - 3q + 4 = 0$

27. $3k^2 + 7k = 4$

28. $2k^2 + 5k = 1$

29. $(x + 3)(x - 1) = 5$

30. $(y - 8)(y + 2) = 24$

31. $(r - 3)(r - 5) = 2$

32. $(k - 1)(k - 7) = 1$

33. $-x^2 + 2x = -5$

34. $-r^2 + 3r = -2$

Solve each problem. See Example 8.

35. If an object is projected upward from ground level on Earth with an initial velocity of 96 ft per sec, its altitude (height) s in feet in t seconds is given by the formula $s = -16t^2 + 96t$. At what times will the object be 80 ft above the ground?

36. At what times will the object described in Exercise 35 be 100 ft above the ground? Round your answers to the nearest tenth.

37. If an object is projected upward on the surface of Mars from ground level with an initial velocity of 104 ft per sec, its altitude (height) s in feet in t seconds is given by the formula $s = -13t^2 + 104t$. At what times will the object be 195 ft above the surface?

38. After how many seconds will the object in Exercise 37 return to the surface? (*Hint:* When it returns to the surface, $s = 0$.)

39. A farmer has a rectangular cattle pen with perimeter 350 ft and area 7500 ft^2. What are the dimensions of the pen? (*Hint:* Use the figure to set up the equation.)

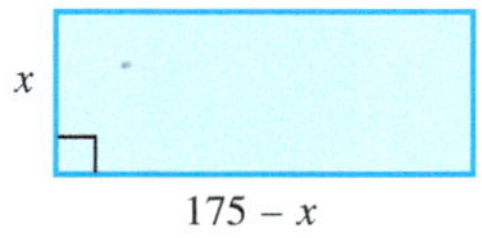

40. The base of a triangle measures 1 m more than three times the height of the triangle. Its area is 15 m^2. Find the lengths of the base and the height.

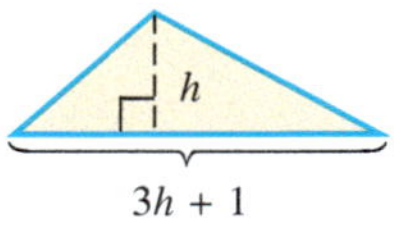

18.3 The Quadratic Formula

In this section, we complete the square to solve the general quadratic equation

$$ax^2 + bx + c = 0,$$

where a, b, and c are complex numbers and $a \neq 0$. The solution of this general equation gives a formula for finding the solution of *any* specific quadratic equation.

OBJECTIVES

1. **Derive the quadratic formula.**
2. **Solve quadratic equations by using the quadratic formula.**
3. **Use the discriminant to determine the number and type of solutions.**

OBJECTIVE 1 Derive the quadratic formula. To solve the general quadratic equation $ax^2 + bx + c = 0$ by completing the square (assuming $a > 0$), we follow the steps given in **Section 10.1.**

$$ax^2 + bx + c = 0$$

$$x^2 + \frac{b}{a}x + \frac{c}{a} = 0 \qquad \text{Divide by } a. \text{ (Step 1)}$$

$$x^2 + \frac{b}{a}x = -\frac{c}{a} \qquad \text{Subtract } \tfrac{c}{a}. \text{ (Step 2)}$$

$$\left[\frac{1}{2}\left(\frac{b}{a}\right)\right]^2 = \left(\frac{b}{2a}\right)^2 = \frac{b^2}{4a^2} \qquad \text{(Step 3)}$$

$$x^2 + \frac{b}{a}x + \frac{b^2}{4a^2} = -\frac{c}{a} + \frac{b^2}{4a^2} \qquad \text{Add } \frac{b^2}{4a^2} \text{ to each side. (Step 4)}$$

Write the left side as a perfect square, and rearrange the right side.

$$\left(x + \frac{b}{2a}\right)^2 = \frac{b^2}{4a^2} + \frac{-c}{a} \qquad \text{(Step 5)}$$

$$\left(x + \frac{b}{2a}\right)^2 = \frac{b^2}{4a^2} + \frac{-4ac}{4a^2} \qquad \text{Write with a common denominator.}$$

$$\left(x + \frac{b}{2a}\right)^2 = \frac{b^2 - 4ac}{4a^2} \qquad \text{Add fractions.}$$

$$x + \frac{b}{2a} = \sqrt{\frac{b^2 - 4ac}{4a^2}} \quad \text{or} \quad x + \frac{b}{2a} = -\sqrt{\frac{b^2 - 4ac}{4a^2}} \qquad \text{Square root property (Step 6)}$$

Since

$$\sqrt{\frac{b^2 - 4ac}{4a^2}} = \frac{\sqrt{b^2 - 4ac}}{\sqrt{4a^2}} = \frac{\sqrt{b^2 - 4ac}}{2a},$$

the right sides of these equations can be expressed as

$$x + \frac{b}{2a} = \frac{\sqrt{b^2 - 4ac}}{2a} \quad \text{or} \quad x + \frac{b}{2a} = \frac{-\sqrt{b^2 - 4ac}}{2a}$$

$$x = \frac{-b}{2a} + \frac{\sqrt{b^2 - 4ac}}{2a} \quad \text{or} \quad x = \frac{-b}{2a} - \frac{\sqrt{b^2 - 4ac}}{2a}$$

$$\boldsymbol{x = \frac{-b + \sqrt{b^2 - 4ac}}{2a}} \quad \text{or} \quad \boldsymbol{x = \frac{-b - \sqrt{b^2 - 4ac}}{2a}}.$$

If $a < 0$, the same two solutions are obtained. The result is the **quadratic formula,** which is abbreviated as shown on the next page.

Quadratic Formula

The solutions of $ax^2 + bx + c = 0$ $(a \neq 0)$ are given by

$$x = \frac{-b \pm \sqrt{b^2 - 4ac}}{2a}.$$

CAUTION

In the quadratic formula, $x = \dfrac{-b \pm \sqrt{b^2 - 4ac}}{2a}$, ***the square root is added to or subtracted from the value of $-b$ before dividing by $2a$.*** A common student error is to fail to divide ***all*** of the expression $-b \pm \sqrt{b^2 - 4ac}$ by $2a$.

OBJECTIVE 2 Solve quadratic equations by using the quadratic formula. To use the quadratic formula, first write the equation in standard form

$$ax^2 + bx + c = 0.$$

Then identify the values of a, b, and c and substitute them into the quadratic formula.

◀ Work Problem 1 at the Side.

1 Identify the values of a, b, and c. (*Hint:* If necessary, first write the equation in standard form with 0 on the right side.) *Do not actually solve.*

(a) $-3x^2 + 9x - 4 = 0$

(b) $3x^2 = 6x + 2$

Example 1 Using the Quadratic Formula (Rational Solutions)

Solve $6x^2 - 5x - 4 = 0$.

Here a, the coefficient of the second-degree term, is 6, while b, the coefficient of the first-degree term, is -5, and the constant c is -4. Substitute these values into the quadratic formula.

$$x = \frac{-b \pm \sqrt{b^2 - 4ac}}{2a} \qquad \text{Quadratic formula}$$

$$x = \frac{-(-5) \pm \sqrt{(-5)^2 - 4(6)(-4)}}{2(6)} \qquad a = 6,\ b = -5,\ c = -4$$

Use parentheses and substitute carefully to avoid errors.

$$x = \frac{5 \pm \sqrt{25 + 96}}{12}$$

$$x = \frac{5 \pm \sqrt{121}}{12}$$

$$x = \frac{5 \pm 11}{12}$$

This last statement leads to two solutions, one from the + sign and one from the − sign.

$$x = \frac{5 + 11}{12} = \frac{16}{12} = \frac{4}{3} \quad \text{or} \quad x = \frac{5 - 11}{12} = \frac{-6}{12} = -\frac{1}{2}$$

Check each solution in the original equation. The solution set is $\left\{-\frac{1}{2}, \frac{4}{3}\right\}$.

◀ Work Problem 2 at the Side.

2 Solve $4x^2 - 11x - 3 = 0$ using the quadratic formula.

Answers

1. (a) -3; 9; -4 (b) 3; -6; -2

2. $\left\{-\frac{1}{4}, 3\right\}$

We could have used factoring to solve the equation in Example 1.

$$6x^2 - 5x - 4 = 0$$

$$(3x - 4)(2x + 1) = 0 \quad \text{Factor.}$$

$$3x - 4 = 0 \quad \text{or} \quad 2x + 1 = 0 \quad \text{Zero-factor property}$$

$$3x = 4 \quad \text{or} \quad 2x = -1 \quad \text{Solve each equation.}$$

$$x = \frac{4}{3} \quad \text{or} \quad x = -\frac{1}{2} \quad \text{Same solutions as in Example 1}$$

When solving quadratic equations, it is a good idea to try factoring first. If the equation cannot be factored or if factoring is difficult, then use the quadratic formula. Later in this section, we will show a way to determine whether factoring can be used to solve a quadratic equation.

Example 2 Using the Quadratic Formula (Irrational Solutions)

Solve $4x^2 = 8x - 1$.

Write the equation in standard form as $4x^2 - 8x + 1 = 0$.

$$x = \frac{-b \pm \sqrt{b^2 - 4ac}}{2a} \quad \text{Quadratic formula}$$

$$x = \frac{-(-8) \pm \sqrt{(-8)^2 - 4(4)(1)}}{2(4)} \quad a = 4, b = -8, c = 1$$

$$x = \frac{8 \pm \sqrt{64 - 16}}{8}$$

$$x = \frac{8 \pm \sqrt{48}}{8}$$

$$x = \frac{8 \pm 4\sqrt{3}}{8} \quad \sqrt{48} = \sqrt{16} \cdot \sqrt{3} = 4\sqrt{3}$$

$$x = \frac{4(2 \pm \sqrt{3})}{4(2)} \quad \text{Factor.}$$

Factor first; then divide out the common factor.

$$x = \frac{2 \pm \sqrt{3}}{2} \quad \text{Lowest terms}$$

The solution set is $\left\{\frac{2 + \sqrt{3}}{2}, \frac{2 - \sqrt{3}}{2}\right\}$.

CAUTION

1. ***Every quadratic equation must be written in standard form $ax^2 + bx + c = 0$ before we begin to solve it,*** whether we use factoring or the quadratic formula.
2. ***When writing solutions in lowest terms, be sure to factor first; then divide out the common factor,*** as shown in the last two steps in Example 2.

Work Problem **3** at the Side. ▶

3 Solve each equation using the quadratic formula.

(a) $6x^2 + 4x - 1 = 0$

(b) $2x^2 + 19 = 14x$

Answers

3. (a) $\left\{\frac{-2 + \sqrt{10}}{6}, \frac{-2 - \sqrt{10}}{6}\right\}$

(b) $\left\{\frac{7 + \sqrt{11}}{2}, \frac{7 - \sqrt{11}}{2}\right\}$

4 Solve each equation using the quadratic formula.

(a) $x^2 + x + 1 = 0$

(b) $(x + 2)(x - 6) = -17$

Example 3 **Using the Quadratic Formula (Nonreal Complex Solutions)**

Solve $(9x + 3)(x - 1) = -8$.

To write this equation in standard form, we first multiply and collect all nonzero terms on the left.

$$(9x + 3)(x - 1) = -8$$

$$9x^2 - 6x - 3 = -8 \quad \text{Multiply.}$$

$$9x^2 - 6x + 5 = 0 \quad \text{Add 8.}$$

From the equation $9x^2 - 6x + 5 = 0$, we identify $a = 9$, $b = -6$, and $c = 5$.

$$x = \frac{-b \pm \sqrt{b^2 - 4ac}}{2a} \quad \text{Quadratic formula}$$

$$x = \frac{-(-6) \pm \sqrt{(-6)^2 - 4(9)(5)}}{2(9)} \quad \text{Substitute.}$$

$$x = \frac{6 \pm \sqrt{-144}}{18}$$

$$x = \frac{6 \pm 12i}{18} \quad \sqrt{-144} = 12i$$

$$x = \frac{6(1 \pm 2i)}{6(3)} \quad \text{Factor.}$$

$$x = \frac{1 \pm 2i}{3} \quad \text{Lowest terms}$$

$$x = \frac{1}{3} \pm \frac{2}{3}i \quad \text{Standard form } a + bi \text{ for a complex number}$$

The solution set is $\left\{\frac{1}{3} + \frac{2}{3}i, \frac{1}{3} - \frac{2}{3}i\right\}$.

◀ Work Problem **4** at the Side.

OBJECTIVE 3 Use the discriminant to determine the number and type of solutions. The solutions of the quadratic equation $ax^2 + bx + c = 0$ are given by

$$x = \frac{-b \pm \sqrt{b^2 - 4ac}}{2a}. \leftarrow \text{Discriminant}$$

If a, b, and c are integers, the type of solutions of a quadratic equation—that is, rational, irrational, or nonreal complex—is determined by the expression under the radical sign, $b^2 - 4ac$. Because it distinguishes among the three types of solutions, $b^2 - 4ac$ is called the *discriminant*. By calculating the discriminant before solving a quadratic equation, we can predict whether the solutions will be rational numbers, irrational numbers, or nonreal complex numbers.

Answers

4. (a) $\left\{-\frac{1}{2} + \frac{\sqrt{3}}{2}i, -\frac{1}{2} - \frac{\sqrt{3}}{2}i\right\}$
(b) $\{2 + i, 2 - i\}$

Discriminant

The **discriminant** of $ax^2 + bx + c = 0$ is $\mathbf{b^2 - 4ac}$. If a, b, and c are integers, then the number and type of solutions are determined as follows.

Discriminant	Number and Type of Solutions
Positive, and the square of an integer	Two rational solutions
Positive, but not the square of an integer	Two irrational solutions
Zero	One rational solution
Negative	Two nonreal complex solutions

Calculating the discriminant can also help you decide whether to solve a quadratic equation by factoring or by using the quadratic formula. ***If the discriminant is a perfect square (including 0), then the equation can be solved by factoring. Otherwise, the quadratic formula (or completing the square) should be used.***

Example 4 Using the Discriminant

Find the discriminant. Use it to predict the number and type of solutions for each equation. Tell whether the equation can be solved by factoring or whether the quadratic formula should be used.

(a) $6x^2 - x - 15 = 0$

First identify the values of a, b, and c. Because $-x = -1x$, the value of b is -1. We find the discriminant by evaluating $b^2 - 4ac$.

$$b^2 - 4ac$$
$$= (-1)^2 - 4(6)(-15) \quad a = 6, b = -1, c = -15$$
$$= 1 + 360$$
$$= 361$$

Use parentheses and substitute carefully.

A calculator shows that $361 = 19^2$, a perfect square. Since a, b, and c are integers and the discriminant is a perfect square, there will be two rational solutions and the equation can be solved by factoring.

(b) $3x^2 - 4x = 5$

Write the equation in standard form as

$$3x^2 - 4x - 5 = 0 \quad \text{Subtract 5.}$$

to find $a = 3$, $b = -4$, and $c = -5$. The discriminant is

$$b^2 - 4ac$$
$$= (-4)^2 - 4(3)(-5)$$
$$= 16 + 60$$
$$= 76.$$

Because 76 is positive but *not* the square of an integer and a, b, and c are integers, the equation will have two irrational solutions and is best solved using the quadratic formula.

Continued on Next Page

5 Find the discriminant. Use it to predict the number and type of solutions for each equation.

(a) $2x^2 + 3x = 4$

(b) $2x^2 + 3x + 4 = 0$

(c) $x^2 + 20x + 100 = 0$

(d) $15x^2 + 11x = 14$

(e) Which of the equations in parts (a)–(d) can be solved by factoring?

(c) $4x^2 + x + 1 = 0$

Since $a = 4$, $b = 1$, and $c = 1$, the discriminant is

$$b^2 - 4ac$$
$$= 1^2 - 4(4)(1)$$
$$= 1 - 16$$
$$= -15.$$

Since the discriminant is negative and a, b, and c are integers, this quadratic equation will have two nonreal complex solutions. The quadratic formula should be used to solve it.

(d) $4x^2 + 9 = 12x$

Write the equation in standard form as

$$4x^2 - 12x + 9 = 0 \qquad \text{Subtract } 12x.$$

to find $a = 4$, $b = -12$, and $c = 9$. The discriminant is

$$b^2 - 4ac$$
$$= (-12)^2 - 4(4)(9)$$
$$= 144 - 144$$
$$= 0.$$

Because the discriminant is 0, the quantity under the radical in the quadratic formula is 0, and there is only one rational solution. Again, the equation can be solved by factoring.

◀ *Work Problem* **5** *at the Side.*

Note

In **Section 18.6** we will see how the discriminant can be used to determine the number of x-intercepts of the graph of a quadratic function.

Answers

5. **(a)** 41; two; irrational
(b) −23; two; nonreal complex
(c) 0; one; rational
(d) 961; two; rational **(e)** (c) and (d)

18.3 Exercises

FOR EXTRA HELP MyMathLab MathXL PRACTICE WATCH DOWNLOAD READ REVIEW

1. A student wrote the following as the quadratic formula for solving $ax^2 + bx + c = 0, a \neq 0$:

$$x = -b \pm \frac{\sqrt{b^2 - 4ac}}{2a}.$$

This is incorrect. ***WHAT WENT WRONG?***

2. A student attempted to solve the equation $5x^2 - 5x + 1 = 0$ as follows.

$$x = \frac{5 \pm \sqrt{25 - 4(5)(1)}}{2(5)} \qquad a = 5, b = -5, c = 1$$

$$x = \frac{5 \pm \sqrt{5}}{10}$$

$$x = \frac{1}{2} \pm \sqrt{5}$$

This is incorrect. ***WHAT WENT WRONG?***

Use the quadratic formula to solve each equation. (All solutions for these equations are real numbers.) See Examples 1 and 2.

3. $x^2 - 8x + 15 = 0$

4. $x^2 + 3x - 28 = 0$

5. $2x^2 + 4x + 1 = 0$

6. $2x^2 + 3x - 1 = 0$

7. $2x^2 - 2x = 1$

8. $9x^2 + 6x = 1$

9. $x^2 + 18 = 10x$

10. $x^2 - 4 = 2x$

11. $4k^2 + 4k - 1 = 0$

12. $4r^2 - 4r - 19 = 0$

13. $2 - 2x = 3x^2$

14. $26r - 2 = 3r^2$

15. $\frac{x^2}{4} - \frac{x}{2} = 1$
(*Hint:* First clear the fractions.)

16. $p^2 + \frac{p}{3} = \frac{1}{6}$
(*Hint:* First clear the fractions.)

17. $-2t(t + 2) = -3$

18. $-3x(x + 2) = -4$

19. $(r - 3)(r + 5) = 2$

20. $(k + 1)(k - 7) = 1$

Use the quadratic formula to solve each equation. (All solutions for these equations are nonreal complex numbers.) See Example 3.

21. $x^2 - 3x + 17 = 0$

22. $x^2 - 5x + 20 = 0$

23. $r^2 - 6r + 14 = 0$

24. $t^2 + 4t + 11 = 0$

25. $4x^2 - 4x = -7$

26. $9x^2 - 6x = -7$

27. $x(3x + 4) = -2$

28. $p(2p + 3) = -2$

Use the discriminant to determine whether the solutions for each equation are

A. *two rational numbers,* **B.** *one rational number,*

C. *two irrational numbers,* **D.** *two nonreal complex numbers.*

Do not actually solve. See Example 4.

29. $25x^2 + 70x + 49 = 0$

30. $4k^2 - 28k + 49 = 0$

31. $x^2 + 4x + 2 = 0$

32. $9x^2 - 12x - 1 = 0$

33. $3x^2 = 5x + 2$

34. $4x^2 = 4x + 3$

35. $3m^2 - 10m + 15 = 0$

36. $18x^2 + 60x + 82 = 0$

37. Using the discriminant, which equations in Exercises 29–36 can be solved by factoring?

38. Based on your answer in Exercise 37, solve the equation given in each exercise.

(a) Exercise 29 **(b)** Exercise 33

18.4 Equations Quadratic in Form

OBJECTIVES

1. **Solve an equation with fractions by writing it in quadratic form.**
2. **Use quadratic equations to solve applied problems.**
3. **Solve an equation with radicals by writing it in quadratic form.**
4. **Solve an equation that is quadratic in form by substitution.**

OBJECTIVE 1 Solve an equation with fractions by writing it in quadratic form. A variety of nonquadratic equations can be written in the form of a quadratic equation and solved by using one of the methods from **Sections 10.1 and 18.3.**

Example 1 Solving an Equation with Fractions That Leads to a Quadratic Equation

Solve $\frac{1}{x} + \frac{1}{x-1} = \frac{7}{12}$.

Clear fractions by multiplying each side by the least common denominator, $12x(x-1)$. (Note that the domain must be restricted to $x \neq 0$ and $x \neq 1$.)

$$12x(x-1)\left(\frac{1}{x} + \frac{1}{x+1}\right) = 12x(x-1)\left(\frac{7}{12}\right)$$

$$12x(x-1)\frac{1}{x} + 12x(x-1)\frac{1}{x-1} = 12x(x-1)\left(\frac{7}{12}\right) \quad \text{Distributive property}$$

$$12(x-1) + 12x = 7x(x-1)$$

$$12x - 12 + 12x = 7x^2 - 7x \quad \text{Distributive property}$$

$$24x - 12 = 7x^2 - 7x \quad \text{Combine like terms.}$$

$$7x^2 - 31x + 12 = 0 \quad \text{Standard form}$$

$$(7x-3)(x-4) = 0 \quad \text{Factor.}$$

$$7x - 3 = 0 \quad \text{or} \quad x - 4 = 0 \quad \text{Zero-factor property}$$

$$x = \frac{3}{7} \quad \text{or} \quad x = 4 \quad \text{Solve each equation.}$$

Check by substituting these solutions in the original equation. The solution set is $\left\{\frac{3}{7}, 4\right\}$.

Work Problem 1 at the Side. ▶

1 Solve each equation. Check your solutions.

(a) $\frac{5}{m} + \frac{12}{m^2} = 2$

(b) $\frac{2}{x} + \frac{1}{x-2} = \frac{5}{3}$

(c) $\frac{4}{m-1} + 9 = -\frac{7}{m}$

OBJECTIVE 2 Use quadratic equations to solve applied problems. In **Sections 2.4 and 8.5** we solved distance-rate-time (or motion) problems that led to linear equations or rational equations. Now we can extend that work to motion problems that lead to quadratic equations. We continue to use the six-step problem-solving method from **Section 2.3.**

Example 2 Solving a Motion Problem

A riverboat for tourists averages 12 mph in still water. It takes the boat 1 hr, 4 min to go 6 mi upstream and return. Find the speed of the current.

Step 1 **Read** the problem carefully.

Step 2 **Assign a variable.** Let $x =$ the speed of the current.

The current slows down the boat when it is going upstream, so the rate (or speed) upstream is the speed of the boat in still water less the speed of the current, or $(12 - x)$ mph. See Figure 1 on the next page.

Continued on Next Page

ANSWERS

1. (a) $\left\{-\frac{3}{2}, 4\right\}$ (b) $\left\{\frac{4}{5}, 3\right\}$ (c) $\left\{\frac{7}{9}, -1\right\}$

Riverboat traveling *upstream*—the current slows it down.

Figure 1

Similarly, the current speeds up the boat as it travels downstream, so its speed downstream is $(12 + x)$ mph. Thus,

$$12 - x = \text{the rate upstream in miles per hour;}$$
$$12 + x = \text{the rate downstream in miles per hour.}$$

This information can be used to complete a table. We use the distance formula, $d = rt$, solved for time t, $t = \frac{d}{r}$, to write expressions for t.

	d	r	t
Upstream	6	$12 - x$	$\dfrac{6}{12 - x}$
Downstream	6	$12 + x$	$\dfrac{6}{12 + x}$

Times in hours

Step 3 **Write an equation.** The total time, 1 hr and 4 min, can be written as

$$1 + \frac{4}{60} = 1 + \frac{1}{15} = \frac{16}{15} \text{ hr.}$$

Because the time upstream plus the time downstream equals $\frac{16}{15}$ hr,

Time upstream + time downstream = total time.

$$\frac{6}{12 - x} + \frac{6}{12 + x} = \frac{16}{15}.$$

Step 4 **Solve** the equation. Multiply each side by $15(12 - x)(12 + x)$, the LCD, and solve the resulting quadratic equation.

$$15(12 + x)6 + 15(12 - x)6 = 16(12 - x)(12 + x)$$
$$90(12 + x) + 90(12 - x) = 16(144 - x^2)$$
$$1080 + 90x + 1080 - 90x = 2304 - 16x^2 \quad \text{Distributive property}$$
$$2160 = 2304 - 16x^2 \quad \text{Combine like terms.}$$
$$16x^2 = 144$$
$$x^2 = 9 \quad \text{Divide by 16.}$$
$$x = 3 \quad \text{or} \quad x = -3 \quad \text{Square root property}$$

Step 5 **State the answer.** The speed of the current cannot be -3, so the answer is 3 mph.

Step 6 **Check** that this value satisfies the original problem.

CAUTION

As shown in Example 2, when a quadratic equation is used to solve an applied problem, sometimes only *one* answer satisfies the application. ***Always check each answer in the words of the original problem.***

Work Problem 2 at the Side. ▶

Recall from **Section 8.5** that a person's work rate is $\frac{1}{t}$ part of the job per hour, where t is the time in hours required to do the complete job. Thus, the part of the job the person will do in x hours is $\frac{1}{t}x$.

Example 3 Solving a Work Problem

It takes two carpet layers 4 hr to carpet a room. If each worked alone, one of them could do the job in 1 hr less time than the other. How long would it take each carpet layer to complete the job alone?

Step 1 **Read** the problem again. There will be two answers.

Step 2 **Assign a variable.** Let $x =$ the number of hours for the slower carpet layer to complete the job alone. Then the faster carpet layer could do the entire job in $(x - 1)$ hours. The slower person's rate is $\frac{1}{x}$, and the faster person's rate is $\frac{1}{x-1}$. Together, they can do the job in 4 hr. Complete a table as shown.

	Rate	Time Working Together	Fractional Part of the Job Done
Slower Worker	$\frac{1}{x}$	4	$\frac{1}{x}(4)$
Faster Worker	$\frac{1}{x-1}$	4	$\frac{1}{x-1}(4)$

Sum is 1 whole job.

Step 3 **Write an equation.** The sum of the fractional parts done by the workers should equal 1 (the whole job).

Part done by slower worker + part done by faster worker = 1 whole job.

$$\frac{4}{x} + \frac{4}{x-1} = 1$$

Step 4 **Solve** the equation. Multiply by $x(x - 1)$, the LCD.

$$x(x-1)\left(\frac{4}{x} + \frac{4}{x-1}\right) = x(x-1)(1) \quad \text{Multiply by the LCD.}$$

$$4(x-1) + 4x = x(x-1) \quad \text{Distributive property}$$

$$4x - 4 + 4x = x^2 - x \quad \text{Distributive property}$$

$$x^2 - 9x + 4 = 0 \quad \text{Standard form}$$

Continued on Next Page

2 Solve each problem.

(a) In 4 hr, Kerrie can go 15 mi upriver and come back. The speed of the current is 5 mph. Complete this table.

	d	r	t
Up			
Down			

(b) Find the speed of the boat from part (a) in still water.

(c) In $1\frac{3}{4}$ hr, Ken rows his boat 5 mi upriver and comes back. The speed of the current is 3 mph. How fast does Ken row?

ANSWERS

2. **(a)** row 1: 15; $x - 5$; $\frac{15}{x-5}$

row 2: 15; $x + 5$; $\frac{15}{x+5}$

(b) 10 mph **(c)** 7 mph

3 Solve each problem. Round answers to the nearest tenth.

(a) Carlos can complete a certain lab test in 2 hr less time than Jaime can. If they can finish the job together in 2 hr, how long would it take each of them working alone?

	Rate	Time Working Together	Fractional Part of the Job Done
Carlos			
Jaime			

(b) Two chefs are preparing a banquet. One chef could prepare the banquet in 2 hr less time than the other. Together, they complete the job in 5 hr. How long would it take the faster chef working alone?

The resulting equation $x^2 - 9x + 4 = 0$ cannot be solved by factoring, so use the quadratic formula.

$$x = \frac{-b \pm \sqrt{b^2 - 4ac}}{2a} \qquad \text{Quadratic formula}$$

$$x = \frac{-(-9) \pm \sqrt{(-9)^2 - 4(1)(4)}}{2(1)} \qquad a = 1, b = -9, c = 4$$

$$x = \frac{9 \pm \sqrt{65}}{2} \qquad \text{Simplify.}$$

$$x = \frac{9 + \sqrt{65}}{2} \approx 8.5 \quad \text{or} \quad x = \frac{9 - \sqrt{65}}{2} \approx 0.5 \qquad \text{Use a calculator.}$$

Step 5 **State the answer.** Only the solution 8.5 makes sense in the original problem. (Why?) Thus, the slower worker can do the job in about 8.5 hr and the faster in about $8.5 - 1 = 7.5$ hr.

Step 6 **Check** that these results satisfy the original problem.

Work Problem 3 at the Side.

OBJECTIVE 3 Solve an equation with radicals by writing it in quadratic form.

Example 4 Solving Radical Equations That Lead to Quadratic Equations

Solve each equation.

(a) $k = \sqrt{6k - 8}$

This equation is not quadratic. However, squaring both sides of the equation gives a quadratic equation that can be solved by factoring.

$$k^2 = 6k - 8 \qquad \text{Square each side.}$$

$$k^2 - 6k + 8 = 0 \qquad \text{Standard form}$$

$$(k - 4)(k - 2) = 0 \qquad \text{Factor.}$$

$$k - 4 = 0 \quad \text{or} \quad k - 2 = 0 \qquad \text{Zero-factor property}$$

$$k = 4 \quad \text{or} \quad k = 2 \qquad \text{Potential solutions}$$

Recall from **Section 9.6** that squaring both sides of a radical equation can introduce extraneous solutions that do not satisfy the original equation. ***All proposed solutions must be checked in the original (not the squared) equation.***

If $k = 4$, then	If $k = 2$, then
$k = \sqrt{6k - 8}$	$k = \sqrt{6k - 8}$
$4 \stackrel{?}{=} \sqrt{6(4) - 8}$	$2 \stackrel{?}{=} \sqrt{6(2) - 8}$
$4 \stackrel{?}{=} \sqrt{16}$	$2 \stackrel{?}{=} \sqrt{4}$
$4 = 4.$ True	$2 = 2.$ True

Both solutions check, so the solution set is $\{2, 4\}$.

Continued on Next Page

Answers

3. **(a)** Jaime: 5.2 hr; Carlos: 3.2 hr **(b)** 9.1 hr

(b) $x + \sqrt{x} = 6$

$(a - b)^2 = a^2 - 2ab + b^2$

$\sqrt{x} = 6 - x$ Isolate the radical on one side.

$x = 36 - 12x + x^2$ Square each side.

$0 = x^2 - 13x + 36$ Standard form

$0 = (x - 4)(x - 9)$ Factor.

$x - 4 = 0$ or $x - 9 = 0$ Zero-factor property

$x = 4$ or $x = 9$ Proposed solutions

Check both proposed solutions in the *original* equation.

If $x = 4$, then	If $x = 9$, then
$x + \sqrt{x} = 6$	$x + \sqrt{x} = 6$
$4 + \sqrt{4} \stackrel{?}{=} 6$	$9 + \sqrt{9} \stackrel{?}{=} 6$
$6 = 6.$ True	$12 = 6.$ False

Only the solution 4 checks, so the solution set is $\{4\}$.

Work Problem **4** at the Side. ▶

4 Solve each equation. Check your solutions.

(a) $x = \sqrt{7x - 10}$

(b) $2x = \sqrt{x} + 1$

OBJECTIVE 4 **Solve an equation that is quadratic in form by substitution.** A nonquadratic equation that can be written in the form

$$au^2 + bu + c = 0,$$

for $a \neq 0$ and an algebraic expression u, is called **quadratic in form.**

Many equations that are quadratic in form can be solved more easily by defining and substituting a "temporary" variable u for an expression involving the variable in the original equation. The first step is to define this temporary variable u.

5 Define a variable u, and write each equation in the form $au^2 + bu + c = 0$.

(a) $2x^4 + 5x^2 - 12 = 0$

(b) $2(x + 5)^2 - 7(x + 5) + 6 = 0$

(c) $x^{4/3} - 8x^{2/3} + 16 = 0$

Example 5 Defining Substitution Variables

Define a variable u, and write each equation in the form $au^2 + bu + c = 0$.

(a) $x^4 - 13x^2 + 36 = 0$

Look at the two terms involving the variable x, ignoring their coefficients. Try to find one variable expression that is the square of the other. We see that $x^4 = (x^2)^2$, so we define $u = x^2$, and rewrite the original equation as the quadratic equation

$$u^2 - 13u + 36 = 0. \quad \text{Here, } u = x^2.$$

(b) $2(4m - 3)^2 + 7(4m - 3) + 5 = 0$

Because this equation involves both $(4m - 3)^2$ and $(4m - 3)$, choose $u = 4m - 3$. Substituting u for $4m - 3$ gives the quadratic equation

$$2u^2 + 7u + 5 = 0. \quad \text{Here, } u = 4m - 3.$$

(c) $2x^{2/3} - 11x^{1/3} + 12 = 0$

Here we apply one of the power rules for exponents from **Section 6.1:** $(a^m)^n = a^{mn}$. Because $(x^{1/3})^2 = x^{2/3}$, we define $u = x^{1/3}$. With this substitution, the original equation becomes

$$2u^2 - 11u + 12 = 0. \quad \text{Here, } u = x^{1/3}.$$

Work Problem **5** at the Side. ▶

Answers

4. (a) $\{2, 5\}$ (b) $\{1\}$
5. (a) $u = x^2$; $2u^2 + 5u - 12 = 0$
(b) $u = x + 5$; $2u^2 - 7u + 6 = 0$
(c) $u = x^{2/3}$; $u^2 - 8u + 16 = 0$

Example 6 Solving Equations That Are Quadratic in Form

Solve each equation.

(a) $x^4 - 13x^2 + 36 = 0$

From Example 5(a), we write this equation in quadratic form by substituting u for x^2.

$$x^4 - 13x^2 + 36 = 0$$

$$(x^2)^2 - 13x^2 + 36 = 0 \qquad x^4 = (x^2)^2$$

$$u^2 - 13u + 36 = 0 \qquad \text{Let } u = x^2.$$

$$(u - 4)(u - 9) = 0 \qquad \text{Factor.}$$

$$u - 4 = 0 \quad \text{or} \quad u - 9 = 0 \qquad \text{Zero-factor property}$$

Don't stop here.

$$u = 4 \quad \text{or} \quad u = 9 \qquad \text{Solve.}$$

$$x^2 = 4 \quad \text{or} \quad x^2 = 9 \qquad \text{Substitute } x^2 \text{ for } u.$$

$$x = \pm 2 \quad \text{or} \quad x = \pm 3 \qquad \text{Square root property}$$

The equation $x^4 - 13x^2 + 36 = 0$, a fourth-degree equation, has four solutions.* The solution set is $\{-3, -2, 2, 3\}$. Check each of the four solutions by substitution.

(b)

$$4x^4 + 1 = 5x^2$$

$$4(x^2)^2 + 1 = 5x^2 \qquad x^4 = (x^2)^2$$

$$4u^2 + 1 = 5u \qquad \text{Let } u = x^2.$$

$$4u^2 - 5u + 1 = 0 \qquad \text{Standard form}$$

$$(4u - 1)(u - 1) = 0 \qquad \text{Factor.}$$

$$4u - 1 = 0 \quad \text{or} \quad u - 1 = 0 \qquad \text{Zero-factor property}$$

$$u = \frac{1}{4} \quad \text{or} \quad u = 1 \qquad \text{Solve.}$$

This is a key step.

$$x^2 = \frac{1}{4} \quad \text{or} \quad x^2 = 1 \qquad \text{Substitute } x^2 \text{ for } u.$$

$$x = \pm\frac{1}{2} \quad \text{or} \quad x = \pm 1 \qquad \text{Square root property}$$

Check that the solution set is $\left\{-1, -\frac{1}{2}, \frac{1}{2}, 1\right\}$.

(c) $x^4 = 6x^2 - 3$

First write the equation as

$$x^4 - 6x^2 + 3 = 0 \quad \text{or} \quad (x^2)^2 - 6x^2 + 3 = 0,$$

which is quadratic in form with $u = x^2$. Substitute u for x^2 and u^2 for x^4 to get

$$u^2 - 6u + 3 = 0.$$

Since this equation cannot be solved by factoring, use the quadratic formula.

Continued on Next Page

*In general, an equation in which an nth-degree polynomial equals 0 has n solutions, although some of them may be repeated.

$$u = \frac{-(-6) \pm \sqrt{(-6)^2 - 4(1)(3)}}{2(1)} \qquad a = 1, b = -6, c = 3$$

$$u = \frac{6 \pm \sqrt{24}}{2} \qquad \text{Simplify.}$$

$$u = \frac{6 \pm 2\sqrt{6}}{2} \qquad \sqrt{24} = \sqrt{4} \cdot \sqrt{6} = 2\sqrt{6}$$

$$u = \frac{2(3 \pm \sqrt{6})}{2} \qquad \text{Factor.}$$

$$u = 3 \pm \sqrt{6} \qquad \text{Lowest terms}$$

$$x^2 = 3 + \sqrt{6} \quad \text{or} \quad x^2 = 3 - \sqrt{6} \qquad \text{Substitute } x^2 \text{ for } u.$$

Find *both* square roots in each case.

$$x = \pm\sqrt{3 + \sqrt{6}} \quad \text{or} \quad x = \pm\sqrt{3 - \sqrt{6}} \qquad \text{Square root property}$$

The solution set contains four numbers:

$$\left\{\sqrt{3 + \sqrt{6}}, -\sqrt{3 + \sqrt{6}}, \sqrt{3 - \sqrt{6}}, -\sqrt{3 - \sqrt{6}}\right\}.$$

Note

Some students prefer to solve equations like those in Examples 6(a) and (b) by factoring directly. For example,

$$x^4 - 13x^2 + 36 = 0 \qquad \text{Example 6(a) equation}$$

$$(x^2 - 9)(x^2 - 4) = 0 \qquad \text{Factor.}$$

$$(x + 3)(x - 3)(x + 2)(x - 2) = 0. \qquad \text{Factor again.}$$

Using the zero-factor property gives the same solutions obtained in Example 6(a). Equations that cannot be solved by factoring (as in Example 6(c)) must be solved by substitution and the quadratic formula.

Work Problem **6** *at the Side.* ▶

The method used in Example 6 can be generalized.

Solving an Equation That Is Quadratic in Form by Substitution

Step 1 Define a temporary variable u, based on the relationship between the variable expressions in the given equation. Substitute u in the original equation and rewrite the equation in the form $au^2 + bu + c = 0$.

Step 2 **Solve the quadratic equation obtained in Step 1** by factoring or the quadratic formula.

Step 3 **Replace u with the expression it defined in Step 1.**

Step 4 **Solve the resulting equations for the original variable.**

Step 5 **Check** all solutions by substituting them in the original equation.

6 Solve each equation. Check your solutions.

(a) $m^4 - 10m^2 + 9 = 0$

(b) $9k^4 - 37k^2 + 4 = 0$

(c) $x^4 - 4x^2 = -2$

Answers

6. **(a)** $\{-3, -1, 1, 3\}$ **(b)** $\left\{-2, -\frac{1}{3}, \frac{1}{3}, 2\right\}$

(c) $\left\{\sqrt{2 + \sqrt{2}}, -\sqrt{2 + \sqrt{2}}, \sqrt{2 - \sqrt{2}}, -\sqrt{2 - \sqrt{2}}\right\}$

7 Solve each equation. Check your solutions.

(a) $5(r+3)^2 + 9(r+3) = 2$

(b) $4m^{2/3} = 3m^{1/3} + 1$

Example 7 **Solving Equations That Are Quadratic in Form**

Solve each equation.

(a) $2(4m-3)^2 + 7(4m-3) + 5 = 0$

Step 1 Because of the repeated quantity $4m-3$, substitute u for $4m-3$ as in Example 5(b).

$$2(4m-3)^2 + 7(4m-3) + 5 = 0$$

$$2u^2 + 7u + 5 = 0 \quad \text{Let } u = 4m-3.$$

Step 2

$$(2u+5)(u+1) = 0 \quad \text{Factor.}$$

$$2u+5=0 \quad \text{or} \quad u+1=0 \quad \text{Zero-factor property}$$

Don't stop here.

$$u = -\frac{5}{2} \quad \text{or} \quad u = -1 \quad \text{Solve for } u.$$

Step 3

$$4m-3 = -\frac{5}{2} \quad \text{or} \quad 4m-3 = -1 \quad \text{Substitute } 4m-3 \text{ for } u.$$

Step 4

$$4m = \frac{1}{2} \quad \text{or} \quad 4m = 2 \quad \text{Solve for } m.$$

$$m = \frac{1}{8} \quad \text{or} \quad m = \frac{1}{2}$$

Step 5 Check that the solution set of the original equation is $\left\{\frac{1}{8}, \frac{1}{2}\right\}$.

(b) $2x^{2/3} - 11x^{1/3} + 12 = 0$

From Example 5(c), substitute u for $x^{1/3}$.

$$2u^2 - 11u + 12 = 0 \quad \text{Let } x^{1/3} = u;\ x^{2/3} = u^2.$$

$$(2u-3)(u-4) = 0 \quad \text{Factor.}$$

$$2u-3=0 \quad \text{or} \quad u-4=0 \quad \text{Zero-factor property}$$

$$u = \frac{3}{2} \quad \text{or} \quad u = 4 \quad \text{Solve for } u.$$

$$x^{1/3} = \frac{3}{2} \quad \text{or} \quad x^{1/3} = 4 \quad u = x^{1/3}$$

$$(x^{1/3})^3 = \left(\frac{3}{2}\right)^3 \quad \text{or} \quad (x^{1/3})^3 = 4^3 \quad \text{Cube each side.}$$

$$x = \frac{27}{8} \quad \text{or} \quad x = 64$$

Check that the solution set is $\left\{\frac{27}{8}, 64\right\}$.

CAUTION

A common error when solving problems like those in Examples 6 and 7 is to stop too soon. ***Once you have solved for u, remember to substitute and solve for the values of the original variable.*** Keep in mind that u is just a temporary variable that helps you solve the given equation. As in any equation, you must solve for the variable in the *original* equation.

◀ Work Problem **7** at the Side.

Answers

7. (a) $\left\{-5, -\frac{14}{5}\right\}$ **(b)** $\left\{-\frac{1}{64}, 1\right\}$

18.4 Exercises

Based on the discussion and examples of this section, write a sentence describing the first step you would take to solve each equation. Do not actually solve.

1. $\frac{14}{x} = x - 5$

2. $\sqrt{1 + x} + x = 5$

3. $(r^2 + r)^2 - 8(r^2 + r) + 12 = 0$

4. $3t = \sqrt{16 - 10t}$

5. Read this incorrect "solution" carefully. ***WHAT WENT WRONG?***

$$\begin{aligned} x &= \sqrt{3x + 4} \\ x^2 &= 3x + 4 \quad \text{Square both sides.} \\ x^2 - 3x - 4 &= 0 \\ (x - 4)(x + 1) &= 0 \\ x - 4 = 0 \quad &\text{or} \quad x + 1 = 0 \\ x = 4 \quad &\text{or} \quad x = -1 \end{aligned}$$

Solution set: $\{4, -1\}$

6. Read this incorrect "solution" carefully. ***WHAT WENT WRONG?***

$$\begin{aligned} 2(m - 1)^2 - 3(m - 1) + 1 &= 0 \\ 2u^2 - 3u + 1 &= 0 \quad \text{Let } u = m - 1. \\ (2u - 1)(u - 1) &= 0 \\ 2u - 1 = 0 \quad &\text{or} \quad u - 1 = 0 \\ u = \frac{1}{2} \quad &\text{or} \quad u = 1 \end{aligned}$$

Solution set: $\left\{\frac{1}{2}, 1\right\}$

Solve each equation. Check your solutions. See Example 1.

7. $1 - \frac{3}{x} - \frac{28}{x^2} = 0$

8. $4 - \frac{7}{r} - \frac{2}{r^2} = 0$

9. $3 - \frac{1}{t} = \frac{2}{t^2}$

10. $1 + \frac{2}{k} = \frac{3}{k^2}$

11. $\frac{1}{x} + \frac{2}{x + 2} = \frac{17}{35}$

12. $\frac{2}{m} + \frac{3}{m + 9} = \frac{11}{4}$

13. $\frac{2}{x + 1} + \frac{3}{x + 2} = \frac{7}{2}$

14. $\frac{4}{3 - p} + \frac{2}{5 - p} = \frac{26}{15}$

15. $\frac{3}{2x} - \frac{1}{2(x + 2)} = 1$

16. $\frac{4}{3x} - \frac{1}{2(x + 1)} = 1$

17. $\frac{6}{p} = 2 + \frac{p}{p + 1}$

18. $\frac{k}{2 - k} + \frac{2}{k} = 5$

19. A boat goes 20 mph in still water, and the rate of the current is t mph.

(a) What is the rate of the boat when it travels upstream?

(b) What is the rate of the boat when it travels downstream?

20. It takes m hours to grade a set of papers.

(a) What is the grader's rate (in job per hour)?

(b) How much of the job will the grader do in 2 hr?

Solve each problem. See Examples 2 and 3.

21. On a windy day Yoshiaki found that he could go 16 mi downstream and then 4 mi back upstream at top speed in a total of 48 min. What was the top speed of Yoshiaki's boat if the current was 15 mph?

	d	r	t
Upstream	4	$x - 15$	
Downstream	16		

22. Lekesha flew her plane for 6 hr at a constant speed. She traveled 810 mi with the wind, then turned around and traveled 720 mi against the wind. The wind speed was a constant 15 mph. Find the speed of the plane.

	d	r	t
With Wind	810		
Against Wind	720		

23. In Canada, Medicine Hat and Cranbrook are 300 km apart. Harry rides his Honda 20 km per hr faster than Yoshi rides his Yamaha. Find Harry's average speed if he travels from Cranbrook to Medicine Hat in $1\frac{1}{4}$ hr less time than Yoshi. (*Source: State Farm Road Atlas.*)

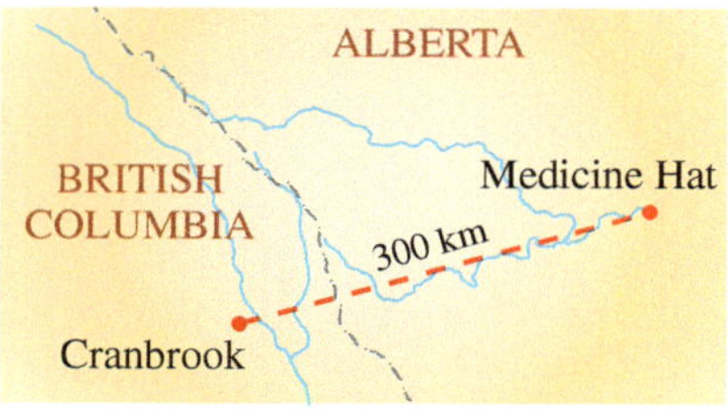

24. In California, the distance from Jackson to Lodi is about 40 mi, as is the distance from Lodi to Manteca. Rico drove from Jackson to Lodi during the rush hour, stopped in Lodi for a root beer, and then drove on to Manteca at 10 mph faster. Driving time for the entire trip was 88 min. Find his speed from Jackson to Lodi. (*Source: State Farm Road Atlas.*)

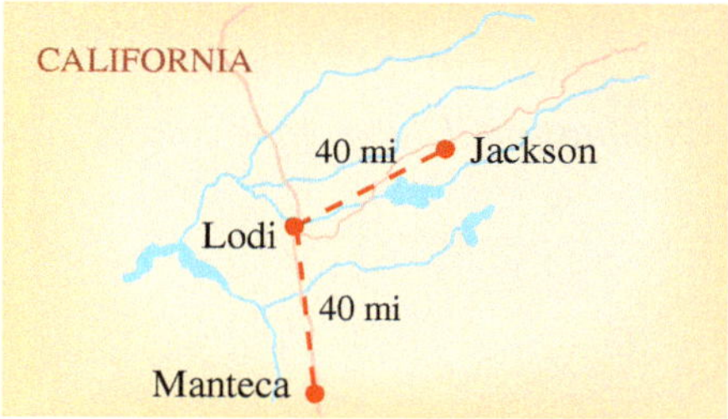

25. Working together, two people can cut a large lawn in 2 hr. One person can do the job alone in 1 hr less time than the other. How long (to the nearest tenth) would it take the faster person to do the job? (*Hint:* x is the time of the faster person.)

	Rate	Time Working Together	Fractional Part of the Job Done
Faster Worker	$\frac{1}{x}$	2	
Slower Worker		2	

26. A janitorial service provides two people to clean an office building. Working together, the two can clean the building in 5 hr. One person is new to the job and would take 2 hr longer than the other person to clean the building alone. How long (to the nearest tenth) would it take the new worker to clean the building alone?

	Rate	Time Working Together	Fractional Part of the Job Done
Faster Worker			
Slower Worker			

27. A washing machine can be filled in 6 min if both the hot and cold water taps are fully opened. Filling the washer with hot water alone takes 9 min longer than filling it with cold water alone. How long does it take to fill the washer with cold water?

28. Two pipes together can fill a large tank in 2 hr. One of the pipes, used alone, takes 3 hr longer than the other to fill the tank. How long would each pipe take to fill the tank alone?

Solve each equation. Check your solutions. See Example 4.

29. $z = \sqrt{5z - 4}$

30. $x = \sqrt{9x - 14}$

31. $2x = \sqrt{11x + 3}$

32. $4x = \sqrt{6x + 1}$

33. $3x = \sqrt{16 - 10x}$

34. $4t = \sqrt{8t + 3}$

35. $p - 2\sqrt{p} = 8$

36. $k + \sqrt{k} = 12$

37. $m = \sqrt{\frac{6 - 13m}{5}}$

38. $r = \sqrt{\frac{20 - 19r}{6}}$

Solve each equation. Check your solutions. See Examples 5–7.

39. $t^4 - 18t^2 + 81 = 0$

40. $x^4 - 8x^2 + 16 = 0$

41. $4k^4 - 13k^2 + 9 = 0$

42. $9x^4 - 25x^2 + 16 = 0$

43. $x^4 + 48 = 16x^2$

44. $z^4 = 17z^2 - 72$

45. $2x^4 - 9x^2 = -2$

46. $8x^4 + 1 = 11x^2$

47. $(x + 3)^2 + 5(x + 3) + 6 = 0$

48. $(k - 4)^2 + (k - 4) - 20 = 0$

49. $(t + 5)^2 + 6 = 7(t + 5)$

50. $3(m + 4)^2 - 8 = 2(m + 4)$

51. $2 + \frac{5}{3k - 1} = \frac{-2}{(3k - 1)^2}$

52. $3 - \frac{7}{2p + 2} = \frac{6}{(2p + 2)^2}$

53. $x^{2/3} + x^{1/3} - 2 = 0$

54. $x^{2/3} - 2x^{1/3} - 3 = 0$

55. $r^{2/3} + r^{1/3} - 12 = 0$

56. $3x^{2/3} - x^{1/3} - 24 = 0$

57. $2\left(1 + \sqrt{r}\right)^2 = 13\left(1 + \sqrt{r}\right) - 6$

58. $(k^2 + k)^2 + 12 = 8(k^2 + k)$

Relating Concepts (Exercises 59–64) For Individual or Group Work

Consider the following equation, which contains variable expressions in the denominators. ***Work Exercises 59–64 in order.***

$$\frac{x^2}{(x-3)^2} + \frac{3x}{x-3} - 4 = 0$$

59. Why must 3 be excluded from the domain of this equation?

60. Multiply each side of the equation by the LCD, $(x - 3)^2$, and solve. There is only one solution—what is it?

61. Write the equation in a different manner so that it is quadratic in form using the expression $\frac{x}{x-3}$.

62. In your own words, explain why the expression $\frac{x}{x-3}$ cannot equal 1.

63. Solve the equation from Exercise 61 by making the substitution $t = \frac{x}{x-3}$. You should get two values for t. Why is one of them impossible for this equation?

64. Solve the equation $x^2(x-3)^{-2} + 3x(x-3)^{-1} - 4 = 0$ by letting $s = (x-3)^{-1}$. You should get two values for s. Why is this impossible for this equation?

Summary Exercises on Solving Quadratic Equations

We have introduced four methods for solving quadratic equations written in standard form $ax^2 + bx + c = 0$. *The following table lists some advantages and disadvantages of each method.*

METHODS FOR SOLVING QUADRATIC EQUATIONS

Method	Advantages	Disadvantages
Factoring	This is usually the fastest method.	Not all polynomials are factorable; some factorable polynomials are hard to factor.
Square root property	This is the simplest method for solving equations of the form $(ax + b)^2 = c$.	Few equations are given in this form.
Completing the square	This method can always be used, although many people prefer the quadratic formula.	It requires more steps than other methods.
Quadratic formula	This method can always be used.	It is more difficult than factoring because of the square root, although calculators can simplify its use.

Refer to the preceding box. Decide whether factoring, the square root property, *or* the quadratic formula *is most appropriate for solving each quadratic equation. Do not actually solve the equations.*

1. $(2x + 3)^2 = 4$

2. $4x^2 - 3x = 1$

3. $z^2 + 5z - 8 = 0$

4. $2k^2 + 3k = 1$

5. $3m^2 = 2 - 5m$

6. $p^2 = 5$

Solve each quadratic equation by the method of your choice. Check your solutions.

7. $p^2 = 47$

8. $6x^2 - x - 15 = 0$

9. $n^2 + 8n + 6 = 0$

10. $(x - 4)^2 = 49$

11. $\frac{9}{m} + \frac{5}{m^2} = 2$

12. $3m^2 = 3 - 8m$

13. $3x^2 - 9x + 4 = 0$

***14.** $x^2 = -12$

15. $x\sqrt{2} = \sqrt{5x - 2}$

16. $12x^4 - 11x^2 + 2 = 0$

17. $(2k + 5)^2 = 12$

18. $\frac{2}{x} + \frac{1}{x - 2} - \frac{5}{3} = 0$

19. $t^4 + 14 = 9t^2$

20. $2x^2 + 4x = 5$

***21.** $z^2 + z + 2 = 0$

22. $x^4 - 8x^2 = -1$

23. $4t^2 - 12t + 9 = 0$

24. $x\sqrt{3} = \sqrt{2 - x}$

25. $r^2 - 72 = 0$

26. $-3x^2 + 4x = -4$

27. $x^2 - 5x - 36 = 0$

28. $w^2 = 169$

***29.** $3p^2 = 6p - 4$

30. $z = \sqrt{\frac{5z + 3}{2}}$

31. $2(3k - 1)^2 + 5(3k - 1) = -2$

***32.** $\frac{4}{r^2} + 3 = \frac{1}{r}$

33. $x - \sqrt{15 - 2x} = 0$

34. $3 = \frac{1}{t + 2} + \frac{2}{(t + 2)^2}$

***35.** $4k^4 + 5k^2 + 1 = 0$

36. $(x + 1)^{2/3} - (x + 1)^{1/3} = 2$

*This exercise requires knowledge of complex numbers.

18.5 Formulas and Further Applications

OBJECTIVES

1. **Solve formulas for variables involving squares and square roots.**
2. **Solve applied problems using the Pythagorean formula.**
3. **Solve applied problems using area formulas.**
4. **Solve applied problems using quadratic functions as models.**

OBJECTIVE 1 Solve formulas for variables involving squares and square roots. The methods presented earlier in this chapter and the previous one can be used to solve formulas with squares and square roots.

Example 1 Solving for Variables Involving Squares or Square Roots

Solve each formula for the given variable.

(a) $w = \dfrac{kFr}{v^2}$ for v

$w = \dfrac{kFr}{v^2}$ (The goal is to isolate v on one side.)

$v^2 w = kFr$ Multiply by v^2.

$v^2 = \dfrac{kFr}{w}$ Divide by w.

$v = \pm\sqrt{\dfrac{kFr}{w}}$ Square root property

$v = \dfrac{\pm\sqrt{kFr}}{\sqrt{w}} \cdot \dfrac{\sqrt{w}}{\sqrt{w}} = \dfrac{\pm\sqrt{kFrw}}{w}$ Rationalize the denominator.

(b) $d = \sqrt{\dfrac{4A}{\pi}}$ for A

$d = \sqrt{\dfrac{4A}{\pi}}$ (The goal is to isolate A on one side.)

$d^2 = \dfrac{4A}{\pi}$ Square both sides.

$\pi d^2 = 4A$ Multiply by π.

$\dfrac{\pi d^2}{4} = A$ Divide by 4.

Work Problem 1 at the Side. ▶

1 Solve each formula for the given variable.

(a) $A = \pi r^2$ for r

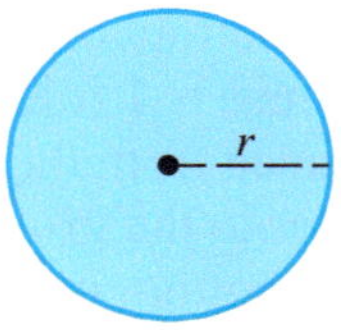

(b) $s = 30\sqrt{\dfrac{a}{p}}$ for a

Note

In formulas like $v = \frac{\pm\sqrt{kFrw}}{w}$ in Example 1(a), we will include both positive and negative values.

Example 2 Solving for a Second-Degree Variable

Solve $s = 2t^2 + kt$ for t.

Since the equation has terms with t^2 and t, write it in standard form $ax^2 + bx + c = 0$, with t as the variable instead of x.

$s = 2t^2 + kt$

$0 = 2t^2 + kt - s$ Subtract s.

$2t^2 + kt - s = 0$ Standard form

Continued on Next Page

ANSWERS

1. (a) $r = \dfrac{\pm\sqrt{A\pi}}{\pi}$ (b) $a = \dfrac{ps^2}{900}$

2 Solve $2t^2 - 5t + k = 0$ for t.

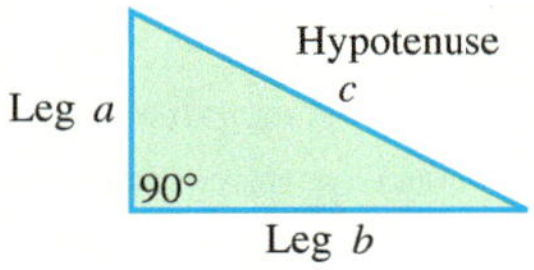

$a^2 + b^2 = c^2$
Pythagorean Formula

3 Solve the problem.

A 13-ft ladder is leaning against a house. The distance from the bottom of the ladder to the house is 7 ft less than the distance from the top of the ladder to the ground. How far is the bottom of the ladder from the house?

Now solve $2t^2 + kt - s = 0$ for t using the quadratic formula.

$$t = \frac{-k \pm \sqrt{k^2 - 4(2)(-s)}}{2(2)} \qquad \text{Let } a = 2,\ b = k, \text{ and } c = -s.$$

$$t = \frac{-k \pm \sqrt{k^2 + 8s}}{4} \qquad \text{Simplify.}$$

The solutions are $t = \dfrac{-k + \sqrt{k^2 + 8s}}{4}$ and $t = \dfrac{-k - \sqrt{k^2 + 8s}}{4}$.

◀ Work Problem **2** at the Side.

OBJECTIVE 2 Solve applied problems using the Pythagorean formula. The Pythagorean formula

$$a^2 + b^2 = c^2,$$

illustrated by the figure in the margin, was introduced in **Section 9.3** and is used to solve applications involving right triangles. Such problems often require solving quadratic equations.

Example 3 Using the Pythagorean Formula

Two cars left an intersection at the same time, one heading due north, the other due west. Some time later, they were exactly 100 mi apart. The car headed north had gone 20 mi farther than the car headed west. How far had each car traveled?

Step 1 **Read** the problem carefully.

Step 2 **Assign a variable.**

Let x = the distance traveled by the car headed west.

Then $x + 20$ = the distance traveled by the car headed north.

See Figure 2. The cars are 100 mi apart, so the hypotenuse of the right triangle equals 100.

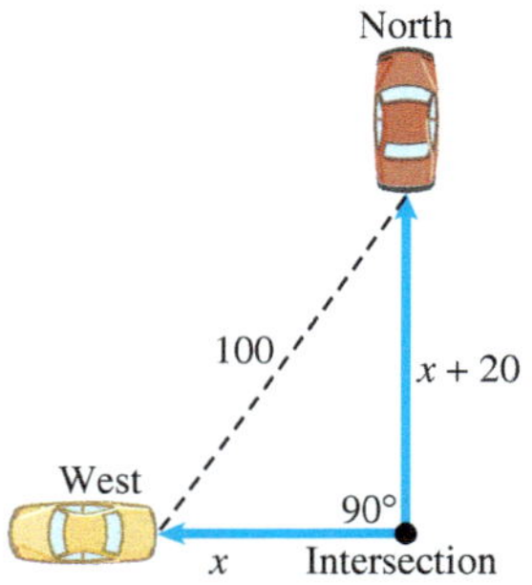

Figure 2

Step 3 **Write an equation.** Use the Pythagorean formula.

$$a^2 + b^2 = c^2$$

$$x^2 + (x + 20)^2 = 100^2$$

$(x + y)^2 = x^2 + 2xy + y^2$

Step 4 **Solve.**

$$x^2 + x^2 + 40x + 400 = 10{,}000 \qquad \text{Square the binomial.}$$

$$2x^2 + 40x - 9600 = 0 \qquad \text{Standard form}$$

$$x^2 + 20x - 4800 = 0 \qquad \text{Divide by 2.}$$

$$(x + 80)(x - 60) = 0 \qquad \text{Factor.}$$

$$x + 80 = 0 \quad \text{or} \quad x - 60 = 0 \qquad \text{Zero-factor property}$$

$$x = -80 \quad \text{or} \quad x = 60 \qquad \text{Solve for } x.$$

Step 5 **State the answer.** Distance cannot be negative, so discard the negative solution. The distances are 60 mi and 60 + 20 = 80 mi.

Step 6 **Check.** Since $60^2 + 80^2 = 100^2$, the answers are correct.

◀ Work Problem **3** at the Side.

ANSWERS

2. $t = \dfrac{5 + \sqrt{25 - 8k}}{4},\ t = \dfrac{5 - \sqrt{25 - 8k}}{4}$

3. 5 ft

OBJECTIVE 3 Solve applied problems using area formulas.

Example 4 Solving an Area Problem

A rectangular reflecting pool in a park is 20 ft wide and 30 ft long. The park gardener wants to plant a strip of grass of uniform width around the edge of the pool. She has enough seed to cover 336 ft^2. How wide will the strip be?

Step 1 **Read** the problem carefully.

Step 2 **Assign a variable.** The pool is shown in Figure 3. If x represents the unknown width of the grass strip, the width of the large rectangle is given by $20 + 2x$ (the width of the pool plus two grass strips), and the length is given by $30 + 2x$.

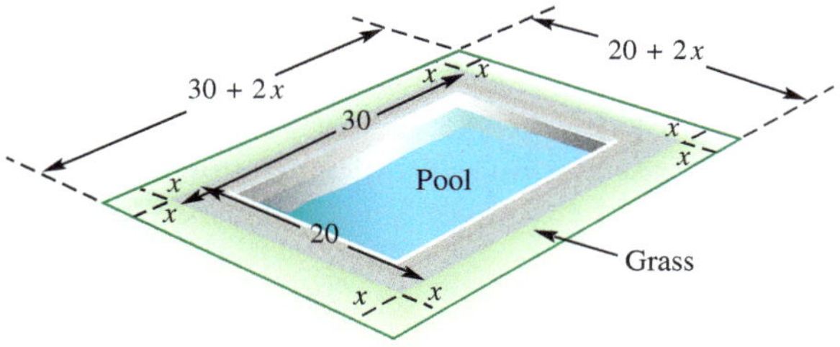

Figure 3

Step 3 **Write an equation.** The area of the large rectangle is given by the product of its length and width, $(30 + 2x)(20 + 2x)$. The area of the pool is $30 \cdot 20 = 600$ ft^2. The area of the large rectangle, minus the area of the pool, should equal the area of the grass strip. Since the area of the grass strip is to be 336 ft^2, the equation is

$$\underbrace{(30 + 2x)(20 + 2x)}_{\text{Area of rectangle}} - \underbrace{600}_{\text{area of pool}} = \underbrace{336.}_{\text{area of grass.}}$$

Step 4 **Solve.**

$$600 + 100x + 4x^2 - 600 = 336 \quad \text{Multiply.}$$
$$4x^2 + 100x - 336 = 0 \quad \text{Standard form}$$
$$x^2 + 25x - 84 = 0 \quad \text{Divide by 4.}$$
$$(x + 28)(x - 3) = 0 \quad \text{Factor.}$$
$$x = -28 \quad \text{or} \quad x = 3 \quad \text{Zero-factor property}$$

Step 5 **State the answer.** The width cannot be -28 ft, so the grass strip should be 3 ft wide.

Step 6 **Check.** If $x = 3$, then the area of the large rectangle is

$$(30 + 2 \cdot 3)(20 + 2 \cdot 3) = 36 \cdot 26 = 936 \text{ ft}^2. \quad \text{Area of pool and strip}$$

The area of the pool is $30 \cdot 20 = 600$ ft^2. So, the area of the grass strip is $936 - 600 = 336$ ft^2, as required. The answer is correct.

Work Problem 4 at the Side. ▶

OBJECTIVE 4 Solve applied problems using quadratic functions as models. Some applied problems can be modeled by *quadratic functions,* which can be written in the form

$$f(x) = ax^2 + bx + c,$$

for real numbers a, b, and c, with $a \neq 0$.

4 Solve the problem.

Suppose the pool in Example 4 is 20 ft by 40 ft and there is enough seed to cover 700 ft^2. How wide should the grass strip be?

ANSWER

4. 5 ft

5 Solve the problem.

A ball is projected vertically upward from the ground. Its distance in feet from the ground at t seconds is

$$s(t) = -16t^2 + 64t.$$

At what times will the ball be 32 ft from the ground? Use a calculator and round answers to the nearest tenth. (*Hint:* There are two answers.)

6 Use a calculator to evaluate

$$\frac{-14.8 \pm \sqrt{14.8^2 - 4(-0.065)(-301)}}{2(-0.065)}$$

for both solutions. Round to the nearest whole number. Which solution is valid for this problem?

ANSWERS

5. at 0.6 sec and at 3.4 sec
6. 23; 205; 23

Example 5 Solving an Applied Problem Using a Quadratic Function

If an object is projected upward from the top of a 144-ft building at 112 ft per sec, its position (in feet above the ground) is given by

$$s(t) = -16t^2 + 112t + 144,$$

where t is time in seconds after it was propelled. When does it hit the ground?

When the object hits the ground, its distance above the ground is 0. We must find the value of t that makes $s(t) = 0$.

$0 = -16t^2 + 112t + 144$ — Let $s(t) = 0$.

$0 = t^2 - 7t - 9$ — Divide by -16.

$$t = \frac{-(-7) \pm \sqrt{(-7)^2 - 4(1)(-9)}}{2(1)} = \frac{7 \pm \sqrt{85}}{2} \approx \frac{7 \pm 9.2}{2}$$ — Use the quadratic formula and a calculator.

The solutions are $t \approx 8.1$ or $t \approx -1.1$. Since time cannot be negative, discard the negative solution. The object will hit the ground about 8.1 sec after it is projected.

◀ Work Problem **5** at the Side.

Example 6 Using a Quadratic Function to Model the CPI

The Consumer Price Index (CPI) is used to measure trends in prices for a "basket" of goods purchased by typical American families. This index uses a base year of 1967, which means that the index number for 1967 is 100. The quadratic function defined by

$$f(x) = -0.065x^2 + 14.8x + 249$$

approximates the CPI for the years 1980–2005, where x is the number of years that have elapsed since 1980. (*Source:* Bureau of Labor Statistics.)

(a) Use the model to approximate the CPI for 1995.
For 1995, $x = 1995 - 1980 = 15$, so find $f(15)$.

$f(x) = -0.065x^2 + 14.8x + 249$ — Given model

$f(15) = -0.065(15)^2 + 14.8(15) + 249$ — Let $x = 15$.

$f(15) \approx 456$ — Nearest whole number

The CPI for 1995 was about 456.

(b) In what year did the CPI reach 550?
Find the value of x that makes $f(x) = 550$.

$f(x) = -0.065x^2 + 14.8x + 249$ — Given model

$550 = -0.065x^2 + 14.8x + 249$ — Let $f(x) = 550$.

$0 = -0.065x^2 + 14.8x - 301$ — Standard form

Now use $a = -0.065$, $b = 14.8$, and $c = -301$ in the quadratic formula.

◀ Work Problem **6** at the Side.

The first solution is $x \approx 23$. Rounding up to the next whole number, the CPI first reached 550 in $1980 + 23 = 2003$. (Reject the solution $x \approx 205$, as this corresponds to a year far beyond the period covered by the model.)

18.5 Exercises

FOR EXTRA HELP MyMathLab MathXL PRACTICE WATCH DOWNLOAD READ REVIEW

1. What is the first step in solving a formula like $gw^2 = 2r$ for w?

2. What is the first step in solving a formula like $gw^2 = kw + 24$ for w?

In Exercises 3 and 4, solve for m in terms of the other variables ($m > 0$).

3.

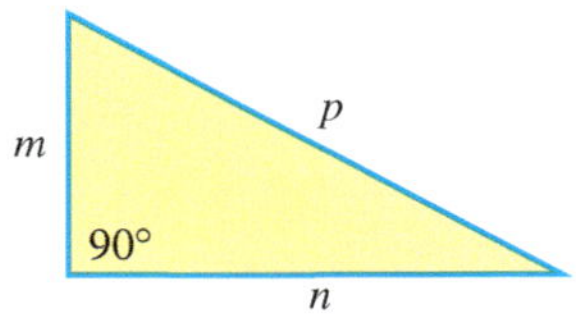

4.

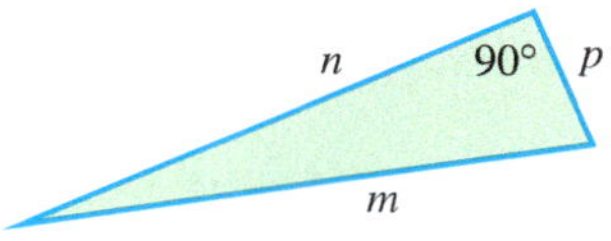

Solve each equation for the indicated variable. (Leave ± in your answers.) See Examples 1 and 2.

5. $d = kt^2$ for t

6. $s = kwd^2$ for d

7. $I = \frac{ks}{d^2}$ for d

8. $R = \frac{k}{d^2}$ for d

9. $F = \frac{kA}{v^2}$ for v

10. $L = \frac{kd^4}{h^2}$ for h

11. $V = \frac{1}{3}\pi r^2 h$ for r

12. $V = \pi(r^2 + R^2)h$ for r

13. $At^2 + Bt = -C$ for t

14. $S = 2\pi rh + \pi r^2$ for r

15. $D = \sqrt{kh}$ for h

16. $F = \frac{k}{\sqrt{d}}$ for d

17. $p = \sqrt{\frac{k\ell}{g}}$ for ℓ

18. $p = \sqrt{\frac{k\ell}{g}}$ for g

Solve each problem. When appropriate, round answers to the nearest tenth. See Example 3.

19. Find the lengths of the sides of the triangle.

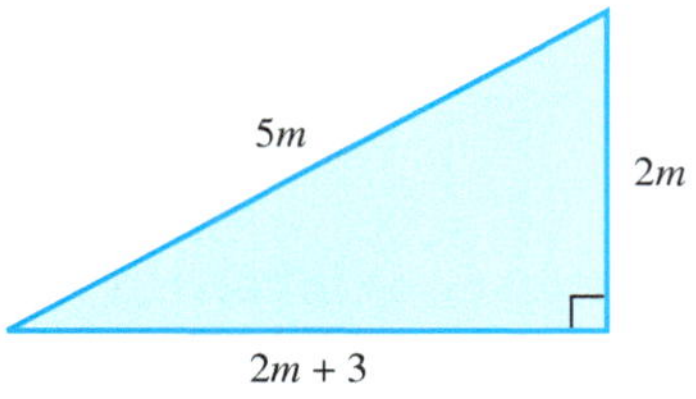

20. Find the lengths of the sides of the triangle.

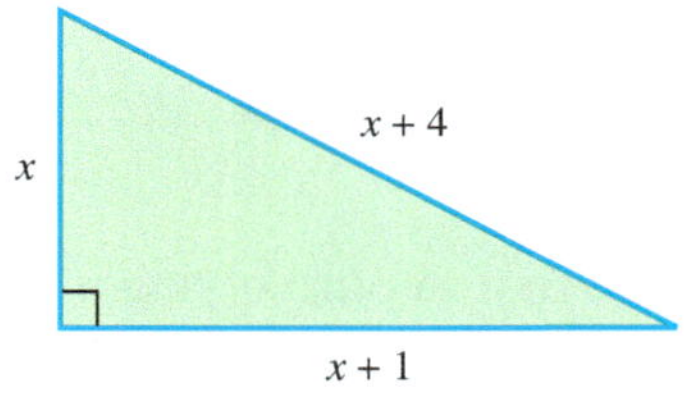

21. Two ships leave port at the same time, one heading due south and the other heading due east. Several hours later, they are 170 mi apart. If the ship traveling south traveled 70 mi farther than the other, how many miles did they each travel?

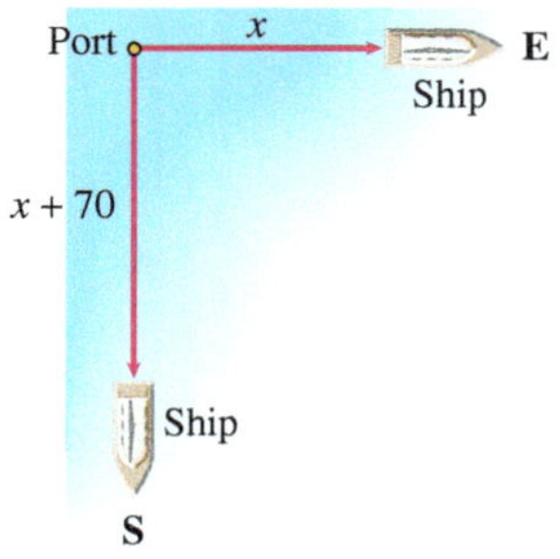

22. Faith Varnado is flying a kite that is 30 ft farther above her hand than its horizontal distance from her. The string from her hand to the kite is 150 ft long. How high is the kite?

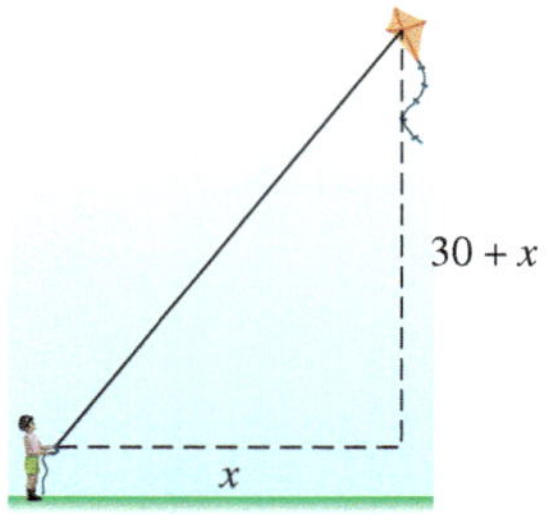

Solve each problem. See Example 4.

23. A couple wants to buy a rug for a room that is 20 ft long and 15 ft wide. They want to leave an even strip of flooring uncovered around the edges of the room. How wide a strip will they have if they buy a rug with an area of 234 ft^2?

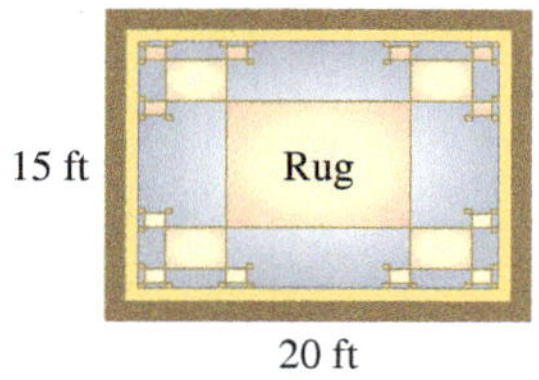

24. A club swimming pool is 30 ft wide and 40 ft long. The club members want an exposed aggregate border in a strip of uniform width around the pool. They have enough material for 296 ft^2. How wide can the strip be?

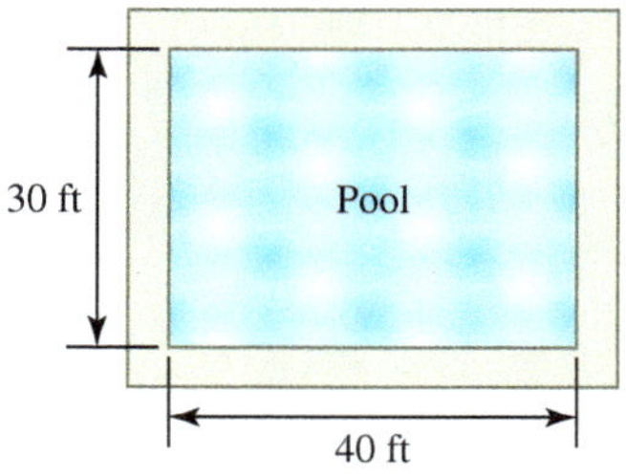

25. A rectangular piece of sheet metal has a length that is 4 in. less than twice the width. A square piece 2 in. on a side is cut from each corner. The sides are then turned up to form an uncovered box of volume 256 $in.^3$. Find the length and width of the original piece of metal.

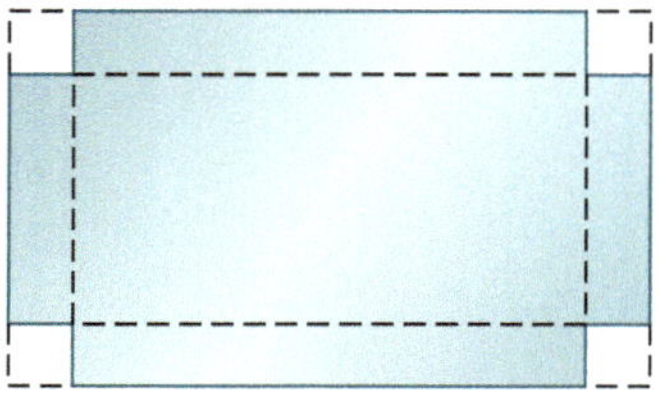

26. Another rectangular piece of sheet metal is 2 in. longer than it is wide. A square piece 3 in. on a side is cut from each corner. The sides are then turned up to form an uncovered box of volume 765 $in.^3$. Find the dimensions of the original piece of metal.

Solve each problem. Round answers to the nearest tenth. See Example 5.

27. A ball is projected upward from the ground. Its distance in feet from the ground in t seconds is given by

$$s(t) = -16t^2 + 128t.$$

At what times will the ball be 213 ft from the ground?

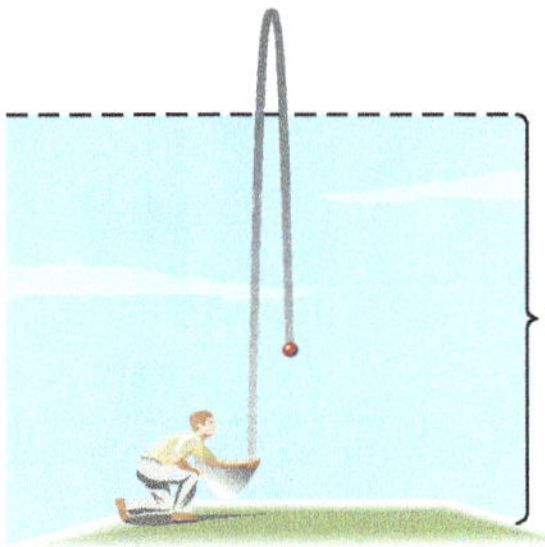

28. A toy rocket is launched from ground level. Its distance in feet from the ground in t seconds is given by

$$s(t) = -16t^2 + 208t.$$

At what times will the rocket be 550 ft from the ground?

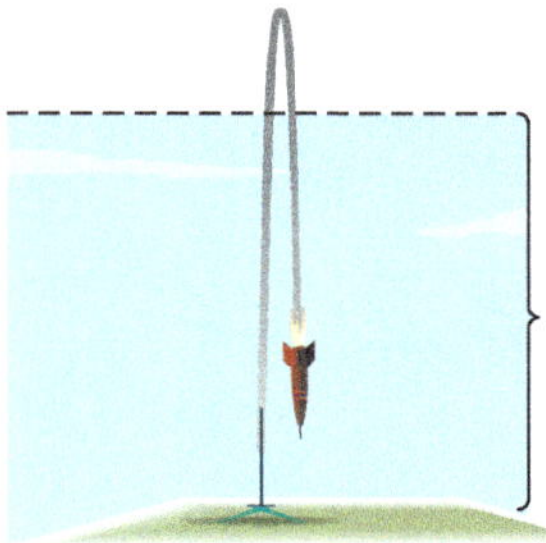

29. The function defined by

$$D(t) = 13t^2 - 100t$$

gives the distance in feet a car going approximately 68 mph will skid in t seconds. Find the time it would take for the car to skid 180 ft.

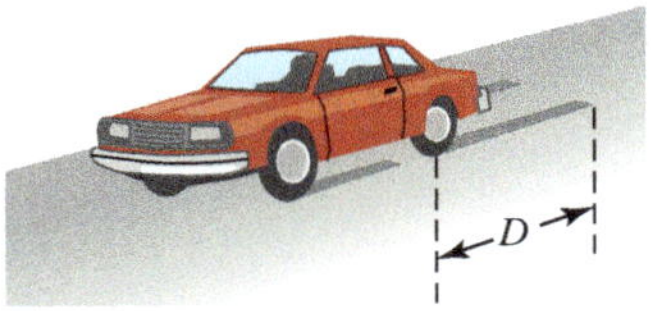

30. The function given in Exercise 29 becomes

$$D(t) = 13t^2 - 73t$$

for a car going 50 mph. Find the time for this car to skid 218 ft.

A ball is projected upward from ground level, and its distance in feet from the ground in t seconds is given by $s(t) = -16t^2 + 160t$. *Use algebra and a short explanation to answer Exercises 31 and 32.*

31. After how many seconds does it reach a height of 400 ft? How would you describe in words its position at this height?

32. After how many seconds does it reach a height of 425 ft? How would you interpret the mathematical result here?

Solve each problem using a quadratic equation.

33. A certain bakery has found that the daily demand for blueberry muffins is $\frac{3200}{p}$, where p is the price of a muffin in cents. The daily supply is $3p - 200$. Find the price at which supply and demand are equal.

34. In one area the demand for compact discs is $\frac{700}{P}$ per day, where P is the price in dollars per disc. The supply is $5P - 1$ per day. At what price, to the nearest cent, does supply equal demand?

The total number of miles traveled by all motor vehicles in the United States for the years 1994–2003 are shown in the bar graph and can be modeled by the quadratic function defined by

$$f(x) = -1.705x^2 + 75.93x + 2351.$$

Here, $x = 0$ represents 1994, $x = 1$ represents 1995, and so on. Use the graph and the model to work Exercises 35–38. See Example 6.

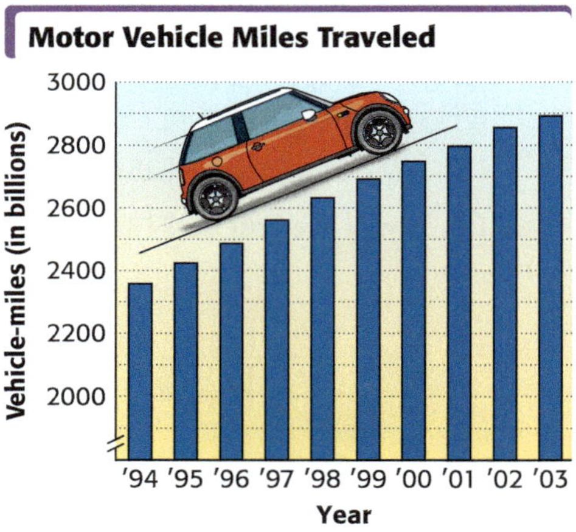

Source: U.S. Federal Highway Administration.

35. **(a)** Use the graph to estimate miles traveled in 2000 to the nearest ten billion.

(b) Use the model to approximate miles traveled in 2000 to the nearest ten billion. How does this result compare to your estimate from part (a)?

36. Based on the model, in what year did miles traveled reach 2600 billion? (Round down to the nearest year.) How does this result compare to the vehicle-miles shown in the graph?

37. Based on the model, in what year did miles traveled reach 2800 billion? (Round down to the nearest year.) How does this result compare to the vehicle-miles shown in the graph?

38. If these data were modeled by a *linear* function defined by $f(x) = ax + b$, would the value of a be positive or negative? Explain.

William Froude was a 19th-century naval architect who used the expression

$$\frac{v^2}{g\ell}$$

in shipbuilding. This expression, known as the ***Froude number,*** *was also used by R. McNeill Alexander in his research on dinosaurs.* (Source: *"How Dinosaurs Ran,"* Scientific American, *April 1991.*) *In Exercises 39 and 40, find to the nearest tenth the value of v (in meters per second), given that $g = 9.8$ m per sec^2.*

39. Rhinoceros: $\ell = 1.2$; Froude number $= 2.57$

40. Triceratops: $\ell = 2.8$; Froude number $= 0.16$

Recall from the ***Section 8.5*** *exercises that corresponding sides of similar triangles are proportional. Use this fact to find the lengths of the indicated sides of each pair of similar triangles. Check all possible solutions in both triangles. Sides of a triangle cannot be negative (and are not drawn to scale here).*

41. Side AC

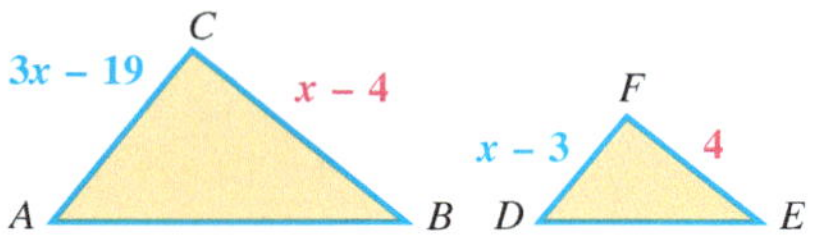

42. Side RQ

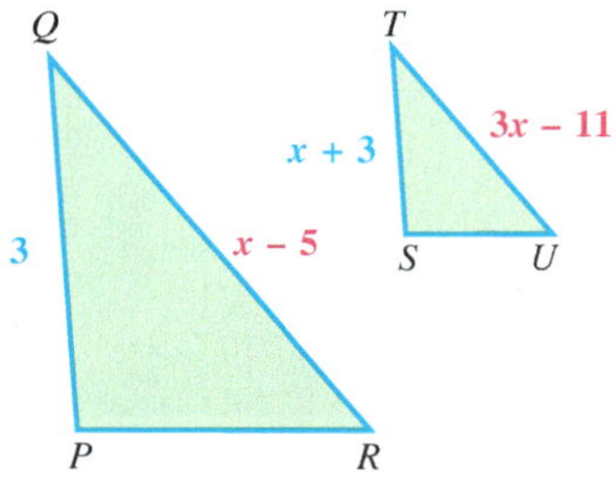

18.6 Graphs of Quadratic Functions

OBJECTIVES

1. Graph a quadratic function.
2. Graph parabolas with horizontal and vertical shifts.
3. Use the coefficient of x^2 to predict the shape and direction in which a parabola opens.
4. Find a quadratic function to model data.

OBJECTIVE 1 Graph a quadratic function. Figure 4 gives a graph of the simplest *quadratic function,* defined by $y = x^2$.

x	y
-2	4
-1	1
0	0
1	1
2	4

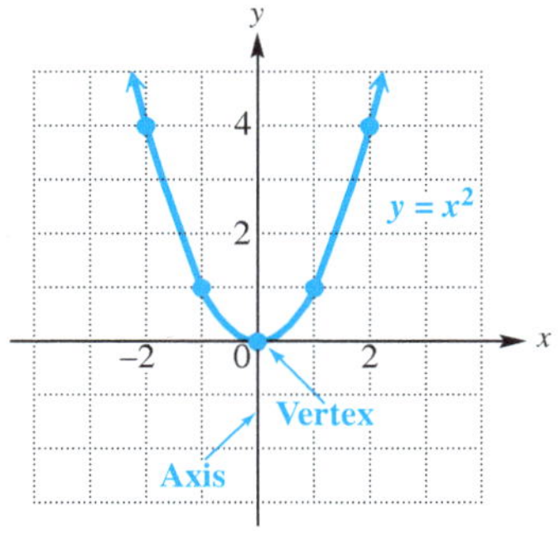

Figure 4

As mentioned in **Section 6.3,** this graph is called a **parabola.** The point (0, 0), the lowest point on the curve, is the **vertex** of this parabola. The vertical line through the vertex is the **axis** of the parabola, here $x = 0$. A parabola is **symmetric about its axis;** that is, if the graph were folded along the axis, the two portions of the curve would coincide. As Figure 4 suggests, x can be any real number, so the domain of the function defined by $y = x^2$ is $(-\infty, \infty)$. Since y is always nonnegative, the range is $[0, \infty)$.

In **Section 18.5,** we solved applications modeled by quadratic functions.

Quadratic Function

A function that can be written in the form

$$f(x) = ax^2 + bx + c$$

for real numbers a, b, and c, with $a \neq 0$, is a **quadratic function.**

The graph of any quadratic function is a parabola with a vertical axis.

Note

We use the variable y and function notation $f(x)$ interchangeably. Although we use the letter f most often to name quadratic functions, other letters can be used. We use the capital letter F to distinguish between different parabolas graphed on the same coordinate axes.

Parabolas, which are a type of *conic section* **(Chapter 12),** have many applications. Cross sections of telescopes, satellite dishes and automobile headlights form parabolas, as do the cables that support suspension bridges.

OBJECTIVE 2 Graph parabolas with horizontal and vertical shifts. Parabolas need not have their vertices at the origin, as does the graph of $f(x) = x^2$. To graph a parabola of the form

$$F(x) = x^2 + k, \qquad \text{Vertical shift } k$$

select sample values of x like those that were used to graph $f(x) = x^2$. The corresponding values of $F(x)$ in $F(x) = x^2 + k$ differ by k from those of $f(x) = x^2$. For this reason, the graph of $F(x) = x^2 + k$ is *shifted,* or *translated,* k units vertically compared with that of $f(x) = x^2$.

1 Graph each parabola. Give the vertex, domain, and range.

(a) $f(x) = x^2 + 3$

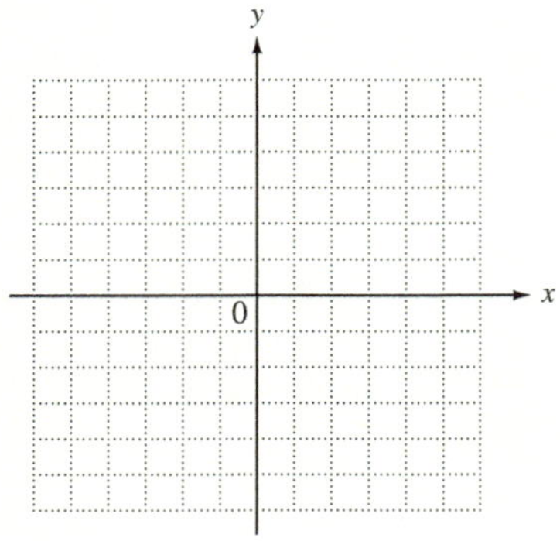

(b) $f(x) = x^2 - 1$

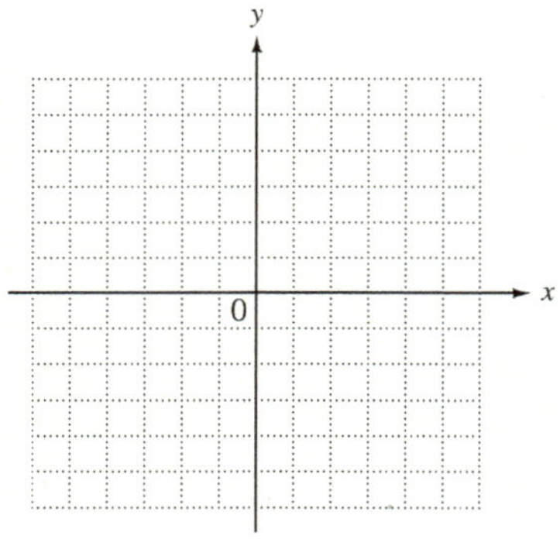

Example 1 Graphing a Parabola with a Vertical Shift

Graph $F(x) = x^2 - 2$.

This graph has the same shape as that of $f(x) = x^2$, but since k here is -2, the graph is shifted 2 units down, with vertex $(0, -2)$. Every function value is 2 less than the corresponding function value of $f(x) = x^2$. Plotting points on both sides of the vertex gives the graph in Figure 5.

x	$f(x) = x^2$	$F(x) = x^2 - 2$
−2	4	2
−1	1	−1
0	0	−2
1	1	−1
2	4	2

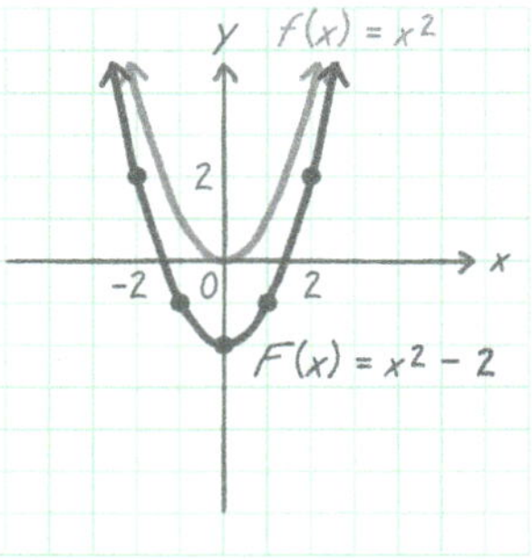

Figure 5

Notice that since the parabola is symmetric about its axis $x = 0$, the plotted points are "mirror images" of each other. Since x can be any real number, the domain is still $(-\infty, \infty)$. The value of y (or $F(x)$) is always greater than or equal to -2, so the range is $[-2, \infty)$. The graph of $f(x) = x^2$ is shown for comparison.

Vertical Shift

The graph of $\boldsymbol{F(x) = x^2 + k}$ is a parabola with the same shape as the graph of $f(x) = x^2$. The parabola is shifted vertically: k units up if $k > 0$, and $|k|$ units down if $k < 0$. The vertex is $(0, k)$.

◀ Work Problem **1** *at the Side.*

The graph of the function defined by

$$\boldsymbol{F(x) = (x - h)^2} \qquad \text{Horizontal shift } h$$

is also a parabola with the same shape as that of $f(x) = x^2$. Because $(x - h)^2 \geq 0$ for all x, the vertex of $F(x) = (x - h)^2$ is the lowest point on the parabola. The lowest point occurs here when $F(x)$ is 0. To get $F(x)$ equal to 0, let $x = h$ so the vertex of $F(x) = (x - h)^2$ is $(h, 0)$. Based on this, the graph of $F(x) = (x - h)^2$ is shifted h units horizontally compared with that of $f(x) = x^2$.

Example 2 Graphing a Parabola with a Horizontal Shift

Graph $F(x) = (x - 2)^2$.

If $x = 2$, then $F(x) = 0$, which gives the vertex $(2, 0)$. The graph of $F(x) = (x - 2)^2$ has the same shape as that of $f(x) = x^2$ but is shifted 2 units to the right. Plotting several points on one side of the vertex and using symmetry about the axis $x = 2$ to find corresponding points on the other side of the vertex gives the graph in Figure 6 on the next page. Again, the domain is $(-\infty, \infty)$; the range is $[0, \infty)$.

Continued on Next Page

Answers

1. (a)

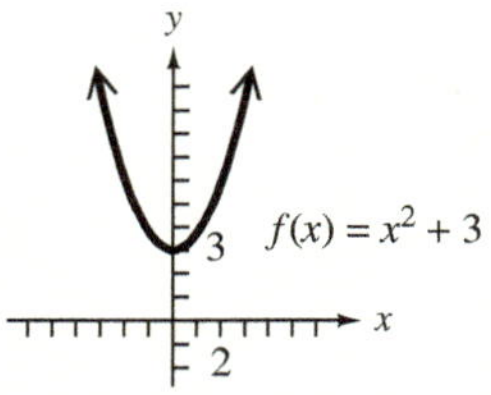

vertex: $(0, 3)$; domain: $(-\infty, \infty)$; range: $[3, \infty)$

(b)

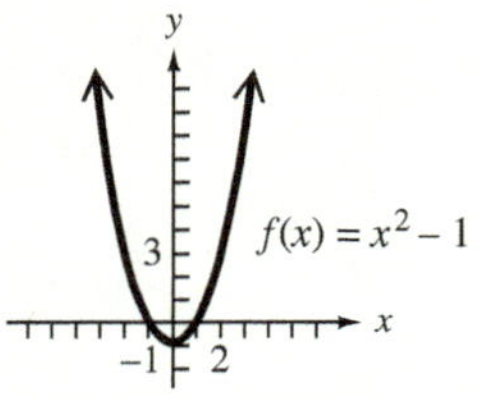

vertex: $(0, -1)$; domain: $(-\infty, \infty)$; range: $[-1, \infty)$

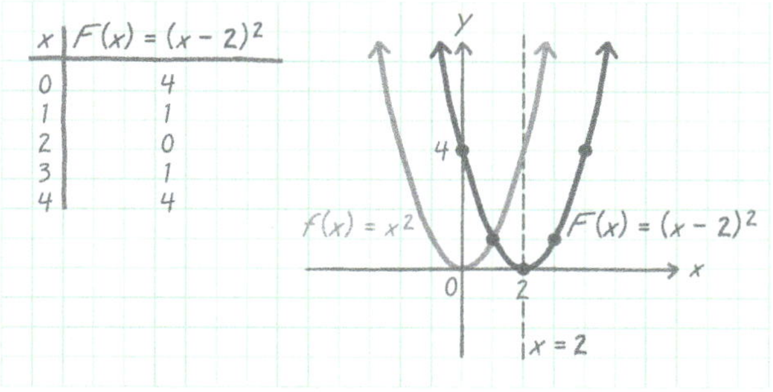

Figure 6

Horizontal Shift

The graph of $\boldsymbol{F(x) = (x - h)^2}$ is a parabola with the same shape as the graph of $f(x) = x^2$. The parabola is shifted h units horizontally: h units to the right if $h > 0$, and $|h|$ units to the left if $h < 0$. The vertex is $(h, 0)$.

CAUTION

Errors frequently occur when horizontal shifts are involved. To determine the direction and magnitude of a horizontal shift, find the value that would cause the expression $x - h$ to equal 0. For example, the graph of $F(x) = (x - 5)^2$ would be shifted 5 units to the *right,* because $+5$ would cause $x - 5$ to equal 0. On the other hand, the graph of $F(x) = (x + 5)^2$ would be shifted 5 units to the *left,* because -5 would cause $x + 5$ to equal 0.

Work Problem **2** at the Side. ▶

A parabola can have both horizontal and vertical shifts.

Example 3 Graphing a Parabola with Horizontal and Vertical Shifts

Graph $F(x) = (x + 3)^2 - 2$.

This graph has the same shape as that of $f(x) = x^2$, but is shifted 3 units to the left (since $x + 3 = 0$ if $x = -3$) and 2 units down (because of the -2). See Figure 7. The vertex is $(-3, -2)$, with axis $x = -3$. This function has domain $(-\infty, \infty)$ and range $[-2, \infty)$.

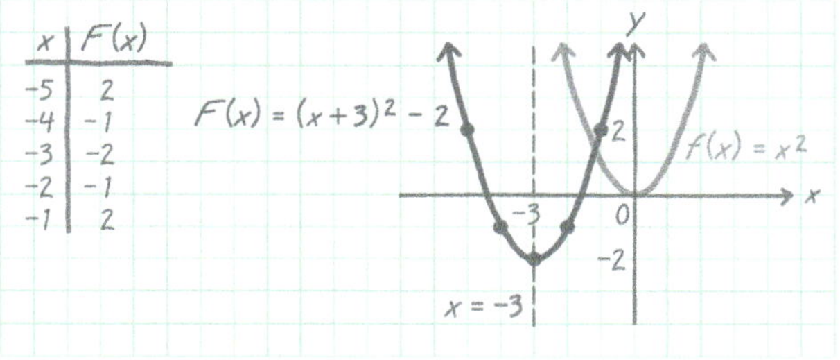

Figure 7

2 Graph each parabola. Give the vertex, axis, domain, and range.

(a) $f(x) = (x - 3)^2$

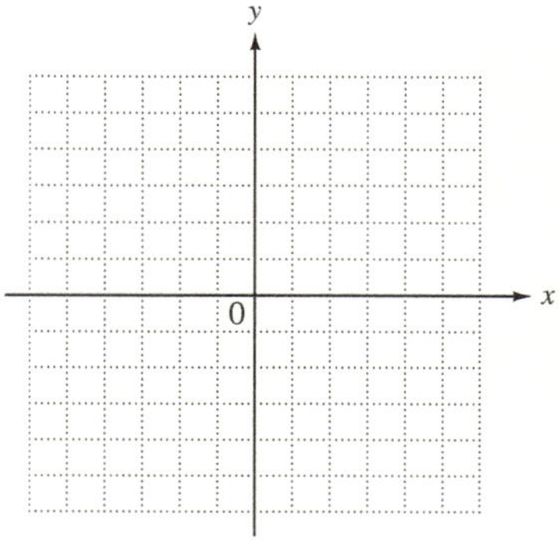

(b) $f(x) = (x + 2)^2$

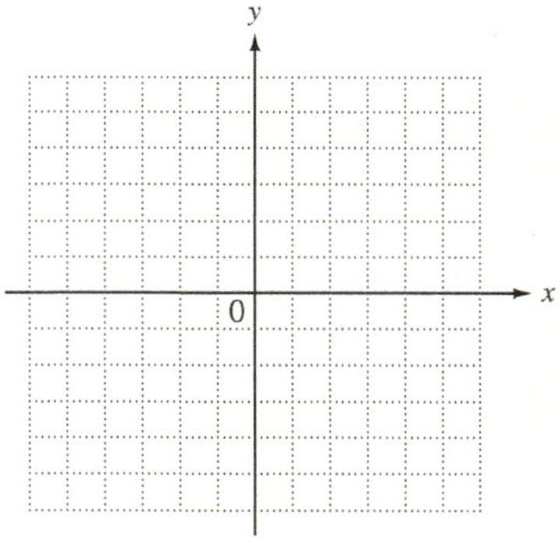

ANSWERS

2. (a)

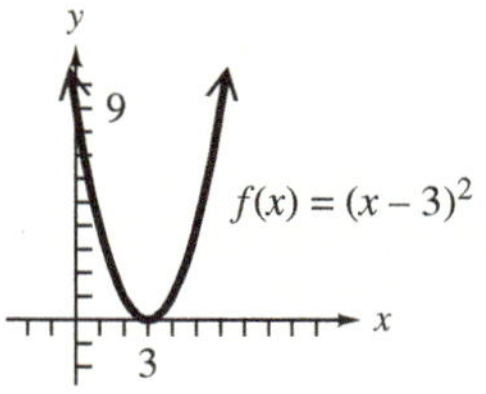

vertex: $(3, 0)$; axis: $x = 3$; domain: $(-\infty, \infty)$; range: $[0, \infty)$

(b)

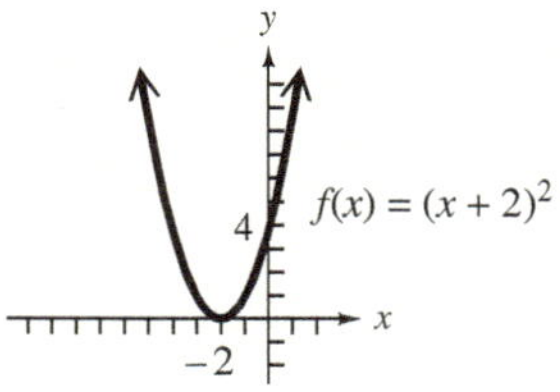

vertex: $(-2, 0)$; axis: $x = -2$; domain: $(-\infty, \infty)$; range: $[0, \infty)$

The characteristics of the graph of a parabola with equation of the form $F(x) = (x - h)^2 + k$ are summarized as follows.

Vertex and Axis of a Parabola

The graph of $\boldsymbol{F(x) = (x - h)^2 + k}$ is a parabola with the same shape as the graph of $f(x) = x^2$ with vertex (h, k). The axis is the vertical line $x = h$.

◀ Work Problem 3 at the Side.

3 Graph each parabola. Give the vertex, axis, domain, and range.

(a) $f(x) = (x + 2)^2 - 1$

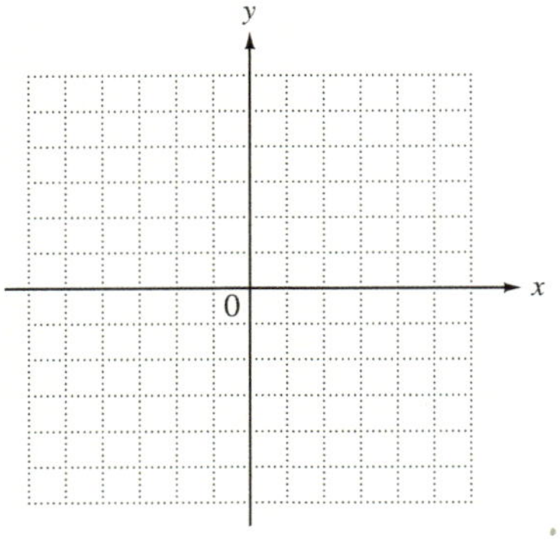

(b) $f(x) = (x - 2)^2 + 5$

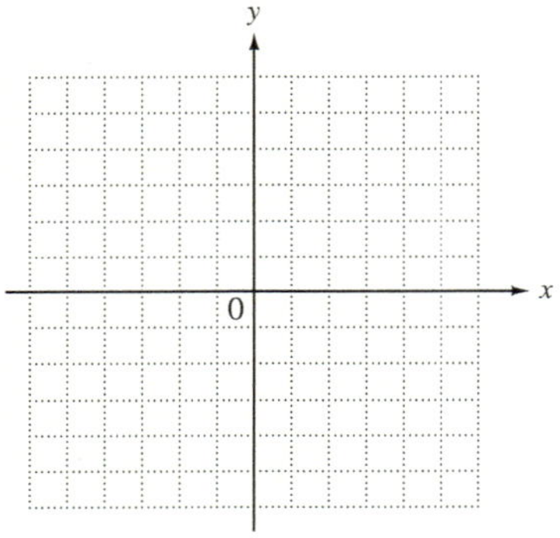

OBJECTIVE 3 Use the coefficient of x^2 to predict the shape and direction in which a parabola opens. Not all parabolas open up, and not all parabolas have the same shape as the graph of $f(x) = x^2$.

Example 4 Graphing a Parabola That Opens Down

Graph $f(x) = -\frac{1}{2}x^2$.

This parabola is shown in Figure 8. The coefficient $-\frac{1}{2}$ affects the shape of the graph; the $\frac{1}{2}$ makes the parabola wider (since the values of $\frac{1}{2}x^2$ increase more slowly than those of x^2), and the negative sign makes the parabola open down. The graph is not shifted in any direction; the vertex is still $(0, 0)$ and the axis has equation $x = 0$. Unlike the parabolas graphed in Examples 1–3, the vertex here has the *greatest* function value of any point on the graph. The domain is $(-\infty, \infty)$; the range is $(-\infty, 0]$.

x	$f(x)$
-2	-2
-1	$-\frac{1}{2}$
0	0
1	$-\frac{1}{2}$
2	-2

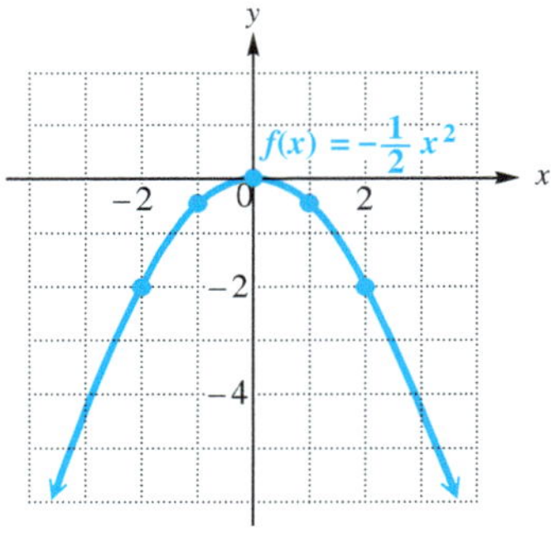

Figure 8

Some general principles concerning the graph of $F(x) = a(x - h)^2 + k$ are summarized as follows.

General Principles

1. The graph of the quadratic function defined by
$$F(x) = a(x - h)^2 + k, \quad a \neq 0,$$
is a parabola with vertex (h, k) and the vertical line $x = h$ as axis.
2. The graph opens up if a is positive and down if a is negative.
3. The graph is wider than that of $f(x) = x^2$ if $0 < |a| < 1$. The graph is narrower than that of $f(x) = x^2$ if $|a| > 1$.

ANSWERS

3. (a)

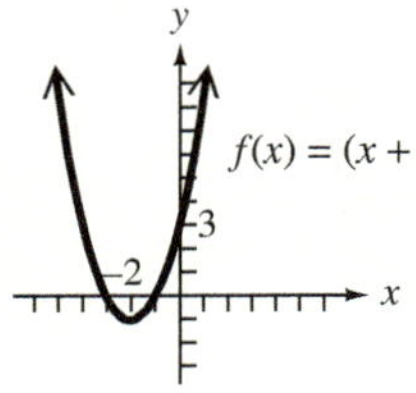

vertex: $(-2, -1)$; axis: $x = -2$; domain: $(-\infty, \infty)$; range: $[-1, \infty)$

(b)

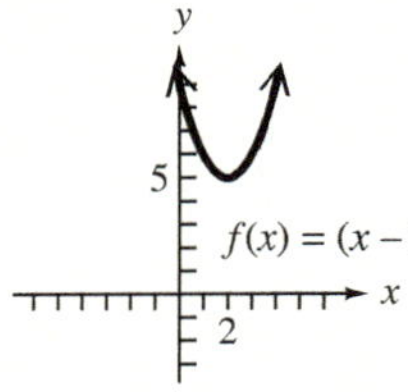

vertex: $(2, 5)$; axis: $x = 2$; domain: $(-\infty, \infty)$; range: $[5, \infty)$

Work Problems **4** *and* **5** *at the Side.* ▶

Example 5 **Using the General Principles to Graph a Parabola**

Graph $F(x) = -2(x + 3)^2 + 4$. Give the domain and the range.

The parabola opens down (because $a < 0$), and is narrower than the graph of $f(x) = x^2$, since $|-2| = 2 > 1$, causing values of $F(x)$ to decrease more quickly than those of $f(x) = -x^2$. This parabola has vertex $(-3, 4)$ as shown in Figure 9. To complete the graph, we plotted the ordered pairs $(-4, 2)$ and, by symmetry, $(-2, 2)$. Symmetry can be used to find additional ordered pairs that satisfy the equation, if desired. The domain is $(-\infty, \infty)$ and the range is $(-\infty, 4]$.

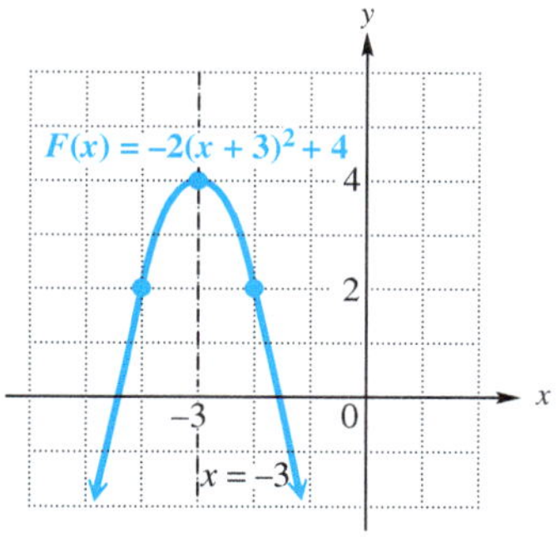

Figure 9

Work Problem **6** *at the Side.* ▶

OBJECTIVE 4 **Find a quadratic function to model data.**

Example 6 **Modeling the Number of Multiple Births**

After rising steadily over several decades, the number of higher-order multiple births (triplets or more) in the United States started to decline during the first decade of the 21st century. Let x represent the number of years since 1995 and y represent the number of higher-order multiple births. Data for selected years are shown in the table.

Year	x	y
1995	0	4973
1996	1	5939
1997	2	6737
1999	4	7321
2001	6	7471
2003	8	7663
2004	9	7275
2005	10	6694

Source: National Center for Health Statistics.

Find a quadratic function that models the data.

A scatter diagram of the ordered pairs (x, y) is shown in Figure 10 on the next page. The general shape suggested by the scatter diagram indicates that a parabola should approximate these points, as shown by the dashed curve in Figure 11. The equation for such a parabola would have a negative coefficient for x^2 since the graph opens down.

Continued on Next Page

4 Decide whether each parabola opens up or down.

(a) $f(x) = -\frac{2}{3}x^2$

(b) $f(x) = \frac{3}{4}x^2 + 1$

(c) $f(x) = -2x^2 - 3$

(d) $f(x) = 3x^2 + 2$

5 Decide whether each parabola in Problem 4 is wider or narrower than the graph of $f(x) = x^2$.

6 Graph

$$f(x) = \frac{1}{2}(x - 2)^2 + 1.$$

Give the domain and the range.

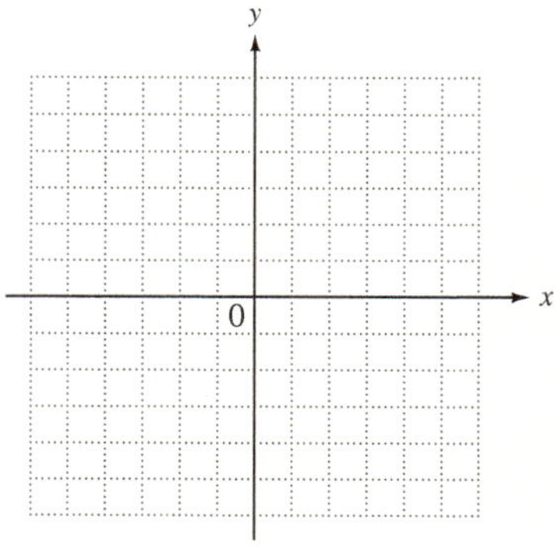

Answers

4. **(a)** down **(b)** up **(c)** down **(d)** up
5. **(a)** wider **(b)** wider **(c)** narrower **(d)** narrower
6.

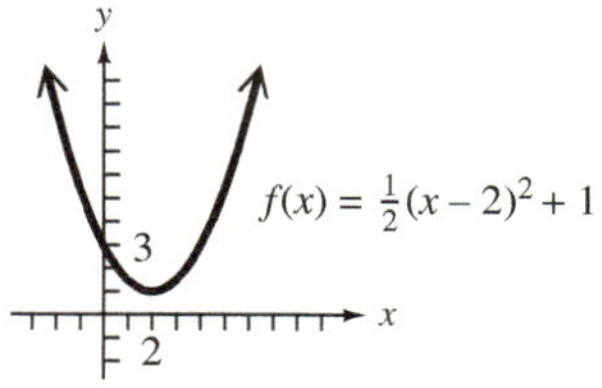

domain: $(-\infty, \infty)$
range: $[1, \infty)$

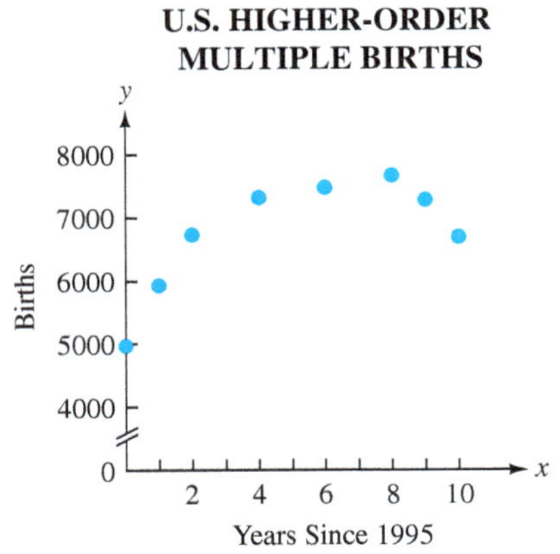

Figure 10

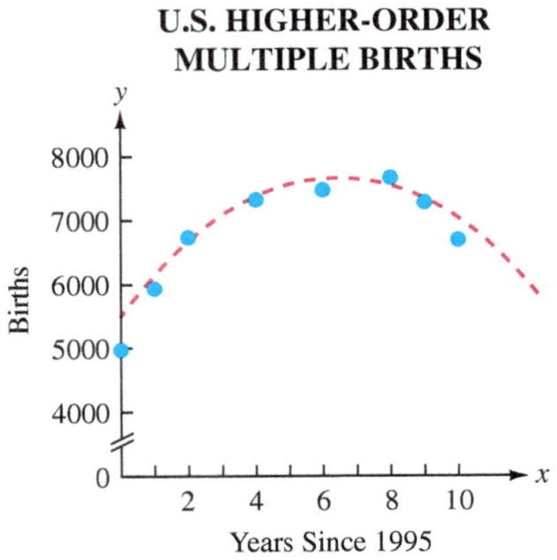

Figure 11

To find a quadratic function of the form

$$y = ax^2 + bx + c$$

that models, or *fits,* these data, we choose three representative ordered pairs and use them to write a system of three equations. Using (0, 4973), (4, 7321), and (10, 6694), we substitute the x- and y-values from the ordered pairs into the quadratic form $y = ax^2 + bx + c$ to get the three equations

$$a(0)^2 + b(0) + c = 4973 \quad \text{or} \quad c = 4973 \qquad (1)$$
$$a(4)^2 + b(4) + c = 7321 \quad \text{or} \quad 16a + 4b + c = 7321 \qquad (2)$$
$$a(10)^2 + b(10) + c = 6694 \quad \text{or} \quad 100a + 10b + c = 6694. \qquad (3)$$

We can find the values of a, b, and c by solving this system of three equations in three variables using the methods of **Section 5.2.** From equation (1), $c = 4973$. Substitute 4973 for c in equations (2) and (3) to obtain the equations

$$16a + 4b + 4973 = 7321, \quad \text{or} \quad 16a + 4b = 2348 \qquad (4)$$
$$100a + 10b + 4973 = 6694, \quad \text{or} \quad 100a + 10b = 1721. \qquad (5)$$

We can eliminate b from this system of equations in two variables by multiplying equation (4) by -5 and equation (5) by 2, and adding the results to get

$$120a = -8298$$
$$a = -69.15. \qquad \text{Divide by 120; use a calculator.}$$

We substitute -69.15 for a in equation (4) or (5) to find that $b = 863.6$. Using the values we have found for a, b, and c, our model is defined by

$$y = -69.15x^2 + 863.6b + 4973.$$

◀ *Work Problems* **7** *and* **8** *at the Side.*

Note

If we had chosen three different ordered pairs of data in Example 6, a slightly different model would have resulted, as in Problem 8 at the side.

Calculator Tip The *quadratic regression* feature on a graphing calculator can be used to generate a quadratic model that fits given data. See your owner's manual for details on how to do this.

7 Tell whether a linear or quadratic function would be a more appropriate model for each set of graphed data. If linear, tell whether the slope should be positive or negative. If quadratic, tell whether the coefficient a of x^2 should be positive or negative.

(a)

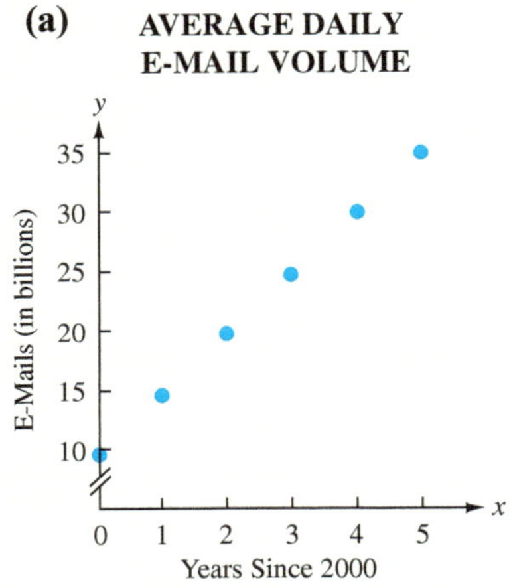

Source: General Accounting Office.

(b)

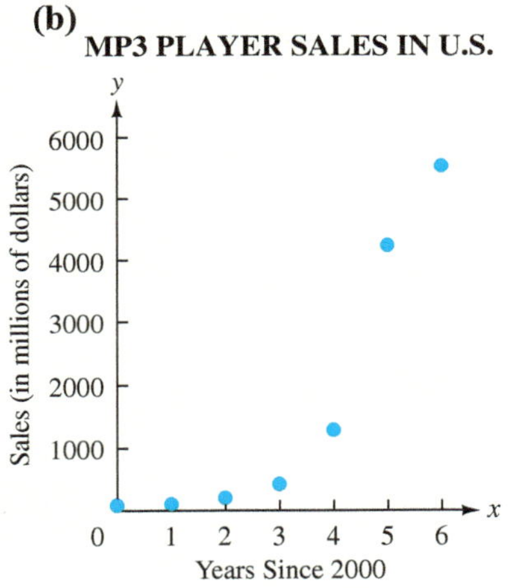

Source: Consumer Electronics Association.

8 Using the points (1, 5939), (6, 7471), and (10, 6694), find another quadratic model for the data on higher-order multiple births in Example 6.

Answers

7. **(a)** linear; positive **(b)** quadratic; positive
8. $y = -55.63x^2 + 695.80x + 5299$

18.6 Exercises

FOR EXTRA HELP MyMathLab — Math XL PRACTICE | WATCH | DOWNLOAD | READ | REVIEW

1. Match each quadratic function with its graph from choices A–D.

(a) $f(x) = (x + 2)^2 - 1$ **(b)** $f(x) = (x + 2)^2 + 1$ **(c)** $f(x) = (x - 2)^2 - 1$ **(d)** $f(x) = (x - 2)^2 + 1$

A.
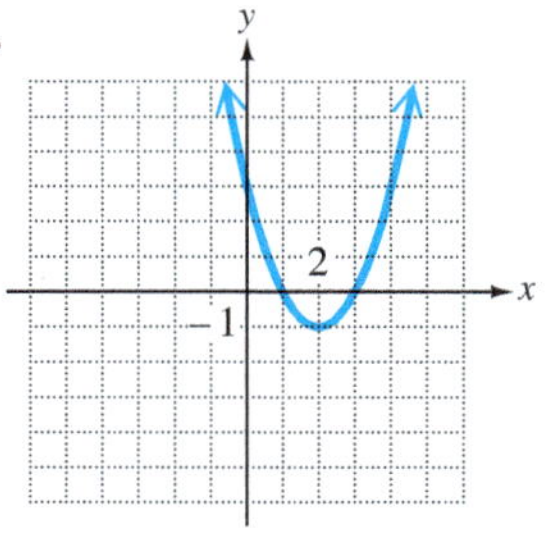

B.
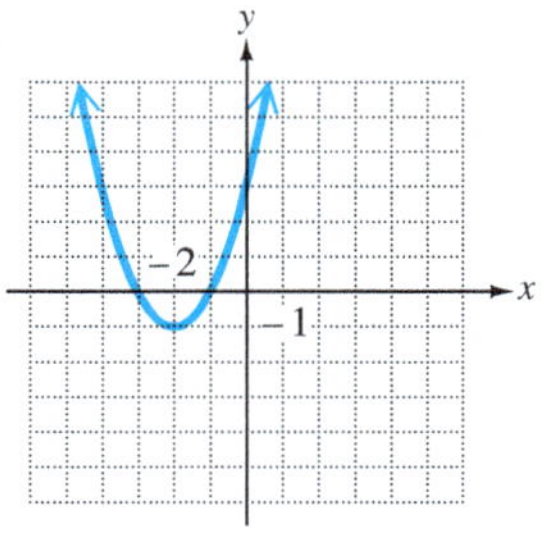

C.
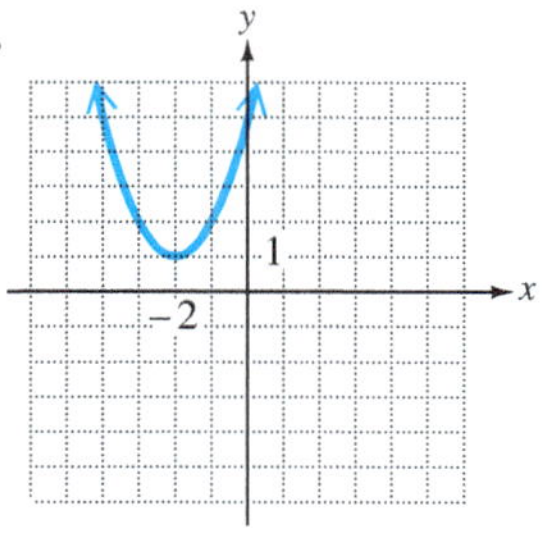

D.
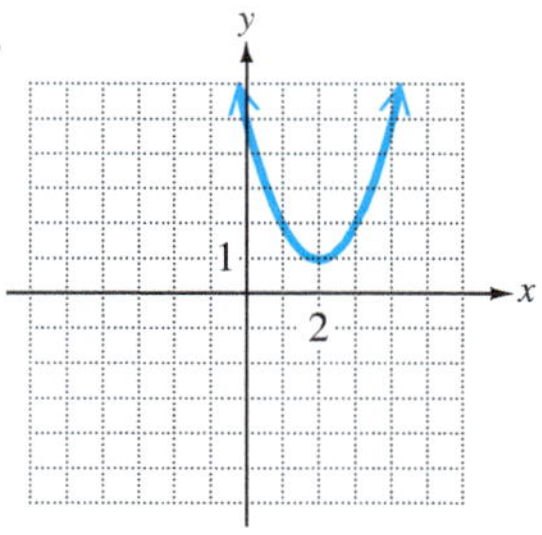

2. Match each quadratic function with its graph from choices A–D.

(a) $f(x) = -x^2 + 2$ **(b)** $f(x) = -x^2 - 2$ **(c)** $f(x) = -(x + 2)^2$ **(d)** $f(x) = -(x - 2)^2$

A.
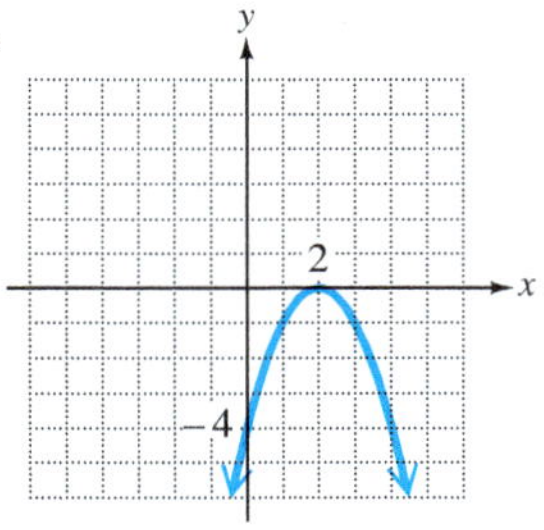

B.
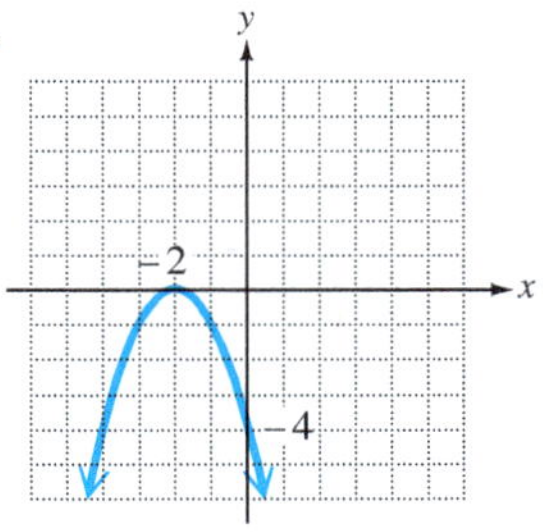

C.
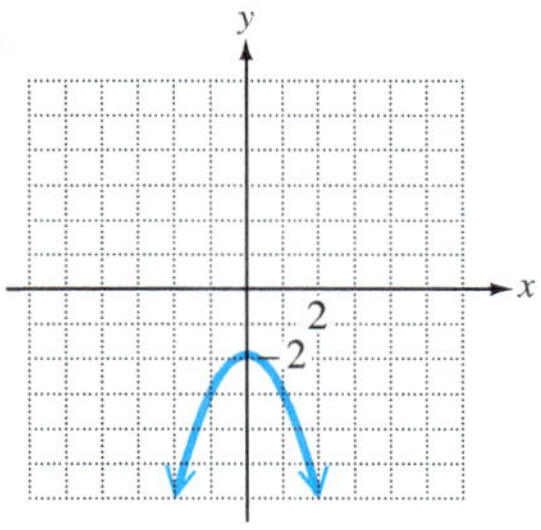

D.
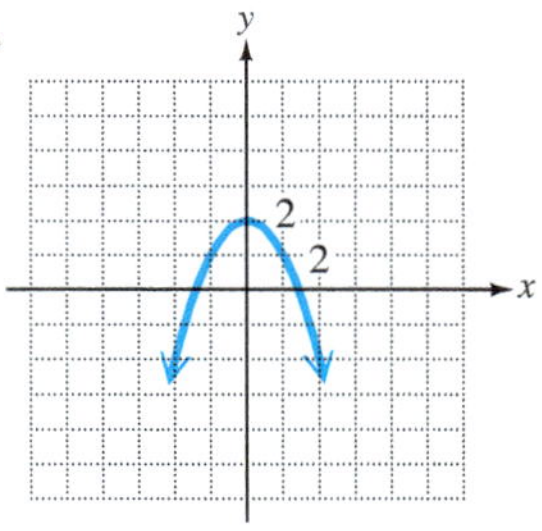

Identify the vertex of each parabola. See Examples 1–4.

3. $f(x) = -3x^2$

4. $f(x) = \frac{1}{2}x^2$

5. $f(x) = x^2 + 4$

6. $f(x) = x^2 - 4$

7. $f(x) = (x - 1)^2$

8. $f(x) = (x + 3)^2$

9. $f(x) = (x + 3)^2 - 4$

10. $f(x) = (x - 5)^2 - 8$

11. Describe how each of the parabolas in Exercises 9 and 10 is shifted compared to the graph of $f(x) = x^2$.

12. What does the value of a in $F(x) = a(x - h)^2 + k$ tell you about the graph of the equation compared to the graph of $f(x) = x^2$?

For each quadratic function, tell whether the graph opens up or down and whether the graph is wider, narrower, or the same shape as the graph of $f(x) = x^2$. See Examples 4 and 5.

13. $f(x) = -\frac{2}{5}x^2$

14. $f(x) = -2x^2$

15. $f(x) = 3x^2 + 1$

16. $f(x) = \frac{2}{3}x^2 - 4$

17. For $f(x) = a(x - h)^2 + k$, in what quadrant is the vertex if

(a) $h > 0, k > 0$;

(b) $h > 0, k < 0$;

(c) $h < 0, k > 0$;

(d) $h < 0, k < 0$?

18. Match each quadratic function with the description of the parabola that is its graph.

(a) $f(x) = (x - 4)^2 - 2$

(b) $f(x) = (x - 2)^2 - 4$

(c) $f(x) = -(x - 4)^2 - 2$

(d) $f(x) = -(x - 2)^2 - 4$

A. Vertex $(2, -4)$, opens down

B. Vertex $(2, -4)$, opens up

C. Vertex $(4, -2)$, opens down

D. Vertex $(4, -2)$, opens up

Sketch the graph of each parabola. Plot at least two points in addition to the vertex. In Exercises 25–32, give the vertex, axis, domain, and range of the parabola. See Examples 1–5.

19. $f(x) = -2x^2$

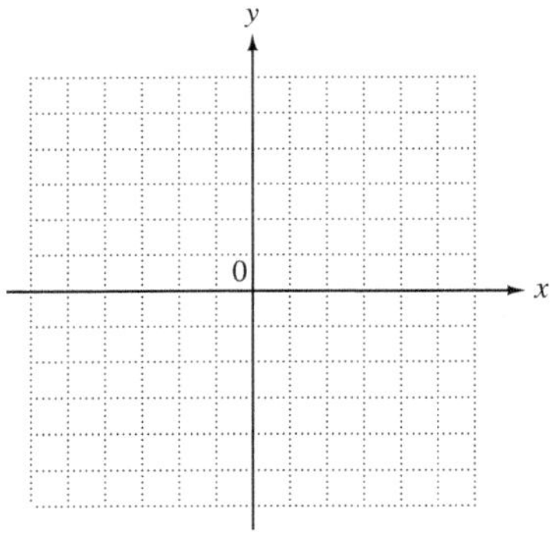

20. $f(x) = \frac{1}{3}x^2$

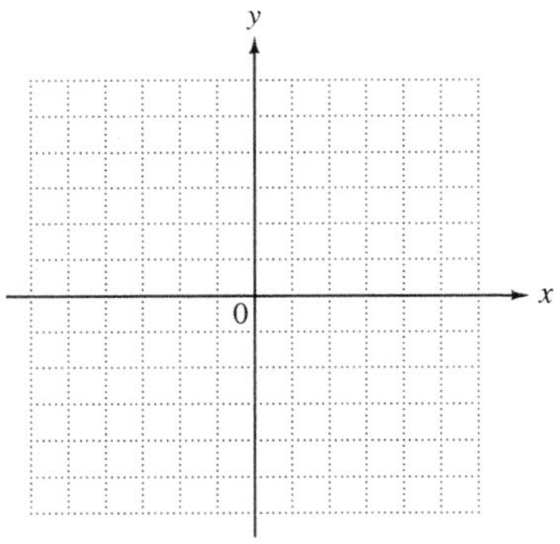

21. $f(x) = x^2 - 1$

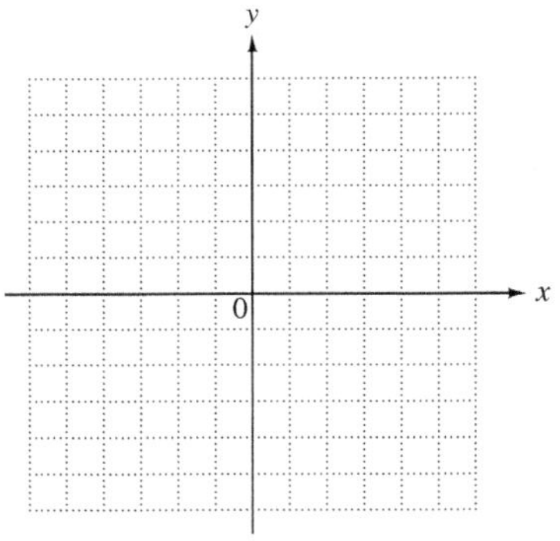

22. $f(x) = x^2 + 3$

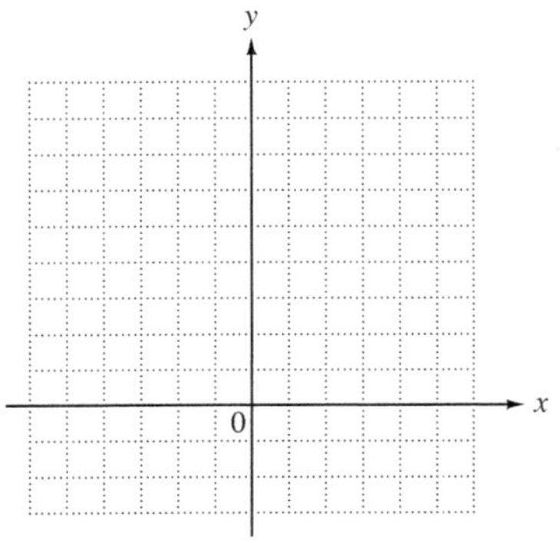

23. $f(x) = -x^2 + 2$

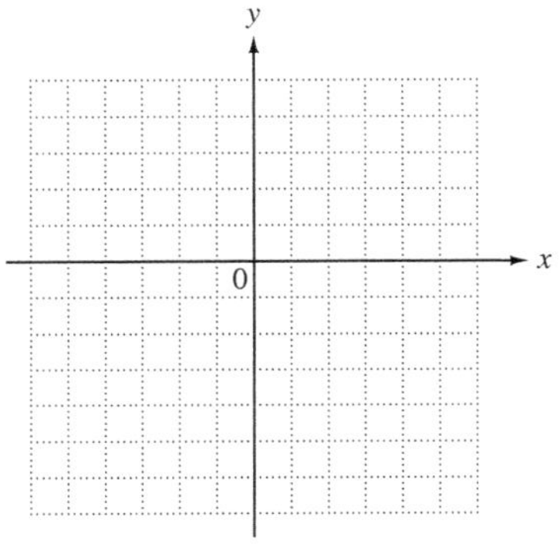

24. $f(x) = 2x^2 - 2$

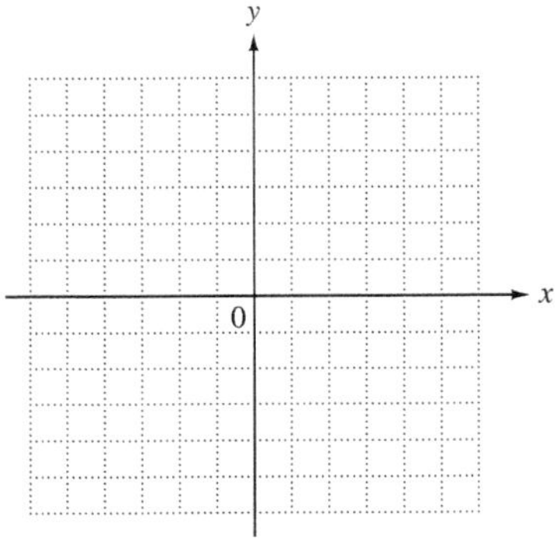

25. $f(x) = \frac{1}{2}(x - 4)^2$

vertex:
axis:
domain:
range:

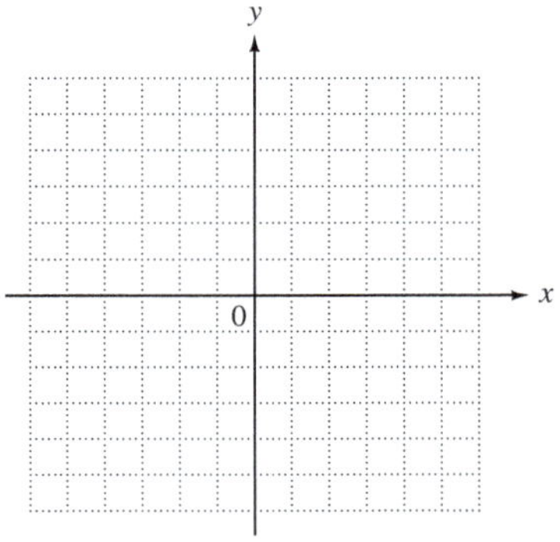

26. $f(x) = -2(x + 1)^2$

vertex:
axis:
domain:
range:

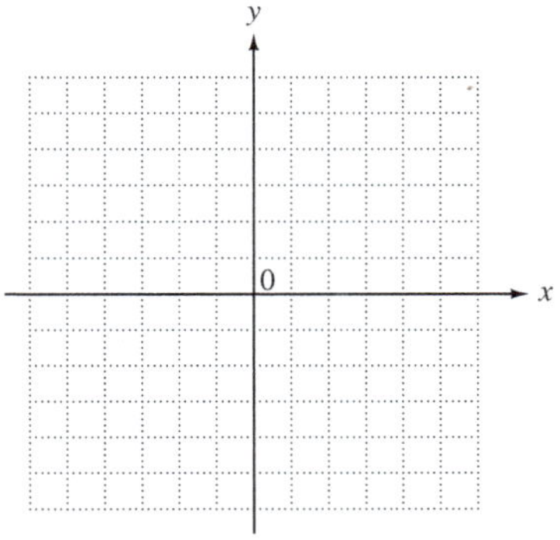

27. $f(x) = (x + 2)^2 - 1$

vertex:
axis:
domain:
range:

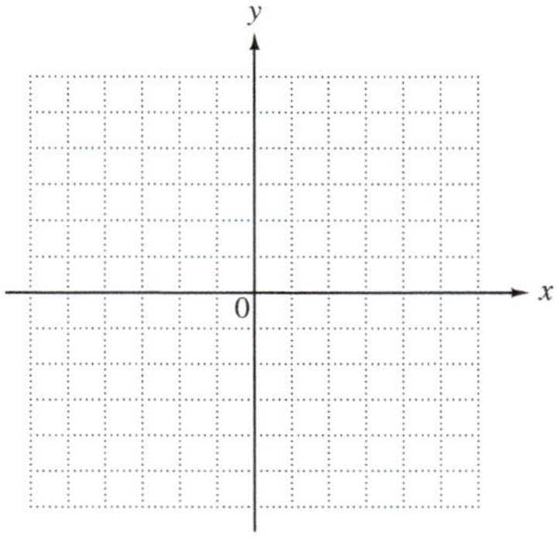

28. $f(x) = (x - 1)^2 + 2$

vertex:
axis:
domain:
range:

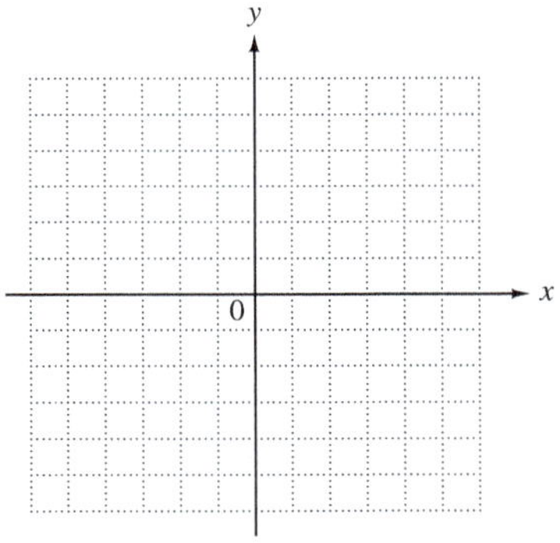

29. $f(x) = -2(x + 3)^2 + 4$

vertex:
axis:
domain:
range:

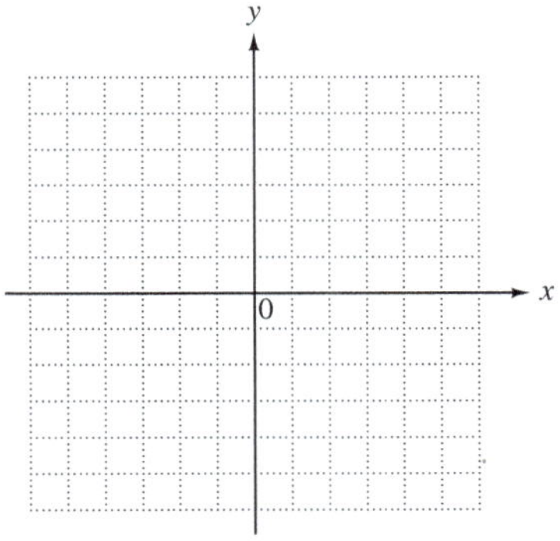

30. $f(x) = 2(x - 2)^2 - 3$

vertex:
axis:
domain:
range:

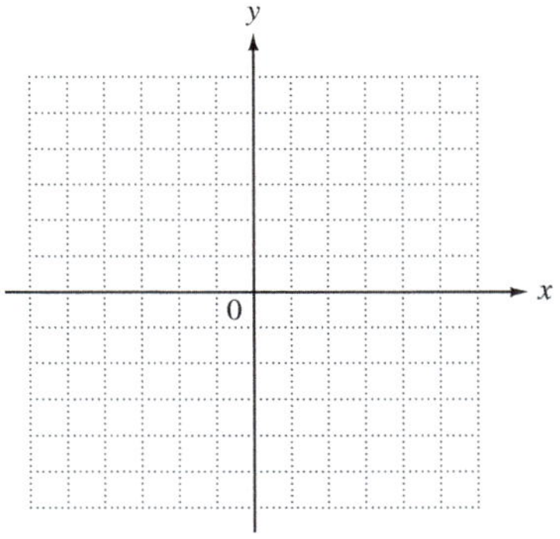

31. $f(x) = -\frac{2}{3}(x + 2)^2 + 1$

vertex:
axis:
domain:
range:

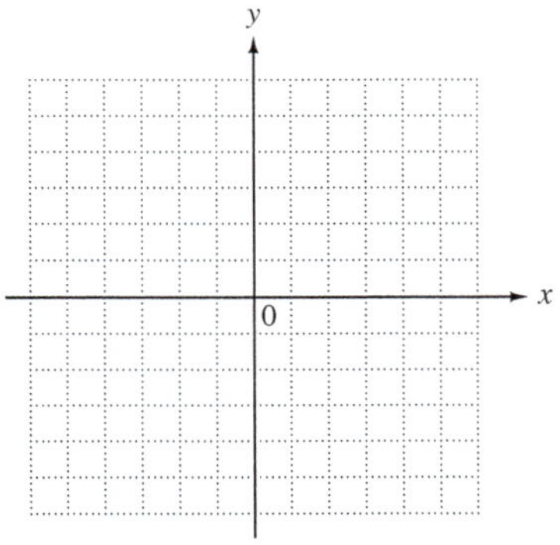

32. $f(x) = -\frac{1}{2}(x + 1)^2 + 2$

vertex:
axis:
domain:
range:

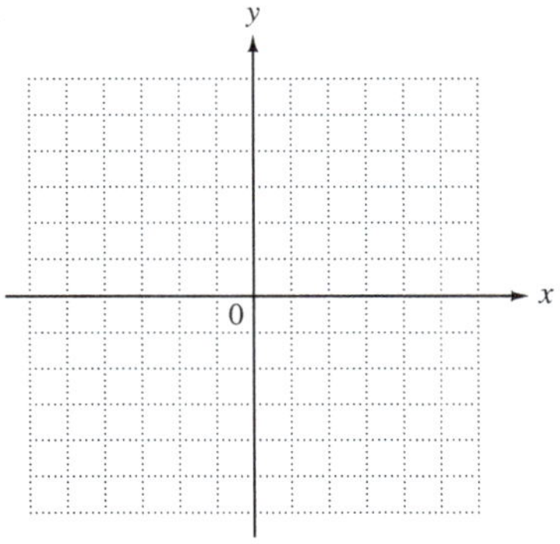

Relating Concepts (Exercises 33–38) For Individual or Group Work

The procedures described in this section that allow the graph of $f(x) = x^2$ to be shifted vertically and horizontally are applicable to other types of functions. In **Section 4.5** *we introduced linear functions of the form $g(x) = ax + b$. Consider the graph of the simplest linear function defined by $g(x) = x$, shown here, and then* **work Exercises 33–38 in order.**

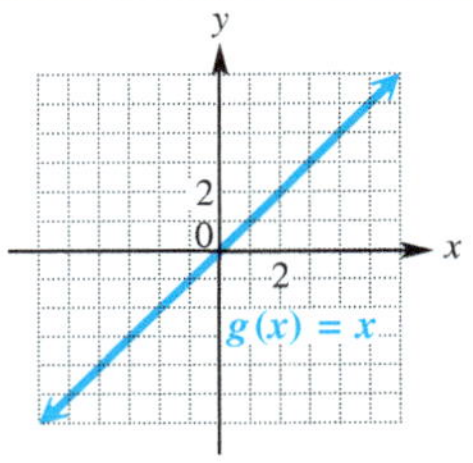

33. Based on the concepts of this section, how does the graph of $F(x) = x^2 + 6$ compare to the graph of $f(x) = x^2$ if a *vertical* shift is considered?

34. Graph the linear function defined by $G(x) = x + 6$.

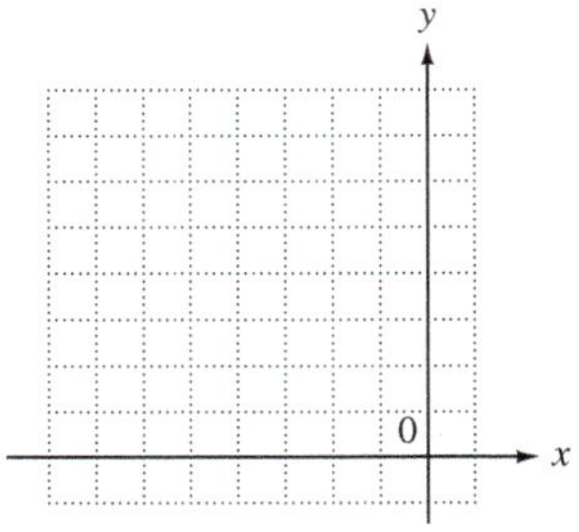

35. Based on the concepts of **Chapter 4,** how does the graph of $G(x) = x + 6$ compare to the graph of $g(x) = x$ if a *vertical* shift is considered? (*Hint:* Look at the *y*-intercept.)

36. Based on the concepts of this section, how does the graph of $F(x) = (x - 6)^2$ compare to the graph of $f(x) = x^2$ if a *horizontal* shift is considered?

37. Graph the linear function defined by $G(x) = x - 6$.

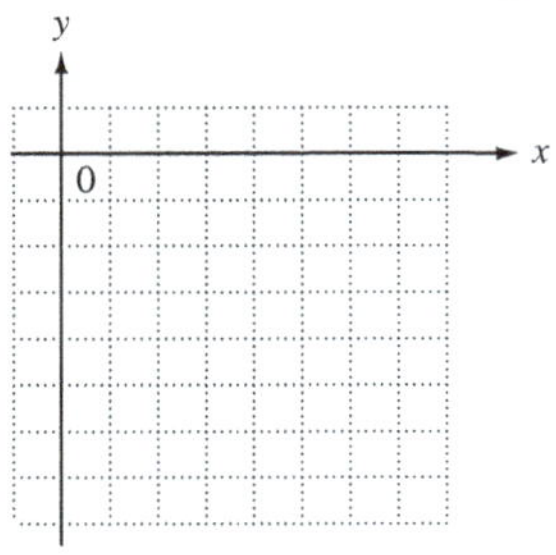

38. Based on the concepts of **Chapter 4,** how does the graph of $G(x) = x - 6$ compare to the graph of $g(x) = x$ if a *horizontal* shift is considered? (*Hint:* Look at the *x*-intercept.)

In Exercises 39–44, tell whether a linear or quadratic function would be a more appropriate model for each set of graphed data. If linear, tell whether the slope should be positive or negative. If quadratic, tell whether the coefficient of x^2 should be positive or negative. See Example 6.

39. **PLASMA TV SALES IN U.S.**

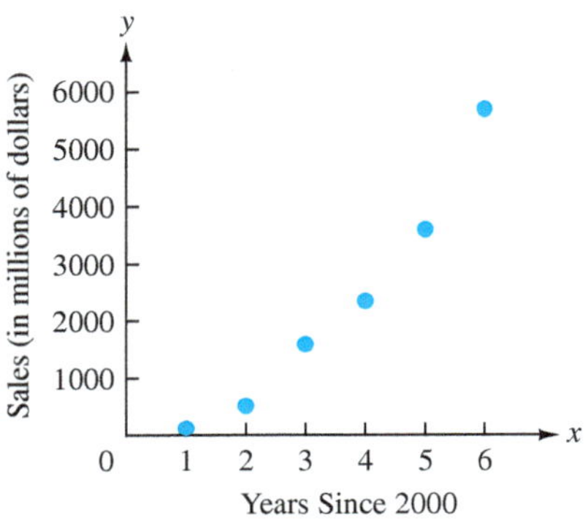

Source: Consumer Electronics Association.

40. **AVERAGE DAILY VOLUME OF FIRST-CLASS MAIL**

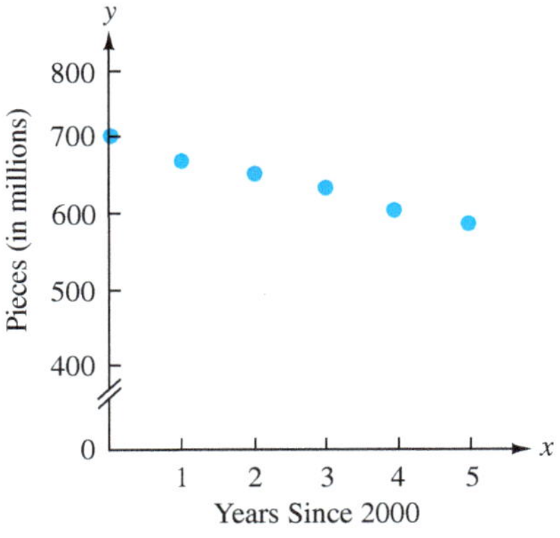

Source: General Accounting Office.

41. **SOCIAL SECURITY ASSETS***

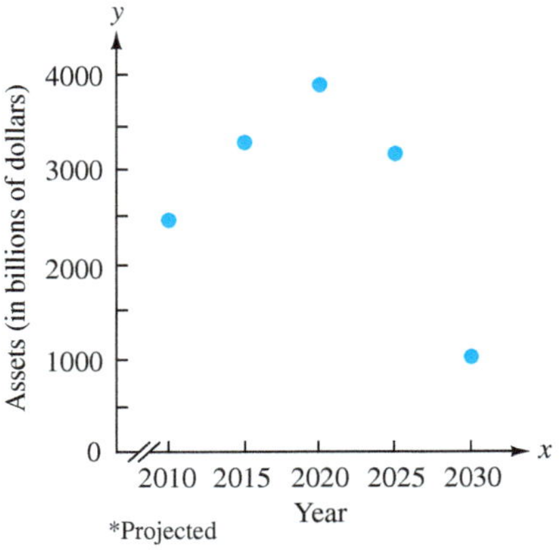

*Projected

Source: Social Security Administration.

42. **FOOD ASSISTANCE SPENDING IN IOWA**

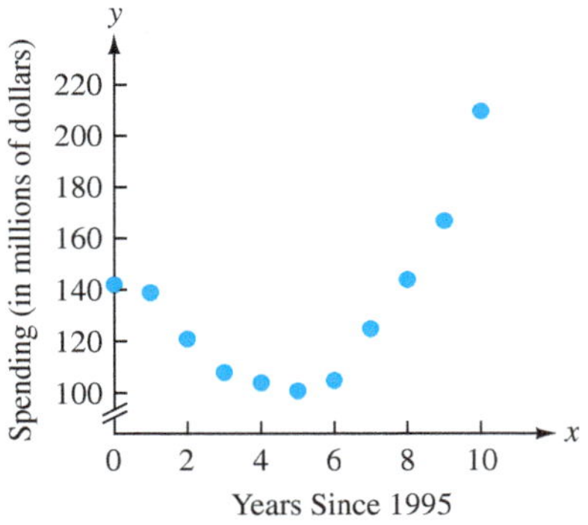

Source: Iowa Department of Human Services.

43. **TIME SPENT PLAYING VIDEO GAMES***

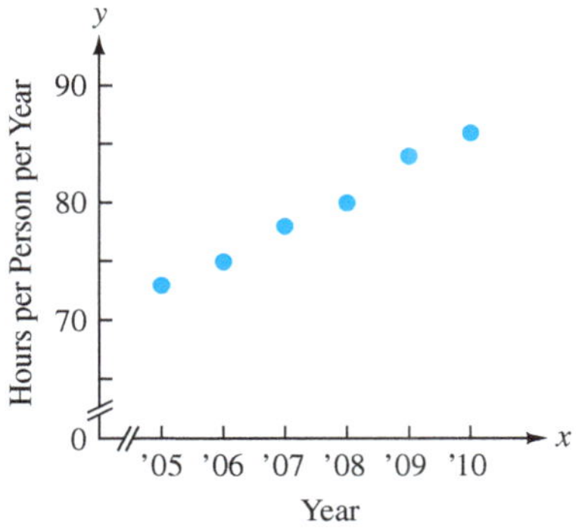

*Later years projected

Source: Veronis Suhler Stevenson.

44. **SALES OF MUSIC CASSETTE TAPES**

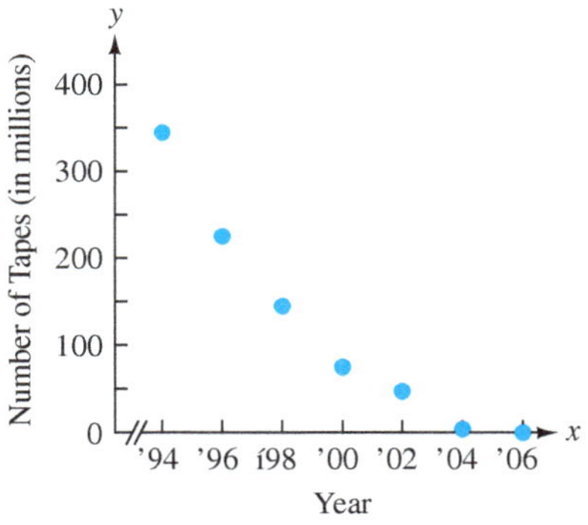

Source: Recording Industry Association of America.

Solve each problem. See Example 6.

45. Sales of digital cameras in the United States (in millions of dollars) between 2000 and 2006 are shown in the table. In the year column, 0 represents 2000, 1 represents 2001, and so on.

SALES OF DIGITAL CAMERAS

Year	Sales
0	1825
1	1972
2	2794
3	3921
4	4739
5	5611
6	7805

Source: Consumer Electronics Association.

(a) Use the ordered pairs (year, sales) to make a scatter diagram of the data.

DIGITAL CAMERA SALES IN U.S.

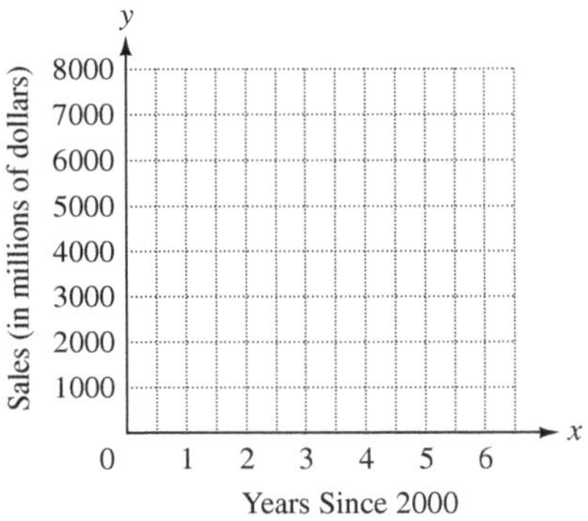

(b) Use the scatter diagram to decide whether a linear or quadratic function would better model the data. If quadratic, should the coefficient a of x^2 be positive or negative?

(c) Use the ordered pairs (0, 1825), (3, 3921), and (6, 7805) to find a quadratic function that models the data. Round the values of a, b, and c in your model to the nearest tenth, as necessary.

(d) Use your model from part (c) to approximate the sales of digital cameras in the United States in 2007. Round your answer to the nearest whole number (of millions).

(e) Sales of digital cameras were projected to be \$6945 million in 2007. Based on this, is the model valid for 2007? Explain.

46. The percent of U.S. high school students in grades 9–12 who smoke is shown in the table for selected years. In the year column, 1 represents 1991, 3 represents 1993, and so on.

HIGH SCHOOL STUDENTS WHO SMOKE

Year	Percent of Students
1	28
3	31
5	35
7	36
9	35
11	29
13	22

Source: Centers for Disease Control and Prevention.

(a) Use the ordered pairs (year, percent of students) to make a scatter diagram of the data.

PERCENT OF HIGH SCHOOL STUDENTS WHO SMOKE

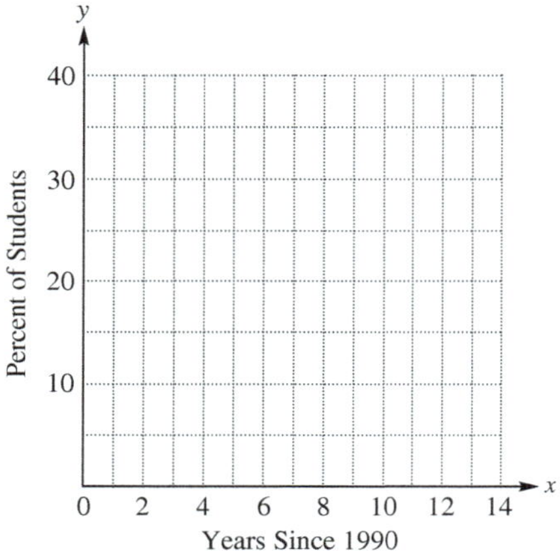

(b) Would a linear or quadratic function better model the data?

(c) Should the coefficient a of x^2 in a quadratic model be positive or negative?

(d) Use the ordered pairs (1, 28), (7, 36), and (11, 29) to find a quadratic function that models the data. Round the values of a, b, and c in your model to the nearest tenth, as necessary.

(e) Use your model from part (d) to approximate the percent of high school students who smoked during 1995 and 2003 to the nearest percent. How well does the model approximate the actual data from the table?

47. In Exercise 46(d), we determined that the quadratic function defined by

$$y = -0.3x^2 + 3.8x + 24.5$$

modeled the percent of U.S. high school students who smoked in the years 1991–2003.

(a) Use this model to approximate the number of high school students who smoked in 2005 and 2007.

(b) The actual smoking rates for high school students were 23% in 2005 and 20% in 2007. (*Source:* Centers for Disease Control and Prevention.) How do the approximations using the model compare to the actual rates for these two years?

48. Should the model from Exercise 46(d) be used to approximate the smoking rate for high school students in years after 2003? Explain.

18.7 More about Parabolas and Their Applications

OBJECTIVES

1. **Find the vertex of a vertical parabola.**
2. **Graph a quadratic function.**
3. **Use the discriminant to find the number of x-intercepts of a parabola with a vertical axis.**
4. **Use quadratic functions to solve problems involving maximum or minimum value.**
5. **Graph parabolas with horizontal axes.**

OBJECTIVE 1 Find the vertex of a vertical parabola. When the equation of a parabola is given in the form $f(x) = ax^2 + bx + c$, we need to locate the vertex in order to sketch an accurate graph. There are two ways to do this:

1. Complete the square, as shown in Examples 1 and 2, or
2. Use a formula derived by completing the square, as shown in Example 3.

Example 1 Completing the Square to Find the Vertex ($a = 1$)

Find the vertex of the graph of $f(x) = x^2 - 4x + 5$.

To find the vertex, we need to write the expression $x^2 - 4x + 5$ in the form $(x - h)^2 + k$. We do this by completing the square on $x^2 - 4x$, as in **Section 10.1.** The process is slightly different here because we want to keep $f(x)$ alone on one side of the equation. Instead of adding the appropriate number to each side, we *add and subtract* it on the right. This is equivalent to adding 0.

$f(x) = x^2 - 4x + 5$

$f(x) = (x^2 - 4x \quad) + 5$ Group the variable terms.

$\left[\frac{1}{2}(-4)\right]^2 = (-2)^2 = 4$

$f(x) = (x^2 - 4x + 4 - 4) + 5$ Add and subtract 4.

$f(x) = (x^2 - 4x + 4) - 4 + 5$ Bring -4 outside the parentheses.

$f(x) = (x - 2)^2 + 1$ Factor; combine like terms.

The vertex of this parabola is (2, 1).

Work Problem 1 at the Side. ▶

Example 2 Completing the Square to Find the Vertex ($a \neq 1$)

Find the vertex of the graph of $f(x) = -3x^2 + 6x - 1$.

We must complete the square on $-3x^2 + 6x$. Because the x^2-term has a coefficient other than 1, we factor that coefficient out of the first two terms and then proceed as in Example 1.

$f(x) = -3x^2 + 6x - 1$

$f(x) = -3(x^2 - 2x) - 1$ Factor out -3.

$\left[\frac{1}{2}(-2)\right]^2 = (-1)^2 = 1$

$f(x) = -3(x^2 - 2x + 1 - 1) - 1$ Add and subtract 1.

***Now bring* −1 *outside the parentheses; be sure to multiply it by* −3.**

$f(x) = -3(x^2 - 2x + 1) + (-3)(-1) - 1$ Distributive property

This is a key step.

$f(x) = -3(x^2 - 2x + 1) + 3 - 1$

$f(x) = -3(x - 1)^2 + 2$ Factor; combine like terms.

The vertex is (1, 2).

Work Problem 2 at the Side. ▶

1 Find the vertex of the graph of each quadratic function.

(a) $f(x) = x^2 - 6x + 7$

(b) $f(x) = x^2 + 4x - 9$

2 Find the vertex of the graph of each quadratic function.

(a) $f(x) = 2x^2 - 4x + 1$

(b) $f(x) = -\frac{1}{2}x^2 + 2x - 3$

ANSWERS

1. (a) (3, −2) (b) (−2, −13)
2. (a) (1, −1) (b) (2, −1)

3 Use the formula to find the vertex of the graph of each quadratic function.

(a) $f(x) = -2x^2 + 3x - 1$

(b) $f(x) = 4x^2 - x + 5$

To derive a formula for the vertex of the graph of the quadratic function defined by $f(x) = ax^2 + bx + c$, complete the square.

$$f(x) = ax^2 + bx + c \quad (a \neq 0) \qquad \text{Standard form}$$

$$f(x) = a\left(x^2 + \frac{b}{a}x\right) + c \qquad \text{Factor } a \text{ from the first two terms.}$$

$$\left[\frac{1}{2}\left(\frac{b}{a}\right)\right]^2 = \left(\frac{b}{2a}\right)^2 = \frac{b^2}{4a^2}$$

$$f(x) = a\left(x^2 + \frac{b}{a}x + \frac{b^2}{4a^2} - \frac{b^2}{4a^2}\right) + c \qquad \text{Add and subtract } \frac{b^2}{4a^2}.$$

$$f(x) = a\left(x^2 + \frac{b}{a}x + \frac{b^2}{4a^2}\right) + a\left(-\frac{b^2}{4a^2}\right) + c \qquad \text{Distributive property}$$

$$f(x) = a\left(x^2 + \frac{b}{a}x + \frac{b^2}{4a^2}\right) - \frac{b^2}{4a} + c \qquad -\frac{ab^2}{4a^2} = -\frac{b^2}{4a}$$

$$f(x) = a\left(x + \frac{b}{2a}\right)^2 + \frac{4ac - b^2}{4a} \qquad \text{Factor; rewrite terms with a common denominator.}$$

$$f(x) = a\left[x - \underbrace{\left(\frac{-b}{2a}\right)}_{h}\right]^2 + \underbrace{\frac{4ac - b^2}{4a}}_{k} \qquad f(x) = (x - h)^2 + k$$

Thus, the vertex (h, k) can be expressed in terms of a, b, and c. It is not necessary to remember the expression for k, since it can be found by replacing x with $\frac{-b}{2a}$. Using function notation, if $y = f(x)$, then the y-value of the vertex is $f\left(\frac{-b}{2a}\right)$.

Vertex Formula

The graph of the quadratic function defined by $f(x) = ax^2 + bx + c$ has vertex

$$\left(\frac{-b}{2a}, f\left(\frac{-b}{2a}\right)\right).$$

The axis of the parabola is the line

$$x = \frac{-b}{2a}.$$

Example 3 **Using the Formula to Find the Vertex**

Use the vertex formula to find the vertex of the graph of $f(x) = x^2 - x - 6$.

For this function, $a = 1$, $b = -1$, and $c = -6$. The x-coordinate of the vertex of the parabola is given by

$$\frac{-b}{2a} = \frac{-(-1)}{2(1)} = \frac{1}{2}.$$

The y-coordinate is $f\left(\frac{-b}{2a}\right) = f\left(\frac{1}{2}\right)$.

$$f\left(\frac{1}{2}\right) = \left(\frac{1}{2}\right)^2 - \frac{1}{2} - 6 = \frac{1}{4} - \frac{1}{2} - 6 = -\frac{25}{4}$$

The vertex is $\left(\frac{1}{2}, -\frac{25}{4}\right)$.

◀ Work Problem 3 at the Side.

Answers

3. (a) $\left(\frac{3}{4}, \frac{1}{8}\right)$ (b) $\left(\frac{1}{8}, \frac{79}{16}\right)$

OBJECTIVE 2 **Graph a quadratic function.**

Graphing a Quadratic Function f

Step 1 **Determine whether the graph opens up or down.** If $a > 0$, the parabola opens up; if $a < 0$, it opens down.

Step 2 **Find the vertex.** Use either the vertex formula or completing the square.

Step 3 **Find any intercepts.** To find the x-intercepts (if any), solve $f(x) = 0$. To find the y-intercept, evaluate $f(0)$.

Step 4 **Complete the graph.** Plot the points found so far. Find and plot additional points as needed, using symmetry about the axis.

Example 4 **Graphing a Quadratic Function**

Graph the quadratic function defined by $f(x) = x^2 - x - 6$.

Step 1 From the equation, $a = 1$, so the graph of the function opens up.

Step 2 The vertex, $\left(\frac{1}{2}, -\frac{25}{4}\right)$, was found in Example 3 by substituting the values $a = 1$, $b = -1$, and $c = -6$ in the vertex formula.

Step 3 Now find any intercepts. Since the vertex, $\left(\frac{1}{2}, -\frac{25}{4}\right)$, is in quadrant IV and the graph opens up, there will be two x-intercepts. To find them, let $f(x) = 0$ and solve the equation.

$$f(x) = x^2 - x - 6$$

$$0 = x^2 - x - 6 \qquad \text{Let } f(x) = 0.$$

$$0 = (x - 3)(x + 2) \qquad \text{Factor.}$$

$$x - 3 = 0 \quad \text{or} \quad x + 2 = 0 \qquad \text{Zero-factor property}$$

$$x = 3 \quad \text{or} \quad x = -2 \qquad \text{Solve each equation.}$$

The x-intercepts are $(3, 0)$ and $(-2, 0)$. Find the y-intercept.

$$f(x) = x^2 - x - 6$$

$$f(0) = 0^2 - 0 - 6 \qquad \text{Let } x = 0.$$

$$f(0) = -6$$

The y-intercept is $(0, -6)$.

Step 4 Plot the points found so far and additional points as needed using symmetry about the axis $x = \frac{1}{2}$. The graph is shown in Figure 12. The domain is $(-\infty, \infty)$, and the range is $\left[-\frac{25}{4}, \infty\right)$.

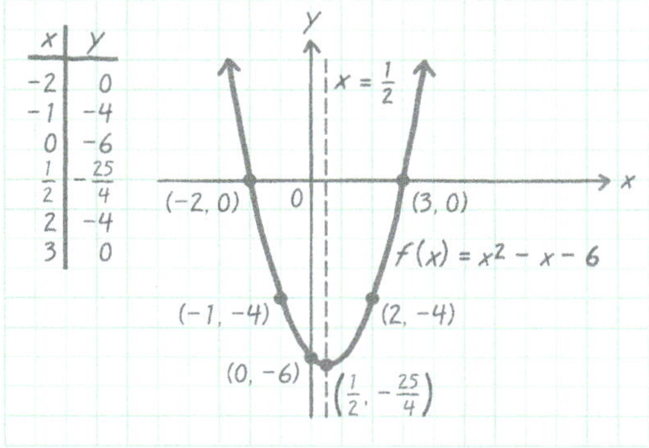

Figure 12

Work Problem **4** *at the Side.* ▶

4 Graph the quadratic function defined by

$$f(x) = x^2 - 6x + 5.$$

Give the vertex, axis, domain, and range.

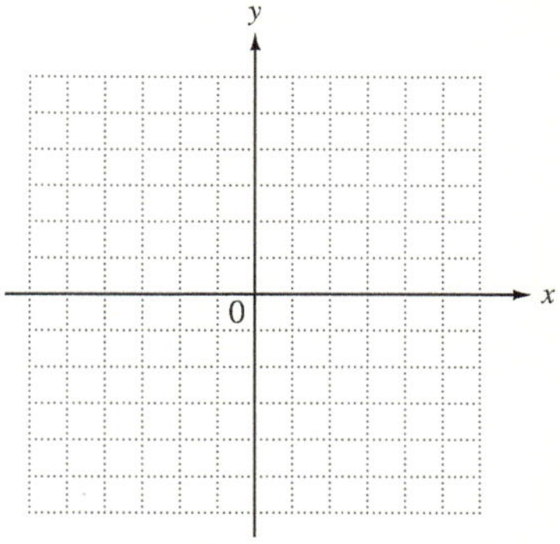

ANSWER

4.

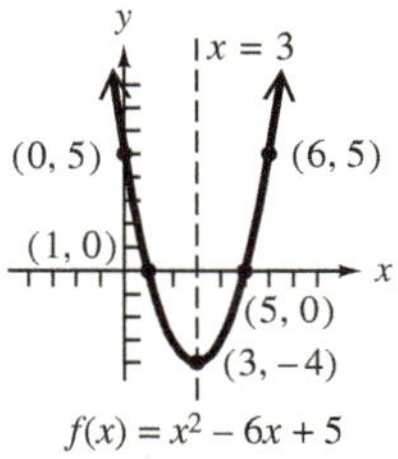

vertex: $(3, -4)$; axis: $x = 3$; domain: $(-\infty, \infty)$; range: $[-4, \infty)$

OBJECTIVE 3 Use the discriminant to find the number of x-intercepts of a parabola with a vertical axis. Recall from **Section 18.3** that the expression $b^2 - 4ac$ is called the **discriminant** of the quadratic *equation* $ax^2 + bx + c = 0$ and that we can use it to determine the number of real solutions of a quadratic equation.

In a similar way, we can use the discriminant of a quadratic *function* to determine the number of x-intercepts of its graph. See Figure 13. If the discriminant is positive, the parabola will have two x-intercepts. If the discriminant is 0, there will be only one x-intercept, and it will be the vertex of the parabola. If the discriminant is negative, the graph will have no x-intercepts.

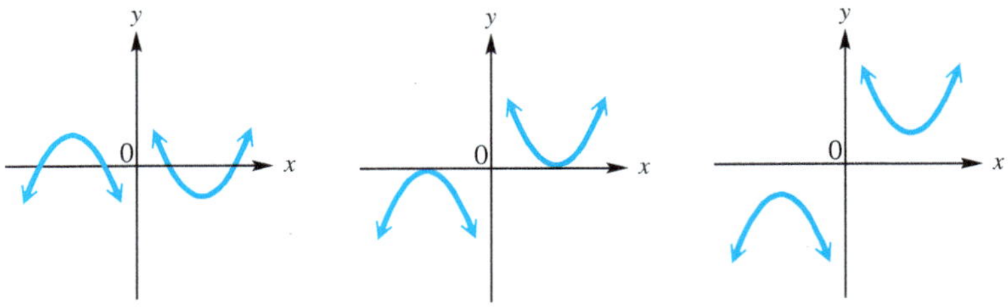

$b^2 - 4ac > 0$	$b^2 - 4ac = 0$	$b^2 - 4ac < 0$
Two x-intercepts	One x-intercept	No x-intercepts

Figure 13

Example 5 Using the Discriminant to Determine the Number of x-Intercepts

Use the discriminant to determine the number of x-intercepts of the graph of each quadratic function.

(a) $f(x) = 2x^2 + 3x - 5$

The discriminant is $b^2 - 4ac$. Here $a = 2$, $b = 3$, and $c = -5$, so

$$\begin{aligned} & b^2 - 4ac \\ &= 3^2 - 4(2)(-5) && \text{Substitute.} \\ &= 9 - (-40) && \text{Apply the exponent; multiply.} \\ &= 49. && \text{Subtract.} \end{aligned}$$

Since the discriminant is positive, the parabola has two x-intercepts.

(b) $f(x) = -3x^2 - 1$

In this equation, $a = -3$, $b = 0$, and $c = -1$. The discriminant is

$$\begin{aligned} & b^2 - 4ac \\ &= 0^2 - 4(-3)(-1) \\ &= -12. \end{aligned}$$

The discriminant is negative, so the graph has no x-intercepts.

(c) $f(x) = 9x^2 + 6x + 1$

Here, $a = 9$, $b = 6$, and $c = 1$. The discriminant is

$$\begin{aligned} & b^2 - 4ac \\ &= 6^2 - 4(9)(1) \\ &= 0. \end{aligned}$$

The parabola has only one x-intercept (its vertex) because the value of the discriminant is 0.

◀ *Work Problem* **5** *at the Side.*

5 Use the discriminant to determine the number of x-intercepts of the graph of each quadratic function.

(a) $f(x) = 4x^2 - 20x + 25$

(b) $f(x) = 2x^2 + 3x + 5$

(c) $f(x) = -3x^2 - x + 2$

ANSWERS

5. **(a)** discriminant is 0; one x-intercept
(b) discriminant is -31; no x-intercepts
(c) discriminant is 25; two x-intercepts

OBJECTIVE 4 Use quadratic functions to solve problems involving maximum or minimum value. The vertex of a parabola is either the highest or the lowest point on the parabola. The y-value of the vertex gives the maximum or minimum value of y, while the x-value tells where that maximum or minimum occurs.

6 Solve Example 6 if the farmer has only 100 ft of fencing.

Problem-Solving Hint

In many applied problems we must find the least or greatest value of some quantity. When we can express that quantity as a quadratic function, the value of k in the vertex (h, k) gives that optimum value.

Example 6 Finding the Maximum Area of a Rectangular Region

A farmer has 120 ft of fencing to enclose a rectangular area next to a building. See Figure 14. Find the maximum area he can enclose and the width required to produce this maximum area.

Figure 14

Let x represent the width of the rectangle. Since he has 120 ft of fencing,

$$x + x + \text{length} = 120 \quad \text{Sum of the sides is 120 ft.}$$
$$2x + \text{length} = 120 \quad \text{Combine like terms.}$$
$$\text{length} = 120 - 2x. \quad \text{Subtract } 2x.$$

The area $A(x)$ is given by the product of the width and length, so

$$A(x) = x(120 - 2x)$$
$$A(x) = 120x - 2x^2.$$

To determine the maximum area, find the vertex of the parabola given by $A(x) = 120x - 2x^2$ using the vertex formula. Writing the equation in standard form as

$$A(x) = -2x^2 + 120x$$

gives $a = -2$, $b = 120$, and $c = 0$, so

$$h = \frac{-b}{2a} = \frac{-120}{2(-2)} = \frac{-120}{-4} = 30;$$

$$A(30) = -2(30)^2 + 120(30) = -2(900) + 3600 = 1800.$$

The graph is a parabola that opens down, and its vertex is $(30, 1800)$. Thus, the maximum area will be 1800 ft^2. This area will occur if x, the width of the rectangle, is 30 ft.

Work Problem 6 at the Side. ▶

ANSWER

6. The rectangle should be 25 ft by 50 ft with a maximum area of 1250 ft^2.

7 Solve the problem.

A toy rocket is launched from the ground so that its distance in feet above the ground after t seconds is

$$s(t) = -16t^2 + 208t.$$

Find the maximum height it reaches and the number of seconds it takes to reach that height.

CAUTION

Be careful when interpreting the meanings of the coordinates of the vertex. The first coordinate, x, gives the value for which the *function value* is a maximum or a minimum. Be sure to read the problem carefully to determine whether you are asked to find the value of the independent variable, the function value, or both.

Example 7 Finding the Maximum Height Attained by a Projectile

If air resistance is neglected, a projectile on Earth shot straight upward with an initial velocity of 40 m per sec will be at a height s in meters given by

$$s(t) = -4.9t^2 + 40t,$$

where t is the number of seconds elapsed after projection. After how many seconds will it reach its maximum height, and what is this maximum height?

For this function, $a = -4.9$, $b = 40$, and $c = 0$. Use the vertex formula.

$$h = \frac{-b}{2a} = \frac{-40}{2(-4.9)} \approx 4.1 \qquad \text{Use a calculator.}$$

Thus, the maximum height is attained at 4.1 sec. To find this maximum height, calculate $s(4.1)$.

$$s(4.1) = -4.9(4.1)^2 + 40(4.1) \approx 81.6 \qquad \text{Use a calculator.}$$

The projectile will attain a maximum height of approximately 81.6 m.

◀ Work Problem **7** at the Side.

OBJECTIVE 5 Graph parabolas with horizontal axes. If x and y are interchanged in the equation $y = ax^2 + bx + c$, the equation becomes

$$x = ay^2 + by + c.$$

Because of the interchange of the roles of x and y, these parabolas are horizontal (with horizontal lines as axes).

Graph of a Horizontal Parabola

The graph of

$$\boldsymbol{x = ay^2 + by + c} \quad \text{or} \quad \boldsymbol{x = a(y - k)^2 + h}$$

is a parabola with vertex (h, k) and the horizontal line $y = k$ as axis. The graph opens to the right if $a > 0$ and to the left if $a < 0$.

Example 8 Graphing a Horizontal Parabola

Graph $x = (y - 2)^2 - 3$. Give the vertex, axis, domain, and range.

This graph has its vertex at $(-3, 2)$, since the roles of x and y are reversed. It opens to the right, the positive x-direction, and has the same shape as $y = x^2$. Plotting a few additional points gives the graph shown in Figure 15 on the next page. Note that the graph is symmetric about its axis, $y = 2$. The domain is $[-3, \infty)$, and the range is $(-\infty, \infty)$.

Continued on Next Page

ANSWER

7. 676 ft; 6.5 sec

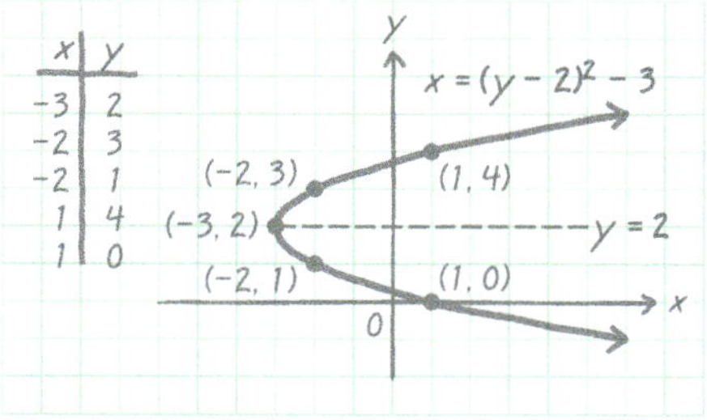

Figure 15

Work Problem **8** at the Side. ▶

8 Graph $x = (y + 1)^2 - 4$. Give the vertex, axis, domain, and range.

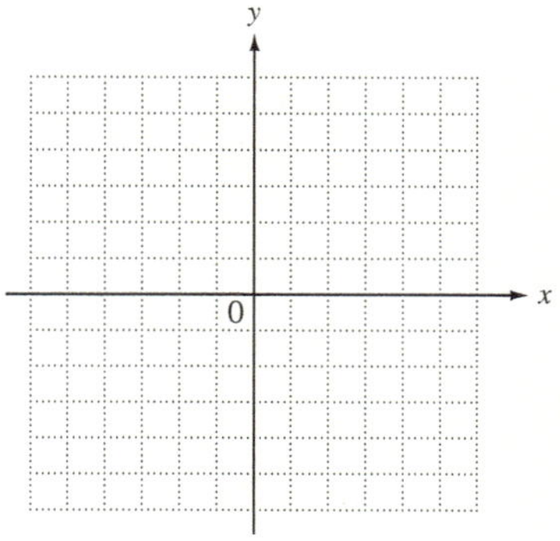

When a quadratic equation is given in the form $x = ay^2 + by + c$, completing the square on y will allow us to find the vertex.

Example 9 **Completing the Square to Graph a Horizontal Parabola**

Graph $x = -2y^2 + 4y - 3$. Give the vertex, axis, domain, and range.

$x = -2y^2 + 4y - 3$

$x = -2(y^2 - 2y) - 3$ — Factor out -2.

$x = -2(y^2 - 2y + 1 - 1) - 3$ — Complete the square within the parentheses; add and subtract 1.

$x = -2(y^2 - 2y + 1) + (-2)(-1) - 3$ — Distributive property

Be careful here.

$x = -2(y - 1)^2 - 1$ — Factor; simplify.

Because of the negative coefficient (-2) in $x = -2(y - 1)^2 - 1$, the graph opens to the left (the negative x-direction). The graph is narrower than the graph of $y = x^2$ because $|-2| > 1$. As shown in Figure 16, the vertex is $(-1, 1)$ and the axis is $y = 1$. The domain is $(-\infty, -1]$, and the range is $(-\infty, \infty)$.

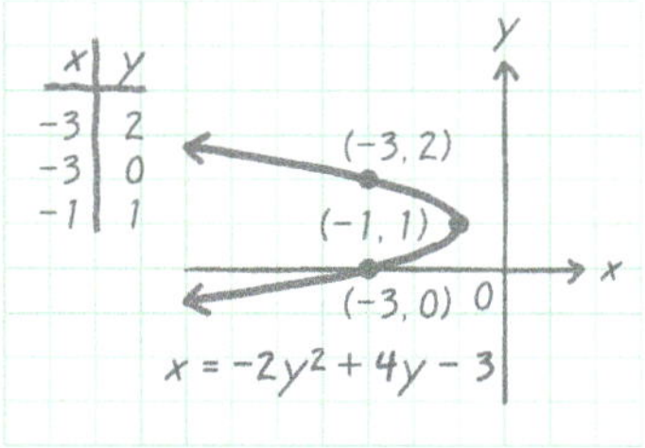

Figure 16

Work Problem **9** at the Side. ▶

9 Graph $x = -y^2 + 2y + 5$. Give the vertex, axis, domain, and range.

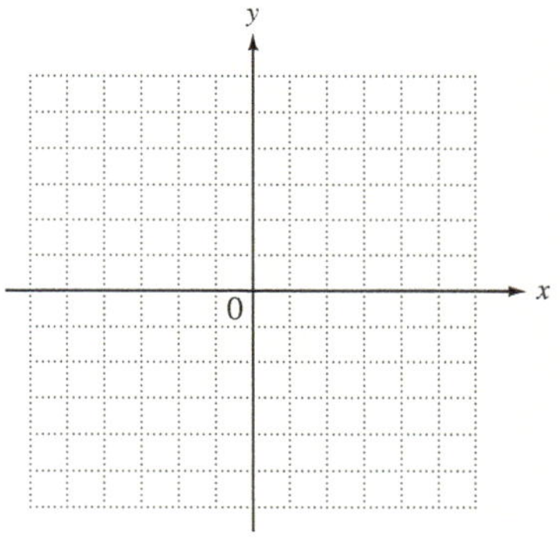

CAUTION

Only quadratic equations solved for y (whose graphs are vertical parabolas) are examples of functions. The horizontal parabolas in Examples 8 and 9 are ***not*** graphs of functions, because they do not satisfy the vertical line test.

ANSWERS

8.

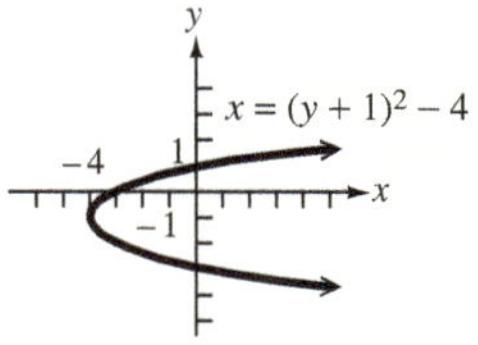

vertex: $(-4, -1)$; axis: $y = -1$; domain: $[-4, \infty)$; range: $(-\infty, \infty)$

9.

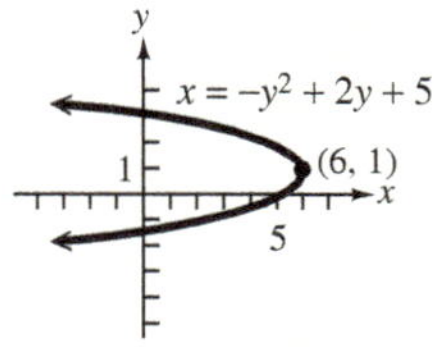

vertex: $(6, 1)$; axis: $y = 1$; domain: $(-\infty, 6]$; range: $(-\infty, \infty)$

In summary, the graphs of parabolas studied in **Sections 18.6 and 18.7** fall into the following categories.

GRAPHS OF PARABOLAS

Equation	Graph
$y = ax^2 + bx + c$ $y = a(x - h)^2 + k$	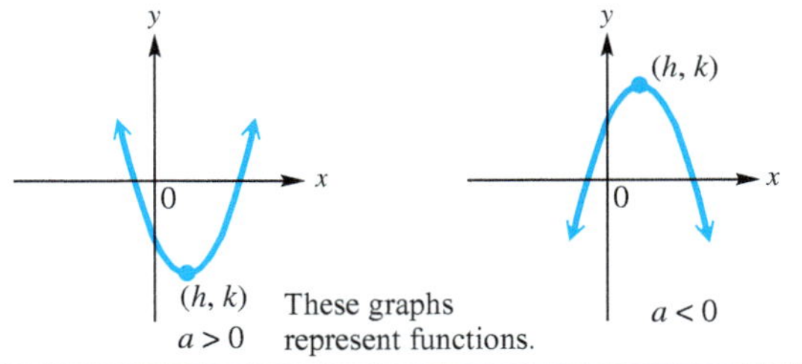
$x = ay^2 + by + c$ $x = a(y - k)^2 + h$	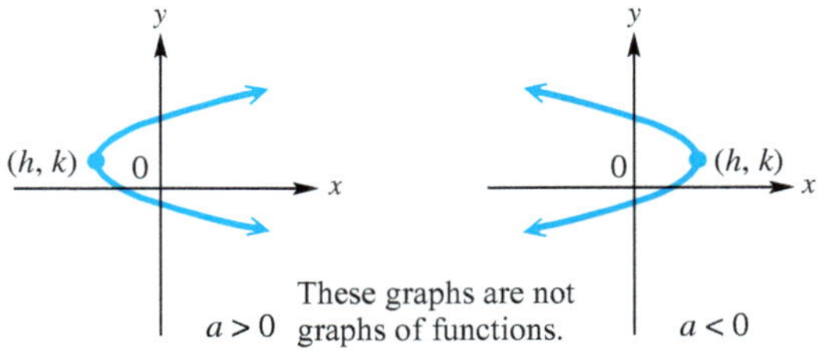

◀ *Work Problems* **10** *and* **11** *at the Side.*

10 Find the vertex of each parabola. Tell whether the graph opens to the right or to the left. Give the domain and range.

(a) $x = 2y^2 - 6y + 5$

(b) $x = -y^2 + 2y + 5$

11 **(a)** Tell whether each of the following equations has a vertical or horizontal parabola as its graph.

A. $y = -x^2 + 20x + 80$

B. $x = 2y^2 + 6y + 5$

C. $x + 1 = (y + 2)^2$

D. $f(x) = (x - 4)^2$

(b) Which of the equations in part (a) represent functions?

ANSWERS

10. **(a)** $\left(\frac{1}{2}, \frac{3}{2}\right)$; right; domain: $\left[\frac{1}{2}, \infty\right)$; range: $(-\infty, \infty)$

(b) $(6, 1)$; left; domain; $(-\infty, 6]$; range: $(-\infty, \infty)$

11. **(a)** A, D are vertical parabolas; B, C are horizontal parabolas.

(b) A, D

18.7 Exercises

FOR EXTRA HELP MyMathLab Math XL PRACTICE WATCH DOWNLOAD READ REVIEW

1. How can you determine just by looking at the equation of a parabola whether it has a vertical or a horizontal axis?

2. Why can't the graph of a quadratic function be a horizontal parabola?

3. How can you determine the number of x-intercepts of the graph of a quadratic function without graphing the function?

4. If the vertex of the graph of a quadratic function is $(1, -3)$ and the graph opens down, how many x-intercepts does the graph have?

Find the vertex of each parabola. For each equation, decide whether the graph opens up, down, to the left, or to the right, and whether it is wider, narrower, or the same shape as the graph of $y = x^2$. If it is a vertical parabola, use the discriminant to determine the number of x-intercepts. See Examples 1–3, 5, 8, and 9.

5. $y = 2x^2 + 4x + 5$

6. $y = 3x^2 - 6x + 4$

7. $y = -x^2 + 5x + 3$

8. $x = -y^2 + 7y - 2$

9. $x = \frac{1}{3}y^2 + 6y + 24$

10. $x = \frac{1}{2}y^2 + 10y - 5$

Graph each parabola. Give the vertex, axis, domain, and range. See Examples 4, 8, and 9.

11. $f(x) = x^2 + 4x + 3$
vertex:
axis:
domain:
range:

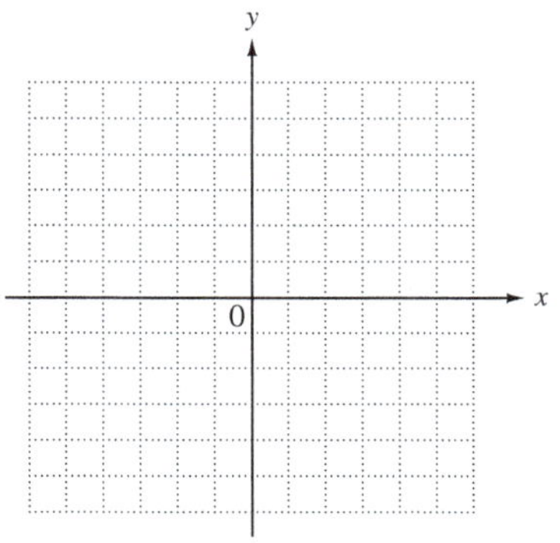

12. $f(x) = x^2 + 2x - 2$
vertex:
axis:
domain:
range:

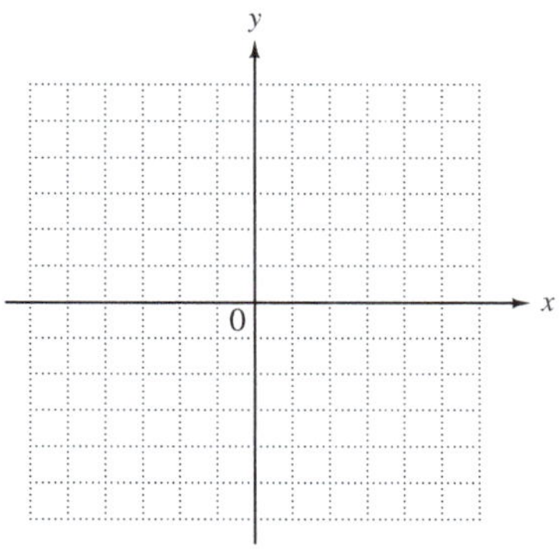

13. $f(x) = -2x^2 + 4x - 5$
vertex:
axis:
domain:
range:

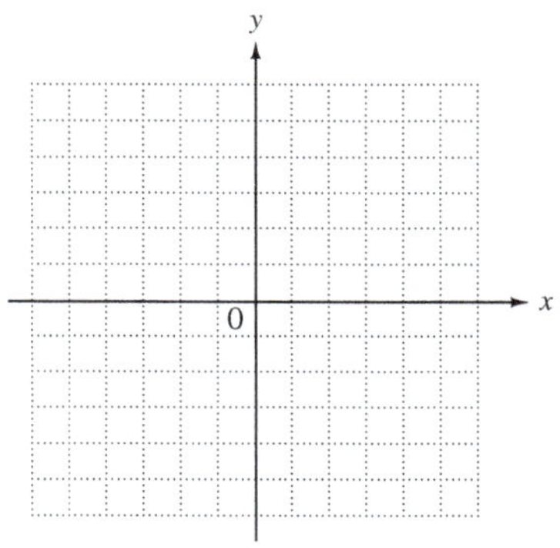

14. $f(x) = -3x^2 + 12x - 8$
vertex:
axis:
domain:
range:

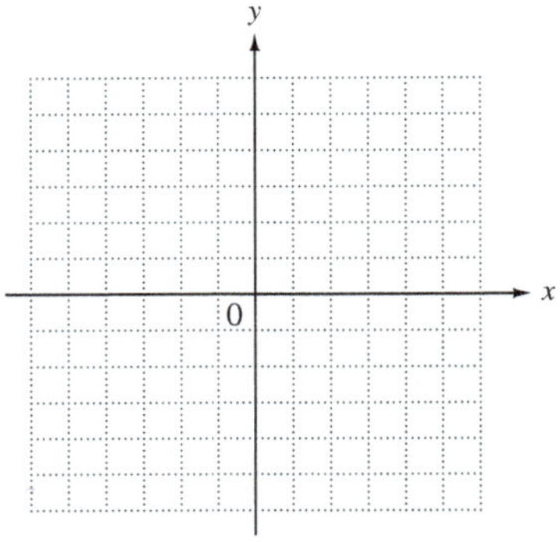

15. $x = -\frac{1}{5}y^2 + 2y - 4$
vertex:
axis:
domain:
range:

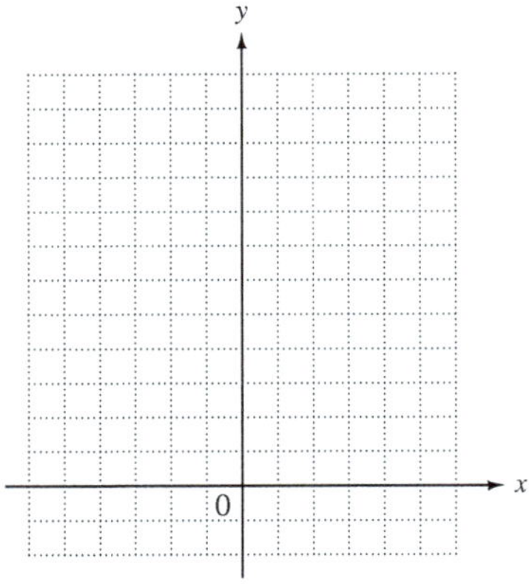

16. $x = -\frac{1}{2}y^2 - 4y - 6$
vertex:
axis:
domain:
range:

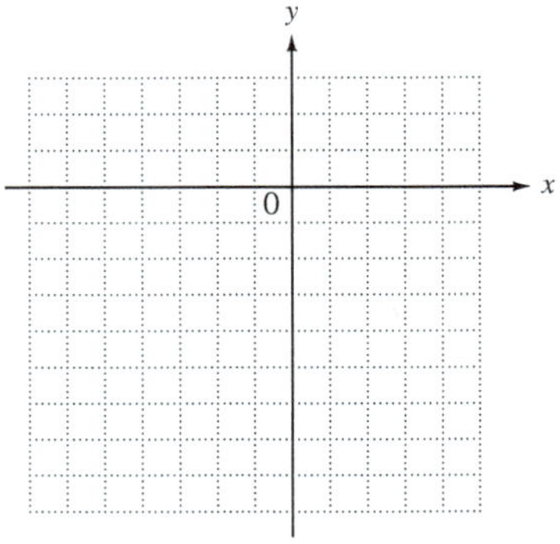

17. $x = 3y^2 + 12y + 5$
vertex:
axis:
domain:
range:

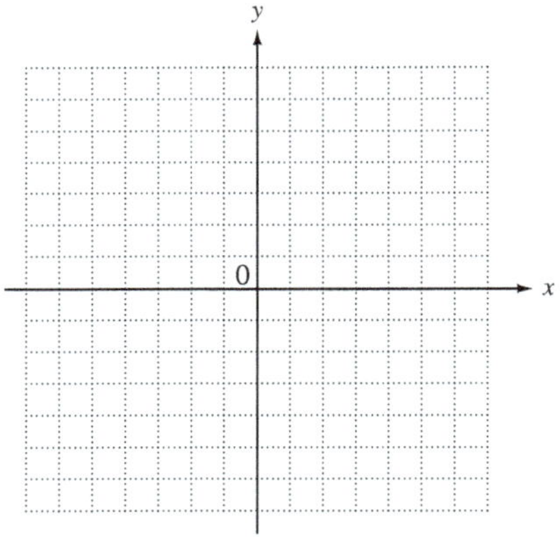

18. $x = 4y^2 + 16y + 11$
vertex:
axis:
domain:
range:

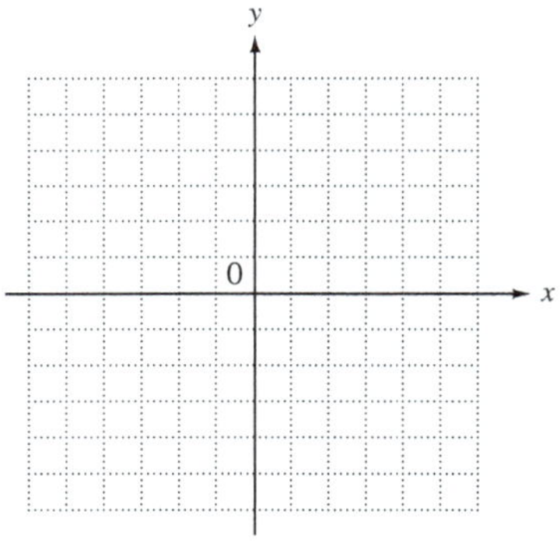

Use the concepts of this section to match each equation with its graph.

19. $y = 2x^2 + 4x - 3$

20. $y = -x^2 + 3x + 5$

21. $y = -\frac{1}{2}x^2 - x + 1$

22. $x = y^2 + 6y + 3$

23. $x = -y^2 - 2y + 4$

24. $x = 3y^2 + 6y + 5$

A.

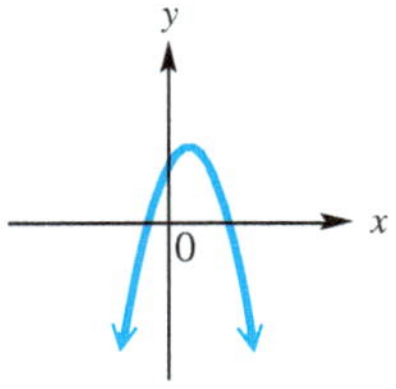

B.

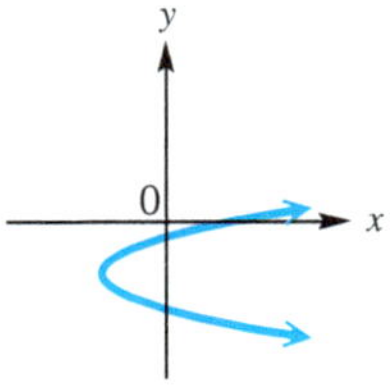

C.

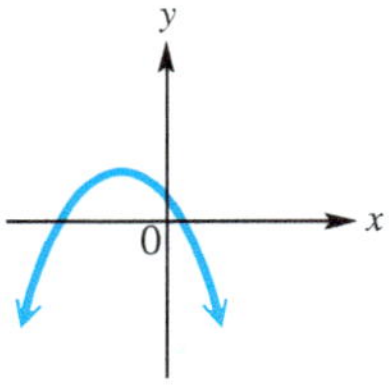

D.

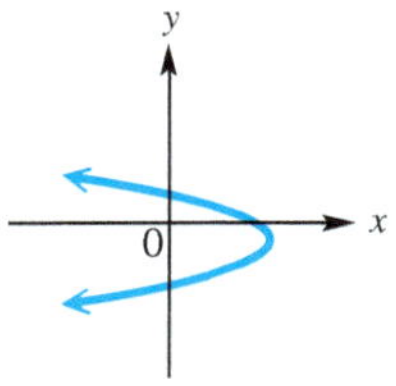

E.

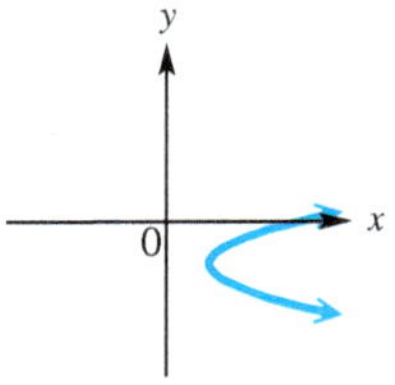

F.

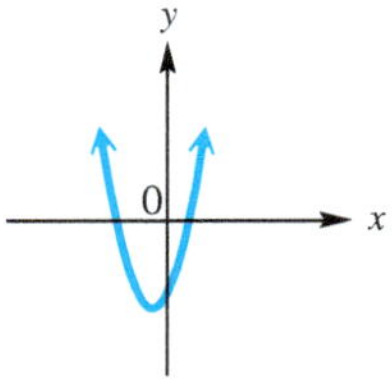

Solve each problem. See Examples 6 and 7.

25. Find the pair of numbers whose sum is 60 and whose product is a maximum. (*Hint:* Let x and $60 - x$ represent the two numbers.)

26. Find the pair of numbers whose sum is 10 and whose product is a maximum.

27. Palo Alto College is planning to construct a rectangular parking lot on land bordered on one side by a highway. The plan is to use 640 ft of fencing to fence off the other three sides. What should the dimensions of the lot be if the enclosed area is to be a maximum?

28. Keisha Hughes has 100 m of fencing material to enclose a rectangular exercise run for her dog. What width will give the enclosure the maximum area?

29. If an object on Earth is projected upward with an initial velocity of 32 ft per sec, then its height (in feet) after t seconds is given by

$$h(t) = 32t - 16t^2.$$

Find the maximum height attained by the object and the number of seconds it takes to hit the ground.

30. A projectile on Earth is fired straight upward so that its distance (in feet) above the ground t seconds after firing is given by

$$s(t) = -16t^2 + 400t.$$

Find the maximum height it reaches and the number of seconds it takes to reach that height.

31. A charter flight charges a fare of \$200 per person, plus \$4 per person for each unsold seat on the plane. If the plane holds 100 passengers and if x represents the number of unsold seats, find the following.

(a) A function defined by $R(x)$ that describes the total revenue received for the flight (*Hint:* Multiply the number of people flying, $100 - x$, by the price per ticket, $200 + 4x$.)

(b) The number of unsold seats that will produce the maximum revenue

(c) The maximum revenue

32. For a trip, a charter bus company charges a fare of \$48 per person, plus \$2 per person for each unsold seat on the bus. If the bus has 42 seats and x represents the number of unsold seats, find the following.

(a) A function defined by $R(x)$ that describes the total revenue from the trip (*Hint:* Multiply the total number riding, $42 - x$, by the price per ticket, $48 + 2x$.)

(b) The number of unsold seats that produces the maximum revenue

(c) The maximum revenue

33. The percent of births in the United States to teenage mothers in the years 1990–2002 can be modeled by the quadratic function defined by

$$f(x) = -0.0334x^2 + 0.2351x + 12.79,$$

where $x = 0$ represents 1990, $x = 1$ represents 1991, and so on. (*Source:* U.S. National Center for Health Statistics.)

(a) Since the coefficient of x^2 in the model is negative, the graph of this quadratic function is a parabola that opens down. Will the y-value of the vertex of this graph be a maximum or a minimum?

(b) In what year during this period was the percent of births in the United States to teenage mothers a maximum? (Round down to the nearest year.) Use the actual y-value of the vertex, to the nearest tenth, to find this percent.

34. The total receipts from individual income taxes by the U.S. Treasury in the years 2002–2006 can be modeled by the quadratic function defined by

$$f(x) = 33.79x^2 - 83.44x + 1036,$$

where $x = 0$ represents 2002, $x = 1$ represents 2003, and so on, and $f(x)$ is in billions of dollars. (*Source:* Internal Revenue Service.)

(a) Since the coefficient of x^2 given in the model is positive, the graph of this quadratic function is a parabola that opens up. Will the y-value of the vertex of this graph be a maximum or minimum?

(b) In what year during this period were total receipts from individual taxes a minimum? Use the actual x-value of the vertex, to the nearest tenth, to find this amount.

35. The graph shows how Social Security assets are expected to change as the number of retirees receiving benefits increases.

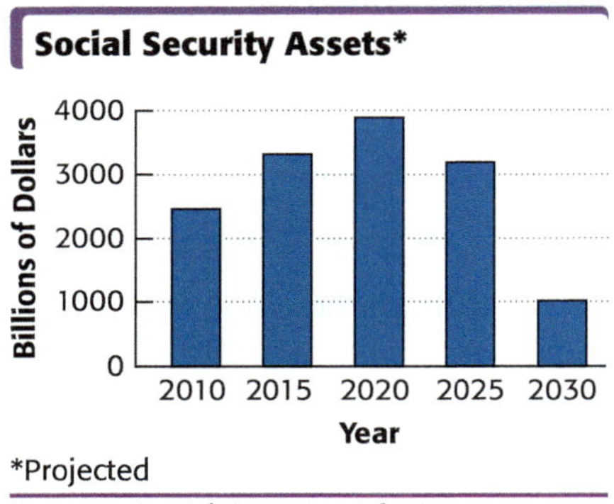

Source: Social Security Administration.

The graph suggests that a quadratic function would be a good fit to the data. The data are approximated by the function defined by

$$f(x) = -20.57x^2 + 758.9x - 3140.$$

In the model, $x = 10$ represents 2010, $x = 15$ represents 2015, and so on, and $f(x)$ is in billions of dollars.

(a) Explain why the coefficient of x^2 in the model is negative, based on the graph.

(b) Algebraically determine the vertex of the graph, with coordinates to four significant digits.

(c) Interpret the answer to part (b) as it applies to the application.

36. The graph shows the performance of investment portfolios with different mixtures of U.S. and foreign investments over a 25-yr period.

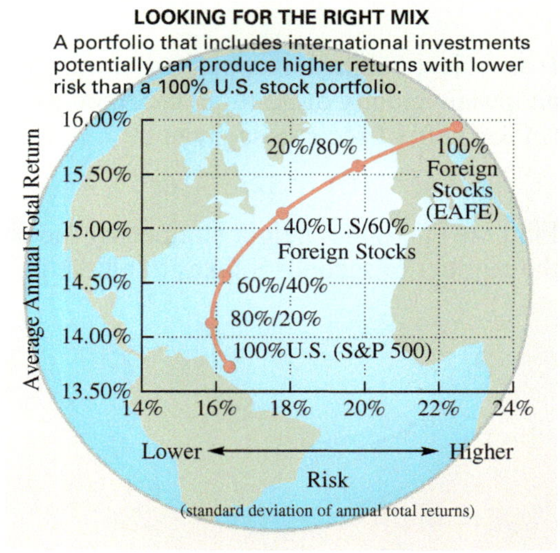

Source: Financial Ink Newsletter, Investment Management and Research, Inc., Feb. 1998. Thanks to David Van Geffen for this information.

(a) Is this the graph of a function? Explain.

(b) What investment mixture shown on the graph appears to represent the vertex? What relative amount of risk does this point represent? What return on investment does it provide?

(c) Which point on the graph represents the riskiest investment mixture? What return on investment does it provide?

Answers to Selected Exercises

Chapter 13 Solving Inequalities/Absolute Values/Intro Functions

Taken from, *Intermediate Algebra* (0321574974)

Section 13.1 (pages 970–983)

1. $\{1, 2, 3, 4, 5\}$ **3.** $\{5, 6, 7, 8, \ldots\}$ **5.** $\{10, 12, 14, 16, \ldots\}$ **7.** $\emptyset$ **9.** $\{-4, 4\}$ **11.** $\{x \mid x$ is an even natural number less than or equal to $8\}$ **13.** $\{x \mid x$ is a multiple of 4 greater than $0\}$

15. 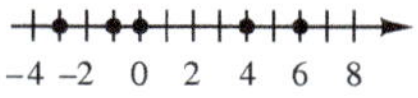

17. 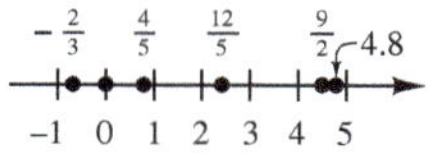

19. (a) $5, 17, \frac{40}{2}$ (or 20) **(b)** $0, 5, 17, \frac{40}{2}$ **(c)** $-8, 0, 5, 17, \frac{40}{2}$ **(d)** $-8, -0.6, 0, \frac{3}{4}, 5, \frac{13}{2}, 17, \frac{40}{2}$ **(e)** $-\sqrt{5}, \sqrt{3}, \pi$ **(f)** All are real numbers. **21.** false; Some are integers, but others, like $\frac{3}{4}$, are not. **23.** false; No irrational number is an integer. **25.** true **27.** true **29.** true **31. (a)** -6 **(b)** 6 **33. (a)** 12 **(b)** 12 **35. (a)** $-\frac{6}{5}$ **(b)** $\frac{6}{5}$ **37.** 8 **39.** $\frac{3}{2}$ **41.** -5 **43.** -2 **45.** -4.5 **47.** 5 **49.** 6 **51.** 0 **53. (a)** Louisiana; It decreased 4.1%. **(b)** West Virginia; It increased 0.6%. **55.** Pacific Ocean, Indian Ocean, Caribbean Sea, South China Sea, Gulf of California **57.** true **59.** true **61.** false **63.** true **65.** true **67.** $7 > y$ **69.** $5 \geq 5$ **71.** $3t - 4 \leq 10$ **73.** $5x + 3 \neq 0$ **75.** $-6 < 10$; true **77.** $10 \geq 10$; true **79.** $-3 \geq -3$; true **81.** $-8 > -6$; false **83.** greater than **85.** California (CA), Florida (FL) **87.** $x < y$

Section 13.2 (pages 984–1000)

1. D **3.** B **5.** F **7. (a)** $131 \leq s \leq 155$ **(b)** $s > 155$ **(c)** $9 \leq x \leq 12$ **(d)** $x > 18$ **9.** Reverse the direction of the inequality symbol only when multiplying or dividing by a *negative* number. The solution set is $[-16, \infty)$.

11. $(-\infty, 7]$ 

13. $[5, \infty)$ 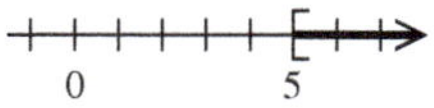

15. $(-5, \infty)$ 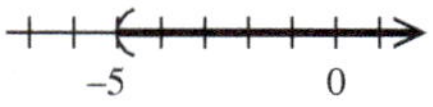

17. $(-4, \infty)$ 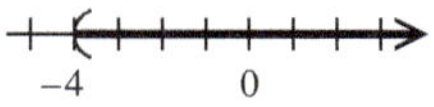

19. $(-\infty, -40]$ 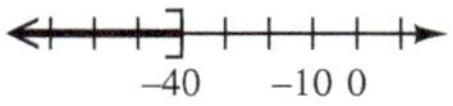

21. $(-\infty, 4]$ 

23. $(7, \infty)$ 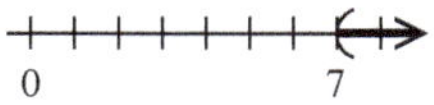

25. $\left(-\infty, -\frac{15}{2}\right)$ 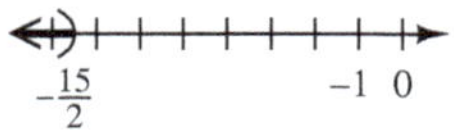

27. $(-\infty, -7)$ 

29. $\left[\frac{1}{2}, \infty\right)$ 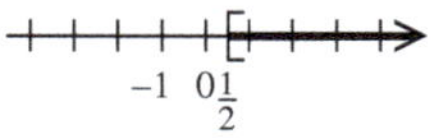

31. $(3, \infty)$ 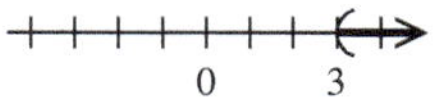

33. $(-\infty, 4)$ 

35. $\left(-\infty, \frac{23}{6}\right]$ 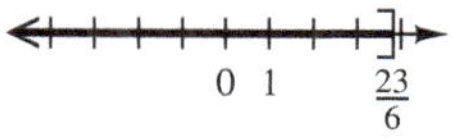

37. $\left(-\infty, \frac{76}{11}\right)$ 

39. $\{-9\}$ 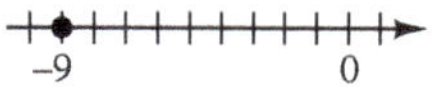

40. $(-9, \infty)$ 

41. $(-\infty, -9)$ 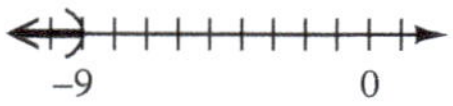

42. We obtain the set of all real numbers. 

43. $(-\infty, -3)$

45. $(1, 11)$ 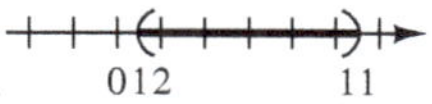

47. $[-14, 10]$ 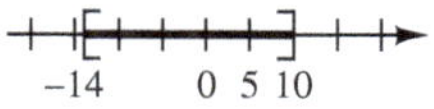

49. $[-5, 6]$ 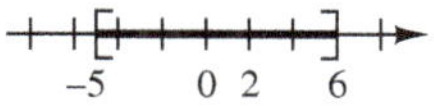

51. $(-6, -4)$ 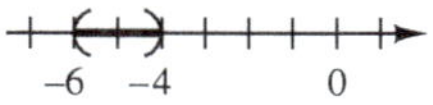

53. $\left[-\frac{1}{3}, \frac{1}{9}\right)$ 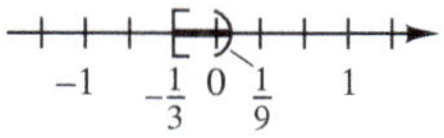

55. $\left[-\frac{1}{2}, \frac{35}{2}\right]$ 

57. from about 2:30 P.M. to 6:00 P.M. **59.** about 84°F–91°F **61.** 26 months **63.** at least 80 **65.** 26 DVDs **67. (a)** 140 to 184 lb **(b)** Answers will vary. **69.** all numbers between -2 and 2, or $(-2, 2)$ **71.** all numbers greater than or equal to 3, or $[3, \infty)$ **73.** all numbers greater than or equal to -9, or $[-9, \infty)$

Section 13.3 (pages 1001–1012)

1. true **3.** false; The union is $(-\infty, 6) \cup (6, \infty)$. **5.** $\{4\}$, or D **7.** $\emptyset$ **9.** $\{1, 2, 3, 4, 5, 6\}$, or A **11.** $\{1, 3, 5, 6\}$

13. 

15. 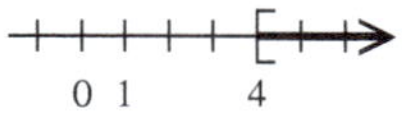

17. 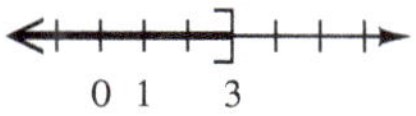

19. Answers will vary. One example is: The intersection of two streets is the region common to *both* streets.

21. $(-3, 2)$ 

23. $(-\infty, 2]$ 

25. $\emptyset$ 

27. $[5, 9]$ 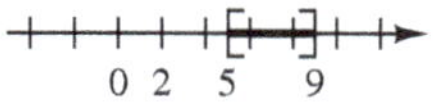

29. $(-\infty, 4]$ 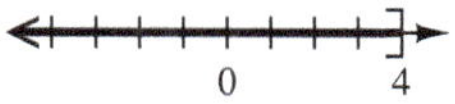

31. $(-\infty, 8]$ 

33. $[-2, \infty)$ 

35. $(-\infty, \infty)$ 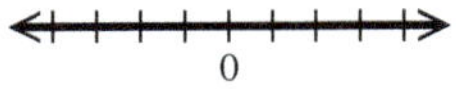

37. $(-\infty, -5) \cup (5, \infty)$ 

39. $(-\infty, -1] \cup (2, \infty)$ 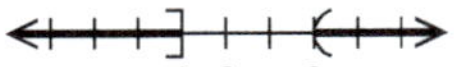

41. $(-\infty, 2) \cup (2, \infty)$ 

43. $[-4, -1]$ **45.** $[-9, -6]$ **47.** $(-\infty, 3)$ **49.** $[3, 9]$

51. intersection; $(-5, -1)$ 

53. union; $(-\infty, 4)$ 

55. intersection; $[4, 12]$ 

57. union; $(-\infty, 0] \cup [2, \infty)$ 

59. Mario, Joe **60.** none of them **61.** none of them **62.** Luigi, Than **63.** Mario, Joe **64.** all of them **65.** {Tuition and fees} **67.** {Tuition and fees, Dormitory charges}

Section 13.4 (pages 1013–1026)

1. E; C; D; B; A **3.** **(a)** one **(b)** two **(c)** none **5.** $\{-12, 12\}$ **7.** $\{-5, 5\}$ **9.** $\{-6, 12\}$ **11.** $\{-5, 4\}$ **13.** $\left\{-3, \frac{11}{2}\right\}$ **15.** $\left\{-\frac{19}{2}, \frac{9}{2}\right\}$ **17.** $\{-10, -2\}$ **19.** $\left\{-8, \frac{32}{3}\right\}$

21. $(-\infty, -3) \cup (3, \infty)$ 

23. $(-\infty, -4] \cup [4, \infty)$ 

25. $(-\infty, -10) \cup (6, \infty)$ 

27. $\left(-\infty, -\frac{7}{3}\right] \cup [3, \infty)$ 

29. $(-\infty, -2) \cup (8, \infty)$ 

31. **(a)** 

(b) 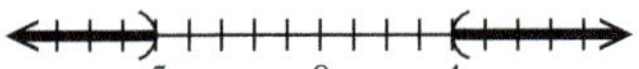

33. $[-3, 3]$ 

35. $(-4, 4)$ 

37. $[-10, 6]$ 

39. $\left(-\frac{7}{3}, 3\right)$ 

41. $[-2, 8]$ 

43. $(-\infty, -2) \cup (10, \infty)$ 

45. $\{-6, -1\}$ 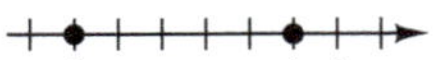

47. $\left[-\frac{10}{3}, 4\right]$ 

49. $\left[-4, -\frac{4}{3}\right]$ 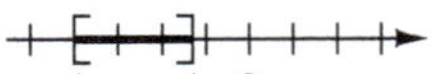

51. $\{-5, 5\}$ **53.** $\{-5, -3\}$ **55.** $(-\infty, -3) \cup (2, \infty)$ **57.** $[-10, 0]$ **59.** $\{-1, 3\}$ **61.** $\left\{-3, \frac{5}{3}\right\}$ **63.** $\left\{-\frac{1}{3}, -\frac{1}{15}\right\}$ **65.** $\left\{-\frac{5}{4}\right\}$ **67.** $(-\infty, \infty)$ **69.** $\emptyset$ **71.** $\left\{-\frac{1}{4}\right\}$ **73.** $\emptyset$ **75.** $(-\infty, \infty)$ **77.** $\left\{-\frac{3}{7}\right\}$ **79.** $(-\infty, \infty)$ **81.** $\left(-\infty, -\frac{7}{10}\right) \cup \left(-\frac{7}{10}, \infty\right)$ **83.** $(-\infty, \infty)$ **85.** $|x - 1000| \le 100$; $900 \le x \le 1100$ **87.** 475.6 ft **88.** 1201 Walnut, City Hall, Fidelity Bank and Trust Building, Kansas City Power and Light, Hyatt Regency Crown Center **89.** City Center Square, Commerce Tower, Federal Office Building, 1201 Walnut, City Hall, Fidelity Bank and Trust Building, Kansas City Power and Light, Hyatt Regency Crown Center **90.** **(a)** $|x - 475.6| \ge 75$ **(b)** $x \ge 550.6$ or $x \le 400.6$ **(c)** Town Pavilion, One Kansas City Place **(d)** It makes sense because it includes all buildings *not* listed in the answer to Exercise 89.

Summary Exercises on Solving Linear and Absolute Value Equations and Inequalities (pages 1025–1026)

1. $\{12\}$ **2.** $\{-5, 7\}$ **3.** $\{7\}$ **4.** $\left\{-\frac{2}{5}\right\}$ **5.** $\emptyset$ **6.** $(-\infty, -1]$ **7.** $\left[-\frac{2}{3}, \infty\right)$ **8.** $\{-1\}$ **9.** $\{-3\}$ **10.** $\left\{1, \frac{11}{3}\right\}$ **11.** $(-\infty, 5]$ **12.** $(-\infty, \infty)$ **13.** $\{2\}$ **14.** $(-\infty, -8] \cup [8, \infty)$ **15.** $\emptyset$ **16.** $(-\infty, \infty)$ **17.** $(-5.5, 5.5)$ **18.** $\left\{\frac{13}{3}\right\}$ **19.** $\left\{-\frac{96}{5}\right\}$ **20.** $(-\infty, 32]$ **21.** $(-\infty, -24)$ **22.** $\left\{\frac{3}{8}\right\}$ **23.** $\left\{\frac{7}{2}\right\}$ **24.** $(-6, 8)$ **25.** $(-\infty, \infty)$ **26.** $(-\infty, 5)$ **27.** $(-\infty, -4) \cup (7, \infty)$ **28.** $\{24\}$ **29.** $\left\{-\frac{1}{5}\right\}$ **30.** $\left(-\infty, -\frac{5}{2}\right]$ **31.** $\left[-\frac{1}{3}, 3\right]$ **32.** $[1, 7]$ **33.** $\left\{-\frac{1}{6}, 2\right\}$ **34.** $\{-3\}$ **35.** $(-\infty, -1] \cup \left[\frac{5}{3}, \infty\right)$ **36.** $\left[\frac{3}{4}, \frac{15}{8}\right]$ **37.** $\left\{-\frac{5}{2}\right\}$ **38.** $\{60\}$ **39.** $\left[-\frac{9}{2}, \frac{15}{2}\right]$ **40.** $(1, 9)$ **41.** $(-\infty, \infty)$ **42.** $\left\{\frac{1}{3}, 9\right\}$ **43.** $(-\infty, \infty)$ **44.** $\left\{-\frac{10}{9}\right\}$ **45.** $\{-2\}$ **46.** $\emptyset$ **47.** $(-\infty, -1) \cup (2, \infty)$ **48.** $[-3, -2]$

Section 13.5 (pages 1027–1034)

1. solid; below **3.** dashed; above **5.** The graph of $Ax + By = C$ divides the plane into two regions. In one of these regions, the ordered pairs satisfy $Ax + By < C$; in the other, they satisfy $Ax + By > C$.

7.

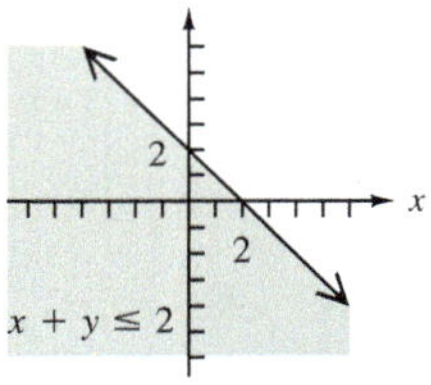

9.

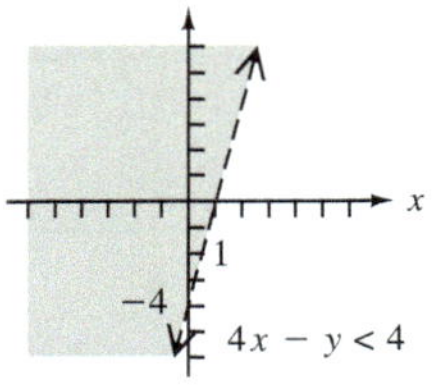

11.

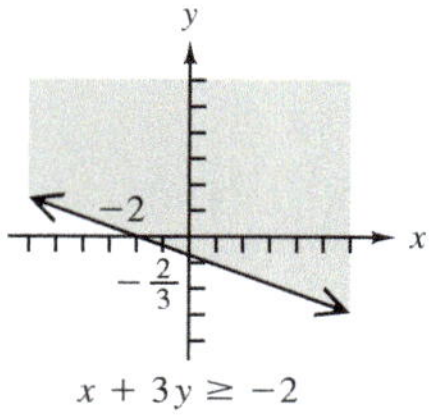

13.

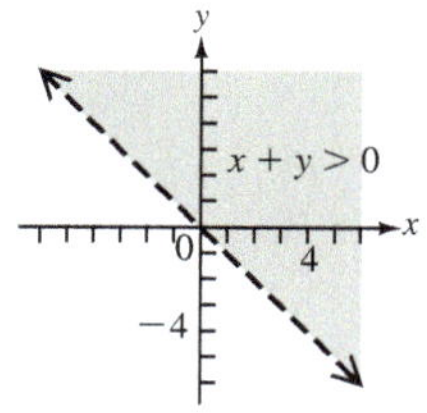

15.

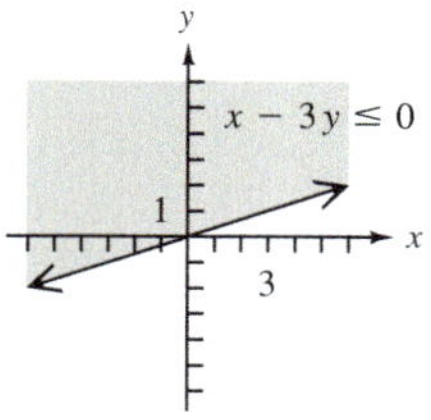

17.

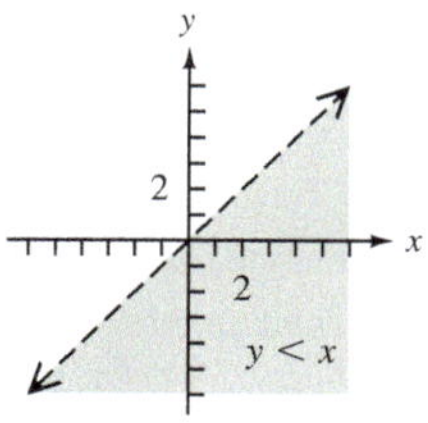

19.

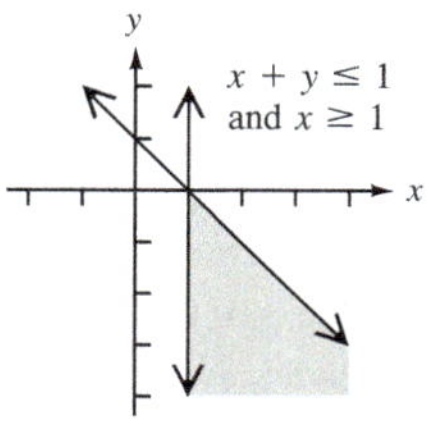

21.

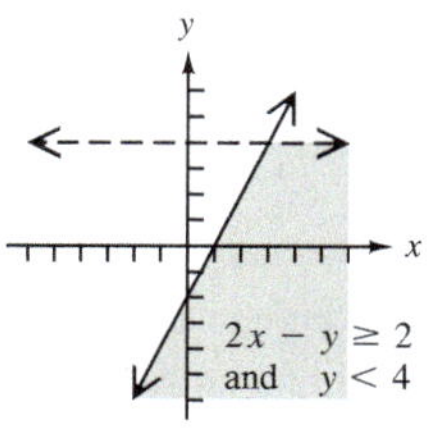

23.

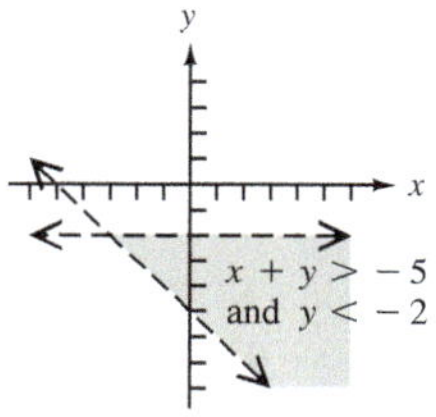

25.

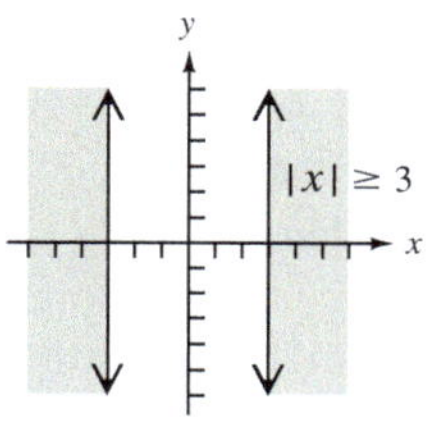

27.

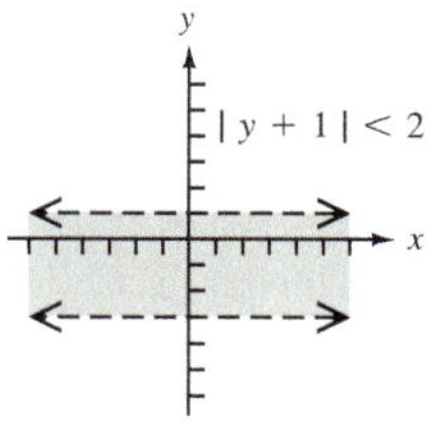

29.

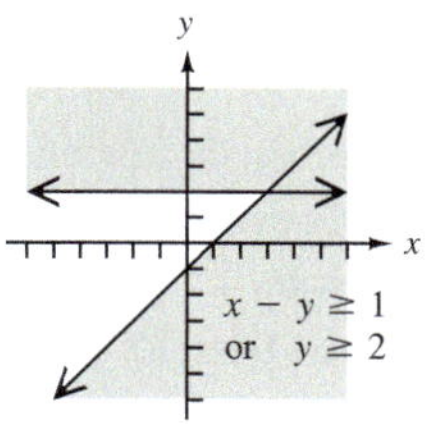

31.

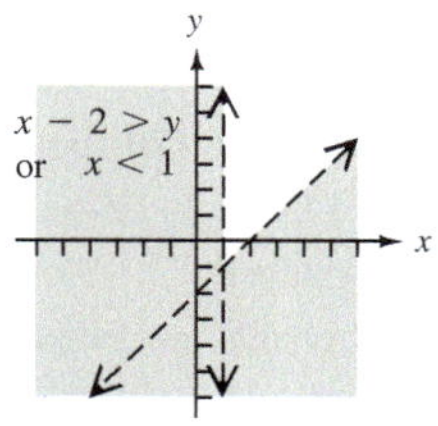

33.

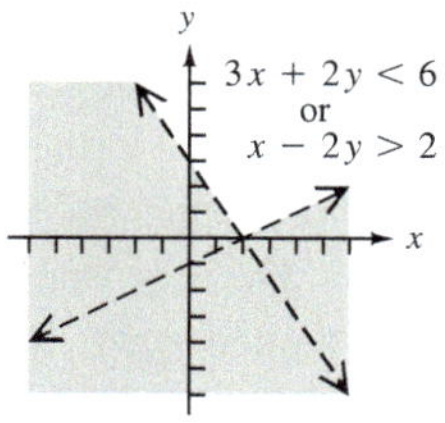

Section 13.6 (pages 1035–1050)

1. independent variable **3. (a)** A relation is a set of ordered pairs. **(b)** The domain is the set of all first components (x-values). **(c)** The range is the set of all second components (y-values). **(d)** A function is a relation in which each domain element is paired with one and only one range element. **5.** function; domain: $\{5, 3, 4, 7\}$; range: $\{1, 2, 9, 3\}$ **7.** not a function; domain: $\{2, 0\}$; range: $\{4, 2, 6\}$ **9.** function; domain: $\{-3, 4, -2\}$; range: $\{1, 7\}$ **11.** not a function; domain: $\{1, 0, 2\}$; range: $\{1, -1, 0, 4, -4\}$ **13.** function; domain: $\{2, 5, 11, 17, 3\}$; range: $\{1, 7, 20\}$ **15.** not a function; domain: $\{1\}$; range: $\{5, 2, -1, -4\}$ **17.** function; domain: $(-\infty, \infty)$; range: $(-\infty, \infty)$ **19.** function; domain: $(-\infty, \infty)$; range: $(-\infty, 4]$ **21.** not a function; domain: $[3, \infty)$; range: $(-\infty, \infty)$ **23.** function; domain: $(-\infty, \infty)$ **25.** not a function; domain: $[0, \infty)$ **27.** function; domain: $(-\infty, \infty)$ **29.** not a function; domain: $(-\infty, \infty)$ **31.** function; domain: $[0, \infty)$ **33.** function; domain: $(-\infty, 0) \cup (0, \infty)$ **35.** function; domain: $\left[-\frac{1}{2}, \infty\right)$ **37.** function; domain: $(-\infty, 9) \cup (9, \infty)$ **39. (a)** $[0, 3000]$ **(b)** 25 hr; 25 hr **(c)** 2000 gal **(d)** $g(0) = 0$; The pool is empty at time 0. **41.** Here is one example. The cost of gasoline; number of gallons purchased; cost; number of gallons **43.** 4 **45.** -11 **47.** $-3p + 4$ **49.** $3x + 4$ **51.** $-3x - 2$ **53.** $-\frac{p^2}{9} + \frac{4p}{3} + 1$ **55. (a)** 2 **(b)** 3 **57. (a)** 15 **(b)** 10 **59. (a)** 3 **(b)** -3 **61.** line; -2; linear; $-2x + 4$; -2; 3; -2 **63. (a)** $f(x) = -\frac{1}{3}x + 4$ **(b)** 3 **65. (a)** $f(x) = 3 - 2x^2$ **(b)** -15 **67. (a)** $f(x) = \frac{4}{3}x - \frac{8}{3}$ **(b)** $\frac{4}{3}$

69. domain: $(-\infty, \infty)$; range: $(-\infty, \infty)$

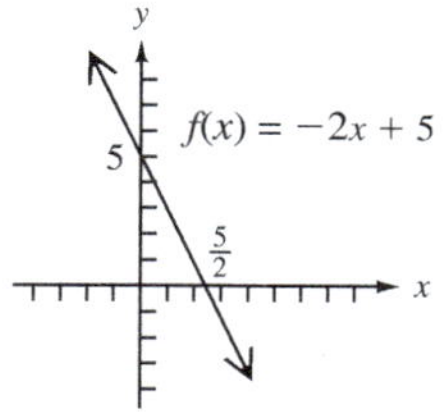

71. domain: $(-\infty, \infty)$; range: $(-\infty, \infty)$

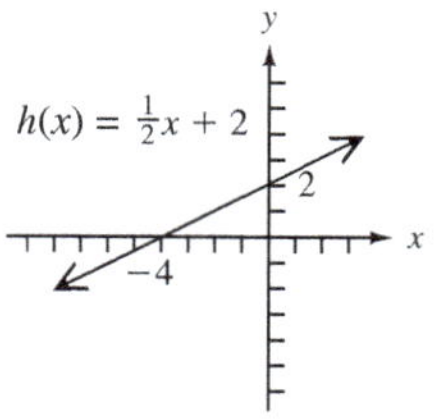

73. domain: $(-\infty, \infty)$; range: $\{-4\}$

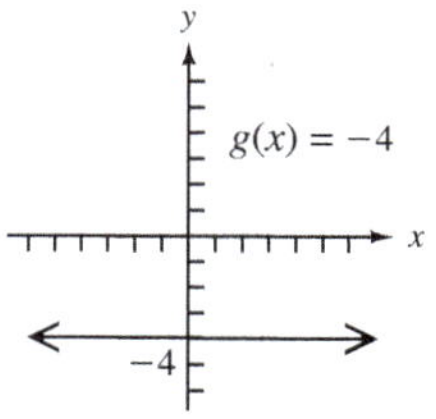

75. (a) \$0; \$2.50; \$5.00; \$7.50 **(b)** $2.50x$
(c)

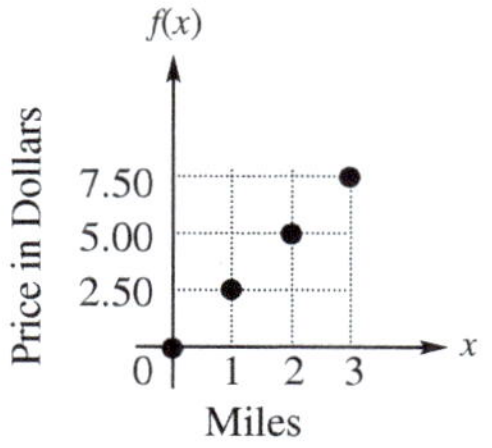

77. 194.53 cm **79.** 177.41 cm

Chapter 14 Systems of Equations

Taken from *Prealgebra & Introductory Algebra* by Lial, et. al (0321578732) and *Intermediate Algebra* (0321574974)

Section 14.1 (pages 1052–1060)

1. B, because the ordered pair must be in quadrant II. **3.** There is no way that the sum of two numbers can be both 2 and 4 at the same time. **5.** no **7.** yes **9.** yes **11.** no

We show the graphs here only for Exercises 13–17.

13. $\{(4, 2)\}$

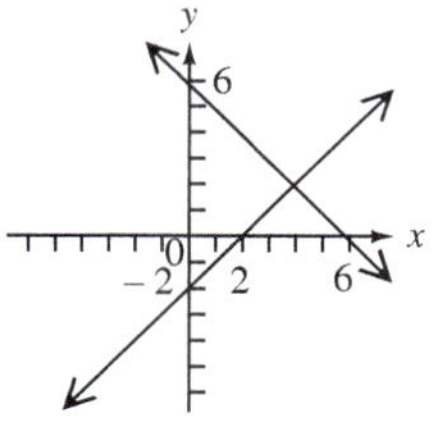

15. $\{(0, 4)\}$

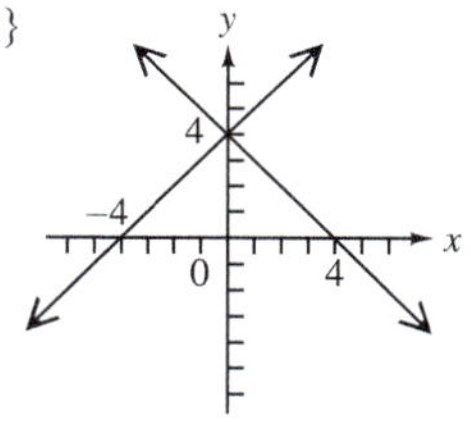

17. $\{(4, -1)\}$

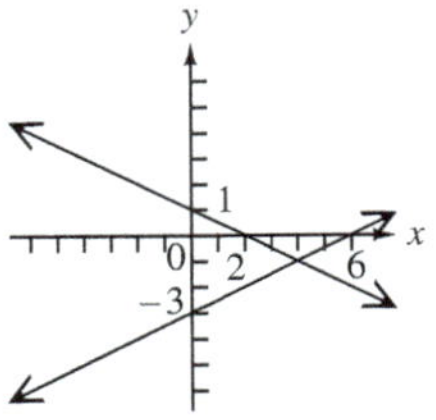

19. $\{(1, 3)\}$ 21. $\{(0, 2)\}$ 23. $\emptyset$ (inconsistent system) 25. $\{(x, y) \mid 2x - y = 4\}$ (dependent equations) 27. $\{(4, -3)\}$ 29. $\emptyset$ (inconsistent system) 31. (a) neither (b) intersecting lines (c) one solution 33. (a) dependent (b) one line (c) infinite number of solutions 35. (a) inconsistent (b) parallel lines (c) no solution 37. 1980–2000 39. (a) 1997–2002 (b) 2001 (c) 2002 (d) (2004, 70) (The y-value is approximate.) (e) During the period 1997–2004, debit card use went from least popular to most popular of the three methods depicted.

Section 14.2 (pages 1061–1068)

1. The y-value must also be determined. The solution set is $\{(3, 0)\}$. 3. $\{(3, 9)\}$ 5. $\{(7, 3)\}$ 7. $\{(-2, 4)\}$ 9. $\{(-4, 8)\}$ 11. $\{(3, -2)\}$ 13. $\{(x, y) \mid 3x - y = 5\}$ 15. $\left\{\left(\frac{1}{3}, -\frac{1}{2}\right)\right\}$ 17. $\emptyset$ 19. $\{(x, y) \mid 3x - 4y = 2\}$ 21. $\{(2, -3)\}$ 23. $\{(10, -12)\}$ 25. $\{(-4, 2)\}$ 27. $\{(7, -3)\}$ 29. $\{(20, 30)\}$ 30. To find the total cost, multiply the number of bicycles (x) by the cost per bicycle (400 dollars) and add the fixed cost (5000 dollars). Thus, $y_1 = 400x + 5000$ gives this total cost (in dollars). 31. $y_2 = 600x$ 32. $y_1 = 400x + 5000$, $y_2 = 600x$; solution set: $\{(25, 15{,}000)\}$ 33. 25; 15,000; 15,000

Section 14.3 (pages 1069–1078)

1. true 3. true 5. $\{(-1, 3)\}$ 7. $\{(-1, -3)\}$ 9. $\{(-2, 3)\}$ 11. $\left\{\left(\frac{1}{2}, 4\right)\right\}$ 13. $\{(3, -6)\}$ 15. $\{(7, 4)\}$ 17. $\{(0, 4)\}$ 19. $\{(-4, 0)\}$ 21. $\{(0, 0)\}$ 23. $\emptyset$ 25. $\{(x, y) \mid x - 3y = -4\}$ 27. $\{(2, 9)\}$ 29. $\{(-6, 5)\}$ 31. $\left\{\left(-\frac{6}{5}, \frac{4}{5}\right)\right\}$ 33. $\left\{\left(\frac{1}{8}, -\frac{5}{6}\right)\right\}$ 35. $\{(11, 15)\}$ 37. $\emptyset$ 39. $\{(x, y) \mid 2x + y = 0\}$ 41. $1339 = 1996a + b$ 42. $1536 = 2004a + b$ 43. $1996a + b = 1339$, $2004a + b = 1536$; solution set: $\{(24.625, -47{,}812.5)\}$ 44. $y = 24.625x - 47{,}812.5$ 45. 1486.8 (million); This is quite a bit less than the actual figure. 46. Since the data do not lie in a perfectly straight line, the quantity obtained from an equation determined in this way will probably be "off" a bit. We cannot put too much faith in models such as this one, because not all sets of data points are linear in nature.

Summary Exercises on Solving Systems of Linear Equations (pages 1077–1078)

1. (a) Use substitution since the second equation is solved for y. (b) Use elimination since the coefficients of the y-terms are opposites. (c) Use elimination since the equations are in standard form with no coefficients of 1 or -1. Solving by substitution would involve fractions. 2. The system on the right is easier to solve by substitution because the second equation is already solved for y. 3. (a) $\{(1, 4)\}$ (b) $\{(1, 4)\}$ (c) Answers will vary. 4. (a) $\{(-5, 2)\}$ (b) $\{(-5, 2)\}$ (c) Answers will vary. 5. $\{(2, 6)\}$ 6. $\{(-3, 2)\}$ 7. $\left\{\left(\frac{1}{3}, \frac{1}{2}\right)\right\}$ 8. $\emptyset$ 9. $\{(3, 0)\}$ 10. $\left\{\left(\frac{3}{2}, -\frac{3}{2}\right)\right\}$ 11. $\{(x, y) \mid 3x + y = 7\}$ 12. $\{(9, 4)\}$ 13. $\left\{\left(-\frac{5}{7}, -\frac{2}{7}\right)\right\}$ 14. $\{(4, -5)\}$ 15. $\emptyset$ 16. $\{(-4, 6)\}$ 17. $\left\{\left(\frac{19}{3}, -5\right)\right\}$ 18. $\left\{\left(\frac{22}{13}, -\frac{23}{13}\right)\right\}$ 19. $\{(-12, -60)\}$ 20. $\{(2, -4)\}$ 21. $\{(18, -12)\}$ 22. $\{(-2, 1)\}$ 23. $\left\{\left(13, -\frac{7}{5}\right)\right\}$ 24. $\{(10, -9)\}$ 25. $\{(0.04, 0.9)\}$

Section 14.4 (pages 1079–1088)

1. The statement means that when -1 is substituted for x, 2 is substituted for y, and 3 is substituted for z in the three equations, the resulting three statements are true. 3. $\{(3, 2, 1)\}$ 5. $\{(1, 4, -3)\}$ 7. $\{(1, 0, 3)\}$ 9. $\left\{\left(1, \frac{3}{10}, \frac{2}{5}\right)\right\}$ 11. $\{(0, 2, -5)\}$ 13. $\left\{\left(\frac{20}{59}, -\frac{33}{59}, \frac{35}{59}\right)\right\}$ 15. $\{(4, 5, 3)\}$ 17. $\{(2, 2, 2)\}$ 19. $\{(-1, 0, 0)\}$ 21. $\left\{\left(\frac{8}{3}, \frac{2}{3}, 3\right)\right\}$ 23. Answers will vary. Some possible answers are (a) two perpendicular walls and the ceiling in a normal room, (b) the floors of three different levels of an office building, and (c) three pages of this book (since they intersect in the spine). 25. $\emptyset$; inconsistent system 27. $\{(x, y, z) \mid x - y + 4z = 8\}$; dependent equations 29. $\{(x, y, z) \mid 2x + y - z = 6\}$; dependent equations 31. $\{(0, 0, 0)\}$ 33. $\emptyset$; inconsistent system 35. $\{(3, 0, 2)\}$ 37. $128 = a + b + c$ 38. $140 = 2.25a + 1.5b + c$ 39. $80 = 9a + 3b + c$ 40. $a + b + c = 128$; $2.25a + 1.5b + c = 140$; $9a + 3b + c = 80$; $\{(-32, 104, 56)\}$ 41. $f(x) = -32x^2 + 104x + 56$ 42. height; time 43. 56 ft 44. 140.5 ft

Section 14.5 (pages 1089–1102)

1. wins: 96; losses: 66 3. length: 78 ft; width: 36 ft 5. Wal-Mart: \$316 billion; ExxonMobil: \$340 billion 7. $x = 40$ and $y = 50$, so the angles measure 40° and 50°. 9. NHL: \$247.32; NBA: \$267.37 11. Junior Roast Beef: \$2.09; Big Montana: \$4.39 13. (a) 6 oz (b) 15 oz (c) 24 oz (d) 30 oz 15. $\$1.29x$ 17. 6 gal of 25%; 14 gal of 35% 19. 6 L of pure acid; 48 L of 10% acid 21. 14 kg of nuts; 16 kg of cereal 23. \$1000 at 2%; \$2000 at 4% 25. (a) $(10 - x)$ mph (b) $(10 + x)$ mph 27. scooter: 25 mph; bicycle: 10 mph 29. boat: 21 mph; current: 3 mph 31. \$0.75-per-lb candy: 5.22 lb; \$1.25-per-lb candy: 3.78 lb 33. 76 general admission; 108 with student ID 35. 8 for a citron; 5 for a wood apple 37. $x + y + z = 180$; angle measures: 70°, 30°, 80° 39. first: 20°; second: 70°; third: 90° 41. shortest: 12 cm; middle: 25 cm; longest: 33 cm 43. Independent: 38; Democrat: 34; Republican: 28 45. \$14 tickets: 300; \$20 tickets: 225; \$50 tickets: 60 47. bookstore A: 140; bookstore B: 280; bookstore C: 380 49. wins: 46; losses: 25; ties: 11

Section 14.6 (pages 1103–1110)

1. (a) $0, 5, -3$ (b) $1, -3, 8$ (c) yes; The number of rows is the same as the number of columns (three). (d) $\begin{bmatrix} 1 & 4 & 8 \\ 0 & 5 & -3 \\ -2 & 3 & 1 \end{bmatrix}$ (e) $\begin{bmatrix} 1 & -\frac{3}{2} & -\frac{1}{2} \\ 0 & 5 & -3 \\ 1 & 4 & 8 \end{bmatrix}$ (f) $\begin{bmatrix} 1 & 15 & 25 \\ 0 & 5 & -3 \\ 1 & 4 & 8 \end{bmatrix}$

3. $\left[\begin{array}{cc|c} 1 & 2 & 11 \\ 2 & -1 & -3 \end{array}\right]$; $\left[\begin{array}{cc|c} 1 & 2 & 11 \\ 0 & -5 & -25 \end{array}\right]$; $\left[\begin{array}{cc|c} 1 & 2 & 11 \\ 0 & 1 & 5 \end{array}\right]$; $x + 2y = 11$; $y = 5$; $\{(1, 5)\}$ **5.** $\{(4, 1)\}$ **7.** $\{(1, 1)\}$ **9.** $\{(-1, 4)\}$ **11.** $\emptyset$ **13.** $\{(x, y) \mid 2x + y = 4\}$ **15.** $\{(0, 0)\}$

17. $\left[\begin{array}{ccc|c} 1 & 1 & -1 & -3 \\ 0 & -1 & 3 & 10 \\ 0 & -6 & 7 & 38 \end{array}\right]$; $\left[\begin{array}{ccc|c} 1 & 1 & -1 & -3 \\ 0 & 1 & -3 & -10 \\ 0 & -6 & 7 & 38 \end{array}\right]$;

$\left[\begin{array}{ccc|c} 1 & 1 & -1 & -3 \\ 0 & 1 & -3 & -10 \\ 0 & 0 & -11 & -22 \end{array}\right]$; $\left[\begin{array}{ccc|c} 1 & 1 & -1 & -3 \\ 0 & 1 & -3 & -10 \\ 0 & 0 & 1 & 2 \end{array}\right]$; $x + y - z = -3$; $y - 3z = -10$; $z = 2$; $\{(3, -4, 2)\}$ **19.** $\{(4, 0, 1)\}$ **21.** $\{(-1, 23, 16)\}$ **23.** $\{(3, 2, -4)\}$ **25.** $\emptyset$ **27.** $\{(x, y, z) \mid x - 2y + z = 4\}$ **29.** $\{(0, 0, 0)\}$

Chapter 15 Factoring

Taken from Prealgebra & *Introductory Algebra* (0321578732) *Intermediate Algebra* (0321574974) and *Prealgebra & Introductory Algebra* (0321578732)

Section 15.1 (pages 1112–1120)

1. 4 **3.** 4 **5.** 6 **7.** 1 **9.** 8 **11.** $10x^3$ **13.** xy^2 **15.** 6 **17.** $3m^2$ **19.** $2z^4$ **21.** $2mn^4$ **23.** $y + 2$ **25.** $a - 2$ **27.** $2 + 3xy$ **29.** $x(x - 4)$ **31.** $3t(2t + 5)$ **33.** $\frac{1}{4}d(d - 3)$ **35.** $-6x^2(2x + 1)$ **37.** $5y^6(13y^4 + 7)$ **39.** no common factor (except 1) **41.** $8m^2n^2(n + 3)$ **43.** $-2x(2x^2 - 5x + 3)$ **45.** $13y^2(y^6 + 2y^2 - 3)$ **47.** $9qp^3(5q^3p^2 + 4p^3 + 9q)$ **49.** $(x + 2)(c + d)$ **51.** $(2a + b)(a^2 - b)$ **53.** $(p + 4)(q - 1)$ **55.** $(5 + n)(m + 4)$ **57.** $(2y - 7)(3x + 4)$ **59.** $(y + 3)(3x + 1)$ **61.** $(z + 2)(7z - a)$ **63.** $(3r + 2y)(6r - x)$ **65.** $(w + 1)(w^2 + 9)$ **67.** $(a + 2)(3a^2 - 2)$ **69.** $(4m - p^2)(4m^2 - p)$ **71.** $(y + 3)(y + x)$ **73.** $(z - 2)(2z - 3w)$ **75.** commutative property **76.** $2x(y - 4) - 3(y - 4)$ **77.** No, because it is not a product. It is the difference between $2x(y - 4)$ and $3(y - 4)$. **78.** $(2x - 3)(y - 4)$; yes

Section 15.2 (pages 1121–1126)

1. a and b must have different signs. **3.** A prime polynomial is one that cannot be factored using only integers in the factors. **5.** 1 and 12, −1 and −12, 2 and 6, −2 and −6, 3 and 4, −3 and −4; The pair with a sum of 7 is 3 and 4. **7.** 1 and −24, −1 and 24, 2 and −12, −2 and 12, 3 and −8, −3 and 8, 4 and −6, −4 and 6; The pair with a sum of −5 is 3 and −8. **9.** C **11.** $x + 11$ **13.** $x - 8$ **15.** $y - 5$ **17.** $x + 11$ **19.** $y - 9$ **21.** $(y + 8)(y + 1)$ **23.** $(b + 3)(b + 5)$ **25.** $(m + 5)(m - 4)$ **27.** $(x + 8)(x - 5)$ **29.** $(y - 5)(y - 3)$ **31.** $(z - 8)(z - 7)$ **33.** $(r - 6)(r + 5)$ **35.** $(a - 12)(a + 4)$ **37.** prime **39.** $(r + 2a)(r + a)$ **41.** $(x + y)(x + 3y)$ **43.** $(t + 2z)(t - 3z)$ **45.** $(v - 5w)(v - 6w)$ **47.** $4(x + 5)(x - 2)$ **49.** $2t(t + 1)(t + 3)$ **51.** $-2x^4(x - 3)(x + 7)$ **53.** $a^3(a + 4b)(a - b)$ **55.** $mn(m - 6n)(m - 4n)$ **57.** The factored form $(2x + 4)(x - 3)$ is incorrect because $2x + 4$ has a common factor of 2, which must be factored out for the trinomial to be *completely* factored.

Section 15.3 (pages 1127–1130)

1. $(m + 6)(m + 2)$ **3.** $(a + 5)(a - 2)$ **5.** $(2t + 1)(5t + 2)$ **7.** $(3z - 2)(5z - 3)$ **9.** $(2s - t)(4s + 3t)$ **11.** $(3a + 2b)(5a + 4b)$ **13.** B **15. (a)** 2; 12; 24; 11 **(b)** 3; 8 (Order is irrelevant.) **(c)** $3m$; $8m$ **(d)** $2m^2 + 3m + 8m + 12$ **(e)** $(2m + 3)(m + 4)$ **(f)** $(2m + 3)(m + 4) = 2m^2 + 11m + 12$ **17.** $(2x + 1)(x + 3)$ **19.** $(4r - 3)(r + 1)$ **21.** $(4m + 1)(2m - 3)$ **23.** $(3m + 1)(7m + 2)$ **25.** $(2b + 1)(3b + 2)$ **27.** $(4y - 3)(3y - 1)$ **29.** $3(4x - 1)(2x - 3)$ **31.** $2m(m - 4)(m + 5)$ **33.** $-4z^3(z - 1)(8z + 3)$ **35.** $(3p + 4q)(4p - 3q)$ **37.** $(3a - 5b)(2a + b)$ **39.** $(5 - x)(1 - x)$ **41.** The student stopped too soon. He needs to factor out the common factor $4x - 1$ to get $(4x - 1)(4x - 5)$ as the correct answer.

Section 15.4 (pages 1131–1136)

1. B **3.** A **5.** A **7.** $2a + 5b$ **9.** $x^2 + 3x - 4$; $x + 4$, $x - 1$, or $x - 1$, $x + 4$ **11.** $2z^2 - 5z - 3$; $2z + 1$, $z - 3$, or $z - 3$, $2z + 1$ **13.** The binomial $2x - 6$ cannot be a factor because it has a common factor of 2, but the polynomial does not. **15.** $(3a + 7)(a + 1)$ **17.** $(2y + 3)(y + 2)$ **19.** $(3m - 1)(5m + 2)$ **21.** $(3s - 1)(4s + 5)$ **23.** $(5m - 4)(2m - 3)$ **25.** $(4w - 1)(2w - 3)$ **27.** $(4y + 1)(5y - 11)$ **29.** prime **31.** $2(5x + 3)(2x + 1)$ **33.** $-q(5m + 2)(8m - 3)$ **35.** $3n^2(5n - 3)(n - 2)$ **37.** $-y^2(5x - 4)(3x + 1)$ **39.** $(5a + 3b)(a - 2b)$ **41.** $(4s + 5t)(3s - t)$ **43.** $m^4n(3m + 2n)(2m + n)$ **45.** $-1(x + 7)(x - 3)$ **47.** $-1(3x + 4)(x - 1)$ **49.** $-1(a + 2b)(2a + b)$ **51.** $5 \cdot 7$ **52.** $(-5)(-7)$ **53.** The product of $3x - 4$ and $2x - 1$ is $6x^2 - 11x + 4$. **54.** The product of $4 - 3x$ and $1 - 2x$ is $6x^2 - 11x + 4$. **55.** The factors in Exercise 53 are the opposites of the factors in Exercise 54. **56.** $(3 - 7t)(5 - 2t)$

Section 15.5 (pages 1137–1146)

1. 1; 4; 9; 16; 25; 36; 49; 64; 81; 100; 121; 144; 169; 196; 225; 256; 289; 324; 361; 400 **3.** 2 **5.** $(y + 5)(y - 5)$ **7.** $\left(p + \frac{1}{3}\right)\left(p - \frac{1}{3}\right)$ **9.** prime **11.** $(3r + 2)(3r - 2)$ **13.** $\left(2m + \frac{3}{5}\right)\left(2m - \frac{3}{5}\right)$ **15.** $4(3x + 2)(3x - 2)$ **17.** $(14p + 15)(14p - 15)$ **19.** $(4r + 5a)(4r - 5a)$ **21.** prime **23.** $(p^2 + 7)(p^2 - 7)$ **25.** $(x^2 + 1)(x + 1)(x - 1)$ **27.** $(p^2 + 16)(p + 4)(p - 4)$ **29.** The teacher was justified, because it was not factored *completely;* $x^2 - 9$ can be factored as $(x + 3)(x - 3)$. The complete factored form is $(x^2 + 9)(x + 3)(x - 3)$. **31.** No, it is not a perfect square since the middle term would have to be $30y$. **33.** $(w + 1)^2$ **35.** $(x - 4)^2$ **37.** $\left(t + \frac{1}{2}\right)^2$ **39.** $(x - 0.5)^2$ **41.** $2(x + 6)^2$ **43.** $(4x - 5)^2$ **45.** $(7x - 2y)^2$ **47.** $(8x + 3y)^2$ **49.** $-2h(5h - 2y)^2$ **51.** $(2x + 3)(5x - 2)$ **52.** $5x - 2$ **53.** Yes. We saw in Exercise 51 that $(2x + 3)(5x - 2)$ is the factored form of $10x^2 + 11x - 6$. **54.** The quotient is $x^2 + x + 1$, so $x^3 - 1$ factors as $(x - 1)(x^2 + x + 1)$.

Summary Exercises on Factoring (pages 1143–1146)

1. F **2.** G **3.** A **4.** B **5.** D **6.** H **7.** C **8.** E **9.** H **10.** D **11.** $8m^3(4m^6 + 2m^2 + 3)$ **12.** $2(m + 3)(m - 8)$ **13.** $7k(2k + 5)(k - 2)$ **14.** prime **15.** $(6z + 1)(z + 5)$ **16.** $(m + n)(m - 4n)$ **17.** $(7z + 4y)(7z - 4y)$ **18.** $10nr(10nr + 3r^2 - 5n)$ **19.** $4x(4x + 5)$ **20.** $(4 + m)(5 + 3n)$ **21.** $(5y - 6z)(2y + z)$ **22.** $(y^2 + 9)(y + 3)(y - 3)$ **23.** $(m - 3)(m + 5)$ **24.** $(2y + 1)(3y - 4)$ **25.** $8z(4z - 1)(z + 2)$ **26.** $(p - 1.2)^2$ **27.** $(z - 6)^2$ **28.** $(3m + 8)(3m - 8)$

29. $(y-6k)(y+2k)$ **30.** $(4z-1)^2$ **31.** $6(y-2)(y+1)$
32. $\left(x+\frac{1}{4}\right)^2$ **33.** $(p-6)(p-11)$ **34.** $(a+8)(a+9)$ **35.** prime
36. $3(6m-1)^2$ **37.** $(z+2a)(z-5a)$ **38.** $(2a+1)(a^2-7)$
39. $(2k-3)^2$ **40.** $(a-7b)(a+4b)$ **41.** $(4r+3m)^2$
42. $(3k-2)(k+2)$ **43.** prime **44.** $(a^2+25)(a+5)(a-5)$
45. $4(2k-3)^2$ **46.** $(4k+1)(2k-3)$ **47.** $6y^4(3y+4)(2y-5)$
48. $5z(z-2)(z-7)$ **49.** $(8p-1)(p+3)$ **50.** $(4k-3h)(2k+h)$
51. $6(3m+2z)(3m-2z)$ **52.** $(2k-5z)^2$ **53.** $2(3a-1)(a+2)$
54. $(3h-2g)(5h+7g)$ **55.** $7(2a+3b)(2a-3b)$
56. $(5z-6)(2z+1)$ **57.** $5m^2(5m-13n)(5m-3n)$
58. $(3y-1)(3y+5)$ **59.** $(3u+11v)^2$ **60.** prime
61. $9p^8(3p+7)(p-4)$ **62.** $5(2m-3)(m+4)$ **63.** $(2-q)(2-3p)$
64. $\left(k+\frac{8}{11}\right)\left(k-\frac{8}{11}\right)$ **65.** $4(4p+5m)(4p-5m)$
66. $(m+4)(m^2-6)$ **67.** $(10a+9y)(10a-9y)$
68. $(8a-b)(a+3b)$ **69.** $(a+4)^2$ **70.** $(2y+5)(2y-5)$ **71.** prime
72. $-3x(x+2y)(x-2y)$ **73.** $(5a-7b)^2$ **74.** $8(t^2+1)(t+1)(t-1)$
75. $-4(x-3y)^2$ **76.** $25(2a+b)(2a-b)$ **77.** $-2(x-9)(x-4)$
78. $(m+3)(2m-5n)$ **79.** $2(2x+5)(3x-2)$ **80.** $(y^2+5)(y^4-3)$
81. $(y+0.8)(y-0.8)$ **82.** $6p(2p+1)(p-5)$

Section 15.6 (pages 1147–1150)

1. $(10a+3b)(10a-3b)$ **3.** $6p^3(3p^2-4+2p^3)$ **5.** $(x+7)(x-5)$
7. prime **9.** $(6b+1)(b-3)$ **11.** $3mn(3m+2n)(2m-n)$
13. $(2p+5q)(p+3q)$ **15.** $(2k+7r)^2$ **17.** $(m-2)(n+5)$
19. $(x+3)^2(x-3)$ **21.** prime **23.** $(3k+1)(2k-1)$
25. $(x^2+25)(x+5)(x-5)$ **27.** $(a+6)(b+c)$ **29.** $4y(y-2)$
31. $(7z+2k)(2z-k)$ **33.** $16(4b+5c)(4b-5c)$
35. $8(5z+4)(25z^2-20z+16)$ **37.** $(5r-s)(2r+5s)$
39. $8x^2(4+2x-3x^3)$ **41.** $(2x-5q)(7x+5q)$ **43.** $(y+5)(y-2)$
45. $2a(a^2+3a-2)$ **47.** $(9p-5r)(2p+7r)$
49. $(x-2y+2)(x-2y-2)$ **51.** $(5r+2s-3)^2$
53. $(z+2)(z-2)(z^2-5)$ **55.** $(p+2)(4+m)$
57. $2(5p+9)(5p-9)$ **59.** $(4a+b)^2$

Section 15.7 (pages 1151–1158)

1. $\{-5, 2\}$ **3.** $\left\{3, \frac{7}{2}\right\}$ **5.** $\left\{-\frac{5}{6}, 0\right\}$ **7.** $\left\{0, \frac{4}{3}\right\}$ **9.** $\left\{-\frac{1}{2}, \frac{1}{6}\right\}$
11. $\{9\}$ **13.** Set each *variable* factor equal to 0, to get $2x = 0$ or $3x - 4 = 0$. The solution set is $\left\{0, \frac{4}{3}\right\}$. **15.** $\{-2, -1\}$ **17.** $\{1, 2\}$
19. $\{-8, 3\}$ **21.** $\{-1, 3\}$ **23.** $\{-2, -1\}$ **25.** $\{-4\}$ **27.** $\left\{-2, \frac{1}{3}\right\}$
29. $\left\{-\frac{4}{3}, \frac{1}{2}\right\}$ **31.** $\left\{-\frac{2}{3}\right\}$ **33.** $\{-3, 3\}$ **35.** $\left\{-\frac{7}{4}, \frac{7}{4}\right\}$ **37.** $\{-11, 11\}$
39. $\{0, 7\}$ **41.** $\left\{0, \frac{1}{2}\right\}$ **43.** $\{2, 5\}$ **45.** $\left\{-4, \frac{1}{2}\right\}$ **47.** $\left\{-12, \frac{11}{2}\right\}$
49. $\{-2, 0, 2\}$ **51.** $\left\{-\frac{7}{3}, 0, \frac{7}{3}\right\}$ **53.** $\left\{-\frac{5}{2}, \frac{1}{3}, 5\right\}$ **55.** $\left\{-\frac{7}{2}, -3, 1\right\}$
57. (a) 64; 144; 4; 6 **(b)** No time has elapsed, so the object hasn't fallen (been released) yet.

Section 15.8 (pages 1159–1170)

1. Read; variable; equation; Solve; answer; Check; original
3. ***Step 3:*** $45 = (2x+1)(x+1)$; ***Step 4:*** $x = 4$ or $x = -\frac{11}{2}$; ***Step 5:*** base: 9 units; height: 5 units; ***Step 6:*** $9 \cdot 5 = 45$
5. ***Step 3:*** $192 = 4x(x+2)$; ***Step 4:*** $x = 6$ or $x = -8$; ***Step 5:*** length: 8 units; width: 6 units; ***Step 6:*** $8 \cdot 6 \cdot 4 = 192$ **7.** length: 14 cm; width: 12 cm **9.** length: 15 in.; width: 12 in. **11.** height: 13 in.; width: 10 in. **13.** mirror: 7 ft; painting: 9 ft **15.** 20, 21 **17.** $-3, -2$ or 4, 5 **19.** $-3, -1$ or 7, 9 **21.** $-2, 0, 2$ or 6, 8, 10 **23.** 12 cm **25.** 12 mi
27. 8 ft **29. (a)** 1 sec **(b)** $\frac{1}{2}$ sec and $1\frac{1}{2}$ sec **(c)** 3 sec **(d)** The negative solution, -1, does not make sense since t represents time, which cannot be negative. **31. (a)** 46 million; The result using the model is a little more than 44 million, the actual number for 1996. **(b)** 14 **(c)** 184 million; The result is a little more than 182 million, the actual number for 2004. **(d)** 318 million **32.** \$58.6 billion; 16% **33.** 2003: \$522.1 billion; 2004: \$610.8 billion; 2005: \$699.5 billion. **34.** The answer using the linear equation is close to the actual data for 2004, but not for the other years. **35.** 2003: \$503.9 billion; 2004: \$601.8 billion; 2005: \$718.2 billion **36.** The answers in Exercise 35 are fairly close to the actual data. The quadratic equation models the data better.
37. (1, 365.1), (2, 423.7), (3, 496.9), (4, 612.1), (5, 714.4)
38. **U.S. TRADE DEFICIT (Goods and Services)**

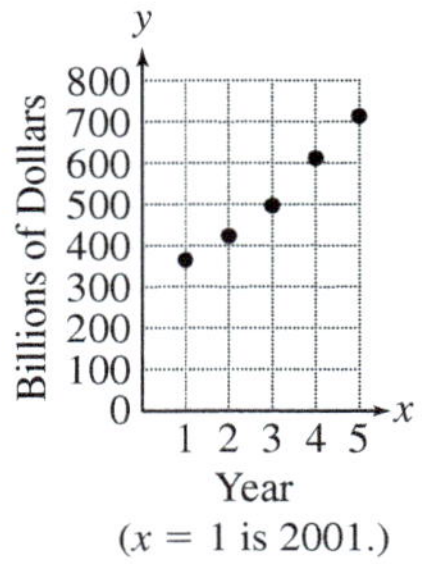

39. \$853.1 billion **40. (a)** The actual deficit is quite a bit less than the estimate. **(b)** No, data for later years might not follow the same pattern.

Chapter 16 Rational Expressions

Taken from *Prealgebra & Introductory Algebra* (0321578732) by Lial, et. al. and *Intermediate Algebra* (0321574974) by Lial, Hornsby, and McGinnis

Section 16.1 (pages 1172–1180)

1. (a) 3; -5 **(b)** q; -1 **3.** A rational expression is a quotient of polynomials, such as $\frac{x+3}{x^2-4}$. **5.** $y \neq 0$ **7.** $x \neq -6$ **9.** $x \neq \frac{5}{3}$
11. $m \neq -3, m \neq 2$ **13.** never undefined **15.** never undefined
17. (a) 1 **(b)** $\frac{17}{12}$ **19. (a)** 0 **(b)** -1 **21. (a)** $\frac{9}{5}$ **(b)** undefined
23. (a) $\frac{2}{7}$ **(b)** $\frac{13}{3}$ **25.** $3r^2$ **27.** $\frac{2}{5}$ **29.** $\frac{x-1}{x+1}$ **31.** $\frac{7}{5}$ **33.** $m-n$
35. $\frac{3(2m+1)}{4}$ **37.** $\frac{3m}{5}$ **39.** $\frac{3r-2s}{3}$ **41.** $\frac{x+1}{x-1}$ **43.** $\frac{z-3}{z+5}$
45. $\frac{a+b}{a-b}$ **47.** -1 **49.** $-(m+1)$ **51.** -1

Answers may vary in Exercises 53–57.

53. $\frac{-(x+4)}{x-3}, \frac{-x-4}{x-3}, \frac{x+4}{-(x-3)}, \frac{x+4}{-x+3}$
55. $\frac{-(2x-3)}{x+3}, \frac{-2x+3}{x+3}, \frac{2x-3}{-(x+3)}, \frac{2x-3}{-x-3}$
57. $\frac{-(3x-1)}{5x-6}, \frac{-3x+1}{5x-6}, \frac{3x-1}{-(5x-6)}, \frac{3x-1}{-5x+6}$ **59.** x^2+3

Section 16.2 (pages 1181–1186)

1. (a) B (b) D (c) C (d) A **3.** $\frac{4m}{3}$ **5.** $\frac{40y^2}{3}$ **7.** $\frac{2}{c+d}$ **9.** $\frac{16q}{3p^3}$ **11.** $\frac{7}{r^2+rp}$ **13.** $\frac{z^2-9}{z^2+7z+12}$ **15.** 5 **17.** $-\frac{3}{2t^4}$ **19.** $\frac{1}{4}$

21. To multiply two rational expressions, multiply the numerators and multiply the denominators. Write the answer in lowest terms. **23.** $\frac{10}{9}$ **25.** $-\frac{3}{4}$ **27.** -1 **29.** $\frac{9(m-2)}{-(m+4)}$, or $\frac{-9(m-2)}{m+4}$ **31.** $\frac{p+4}{p+2}$ **33.** $\frac{(k-1)^2}{(k+1)(2k-1)}$ **35.** $\frac{4k-1}{3k-2}$ **37.** $\frac{m+4p}{m+p}$ **39.** $\frac{10}{x+10}$ **41.** $\frac{5xy^2}{4q}$

Section 16.3 (pages 1187–1192)

1. C **3.** C **5.** 30 **7.** x^7 **9.** $72q$ **11.** $84r^5$ **13.** $2^3 \cdot 3 \cdot 5$ **15.** The least common denominator is their product. **17.** $28m^2(3m-5)$ **19.** $30(b-2)$ **21.** $c-d$ or $d-c$ **23.** $k(k+5)(k-2)$ **25.** $(p+3)(p+5)(p-6)$ **27.** $\frac{20}{55}$ **29.** $\frac{-45}{9k}$ **31.** $\frac{26y^2}{80y^3}$ **33.** $\frac{35t^2r^3}{42r^4}$ **35.** $\frac{20}{8(m+3)}$ **37.** $\frac{8t}{12-6t}$ **39.** $\frac{14(z-2)}{z(z-3)(z-2)}$ **41.** $\frac{2(b-1)(b+2)}{b^3+3b^2+2b}$

Section 16.4 (pages 1193–1202)

1. E **3.** C **5.** B **7.** G **9.** $\frac{11}{m}$ **11.** b **13.** $\frac{4}{y+4}$ **15.** $\frac{m-1}{m+1}$ **17.** x **19.** $y-6$ **21.** Combine the numerators and keep the same denominator. For example, $\frac{3x+2}{x-6}+\frac{-2x-8}{x-6}=\frac{x-6}{x-6}$. Then write in lowest terms: $\frac{x-6}{x-6}=1$. **23.** $\frac{3z+5}{15}$ **25.** $\frac{10-7r}{14}$ **27.** $\frac{-3x-2}{4x}$ **29.** $\frac{57}{20x}$ **31.** $\frac{x+1}{2}$ **33.** $\frac{5x+9}{6x}$ **35.** $\frac{3x+3}{x(x+3)}$ **37.** $\frac{-k-10}{k(k+5)}$ **39.** $\frac{x+4}{x+2}$ **41.** $\frac{x^2+6x-8}{(x-2)(x+2)}$ **43.** $\frac{3}{t}$ **45.** $m-2$ or $2-m$ **47.** $\frac{-2}{x-5}$, or $\frac{2}{5-x}$ **49.** -4 **51.** $\frac{-5}{x-y^2}$, or $\frac{5}{y^2-x}$ **53.** $\frac{x+y}{5x-3y}$, or $\frac{-x-y}{3y-5x}$ **55.** $\frac{-6}{4p-5}$, or $\frac{6}{5-4p}$ **57.** $\frac{-(m+n)}{2(m-n)}$ **59.** $\frac{-x^2+6x+11}{(x+3)(x-3)(x+1)}$ **61.** $\frac{-5q^2-13q+7}{(3q-2)(q+4)(2q-3)}$ **63.** $\frac{9r+2}{r(r+2)(r-1)}$ **65.** $\frac{2x^2+6xy+8y^2}{(x+y)(x+y)(x+3y)}$, or $\frac{2x^2+6xy+8y^2}{(x+y)^2(x+3y)}$ **67.** $\frac{15r^2+10ry-y^2}{(3r+2y)(6r-y)(6r+y)}$ **69.** (a) $\frac{9k^2+6k+26}{5(3k+1)}$ (b) $\frac{1}{4}$

Section 16.5 (pages 1203–1210)

1. (a) 6; $\frac{1}{6}$ (b) 12; $\frac{3}{4}$ (c) $\frac{1}{6} \div \frac{3}{4}$ (d) $\frac{2}{9}$ **3.** -6 **5.** $\frac{1}{pq}$ **7.** $\frac{1}{xy}$ **9.** $\frac{2a^2b}{3}$ **11.** $\frac{m(m+2)}{3(m-4)}$ **13.** $\frac{2}{x}$ **15.** $\frac{8}{x}$ **17.** $\frac{a^2-5}{a^2+1}$ **19.** $\frac{3(p+2)}{2(2p+3)}$ **21.** $\frac{40-12p}{85p}$ **23.** $\frac{t(t-2)}{4}$ **25.** $\frac{-k}{2+k}$ **27.** $\frac{2x-7}{3x+1}$ **29.** $\frac{3m(m-3)}{(m-1)(m-8)}$ **31.** $\frac{6}{5}$

Section 16.6 (pages 1211–1220)

1. (a) $-1, 2$ (b) $\{x \mid x \neq -1, 2\}$ **3.** (a) $-\frac{5}{3}, 0, -\frac{3}{2}$ (b) $\left\{x \mid x \neq -\frac{5}{3}, 0, -\frac{3}{2}\right\}$ **5.** (a) 0 (b) $\{x \mid x \neq 0\}$ **7.** (a) $4, \frac{7}{2}$ (b) $\left\{x \mid x \neq 4, \frac{7}{2}\right\}$ **9.** (a) $0, 1, -3, 2$ (b) $\{x \mid x \neq 0, 1, -3, 2\}$ **11.** $\{1\}$ **13.** $\{-6, 4\}$ **15.** $\left\{-\frac{7}{12}\right\}$ **17.** $\emptyset$ **19.** $\{-3\}$ **21.** $\{5\}$ **23.** $\{5\}$ **25.** $\emptyset$ **27.** $\left\{\frac{27}{56}\right\}$ **29.** $\emptyset$ **31.** $\{-10\}$ **33.** $\emptyset$ **35.** $\{0\}$ **37.** $\left\{x \mid x \neq -\frac{3}{2}, \frac{3}{2}\right\}$ **39.** $x=0$; $y=0$ **41.** $x=2$; $y=0$

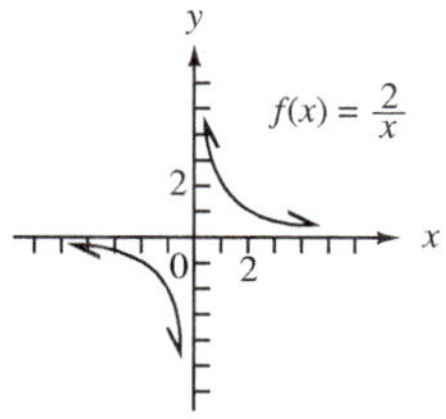

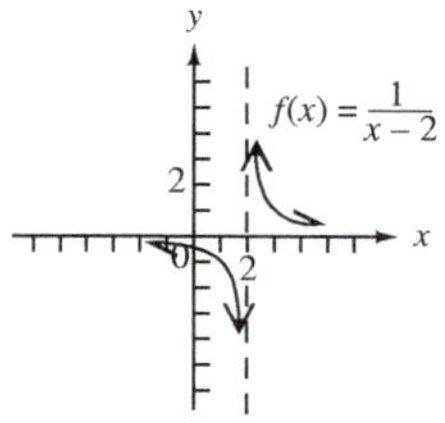

43. (a) 0 (b) 1.6 (c) 4.1 (d) The waiting time also increases. **45.** Substituting -1 for x gives a true statement, $\frac{4}{3}=\frac{4}{3}$. Substituting -2 for x leads to 0 in the first and third denominators. **46.** $C=-4$; $\{-2\}$; -1 is rejected. **47.** $C=24$; $\{-4\}$; 3 is rejected. **48.** Answers will vary.

Summary Exercises on Rational Expressions and Equations (pages 1219–1220)

1. equation; $\{20\}$ **2.** expression; $\frac{2(x+5)}{5}$ **3.** expression; $-\frac{22}{7x}$ **4.** expression; $\frac{y+x}{y-x}$ **5.** equation; $\left\{\frac{1}{2}\right\}$ **6.** equation; $\{7\}$ **7.** expression; $\frac{43}{24x}$ **8.** equation; $\{1\}$ **9.** expression; $\frac{5x-1}{-2x+2}$, or $\frac{5x-1}{-2(x-1)}$ **10.** expression; $\frac{25}{4(r+2)}$ **11.** expression; $\frac{x^2+xy+2y^2}{(x+y)(x-y)}$ **12.** expression; $\frac{24p}{p+2}$ **13.** expression; $-\frac{5}{36}$ **14.** equation; $\{0\}$ **15.** expression; $\frac{b+3}{3}$ **16.** expression; $\frac{5}{3z}$ **17.** expression; $\frac{2x+10}{x(x-2)(x+2)}$ **18.** equation; $\{2\}$ **19.** expression; $\frac{-x}{3x+5y}$ **20.** equation; $\{-13\}$ **21.** expression; $\frac{3y+2}{y+3}$ **22.** equation; $\left\{\frac{5}{4}\right\}$ **23.** equation; $\emptyset$ **24.** expression; $\frac{2z-3}{2z+3}$ **25.** expression; $\frac{-1}{x-3}$, or $\frac{1}{3-x}$ **26.** expression; $\frac{t-2}{8}$ **27.** equation; $\{-10\}$ **28.** expression; $\frac{13x+28}{2x(x+4)(x-4)}$ **29.** equation; $\emptyset$ **30.** expression; $\frac{k(2k^2-2k+5)}{(k-1)(3k^2-2)}$

Section 16.7 (pages 1221–1234)

1. A **3.** D **5.** 65.625 **7.** $\frac{25}{4}$ **9.** $G=\frac{Fd^2}{Mm}$ **11.** $a=\frac{bc}{c+b}$ **13.** $v=\frac{PVt}{pT}$ **15.** $r=\frac{nE-IR}{In}$ **17.** $b=\frac{2A}{h}-B$, or $b=\frac{2A-Bh}{h}$ **19.** $r=\frac{eR}{E-e}$ **21.** Multiply each side by $a-b$.

23. 21 girls, 7 boys **25.** $\frac{1}{2}$ job per hr **27.** 5.4 in. **29.** 7.6 in. **31.** 100 games **33.** 25,000 fish **35.** 6.6 more gallons **37.** 2.4 mL **39.** $x = \frac{7}{2}$; $AC = 8$; $DF = 12$ **41.** 3 mph **43.** 1020 mi **45.** 480 mi **47.** 190 mi **49.** $6\frac{2}{3}$ min **51.** 30 hr **53.** 20 hr **55.** $2\frac{4}{5}$ hr

Chapter 17 Roots and Radicals

Taken from *Prealgebra & Introductory Algebra* by Lial, et. al. (0321578732)

Section 17.1 (pages 1236–1246)

1. true **3.** false; Zero has only one square root. **5.** true **7.** $-3, 3$ **9.** $-8, 8$ **11.** $-13, 13$ **13.** $-\frac{5}{14}, \frac{5}{14}$ **15.** $-30, 30$ **17.** 1 **19.** 7 **21.** -16 **23.** $-\frac{12}{11}$ **25.** 0.8 **27.** not a real number **29.** not a real number **31.** 100 **33.** 19 **35.** $\frac{2}{3}$ **37.** $3x^2 + 4$ **39.** a must be positive. **41.** a must be negative. **43.** rational; 5 **45.** irrational; 5.385 **47.** rational; -8 **49.** irrational; -17.321 **51.** not a real number **53.** irrational; 34.641 **55.** C **57.** $c = 17$ **59.** $b = 8$ **61.** $c = 11.705$ **63.** 24 cm **65.** 80 ft **67.** 195 ft **69.** 158.6 ft **71.** 11.1 ft **73.** 9.434 **75.** 1 **77.** 5 **79.** -3 **81.** -6 **83.** 2 **85.** 4 **87.** 6 **89.** not a real number **91.** -5 **93.** -4

ANSWERS

Section 17.2 (pages 1248–1256)

1. E **3.** D **5.** C **7.** C **9.** C **11.** **(a)** not a real number **(b)** negative **(c)** 0 **13.** -9 **15.** 6 **17.** -4 **19.** -8 **21.** 6 **23.** -2 **25.** not a real number **27.** 3 **29.** not a real number **31.** $\frac{8}{9}$ **33.** 0.7 **35.** $\frac{4}{3}$ **37.** $-\frac{1}{2}$ **39.** 0.1

41. domain: $[-3, \infty)$; range: $[0, \infty)$ **43.** domain: $[0, \infty)$; range: $[-2, \infty)$

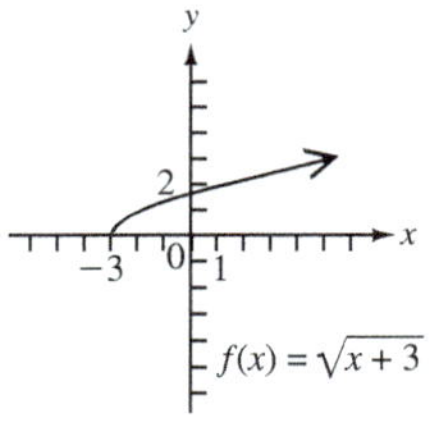

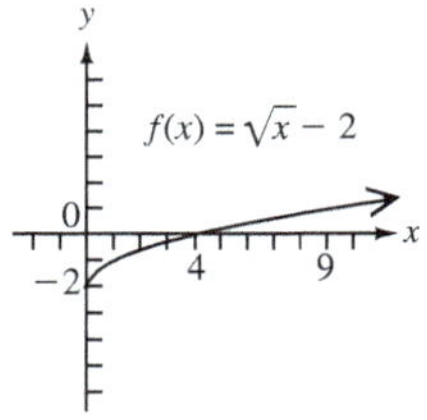

45. domain: $(-\infty, \infty)$; range: $(-\infty, \infty)$

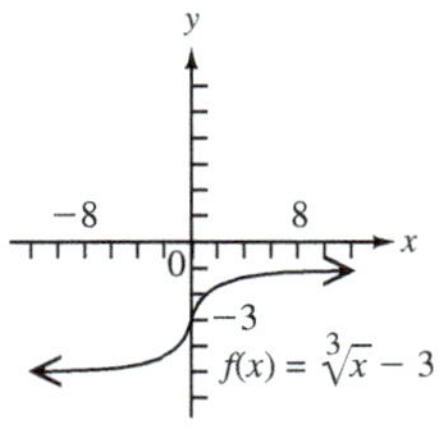

47. 12 **49.** 10 **51.** 2 **53.** -9 **55.** -5 **57.** $|x|$ **59.** $|z|$ **61.** x **63.** x^5 **65.** $|x|^5$ (or $|x^5|$) **67.** 97.381 **69.** 16.863 **71.** -9.055 **73.** 7.507 **75.** 3.162 **77.** 1.885 **79.** 1,183,000 cycles per sec **81.** 10 mi **83.** 392,000 mi² **85.** 1.732 amps

Section 17.3 (pages 1257–1266)

1. C **3.** A **5.** H **7.** B **9.** D **11.** 13 **13.** 9 **15.** 2 **17.** $\frac{8}{9}$ **19.** -3 **21.** not a real number **23.** 1000 **25.** 27 **27.** -1024 **29.** 16 **31.** $\frac{1}{8}$ **33.** $\frac{1}{512}$ **35.** $\frac{9}{25}$ **37.** $\sqrt{12}$ **39.** $(\sqrt[4]{8})^3$ **41.** $(\sqrt[8]{9q})^5 - (\sqrt[3]{2x})^2$ **43.** $\frac{1}{(\sqrt{2m})^3}$ **45.** $(\sqrt[3]{2y + x})^2$ **47.** $\frac{1}{(\sqrt[3]{3m^4 + 2k^2})^2}$ **49.** $\sqrt{a^2 + b^2} = \sqrt{3^2 + 4^2} = 5$; $a + b = 3 + 4 = 7$; $5 \neq 7$ **51.** 64 **53.** 64 **55.** x^{10} **57.** $\sqrt[6]{x^5}$ **59.** $\sqrt[15]{t^8}$ **61.** 9 **63.** 4 **65.** y **67.** $x^{5/12}$ **69.** $k^{2/3}$ **71.** x^3y^8 **73.** $\frac{1}{x^{10/3}}$ **75.** $\frac{1}{m^{1/4}n^{3/4}}$ **77.** p^2 **79.** $\frac{c^{11/3}}{b^{11/4}}$ **81.** $\frac{q^{5/3}}{9p^{7/2}}$ **83.** $p + 2p^2$ **85.** $k^{7/4} - k^{3/4}$ **87.** $6 + 18a$ **89.** $5 + \frac{5}{m^3}$ **91.** $y^{3/2}$ **93.** $\frac{1}{k^{2/3}}$ **95.** $x^{1/3}z^{5/6}$ **97.** $k^{1/6}$ **99.** $y^{1/30}$ **101.** $x^{5/27}$ **103.** 72 in.; 6.0 ft **105.** $-12.3°$; The table gives $-12°$.

Section 17.4 (pages 1267–1280)

1. true; Both are equal to $4\sqrt{3}$ and approximately 6.92820323. **3.** true; Both are equal to $6\sqrt{2}$ and approximately 8.485281374. **5.** Because there are only two factors of $\sqrt[3]{x}$, $\sqrt[3]{x} \cdot \sqrt[3]{x} = (\sqrt[3]{x})^2$, or $\sqrt[3]{x^2}$. **7.** D **9.** $\sqrt{30}$ **11.** $\sqrt{14x}$ **13.** $\sqrt{42pqr}$ **15.** $\sqrt[3]{14xy}$ **17.** $\sqrt[4]{33}$ **19.** $\sqrt[4]{6xy^2}$ **21.** This product cannot be simplified using the product rule. **23.** $\frac{8}{11}$ **25.** $\frac{\sqrt{3}}{5}$ **27.** $\frac{\sqrt{x}}{5}$ **29.** $\frac{p^3}{9}$ **31.** $-\frac{3}{4}$ **33.** $\frac{\sqrt[3]{r^2}}{2}$ **35.** $-\frac{3}{x}$ **37.** $\frac{1}{x^3}$ **39.** $2\sqrt{3}$ **41.** $12\sqrt{2}$ **43.** $-4\sqrt{2}$ **45.** $-2\sqrt{7}$ **47.** cannot be simplified further **49.** $4\sqrt[3]{2}$ **51.** $-2\sqrt[3]{2}$ **53.** $2\sqrt[3]{5}$ **55.** $-4\sqrt[4]{2}$ **57.** $2\sqrt[5]{2}$ **59.** His reasoning was incorrect. Here 8 is a term, not a factor. **61.** $6k\sqrt{2}$ **63.** $12xy^4\sqrt{xy}$ **65.** $11x^3$ **67.** $-3t^4$ **69.** $-10m^4z^2$ **71.** $5a^2b^3c^4$ **73.** $\frac{1}{2}r^2t^5$ **75.** $5x\sqrt{2x}$ **77.** $-10r^5\sqrt{5r}$ **79.** $x^3y^4\sqrt{13x}$ **81.** $2z^2w^3$ **83.** $-2zt^2\sqrt[3]{2z^2t}$ **85.** $3x^3y^4$ **87.** $-3r^3s^2\sqrt[4]{2r^3s^2}$ **89.** $\frac{y^5\sqrt{y}}{6}$ **91.** $\frac{x^5\sqrt[3]{x}}{3}$ **93.** $4\sqrt{3}$ **95.** $\sqrt{5}$ **97.** $x^2\sqrt{x}$ **99.** $\sqrt[6]{432}$ **101.** $\sqrt[12]{6912}$ **103.** $\sqrt[6]{x^5}$ **105.** 5 **107.** $8\sqrt{2}$ **109.** $2\sqrt{14}$ **111.** 13 **113.** $9\sqrt{2}$ **115.** $\sqrt{17}$ **117.** 5 **119.** $6\sqrt{2}$ **121.** $\sqrt{5y^2 - 2xy + x^2}$ **123.** 27.0 in. **125.** 0.003 **127.** 15.3 mi

Section 17.5 (pages 1281–1284)

1. B **3.** 15; Each radical expression simplifies to a whole number. **5.** -4 **7.** $7\sqrt{3}$ **9.** $14\sqrt[3]{2}$ **11.** $5\sqrt[4]{2}$ **13.** $24\sqrt{2}$ **15.** cannot be simplified further **17.** $20\sqrt{5}$ **19.** $12\sqrt{2x}$ **21.** $-2m\sqrt{2}$ **23.** $\sqrt[3]{2}$ **25.** $2\sqrt[3]{x}$ **27.** $-\sqrt[3]{x^2y}$ **29.** $-x\sqrt[3]{xy^2}$ **31.** $19\sqrt[4]{2}$ **33.** $x\sqrt[4]{xy}$ **35.** $9\sqrt[4]{2a^3}$ **37.** $\frac{5\sqrt{5}}{6}$ **39.** $\frac{7\sqrt{2}}{6}$ **41.** $\frac{5\sqrt{2}}{3}$ **43.** $5\sqrt{2} + 4$ **45.** $\frac{5 - 3x}{x^4}$ **47.** $\frac{m\sqrt[3]{m^2}}{2}$ **49.** $\frac{3x\sqrt[3]{2} - 4\sqrt[3]{5}}{x^3}$ **51.** $(12\sqrt{5} + 5\sqrt{3})$ in. **53.** $(24\sqrt{2} + 12\sqrt{3})$ in.

Section 17.6 (pages 1285–1294)

1. E **3.** A **5.** D **7.** $6 - 4\sqrt{3}$ **9.** $6 - \sqrt{6}$ **11.** 2 **13.** 9 **15.** $3\sqrt{2} - 5\sqrt{3} + 2\sqrt{6} - 10$ **17.** $3x - 4$ **19.** $4x - y$ **21.** $16x + 24\sqrt{x} + 9$ **23.** $81 - \sqrt[3]{4}$ **25.** Because 6 and $4\sqrt{3}$ are

not like terms, they cannot be combined. **27.** $\sqrt{7}$ **29.** $5\sqrt{3}$ **31.** $\frac{\sqrt{6}}{2}$ **33.** $\frac{9\sqrt{15}}{5}$ **35.** $-\sqrt{2}$ **37.** $\frac{-8\sqrt{3k}}{k}$ **39.** $\frac{6\sqrt{3}}{y}$
41. To rationalize a cube root, three factors of the quantity under the radical sign are needed. We must multiply by $\sqrt[3]{2^2}$, or $\sqrt[3]{4}$, to rationalize $\sqrt[3]{2}$.
43. $\frac{\sqrt{14}}{2}$ **45.** $-\frac{\sqrt{14}}{10}$ **47.** $\frac{2\sqrt{6x}}{x}$ **49.** $-\frac{7r\sqrt{2rs}}{s}$
51. $\frac{12x^3\sqrt{2xy}}{y^5}$ **53.** $\frac{\sqrt[3]{18}}{3}$ **55.** $\frac{\sqrt[3]{12}}{3}$ **57.** $-\frac{\sqrt[3]{2pr}}{r}$
59. $\frac{2\sqrt[4]{x^3}}{x}$ **61.** $\frac{\sqrt[4]{2yz^3}}{z}$ **63.** $\frac{2(4-\sqrt{3})}{13}$ **65.** $3(\sqrt{5}-\sqrt{3})$
67. $\sqrt{3}+\sqrt{7}$ **69.** $\sqrt{7}-\sqrt{6}-\sqrt{14}+2\sqrt{3}$
71. $2\sqrt{3}+\sqrt{10}-3\sqrt{2}-\sqrt{15}$ **73.** $\frac{4(\sqrt{x}+2\sqrt{y})}{x-4y}$
75. $\frac{x\sqrt{2}-\sqrt{3xy}-\sqrt{2xy}+y\sqrt{3}}{2x-3y}$ **77.** $\frac{5+2\sqrt{6}}{4}$ **79.** $\frac{4+2\sqrt{2}}{3}$
81. $\frac{6+2\sqrt{6x}}{3}$ **83.** $\frac{319}{6(8\sqrt{5}+1)}$ **84.** $\frac{9a-b}{(\sqrt{b}-\sqrt{a})(3\sqrt{a}-\sqrt{b})}$
85. $\frac{(3\sqrt{a}+\sqrt{b})(\sqrt{b}+\sqrt{a})}{b-a}$ **86.** In Exercise 84, we multiplied the numerator and denominator by the conjugate of the numerator, while in Exercise 85 we multiplied by the conjugate of the denominator.

Summary Exercises on Operations with Radicals and Rational Exponents (pages 1295–1296)

1. $-6\sqrt{10}$ **2.** $7-\sqrt{14}$ **3.** $2+\sqrt{6}-2\sqrt{3}-3\sqrt{2}$ **4.** $4\sqrt{2}$
5. $73+12\sqrt{35}$ **6.** $\frac{-\sqrt{6}}{2}$ **7.** $4(\sqrt{7}-\sqrt{5})$ **8.** $3\sqrt[3]{2x^2}$
9. $-3+2\sqrt{2}$ **10.** -2 **11.** -44 **12.** $\frac{\sqrt{x}+\sqrt{5}}{x-5}$ **13.** $2abc^3\sqrt[3]{b^2}$
14. $5\sqrt[3]{3}$ **15.** $3(\sqrt{5}-2)$ **16.** $\frac{\sqrt{15x}}{5x}$ **17.** $\frac{8}{5}$ **18.** $\frac{\sqrt{2}}{8}$
19. $-\sqrt[3]{100}$ **20.** $11+2\sqrt{30}$ **21.** $-3\sqrt{3x}$ **22.** $52-30\sqrt{3}$
23. 1 **24.** $\frac{\sqrt[3]{117}}{9}$ **25.** $t^2\sqrt[4]{t}$ **26.** $3\sqrt{2}+\sqrt{15}+\sqrt{42}+\sqrt{35}$
27. $2\sqrt[4]{27}$ **28.** $\frac{1+\sqrt[3]{3}+\sqrt[3]{9}}{-2}$ **29.** $\frac{x\sqrt[3]{x^2}}{y}$ **30.** $-4\sqrt{3}-3$
31. $xy^{6/5}$ **32.** $x^{10}y$ **33.** $\frac{1}{25x^2}$ **34.** $\frac{-6y^{1/6}}{x^{1/24}}$ **35.** $7+4\cdot 3^{1/2}$, or $7+4\sqrt{3}$ **36.** 1

Section 17.7 (pages 1297–1304)

1. (a) yes **(b)** no **3. (a)** yes **(b)** no **5.** no; There is no solution. The radical expression, which is positive, cannot equal a negative number. **7.** $\{11\}$ **9.** $\left\{\frac{1}{3}\right\}$ **11.** $\emptyset$ **13.** $\{5\}$ **15.** $\{18\}$
17. $\{5\}$ **19.** $\{4\}$ **21.** $\{17\}$ **23.** $\{5\}$ **25.** $\emptyset$ **27.** $\{0\}$ **29.** $\{0\}$
31. $\left\{-\frac{1}{3}\right\}$ **33.** $\emptyset$ **35.** We cannot just square each term. The right side should be $(8-x)^2 = 64-16x+x^2$. The correct first step is $3x+4 = 64-16x+x^2$, and the solution set is $\{4\}$. **37.** $\{1\}$
39. $\{-1\}$ **41.** $\{14\}$ **43.** $\{8\}$ **45.** $\{0\}$ **47.** $\emptyset$ **49.** $\{7\}$ **51.** $\{7\}$
53. $\{4, 20\}$ **55.** $\emptyset$ **57.** $\left\{\frac{5}{4}\right\}$ **59.** $\{9, 17\}$ **61.** $\left\{\frac{1}{4}, 1\right\}$
63. $K = \frac{V^2m}{2}$ **65.** $C = \frac{L}{Z^2}$ **67.** $L = \frac{1}{4\pi^2f^2C}$ **69.** $r = \frac{a}{4\pi^2N^2}$

Section 17.8 (pages 1305–1315)

1. i **3.** $-i$ **5.** Any real number a can be written as $a+0i$, a complex number with imaginary part 0. **7.** $13i$ **9.** $-12i$ **11.** $i\sqrt{5}$
13. $4i\sqrt{3}$ **15.** -15 **17.** $-\sqrt{57}$ **19.** -10 **21.** $i\sqrt{33}$
23. $\sqrt{3}$ **25.** $5i$ **27.** $-1+7i$ **29.** 0 **31.** $7+3i$ **33.** -2
35. $1+13i$ **37.** $6+6i$ **39.** $4+2i$ **41.** -81 **43.** -16
45. $-10-30i$ **47.** $10-5i$ **49.** $-9+40i$ **51.** 153
53. (a) $a-bi$ **(b)** $a^2; b^2$ **55.** $1+i$ **57.** $-1+2i$
59. $2+2i$ **61.** $-\frac{5}{13}-\frac{12}{13}i$ **63. (a)** $4x+1$ **(b)** $4+i$
64. (a) $-2x+3$ **(b)** $-2+3i$ **65. (a)** $3x^2+5x-2$
(b) $5+5i$ **66. (a)** $-\sqrt{3}+\sqrt{6}+1-\sqrt{2}$ **(b)** $\frac{1}{5}-\frac{7}{5}i$
67. Because $i^2 = -1$, two pairs of like terms can be combined in Exercise 65(b). **68.** Because $i^2 = -1$, additional terms can be combined in the numerator and denominator. **69.** $\frac{5}{41}+\frac{4}{41}i$
71. -1 **73.** i **75.** 1 **77.** $-i$ **79.** Since $i^{20} = (i^4)^5 = 1^5 = 1$, the student multiplied by 1, which is justified by the identity property for multiplication. **81.** $\frac{1}{2}+\frac{1}{2}i$ **83.** $(1+5i)^2-2(1+5i)+26$ will simplify to 0 when the operations are applied.

Chapter 18 Quadratics

Taken from *Intermediate Algebra* (0321574974) by Lial Hornsby, and McGinnis and *Prealgebra & Introductory Algebra* (0321578732) by Lial, et. al.

Section 18.1 (pages 1318–1324)

1. true **3.** false; If k is a positive integer that is not a perfect square, then the solutions will be irrational. **5.** false; For values of k that satisfy $0 \le k < 10$, there are real solutions. **7.** $\{-9, 9\}$
9. $\{-\sqrt{14}, \sqrt{14}\}$ **11.** $\{-4\sqrt{3}, 4\sqrt{3}\}$ **13.** $\left\{-\frac{5}{2}, \frac{5}{2}\right\}$
15. $\emptyset$ **17.** $\{-1.5, 1.5\}$ **19.** $\{-\sqrt{3}, \sqrt{3}\}$ **21.** $\left\{-\frac{2\sqrt{7}}{7}, \frac{2\sqrt{7}}{7}\right\}$
23. $\{-3\sqrt{2}, 3\sqrt{2}\}$ **25.** $\{-2\sqrt{6}, 2\sqrt{6}\}$ **27.** $\left\{-\frac{2\sqrt{5}}{5}, \frac{2\sqrt{5}}{5}\right\}$
29. $\{-2, 8\}$ **31.** $\emptyset$ **33.** $\{8+3\sqrt{3}, 8-3\sqrt{3}\}$ **35.** $\left\{-3, \frac{5}{3}\right\}$
37. $\left\{0, \frac{3}{2}\right\}$ **39.** $\left\{\frac{5+\sqrt{30}}{2}, \frac{5-\sqrt{30}}{2}\right\}$
41. $\left\{\frac{-1+3\sqrt{2}}{3}, \frac{-1-3\sqrt{2}}{3}\right\}$ **43.** $\{-10+4\sqrt{3}, -10-4\sqrt{3}\}$
45. $\left\{\frac{1+4\sqrt{3}}{4}, \frac{1-4\sqrt{3}}{4}\right\}$ **47.** The answers are equivalent. If the answer of either student is multiplied by $\frac{-1}{-1}$, it will look like the answer of the other student. **49.** about $\frac{1}{2}$ sec **51.** 9 in. **53.** 5% **55.** 2%

Section 18.2 (pages 1323–1332)

1. 25; $(x+5)^2$ **3.** 1; $(x+1)^2$ **5.** $\frac{25}{4}$; $\left(p-\frac{5}{2}\right)^2$ **7.** D

9. $\{1, 3\}$ **11.** $\{-3, -2\}$ **13.** $\{-1+\sqrt{6}, -1-\sqrt{6}\}$

15. $\{4+2\sqrt{3}, 4-2\sqrt{3}\}$ **17.** $\{-3\}$

19. $\left\{\frac{-1+\sqrt{5}}{2}, \frac{-1-\sqrt{5}}{2}\right\}$ **21.** $\left\{-\frac{3}{2}, \frac{1}{2}\right\}$

23. $\left\{\frac{2+\sqrt{14}}{2}, \frac{2-\sqrt{14}}{2}\right\}$ **25.** $\emptyset$

27. $\left\{\frac{-7+\sqrt{97}}{6}, \frac{-7-\sqrt{97}}{6}\right\}$ **29.** $\{-4, 2\}$

31. $\{4+\sqrt{3}, 4-\sqrt{3}\}$ **33.** $\{1+\sqrt{6}, 1-\sqrt{6}\}$

35. 1 sec and 5 sec **37.** 3 sec and 5 sec **39.** 75 ft by 100 ft

Section 18.3 (pages 1333–1340)

1. The fraction bar should extend under the term $-b$. **3.** $\{3, 5\}$

5. $\left\{\frac{-2+\sqrt{2}}{2}, \frac{-2-\sqrt{2}}{2}\right\}$ **7.** $\left\{\frac{1+\sqrt{3}}{2}, \frac{1-\sqrt{3}}{2}\right\}$

9. $\{5+\sqrt{7}, 5-\sqrt{7}\}$ **11.** $\left\{\frac{-1+\sqrt{2}}{2}, \frac{-1-\sqrt{2}}{2}\right\}$

13. $\left\{\frac{-1+\sqrt{7}}{3}, \frac{-1-\sqrt{7}}{3}\right\}$ **15.** $\{1+\sqrt{5}, 1-\sqrt{5}\}$

17. $\left\{\frac{-2+\sqrt{10}}{2}, \frac{-2-\sqrt{10}}{2}\right\}$ **19.** $\{-1+3\sqrt{2}, -1-3\sqrt{2}\}$

21. $\left\{\frac{3}{2}+\frac{\sqrt{59}}{2}i, \frac{3}{2}-\frac{\sqrt{59}}{2}i\right\}$ **23.** $\{3+i\sqrt{5}, 3-i\sqrt{5}\}$

25. $\left\{\frac{1}{2}+\frac{\sqrt{6}}{2}i, \frac{1}{2}-\frac{\sqrt{6}}{2}i\right\}$ **27.** $\left\{-\frac{2}{3}+\frac{\sqrt{2}}{3}i, -\frac{2}{3}-\frac{\sqrt{2}}{3}i\right\}$

29. B **31.** C **33.** A **35.** D **37.** The equations in Exercises 29, 30, 33, and 34 can be solved by factoring.

Section 18.4 (pages 1341–1354)

1. Multiply by the LCD, x. **3.** Substitute a variable for r^2+r.
5. The proposed solution -1 does not check. The solution set is $\{4\}$.

7. $\{-4, 7\}$ **9.** $\left\{-\frac{2}{3}, 1\right\}$ **11.** $\left\{-\frac{14}{17}, 5\right\}$ **13.** $\left\{-\frac{11}{7}, 0\right\}$

15. $\left\{\frac{-1+\sqrt{13}}{2}, \frac{-1-\sqrt{13}}{2}\right\}$ **17.** $\left\{\frac{2+\sqrt{22}}{3}, \frac{2-\sqrt{22}}{3}\right\}$

19. **(a)** $(20-t)$ mph **(b)** $(20+t)$ mph **21.** 25 mph
23. 80 km per hr **25.** 3.6 hr **27.** 9 min **29.** $\{1, 4\}$

31. $\{3\}$ **33.** $\left\{\frac{8}{9}\right\}$ **35.** $\{16\}$ **37.** $\left\{\frac{2}{5}\right\}$ **39.** $\{-3, 3\}$

41. $\left\{-\frac{3}{2}, -1, 1, \frac{3}{2}\right\}$ **43.** $\{-2\sqrt{3}, -2, 2, 2\sqrt{3}\}$

45. $\left\{\frac{\sqrt{9+\sqrt{65}}}{2}, -\frac{\sqrt{9+\sqrt{65}}}{2}, \frac{\sqrt{9-\sqrt{65}}}{2}, -\frac{\sqrt{9-\sqrt{65}}}{2}\right\}$

47. $\{-6, -5\}$ **49.** $\{-4, 1\}$ **51.** $\left\{-\frac{1}{3}, \frac{1}{6}\right\}$ **53.** $\{-8, 1\}$

55. $\{-64, 27\}$ **57.** $\{25\}$ **59.** It would cause both denominators to equal 0, and division by 0 is undefined. **60.** $\frac{12}{5}$

61. $\left(\frac{x}{x-3}\right)^2+3\left(\frac{x}{x-3}\right)-4=0$ **62.** The numerator can never equal the denominator, since the denominator is 3 less than the numerator. **63.** $\left\{\frac{12}{5}\right\}$; The values for t are -4 and 1. The value 1 is impossible because it leads to a contradiction $\left(\text{since } \frac{x}{x-3} \text{ is never equal to } 1\right)$. **64.** $\left\{\frac{12}{5}\right\}$; The values for s are $\frac{1}{x}$ and $\frac{-4}{x}$. The value $\frac{1}{x}$ is impossible, since $\frac{1}{x} \neq \frac{1}{x-3}$ for all x.

Summary Exercises on Solving Quadratic Equations 1353–1354

1. square root property **2.** factoring **3.** quadratic formula
4. quadratic formula **5.** factoring **6.** square root property

7. $\{\sqrt{47}, -\sqrt{47}\}$ **8.** $\left\{-\frac{3}{2}, \frac{5}{3}\right\}$ **9.** $\{-4+\sqrt{10}, -4-\sqrt{10}\}$

10. $\{-3, 11\}$ **11.** $\left\{-\frac{1}{2}, 5\right\}$ **12.** $\left\{-3, \frac{1}{3}\right\}$

13. $\left\{\frac{9+\sqrt{33}}{6}, \frac{9-\sqrt{33}}{6}\right\}$ **14.** $\{2i\sqrt{3}, -2i\sqrt{3}\}$ **15.** $\left\{\frac{1}{2}, 2\right\}$

16. $\left\{-\frac{\sqrt{6}}{3}, -\frac{1}{2}, \frac{1}{2}, \frac{\sqrt{6}}{3}\right\}$ **17.** $\left\{\frac{-5+2\sqrt{3}}{2}, \frac{-5-2\sqrt{3}}{2}\right\}$

18. $\left\{\frac{4}{5}, 3\right\}$ **19.** $\{-\sqrt{7}, -\sqrt{2}, \sqrt{2}, \sqrt{7}\}$

20. $\left\{\frac{-2+\sqrt{14}}{2}, \frac{-2-\sqrt{14}}{2}\right\}$ **21.** $\left\{-\frac{1}{2}+\frac{\sqrt{7}}{2}i, -\frac{1}{2}-\frac{\sqrt{7}}{2}i\right\}$

22. $\left\{\sqrt{4+\sqrt{15}}, -\sqrt{4+\sqrt{15}}, \sqrt{4-\sqrt{15}}, -\sqrt{4-\sqrt{15}}\right\}$

23. $\left\{\frac{3}{2}\right\}$ **24.** $\left\{\frac{2}{3}\right\}$ **25.** $\{6\sqrt{2}, -6\sqrt{2}\}$ **26.** $\left\{-\frac{2}{3}, 2\right\}$

27. $\{-4, 9\}$ **28.** $\{13, -13\}$ **29.** $\left\{1+\frac{\sqrt{3}}{3}i, 1-\frac{\sqrt{3}}{3}i\right\}$

30. $\{3\}$ **31.** $\left\{-\frac{1}{3}, \frac{1}{6}\right\}$ **32.** $\left\{\frac{1}{6}+\frac{\sqrt{47}}{6}i, \frac{1}{6}-\frac{\sqrt{47}}{6}i\right\}$

33. $\{3\}$ **34.** $\left\{-\frac{8}{3}, -1\right\}$ **35.** $\left\{-i, i, -\frac{1}{2}i, \frac{1}{2}i\right\}$ **36.** $\{-2, 7\}$

Section 18.5 (pages 1355–1362)

1. Solve for w^2 by dividing each side by g. **3.** $m=\sqrt{p^2-n^2}$

5. $t=\frac{\pm\sqrt{dk}}{k}$ **7.** $d=\frac{\pm\sqrt{skI}}{I}$ **9.** $v=\frac{\pm\sqrt{kAF}}{F}$

11. $r=\frac{\pm\sqrt{3\pi Vh}}{\pi h}$ **13.** $t=\frac{-B\pm\sqrt{B^2-4AC}}{2A}$ **15.** $h=\frac{D^2}{k}$

17. $\ell=\frac{p^2g}{k}$ **19.** 2.3, 5.3, 5.8 **21.** eastbound ship: 80 mi; southbound ship: 150 mi **23.** 1 ft **25.** 20 in. by 12 in.
27. 2.4 sec and 5.6 sec **29.** 9.2 sec **31.** It reaches its *maximum* height at 5 sec because this is the only time it reaches 400 ft. **33.** \$0.80
35. **(a)** 2750 billion **(b)** 2750 billion; They are the same.
37. 2001; The graph indicates that vehicle-miles reached 2800 in 2001.
39. 5.5 m per sec **41.** 5 or 14

ANSWERS

Section 18.6 (pages 1363–1374)

1. (a) B **(b)** C **(c)** A **(d)** D **3.** (0, 0) **5.** (0, 4) **7.** (1, 0) **9.** (−3, −4) **11.** In Exercise 9, the parabola is shifted 3 units to the left and 4 units down. The parabola in Exercise 10 is shifted 5 units to the right and 8 units down. **13.** down; wider **15.** up; narrower **17. (a)** I **(b)** IV **(c)** II **(d)** III

19.

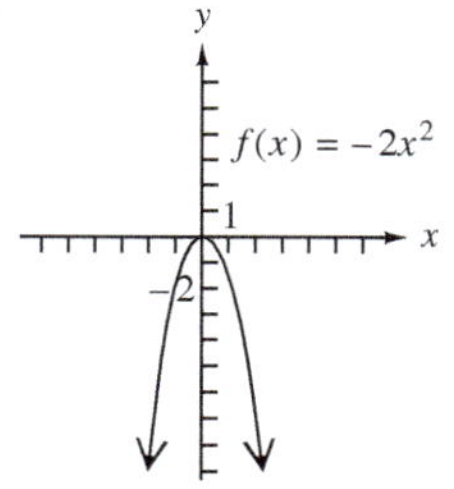

21.

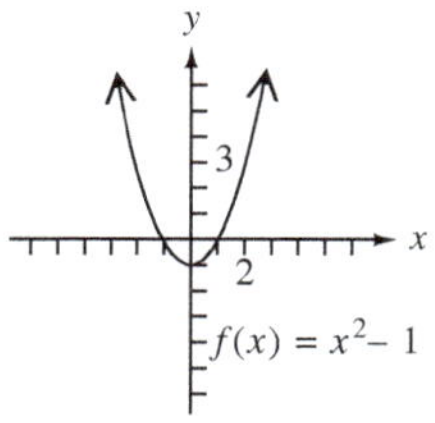

23.

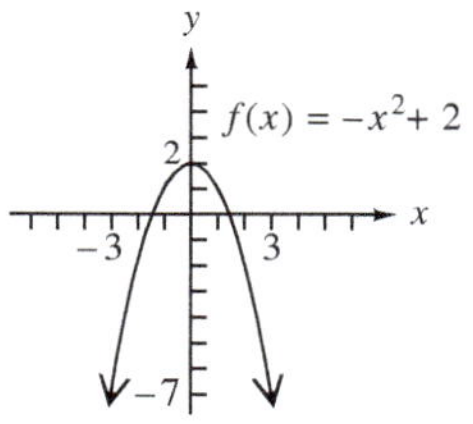

25. vertex: (4, 0); axis: $x = 4$; domain: $(-\infty, \infty)$; range: $[0, \infty)$

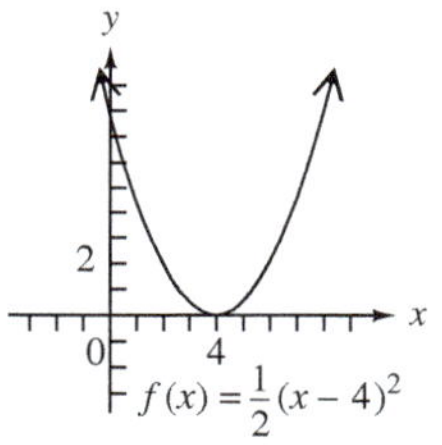

27. vertex: (−2, −1); axis: $x = -2$; domain: $(-\infty, \infty)$; range: $[-1, \infty)$

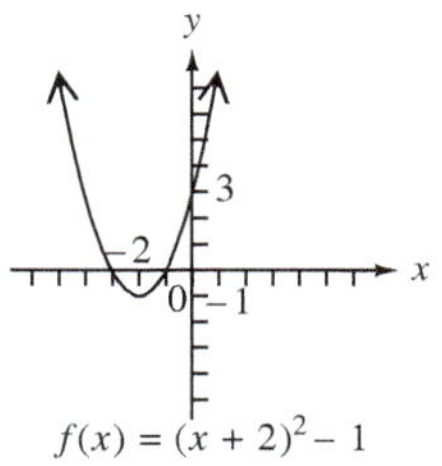

29. vertex: (−3, 4); axis: $x = -3$; domain: $(-\infty, \infty)$; range: $(-\infty, 4]$

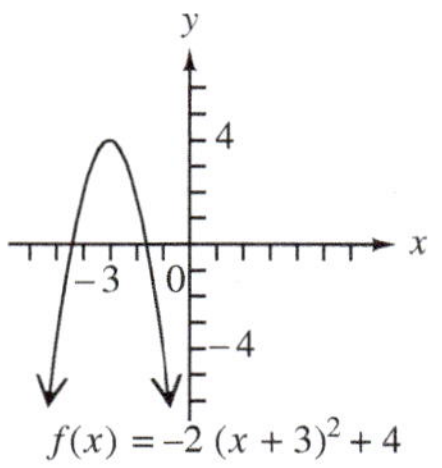

31. vertex: (−2, 1); axis: $x = -2$; domain: $(-\infty, \infty)$; range: $(-\infty, 1]$

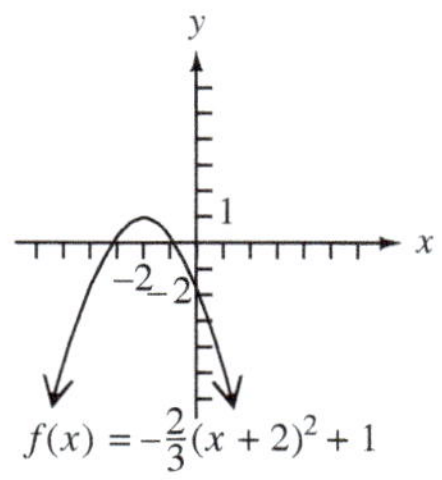

33. It is shifted 6 units up.

34.

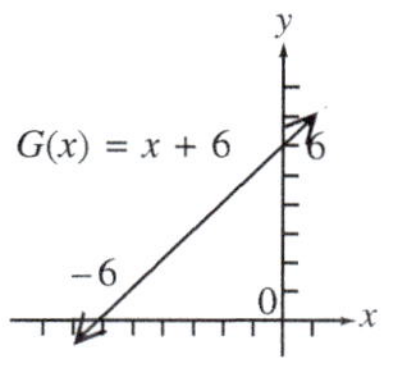

35. It is shifted 6 units up. **36.** It is shifted 6 units to the right.

37.

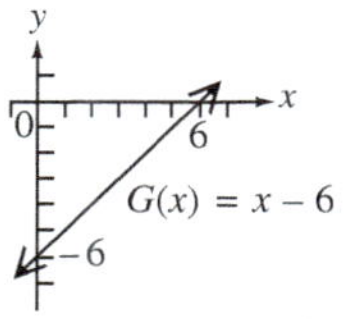

38. It is shifted 6 units to the right. **39.** quadratic; positive **41.** quadratic; negative **43.** linear; positive

45. (a) DIGITAL CAMERA SALES IN U.S.

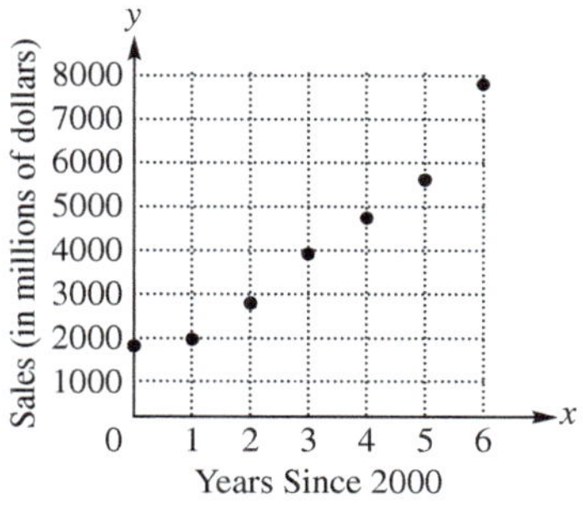

(b) quadratic; positive **(c)** $f(x) = 99.3x^2 + 400.7x + 1825$ **(d)** \$9496 million **(e)** No. The number of digital cameras sold in 2007 is far below the number approximated by the model. Rather than continuing to increase, sales of digital cameras fell in 2007.

47. (a) 2005: 14%; 2007: 2.4% **(b)** The approximations using the model are far too low.

Section 18.7 (pages 1375–1386)

1. If x is squared, it has a vertical axis; if y is squared, it has a horizontal axis. **3.** Use the discriminant of the corresponding quadratic equation. If it is positive, there are two x-intercepts. If it is 0, there is just one x-intercept (the vertex), and if it is negative, there are no x-intercepts. **5.** (−1, 3); up; narrower; no x-intercepts

7. $\left(\frac{5}{2}, \frac{37}{4}\right)$; down; same; two x-intercepts

9. (−3, −9); to the right; wider

11. vertex: (−2, −1); axis: $x = -2$; domain: $(-\infty, \infty)$; range: $[-1, \infty)$

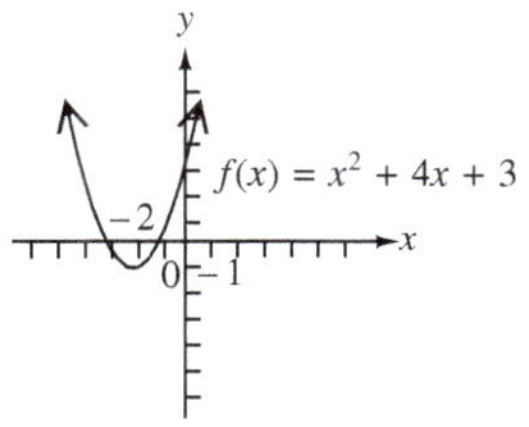

13. vertex: $(1, -3)$; axis: $x = 1$; domain: $(-\infty, \infty)$; range: $(-\infty, -3]$

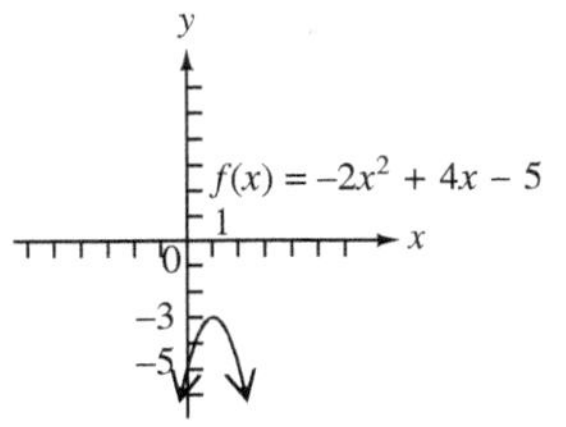

15. vertex: $(1, 5)$; axis: $y = 5$; domain: $(-\infty, 1]$; range: $(-\infty, \infty)$

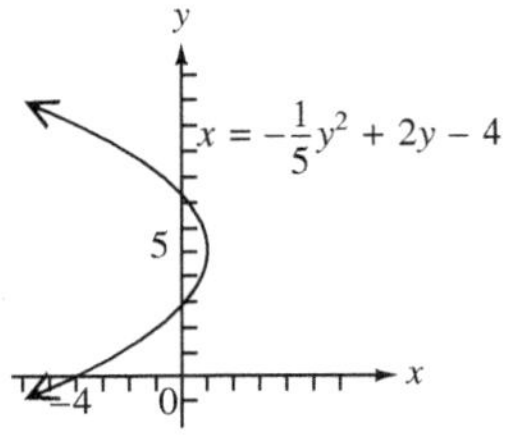

17. vertex: $(-7, -2)$; axis: $y = -2$; domain: $[-7, \infty)$; range: $(-\infty, \infty)$

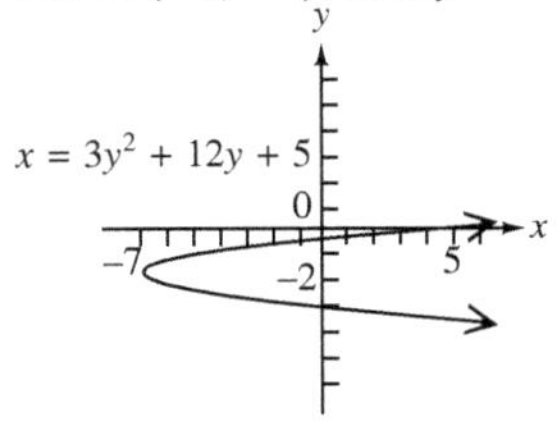

19. F **21.** C **23.** D **25.** 30 and 30 **27.** 160 ft by 320 ft **29.** 16 ft; 2 sec **31.** **(a)** $R(x) = 20{,}000 + 200x - 4x^2$ **(b)** 25 **(c)** \$22,500 **33.** **(a)** maximum **(b)** 1993; 13.2% **35.** **(a)** The coefficient of x^2 is negative because the parabola opens down. **(b)** (18.45, 3860) **(c)** In 2018 Social Security assets will reach their maximum value of \$3860 billion.

INDEX

G

H

I

S

T

GEOMETRY FORMULAS

Square

Perimeter: $P = 4s$

Area: $A = s^2$

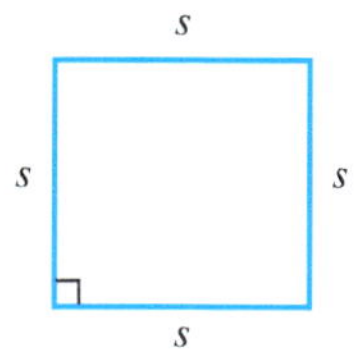

Rectangle

Perimeter: $P = 2L + 2W$

Area: $A = LW$

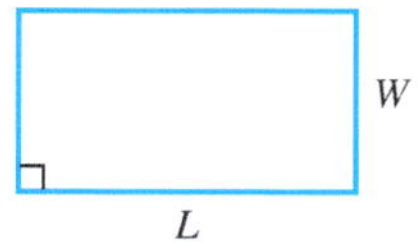

Triangle

Perimeter: $P = a + b + c$

Area: $A = \frac{1}{2}bh$

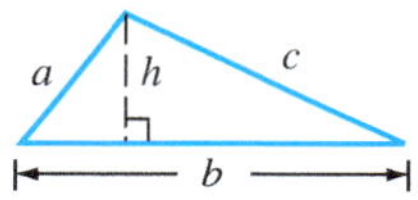

Parallelogram

Perimeter: $P = 2a + 2b$

Area: $A = bh$

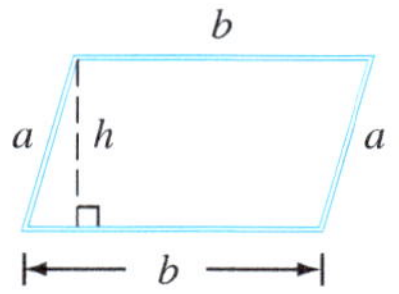

Trapezoid

Perimeter: $P = a + b + c + B$

Area: $A = \frac{1}{2}h(b + B)$

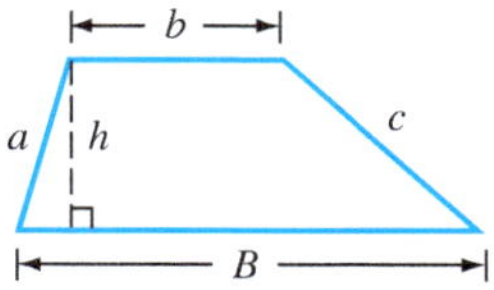

Circle

Diameter: $d = 2r$

Circumference: $C = 2\pi r$

$C = \pi d$

Area: $A = \pi r^2$

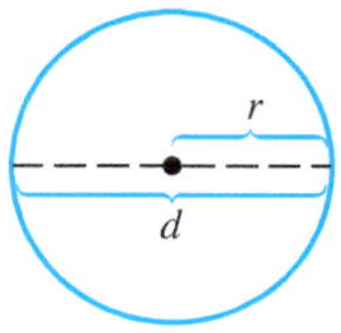

Rectangular Solid

Volume: $V = LWH$

Surface area: $A = 2HW + 2LW + 2LH$

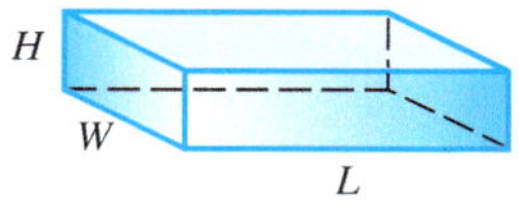

Cube

Volume: $V = e^3$

Surface area: $S = 6e^2$

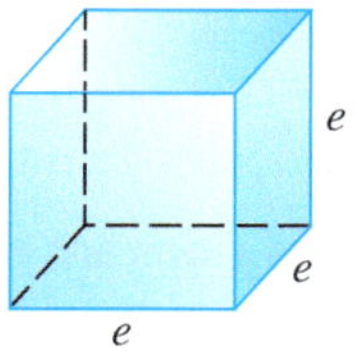

Right Circular Cylinder

Volume: $V = \pi r^2 h$

Surface area: $S = 2\pi rh + 2\pi r^2$

(Includes both circular bases)

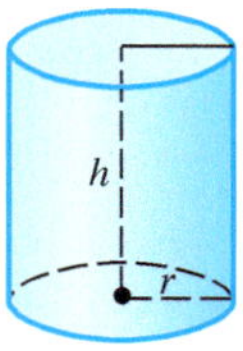

Cone

Volume: $V = \frac{1}{3}\pi r^2 h$

Surface area: $S = \pi r\sqrt{r^2 + h^2} + \pi r^2$

(Includes circular base)

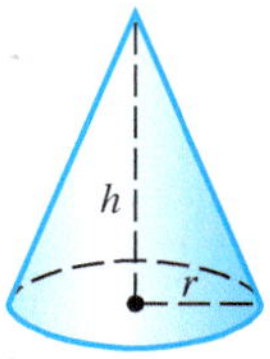

Right Pyramid

Volume: $V = \frac{1}{3}Bh$

B = area of the base

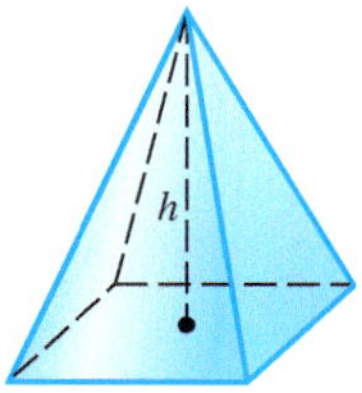

Sphere

Volume: $V = \frac{4}{3}\pi r^3$

Surface area: $S = 4\pi r^2$

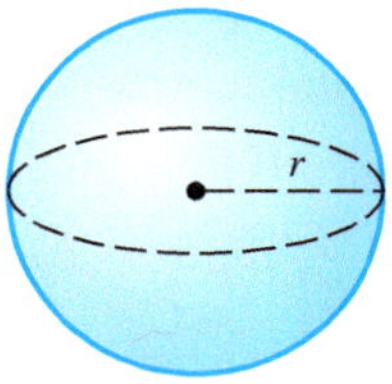

TRIANGLES AND ANGLES

Right Triangle

Triangle has one 90° (right) angle.

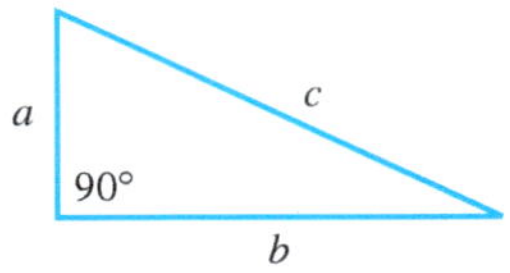

Pythagorean Formula (for right triangles)

$a^2 + b^2 = c^2$

Isosceles Triangle

Two sides are equal.

$AB = BC$

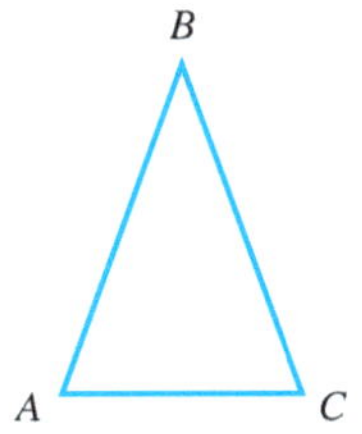

Equilateral Triangle

All sides are equal.

$AB = BC = CA$

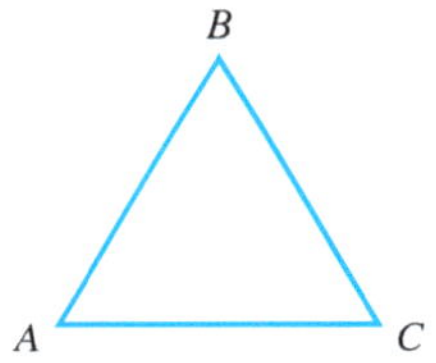

Sum of the Angles of Any Triangle

$A + B + C = 180°$

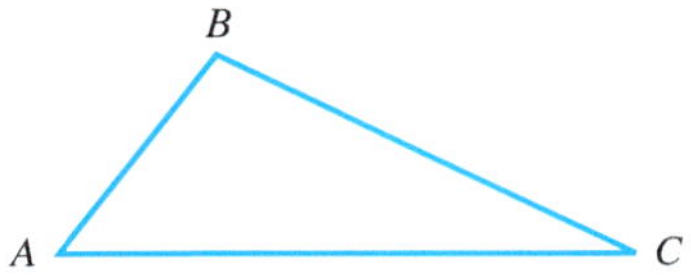

Similar Triangles

Corresponding angles are equal; corresponding sides are proportional.

$A = D, B = E, C = F$

$$\frac{AB}{DE} = \frac{AC}{DF} = \frac{BC}{EF}$$

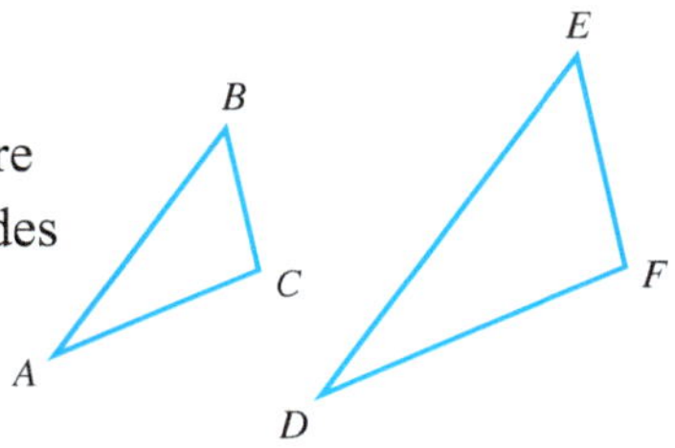

Right Angle

Measure is 90°.

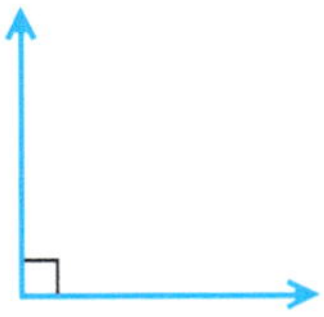

Straight Angle

Measure is 180°.

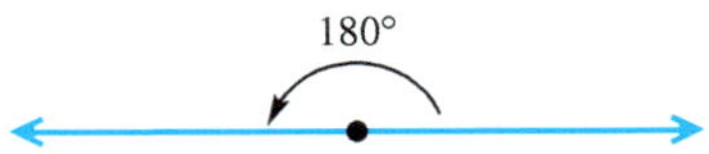

Complementary Angles

The sum of the measures of two complementary angles is 90°.

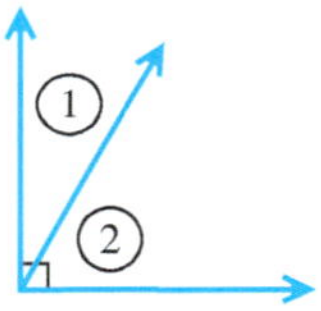

Angles ① and ② are complementary.

Supplementary Angles

The sum of the measures of two supplementary angles is 180°.

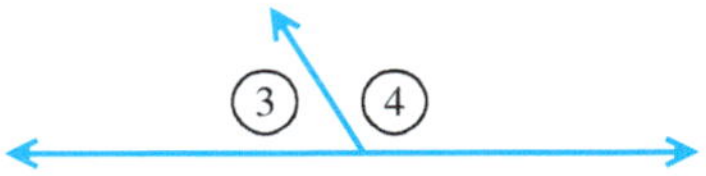

Angles ③ and ④ are supplementary.

Vertical Angles

Vertical angles have equal measures.

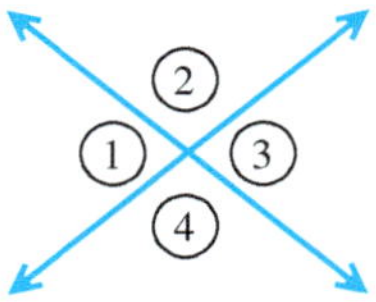

Angle ① = Angle ③

Angle ② = Angle ④